CCEA GCSE

MATHEMATICS
M3 M7

COLOURPOINT EDUCATIONAL

© Luke Robinson, Sam Stevenson and Colourpoint Creative Ltd 2025

First Edition

Print ISBN: 978-1-78073-388-3
eBook ISBN: 978-1-78073-389-0

Layout and design: April Sky Design
Printed by: GPS Colour Graphics Ltd, Belfast

All rights reserved. No part of this publication may be reproduced, stored in a retrieval system or transmitted in any form or by any means, electronic, mechanical, photocopying, scanning, recording or otherwise, without the prior written permission of the copyright owners and publisher of this book.

Copyright has been acknowledged to the best of our ability. If there are any inadvertent errors or omissions, we shall be happy to correct them in any future editions.

Colourpoint Educational
An imprint of Colourpoint Creative Ltd
Colourpoint House
Jubilee Business Park
21 Jubilee Road
Newtownards
County Down
Northern Ireland
BT23 4YH

Tel: 028 9182 6339
E-mail: sales@colourpoint.co.uk
Web site: www.colourpointeducational.com

The Authors

Luke Robinson took a mathematics degree, followed by an MSc and PhD in meteorology. He taught at Northwood College in London before becoming a freelance Mathematics tutor and writer. He now lives in County Down with his wife and son.

Sam Stevenson has taught Mathematics and Informatics in grammar schools to a generation. He has a passion for developing and delivering Further Mathematics at both GCSE and GCE levels. He will be known to many from the Northern Ireland Mathematics educational circuit. His hobbies include family, cycling, keeping tortoises, skiing and combinatorics, in no particular order.

Note: This book has been written to meet the requirements of the GCSE Mathematics specification from CCEA. While the authors and Colourpoint Creative Limited have taken all reasonable care in the preparation of this book, it is not possible to guarantee that is completely error-free. In addition, it is the responsibility of each candidate to satisfy themselves that they have covered all necessary material before sitting an examination based on the CCEA specification. The publishers will therefore accept no legal responsibility or liability for any errors or omissions from this book or the consequences thereof.

Contents

Introduction 5
1 Working With Integers 6
2 Approximation and Estimation 15
3 Percentages and Finance 22
4 Bounds 36
   Progress Review 42
5 Brackets 46
6 Factorisation 53
7 Algebraic Fractions 57
8 Solving Equations and Identities 63
9 Quadratic Expressions and Equations 74
10 Trial and Improvement 80
   Progress Review 84
11 Straight Lines 87
12 Angles 102
13 Circles 109
14 Pythagoras' Theorem and Trigonometry 112
15 Compound Measure 124
16 Perimeter, Area and Volume 133
   Progress Review 146
17 Collecting Data 151
18 Statistical Averages and Spread 160
19 Statistical Diagrams 173
20 Scatter Graphs 183
21 Cumulative Frequency 190
   Progress Review 200
22 Number Systems 205
23 Surds 210
24 Indices 215
25 Standard Form 221
26 Simultaneous Equations 229
27 Inequalities 235
   Progress Review 247
28 Formulae 251
29 Sequences 258
30 Non-Linear Graphs 268
31 Real-Life Graphs 275
32 Graphical Solutions 284
33 Proportion and Variation 290
   Progress Review 299
34 Transformations 306
35 Enlargement and Similarity 317
36 Constructions and Loci 335
37 Probability 344
   Progress Review 361

See page 5 for information on how to obtain the Answers.

Introduction

This book covers the Higher Tier Option 1 (M3 and M7) pathway of the CCEA specification for Mathematics (first teaching in September 2017). Specifically, it covers:
- the material required at level M3 and level M7, and
- the material required at M1, M2, M5 and M6 at a depth that is appropriate for an M3 and M7 student.

This book has undergone a detailed quality assurance check by experienced mathematician Joe McGurk prior to publication.

Feedback from teachers is that highlighting the different units in a Mathematics textbook is overly complex and potentially confusing. As a result, this book has been deliberately designed to cover the course without making any reference to specific units within the text.

Teachers and students can be assured that, if they are on the M3 and M7 pathway, then they simply need to teach or study the whole contents of this textbook. The authors have chosen a thematic approach, where similar material that appears in both M3 and M7 is treated together in the book. Thus, while the first part of the book generally covers M3 material, and the second part focuses on M7 material, there are some chapters that contains both M3 and M7 material.

The book also contains seven 'progress review' sections which act as checkpoints for students. Students can use these to assess their understanding of material at regular intervals throughout the course.

Finally, the book is supplemented by two Revision Booklets, one for M3 and one for M7, which students can use to prepare for the exam. These are two of eight books that Colourpoint publishes for M1 to M8. They are available from Colourpoint.

Answers

The answers to all the exercises are available in PDF format and can be downloaded from our web site. Go to www.colourpointeducational.com and search for the title of this book. Once you reach the page about this book, you will find a link to download the answers.

If you have any problems, please contact Colourpoint directly. Our contact details are on page 2.

Chapter 1
Working With Integers

1.1 Introduction

This chapter is about **positive integers**, sometimes called the **natural numbers** or the **counting numbers**: 1, 2, 3, 4, …

Before you start you should:
- Know and understand the following terms: factor, multiple, prime.
- Understand index notation, for example that 4^2 means 4×4 and that 4^3 means $4 \times 4 \times 4$

In this chapter you will learn how to:
- Find a prime factorisation for any positive integer.
- Find the Highest Common Factor for two positive integers.
- Find the Lowest Common Multiple for two positive integers.

1.2 Factors and Multiples

The **factors** of a number are all the integers that divide into that number. This includes one and the number itself.
The **multiples** of a number are all the integers that the number divides into.

Example 1

Write down the first five multiples of 8

The **multiples** of 8 are the numbers in the 8 times table:
$1 \times 8 = 8$ $2 \times 8 = 16$ $3 \times 8 = 24$ $4 \times 8 = 32$ $5 \times 8 = 40$ …
So, the first five multiples of 8 are: 8, 16, 24, 32 and 40
There are an infinite number of multiples of 8

Example 2

(a) Write down all the factors of 12
(b) Write down the first 4 multiples of 12

(a) The factors of 12 are the integers that divide into 12
You may find it helpful to find these in **factor pairs** that multiply to make 12:

1	2	3
12	6	4

So the factors of 12 are 1, 2, 3, 4, 6 and 12

> **Note:** Don't forget to include 1 and 12 itself!

(b) The first 4 multiples of 12 are 12, 24, 36 and 48.

> **Note:** Don't forget to include 12 itself! It is both a factor and a multiple of 12

Exercise 1A

1. Write down the first 5 multiples of
   - (a) 2
   - (b) 5
   - (c) 3
   - (d) 7
   - (e) 11
   - (f) 10
   - (g) 1
   - (h) 6
   - (i) 9
   - (j) 8

2. List all the factors of:
   - (a) 36
   - (b) 45
   - (c) 20
   - (d) 18
   - (e) 60
   - (f) 80
   - (g) 84
   - (h) 96
   - (i) 100
   - (j) 105

3. (a) List the first ten multiples of 6
   (b) List the first five multiples of 12
   (c) James spots a connection between the multiples of 6 and the multiples of 12
   What is it?

4. (a) Write down all the multiples of 9 between 15 and 95
   (b) What pattern do you notice?

5. Ciara thinks of a number. She writes down some multiples of her number.

   12   16   24

   What number could Ciara be thinking of? Find three different answers.

6. Find the answers to the following calculations.
   (a) The third multiple of 3 plus the second multiple of 12
   (b) The fifth multiple of 4 multiplied by the third multiple of 5
   (c) The fifth multiple of 12 minus the fifth multiple of 8
   (d) The ninth multiple of 8 divided by the sixth multiple of 12

7. Copy all the numbers from the grid on the right.
   (a) Draw a square around each multiple of 4
   (b) Draw a triangle around each multiple of 3
   (c) Draw a circle around each multiple of 10
   (d) Which numbers need two shapes?
   (e) Which number needs 3 shapes?
   (f) Which number has no shapes around it?

20	40	45	35
36	24	18	16
60	64	8	12
30	15	6	10

8. Jamie thinks of a number. He tells his sister Delyth:

   My number is a multiple of 4 and a multiple of 6   It is a factor of 96

   What could Jamie's number be? How many possible answers are there?

1.3 Prime Numbers

A **prime number** has only two factors: 1 and itself. A number that is not prime is called a **composite number**.

Example 3

(a) Is 7 a prime number?
(b) Is 9 a prime number?
(c) Is 11 a prime number?
(d) How many prime numbers are there between 20 and 30?

(a) 7 is a prime number. The only factors of 7 are 1 and 7
(b) 9 is **not** a prime number because 3 is a factor of 9
   It is a composite number.
(c) 11 is a prime number. The only factors of 11 are 1 and 11
(d) There are 2 prime numbers between 20 and 30: 23 and 29

Exercise 1B

1. Charlie says 'All prime numbers are odd!' Is Charlie right or wrong? Explain your answer.
2. List the first 10 prime numbers.

3. Copy the grid of numbers on the right.
   Circle all the prime numbers.

39	26	11	8	7
33	25	16	40	20
5	31	21	12	17

4. Here is a list of twelve numbers.

   29, 6, 25, 20, 2, 23, 8, 17, 3, 4, 9, 31

   From the list, write down:
   (a) The smallest prime number.
   (b) A prime number greater than 20.
   (c) An even prime number.
   (d) The largest prime number.
   (e) Three numbers that are not prime.

5. Goldbach's Conjecture states that:

   'Every even number greater than 2 can be written as the sum of two primes.'

   This has been tested for several billion even numbers, but, so far, mathematicians have not proved it is *always* true. Show that Goldbach's Conjecture is true for the even numbers from 4 to 12.
   The first one, 4, has been done for you:

   $$4 = 2 + 2$$
   ...

1.4 Prime Factorisation

Every positive integer can be written as a **product of prime factors**.

The **product of prime factors** is also called the **prime factorisation** for that integer.

Example 4

Write 120 as a product of prime factors.

Method 1: Factor trees

Find two numbers that multiply to make 120. The numbers 4 and 30 have been chosen, as shown in the diagram on the right.

Repeat the process for 4 and 30.

At each stage, circle the prime numbers. These cannot be factorised further.

Repeat until you are only left with prime numbers. These primes make up the product of prime factors.

So: $120 = 2 \times 2 \times 2 \times 3 \times 5$

or, using index notation: $120 = 2^3 \times 3 \times 5$

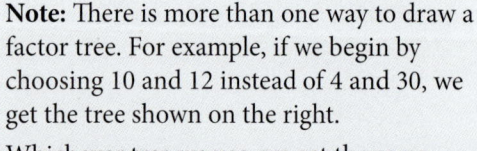

> **Note:** There is more than one way to draw a factor tree. For example, if we begin by choosing 10 and 12 instead of 4 and 30, we get the tree shown on the right.
>
> Whichever tree we use, we get the same prime factorisation.

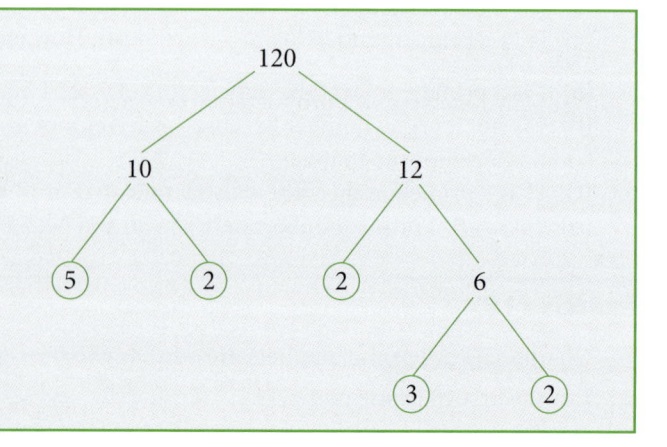

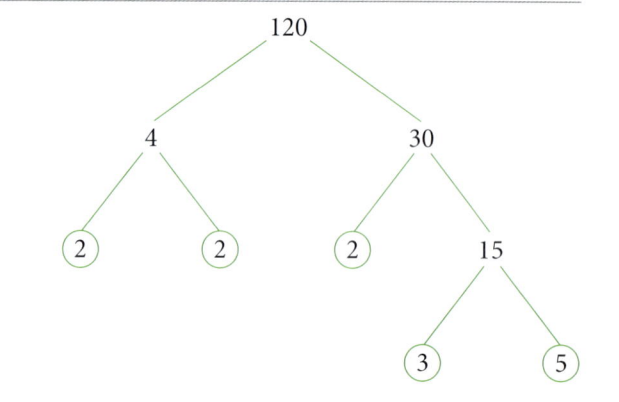

Method 2: Repeated division

Choose the smallest prime number that is a factor of 120, which is 2

Divide 120 by 2, giving 60

Repeat the process until you reach 1

The prime factorisation for 120 is the product of the prime numbers you divide by.

So: $\qquad 120 = 2 \times 2 \times 2 \times 3 \times 5$

or, using index notation: $\qquad 120 = 2^3 \times 3 \times 5$

```
2 | 120
2 |  60
2 |  30
3 |  15
5 |   5
        1
```

> **Note:** For any integer, there is only one way to write it using a product of prime factors.

Calculator Tip

You can check a prime factorisation on your calculator. This tip is for the Casio fx-83GT CW, but similar functionality is available on most GCSE calculators.

For example:

Enter: $\qquad\qquad\qquad\qquad\qquad\qquad\qquad$ 5×9 and press EXE

The calculator displays: $\qquad\qquad\qquad\qquad\qquad$ 45

Then press: $\qquad\qquad\qquad\qquad\qquad\qquad\qquad$ FORMAT

Move down with the down arrow key to the option 'Prime Factor' and press EXE

The calculator displays: $\qquad\qquad\qquad\qquad\qquad$ $3^2 \times 5$

Alternatively, just type in a number:

Enter: $\qquad\qquad\qquad\qquad\qquad\qquad\qquad$ $100 =$

Then use the FORMAT button in the same way to get: $\quad 2^2 \times 5^2$

You may use this calculator function only to **check** a prime factorisation. In exercises and exam questions, you must show your working.

Example 5

The prime factorisation for 675 is $3^3 \times 5^2$

Are the following numbers factors of 675?

(a) 2 (b) 3 (c) 6 (d) 11 (e) 15 (f) 25

(a) No, because 2 is not a part of the prime factorisation.
(b) Yes, because 3 is a part of the prime factorisation.
(c) No, because $6 = 2 \times 3$ and the prime factorisation does not include 2
(d) No, because 11 is a prime that does not appear in the prime factorisation.
(e) Yes, because $15 = 3 \times 5$ and the prime factorisation includes both 3 and 5
(f) Yes, because $25 = 5^2$ and the prime factorisation includes 5^2

Example 6

Write the following as a product of prime factors: $14^2 \times 20$

Begin with a prime factorisation of 14 and 20:

$\qquad 14 = 2 \times 7$, so: $14^2 = 2 \times 7 \times 2 \times 7$
$\qquad 20 = 2 \times 2 \times 5$

So:
$14^2 \times 20 = 2 \times 7 \times 2 \times 7 \times 2 \times 2 \times 5$
$\qquad\qquad = 2^4 \times 5 \times 7^2$

If the powers in the prime factorisation of a number are all multiples of 2, the number is a square number.

If the powers are all multiples of 3, the number is a cube number.

Example 7

(a) Write 2400 as a product of prime factors.
(b) Using your answer to part (a), find the smallest number that, when multiplied by 2400, gives a square number.
(c) Find the smallest number that, when multiplied by 2400, gives a cube number.

(a) Prime factorisation: $2400 = 2^5 \times 3 \times 5^2$

(b) For a square number, all powers must be multiples of 2
We must multiply by 2^1 (to change 2^5 into 2^6) and by 3^1 (to change 3^1 into 3^2).
Overall, we multiply by $2 \times 3 = 6$
Check: $2400 \times 6 = 14\,400$ This is a square number, as $\sqrt{14\,400} = 120$

(c) For a cube number, all powers must be multiples of 3
We must multiply by 2^1 (to change 2^5 into 2^6) and by 3^2 (to change 3^1 into 3^3) and by 5^1 (to change 5^2 into 5^3).
Overall, we multiply by $2 \times 3^2 \times 5 = 90$
Check: $2400 \times 90 = 216\,000$ This is a cube number, as $\sqrt[3]{216\,000} = 60$

Exercise 1C

1. Write each of these numbers as a product of prime factors.
   (a) 30   (b) 42   (c) 70   (d) 105   (e) 33
   (f) 25   (g) 20   (h) 18   (i) 45   (j) 75

2. Match each number to its prime factorisation.

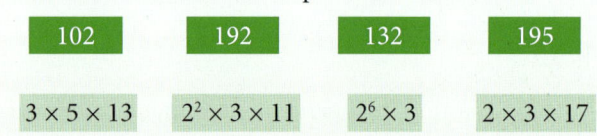

   | 102 | 192 | 132 | 195 |

   | $3 \times 5 \times 13$ | $2^2 \times 3 \times 11$ | $2^6 \times 3$ | $2 \times 3 \times 17$ |

3. Find the prime factorisation for each of these numbers, giving your answers in index form.
   (a) 60   (b) 152   (c) 162   (d) 810

4. The prime factorisation for 3300 is $2^2 \times 3 \times 5^2 \times 11$
   Using this, determine whether the following numbers are factors of 3300
   (a) 2   (b) 11   (c) 22   (d) 20   (e) 24

5. The prime factorisation for 975 is $3 \times 5^2 \times 13$
   Using this, state whether the following numbers are factors of 975, giving a reason in each case.
   (a) 3   (b) 5   (c) 7   (d) 25   (e) 9   (f) 125

6. Write these as a product of prime factors, giving your answers in index form.
   (a) 6×10^2 (b) $6^2 \times 9^3$ (c) $10^2 \times 14^3$

7. (a) Write 5850 as a product of prime factors, giving your answer in index form.
   (b) Find the smallest number you could multiply 5850 by to give a square number.

8. (a) Write 5292 as a product of prime factors, giving your answer in index form.
   (b) Find the smallest number you could multiply 5292 by to give a cube number.

1.5 Highest Common Factor and Lowest Common Multiple

A **common factor** is an integer that is a factor of two or more numbers.
A **common multiple** is an integer that is a multiple of two or more numbers.

Example 8

(a) Write down a common factor of 15 and 20
(b) Is 30 a common multiple of 6 and 12?

(a) The factors of 15 are: ①　3　⑤　15

The factors of 20 are: ①　2　4　⑤　10　20

The **common factors** have been circled: 1 and 5
(The question just asks for one common factor, so either 1 or 5 would be a correct answer.)

(b) 30 is not a common multiple of 6 and 12; it is a multiple of 6, but not a multiple of 12

The **highest common factor** (or **HCF**) of two numbers is the biggest number that is a factor of both.

The **lowest common multiple** (or **LCM**) of two numbers is the smallest number that is a multiple of both.

Example 9

Find the highest common factor of 48 and 84

Method 1: List the factors

We must find the **highest** number that is a factor of both 48 and 84
First write down all the factors of 48 and 84
Then circle the numbers that appear in both lists.

The factors of 48 are: ①　②　③　④　⑥　8　⑫　16　24　48

The factors of 84 are: ①　②　③　④　⑥　7　⑫　14　21　28　42　84

1, 2, 3, 4, 6 and 12 are the **common factors** of 48 and 84
So, 12 is the **highest common factor**.

Method 2: Use the prime factorisations

$48 = 2^4 \times 3$
$84 = 2^2 \times 3 \times 7$

To find the HCF, look at both factorisations. Take the **lowest** power of each prime, but do not include a prime if it only appears in one factorisation.

The lowest power of 2 is 2^2 (in the factorisation of 84).
The lowest power of 3 is 3^1 (in both factorisations).
Do not include 7 (as it only appears in the factorisation of 84).
So, the HCF is $2^2 \times 3^1 = 12$

Method 3: Use a Venn diagram

$48 = 2^4 \times 3$
$84 = 2^2 \times 3 \times 7$

Draw a Venn diagram with a circle for each of the starting numbers, 48 and 84

In each circle, place the number's prime factors. Any prime factors in common appear in the intersection region, as shown on the right, with the remaining factors appearing in the rest of the circle.

The HCF can then be found by multiplying **all the numbers that appear in the intersection**.

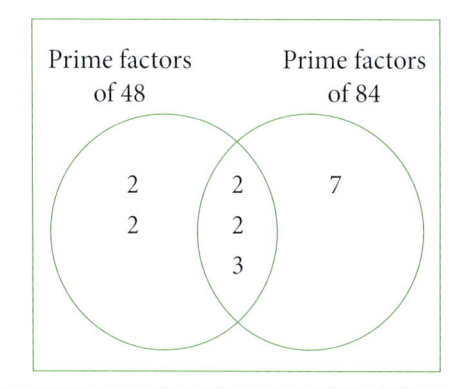

So: HCF = 2 × 2 × 3
　　　　= 12

Example 10

Find the lowest common multiple of 20 and 44

We must find the **lowest** number that is a multiple of both 20 and 44

Method 1: List the multiples

First write down the first few multiples of 20 and 44
Then find the first number that appears in both lists.

Multiples of 20 are: 20   40   60   80   100   120   140   160   180   200   (220)   ...

Multiples of 44 are: 44   88   132   176   (220)   ...

220 is the first number to appear in both lists.
So, 220 is the **lowest common multiple** of 20 and 44

Method 2: Use the prime factorisations

$20 = 2^2 \times 5$ $44 = 2^2 \times 11$

To find the LCM, look at both factorisations. Take the highest power of each prime.
The highest power of 2 is 2^2 (this appears in both factorisations).
The highest power of 5 is 5
The highest power of 11 is 11
So the LCM is $2^2 \times 5 \times 11 = 220$

Method 3: Use a Venn diagram

$20 = 2^2 \times 5$ $44 = 2^2 \times 11$

Draw a Venn diagram with a circle for each of the starting numbers, 20 and 44.

In each circle, place the number's prime factors. Any prime factors in common appear in the intersection region, as shown on the right, with the remaining factors appearing in the rest of the circle.

The LCM can then be found by multiplying **all the numbers that appear in the diagram**.

So: LCM = $5 \times 2 \times 2 \times 11$
       = 220

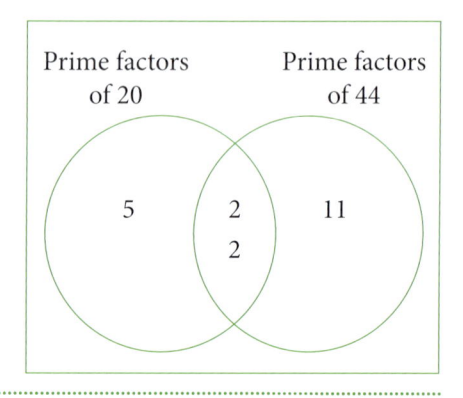

Some worded problems require calculation of the highest common factor or lowest common multiple.

Example 11

Copeland lighthouse flashes every 3 minutes. Portpatrick lighthouse flashes every 5 minutes. If both lighthouses flash together at 13:06, when will they next flash at the same time?

In minutes after 13:06:

Copeland flashes at: 3   6   9   12   (15)   ...

Portpatrick flashes at: 5   10   (15)   20   25   ...

The first time both lighthouses flash together is at 15 minutes.

The time will be 15 minutes after 13:06, which is 13:21

Note: This question requires you to find the LCM of 3 and 5; this has been done by listing multiples of both numbers.

Example 12

Terry runs a café. He buys big boxes of ketchup sachets and big boxes of mustard sachets. There are 100 ketchup sachets in a box and 80 mustard sachets in a box. Terry wants to buy the same number of ketchup sachets and mustard sachets.

(a) What is the smallest number of ketchup and mustard sachets he could buy?
(b) How many **boxes** of each must he buy?

(c) If a box of ketchup sachets costs £12 and a box of mustard sachets costs £9.50, how much does he spend?

(a) If Terry buys the same number of ketchup and mustard sachets, this number must be a multiple of 100 and a multiple of 80, in other words a common multiple of 100 and 80. If he buys the smallest possible number, it is the lowest common multiple (LCM).

The prime factorisations are:
$100 = 2^2 \times 5^2$
$80 = 2^4 \times 5$
$\text{LCM} = 2^4 \times 5^2$
$= 400$

The smallest number of ketchup and mustard sachets Terry could buy is 400 of each.

(b) There are 100 ketchup sachets in a box. So 400 ketchup sachets is 4 boxes.
There are 80 mustard sachets in a box. So 400 mustard sachets is 5 boxes.

(c) Total cost = 4 × £12 + 5 × £9.50
= £95.50

Example 13

David is tiling a bathroom floor, which measures 360 cm by 280 cm. He would like to use square tiles, and he does not want to cut any tiles. The length and width of each square tile is a whole number of centimetres.
(a) What is the largest tile David could use?
(b) If David uses the largest possible tile, how many tiles would he need to cover the floor?

(a) The side length of each square must be a factor of 360 so that a whole number of them fit along the length. The side length of each square must also be a factor of 280 so that a whole number of them fit along the width.

So, the tile size is a common factor of 360 and 280

Since David wants the largest possible square, we must find the highest common factor (HCF):
$360 = 2^3 \times 3^2 \times 5$
$280 = 2^3 \times 5 \times 7$
$\text{HCF} = 2^3 \times 5 = 40$
David must use square tiles with a side length of 40 cm.

(b) If each tile is to be 40 cm, David needs 9 of them along the length and 7 of them along the width. In total he needs 63 tiles.

Exercise 1D

1. Find the lowest common multiple (LCM) of:
   (a) 6 and 8   (b) 4 and 10   (c) 5 and 15   (d) 8 and 20   (e) 10 and 14
   (f) 70 and 56 (g) 80 and 60 (h) 90 and 48 (i) 60 and 96 (j) 120 and 150

2. Find the highest common factor (HCF) of:
   (a) 6 and 8   (b) 12 and 16 (c) 15 and 27 (d) 22 and 42 (e) 30 and 36
   (f) 60 and 114 (g) 84 and 120 (h) 72 and 90 (i) 90 and 126 (j) 148 and 120

3. (a) By drawing a prime factor tree diagram for 20, find the prime factorisation of 20
   (b) By drawing a prime factor tree diagram for 28, find the prime factorisation of 28
   (c) Using your answers to parts (a) and (b), draw a Venn diagram to help you find
       (i) the highest common factor and (ii) the lowest common multiple of 20 and 28

4. Suzie says that the lowest common multiple of 6 and 9 is 54 as 6 × 9 = 54
   Is Suzie right? Explain your answer fully.

5. (a) Using the method of repeated division, find the prime factorisations for both 40 and 72
   (b) Hence find both the HCF and LCM of 40 and 72

6. $p = 42$ and $q = 54$
   (a) Find the highest common factor of p and q

(b) Find the lowest common multiple of p and q

7. $m = 72$ and $n = 9$
   (a) Find the highest common factor of $\frac{m}{n}$ and mn
   (b) Find the lowest common multiple of $\frac{m}{n}$ and mn

8. Find the lowest common multiple of:
   (a) 2, 3 and 5   (b) 2, 3 and 6

9. Farmer McDonald sends eggs to the supermarket in crates of 48
   Farmer Shannon sends eggs to the supermarket in crates of 36
   The supermarket needs the same number of eggs from each farmer. Find the smallest number of eggs they could deliver.

10. Trains arriving in Belfast's Grand Central Station from Dublin carry 480 passengers when full. Trains arriving from Derry~Londonderry carry up to 300 passengers. If, one day, the same number of passengers arrive on full trains from both places, what is the smallest possible number of trains that come from
    (a) Dublin   (b) Derry~Londonderry?

11. Find two numbers, both greater than 1, that have a lowest common multiple of 36

1.6 Summary

In this chapter you have learnt about prime numbers, factors, multiples, prime factorisations, lowest common multiples (LCM) and highest common factors (HCF).

You have also learned that:

- A **factor** of a number is a number that divides into it. For example, 3 is a factor of 12
- A **multiple** of a number is in the times table of that number. For example, 15 is a multiple of 5
- A **prime number** has no factors apart from 1 and itself. For example, 7 is a prime number because its only factors are 1 and 7
- A **prime factorisation** is also called a **product of prime factors**. You can find the prime factorisation of a number using a tree diagram or repeated division by primes.
- For any number, there is only one prime factorisation, for example $12 = 2^2 \times 3$
- A **common factor** is a factor of two or more numbers. For example, 2 is a common factor of 8 and 12
- The **highest common factor (HCF)** of two or more numbers is the largest number that is a common factor of the starting numbers. For example, 4 is the highest common factor of 8 and 12
- A **common multiple** is a multiple of two or more numbers. For example, 40 is a common multiple of 4 and 10
- The **lowest common multiple (LCM)** of two or more numbers is the smallest number that is a common multiple of the starting numbers. For example, 20 is the lowest common multiple of 4 and 10
- To find the HCF or LCM of two numbers, various methods are available. If you first find the prime factorisation for each number, then a common method is to use a Venn diagram to find the HCF and LCM.

Chapter 2
Approximation and Estimation

2.1 Introduction

There are many situations in which calculations result in answers such as 43.2198 metres, or 3.12456 seconds.

It is not always sensible to record so many digits, so we need a method of **approximating**. You have already come across rounding to the nearest unit, ten or hundred, for example.

In this chapter, we consider rounding to a given number of decimal places or significant figures.

Estimation is a useful skill when a calculation involves more than one number or measurement given to a high level of accuracy. An estimate can be found by rounding every number in the calculation, usually to one significant figure. The aim is to estimate an answer without using a calculator.

Key words

- **Approximation**: A key skill to round a number. In this chapter we consider approximating to some number of decimal places or significant figures.
- **Decimal places**: The digits after the decimal point are the decimal places.
- **Significant figures**: The significant figures begin on the left, with the first non-zero digit. For example, in the number 1234.5, the first two significant figures are the 1 and 2 in the thousands and hundreds columns.
- **Estimation**: Finding an approximate value for a calculation that involves more than one number. Each number in the calculation is rounded, usually to one significant figure.

Before you start you should know:

- The meaning of **place value**.
- How to add, subtract, multiply and divide integers and decimal numbers.
- Remember the correct order of operations (BIDMAS).
- How to do calculations with money.

In this chapter you will learn how to:

- Round to a given number of decimal places.
- Round to a given number of significant figures.
- Estimate the answer to a complicated calculation.

Exercise 2A (Revision)

1. Write down the value of the underlined digits in each of the following numbers.
   (a) 5901.4   (b) 234.05   (c) 21.067
2. Carry out the following calculations without a calculator.
   (a) 6182 + 597   (b) 823.4 − 97.5   (c) 36.7 × 8   (d) 1320 ÷ 1.2
3. Use the correct order of operations to do the following calculations.
   (a) 7 × 20 − 10 ÷ 5   (b) 1.3 × (6.9 + 3.1)   (c) 5 + 5 × 5 − 5 ÷ 5
   (d) 5×2^2 (e) $3 \times 2 + (5 - 4)^2$
4. At the cinema Nicky buys 4 child tickets and 2 adult tickets. The child tickets are two thirds of the price of the adult tickets. If an adult ticket is £9.60, how much does Nicky spend altogether?

2.2 Approximation

In this section we will consider two ways of rounding:
- rounding to a given number of decimal places; and
- rounding to a given number of significant figures.

Rounding to a given number of decimal places

When rounding to one decimal place, look at the digit in the second decimal place (the hundredths column). If that is 5 or more, the digit in the tenths column must round up. Otherwise, it stays the same.

Likewise, when rounding to two decimal places, look at the digit in the third decimal place (the thousandths column) to determine whether the digit in the hundredths column rounds up or stays the same.

This process is illustrated in the following example.

Example 1

Round each of the following numbers to:
(i) one decimal place (ii) two decimal places (iii) three decimal places.
(a) 0.42765739 (b) 62.004596 (c) 89.9979

(a) (i) 0.4 (ii) 0.43 (iii) 0.428

(b) If you are asked to round to one decimal place, you must give an answer with exactly one decimal place. So, in part (i), we must write 62.0 and not just 62
Likewise, in part (ii), when giving an answer to 2 decimal places, you must write 62.00
(i) 62.0 (ii) 62.00 (iii) 62.005

(c) (i) 90.0
The 9 in the tenths column rounds up because there is a 9 in the hundredths column. The tenths column becomes a zero, the units a zero and the number in the tens column becomes a 9
(ii) 90.00
Like in part (i), the 9 in the hundredths column rounds up because it is followed by 7 in the thousandths column. The hundredths, tenths and units columns all become zeroes, and the number in the tens column becomes a 9
(iii) 89.998
When rounding to three decimal places, the 7 in the thousandths column rounds up from 7 to 8 because it is followed by a 9

Exercise 2B

1. Round each of the following to (i) one decimal place (ii) 2 decimal places (iii) 3 decimal places:
    (a) 40.2478   (b) 258.245498   (c) 934.2789   (d) 0.77160   (e) 468.56698
    (f) 1492.1995298   (g) 39.969   (h) 101.0101   (i) 596.9696   (j) 23.7894

2. The distances in kilometres by road between some towns in Northern Ireland are shown in the distance table below. For example, it is 67.202 km from Belfast to Dungannon.

Antrim				
27.681	Belfast			
46.242	73.284	Cookstown		
61.761	67.202	15.525	Dungannon	
136.489	132.163	84.486	69.287	Enniskillen

(a) Round the distance from Dungannon to Enniskillen to:
   (i) one decimal place (ii) two decimal places.
(b) Paul the plumber travels from Belfast to Cookstown. From there he travels to Dungannon. How far does he travel altogether? Give your answer in kilometres to (i) 1 decimal place (ii) 2 decimal places.
(c) How much further is it from Belfast to Dungannon than from Belfast to Antrim? Give your answer in kilometres to (i) one decimal place (ii) two decimal places.

CHAPTER 2: APPROXIMATION AND ESTIMATION

Rounding to a given number of significant figures

You can also round to a given number of significant figures. The first significant figure is the first non-zero digit, either before or after the decimal point.

Example 2

The area of Scotland is 77 910 km². Round this to:
(a) one (b) two (c) three significant figures.

We are rounding 77 910 km²
(a) The first significant figure is the first 7
It rounds up to 8 because it is followed by another 7
The other digits become zeroes.
To one significant figure the answer is 80 000 km²
(b) The first two significant figures are 7 and 7
The second 7 rounds up to 8 because it is followed by 9
To two significant figures the answer is 78 000 km²
(c) The first three significant figures are 7, 7 and 9
The 9 does not change as it is followed by a 1
The answer is 77 900 km²

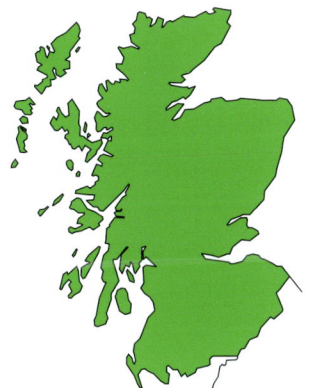

Example 3

A biology student measures the mass of an ant on precision scales as 0.0020475 g. Round this to:
(a) one (b) two (c) three significant figures.

We are rounding 0.0020475 g

Remember to begin counting the significant figures from the first non-zero digit. The zero after the two is a significant figure.

(a) To one significant figure this is 0.002 g
(b) To two significant figures this is 0.0020 g
(c) To three significant figures this is 0.00205 g

Exercise 2C

1. In 2019 the population of Northern Ireland was approximately 1 885 000
   The population of the Republic of Ireland was approximately 5 033 000
   What was the combined population of the Republic of Ireland and Northern Ireland? Give your answer to:
   (a) one (b) two (c) three significant figures.

2. Round each of these numbers to:
   (i) one significant figure (ii) two significant figures (iii) three significant figures.
   (a) 557.9303 (b) 89.9033 (c) 475.0452 (d) 3.805874 (e) 996.5346247

3. The Belfast Giants ice hockey team plays its games in a venue that can hold 8700 spectators. The table below shows the attendance for its home games against various teams in the 2022-2023 season.

Match	Attendance
vs Cardiff Devils	6167
vs Coventry Blaze	5605
vs Dundee Stars	4709
vs Glasgow Clan	8696

   (a) Which match was the best attended?
   (b) Which match was the least well attended?
   (c) Round each attendance figure to:
   (i) one significant figure (ii) two significant figures (iii) three significant figures.

4. The bar chart on the next page shows the number of items, in millions, that Royal Mail delivered across the UK for each year from 2017 to 2022.

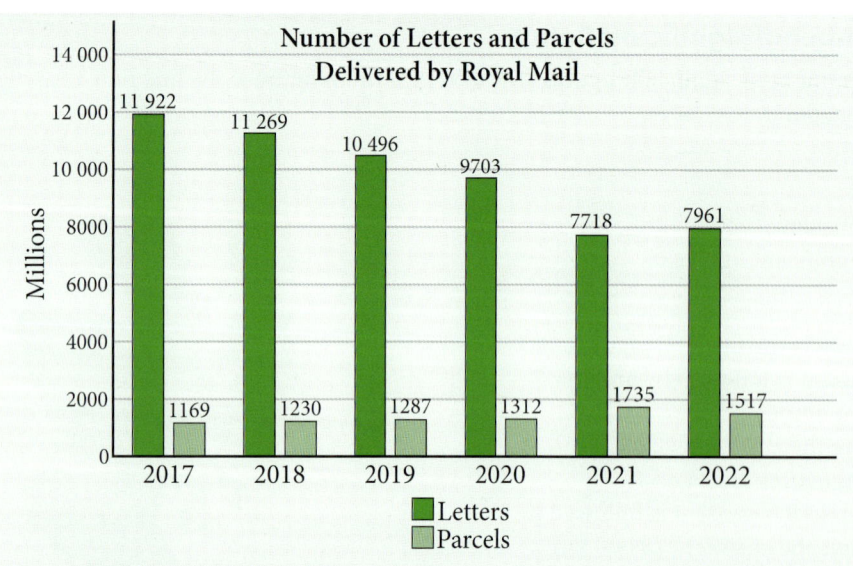

Number of Letters and Parcels Delivered by Royal Mail

Year	Letters (Millions)	Parcels (Millions)
2017	11 922	1169
2018	11 269	1230
2019	10 496	1287
2020	9703	1312
2021	7718	1735
2022	7961	1517

For example, in 2017 Royal Mail delivered 11 922 million letters and 1169 million parcels.

(a) How many parcels did Royal Mail deliver during 2019? Give your answer rounded to:
   (i) one significant figure (ii) two significant figures (iii) three significant figures.
(b) How many letters did Royal Mail deliver during 2022? Give your answer rounded to:
   (i) one significant figure (ii) two significant figures (iii) three significant figures.
(c) How many items did Royal Mail deliver in total during 2018? Give your answer rounded to:
   (i) one significant figure (ii) two significant figures (iii) three significant figures.

5. Henry is growing four tomato plants from seed on his window sill. After 2 weeks, he measures the height of each plant:
   (a) 3.55 cm   (b) 2.79 cm   (c) 4.97 cm   (d) 3.89 cm
   Round these measurements to: (i) one significant figure (ii) two significant figures.

6. A field has an area of 0.005274 km². Round this to:
   (a) one significant figure (b) two significant figures (c) three significant figures.

7. In each part (a) to (f), round both numbers to 3 significant figures. In which part do the two numbers round to the same value?
   (a) 5 643 000   5 612 000
   (b) 2.71828   2.719
   (c) 3.738596   3.698034
   (d) 212.9091   212.333
   (e) 43 936   43 983
   (f) 19.938   19.965

2.3 Estimation

Estimation is useful when we want to get some idea of the approximate size of a calculation, rather than an exact answer. It can serve as a check for many calculations, whether they are done mentally or using a calculator.

For example, 9.24 times 26.3 is approximately 9 × 30 and so, as a rough guide, the answer should be somewhere in the region of 270

Estimation is useful in many practical situations where an exact answer is not needed and a rough answer is sufficient. For example, suppose you need to buy 5.8 metres of piping priced at £6.75 per metre. A rough estimate of the cost would be 6 × £7 = £42

The symbol ≈ means 'is approximately equal to'. You will come across this symbol in this chapter and elsewhere.

When estimating the answer to a calculation, you will usually round each number in the calculation to one significant figure. However, in some cases, you may find that using a different level of accuracy allows you to estimate more easily. For example, you may choose to round to two significant figures or to the nearest whole number.

Example 4

For each of these calculations, estimate the answer, showing your working. Round each number in the calculation to 1 significant figure.

(a) 2.96×5.17 (b) $31.77 \div 6.484$ (c) $\dfrac{0.846 \times 129.3}{2.258}$ (d) $\dfrac{30.9857 + 68.213}{2(23.27 - 11.98)}$

(e) $\dfrac{47.6 \times 0.18}{4.75}$ (f) $\dfrac{98.75 \times 23.41}{0.506}$

(a) 2.96×5.17
$\approx 3 \times 5$ by rounding both numbers to one significant figure.
≈ 15

(b) $31.77 \div 6.484$
$\approx 30 \div 6$
≈ 5

(c) $\dfrac{0.846 \times 129.3}{2.258}$
$\approx \dfrac{0.8 \times 100}{2}$
$\approx \dfrac{80}{2}$
≈ 40

(d) $\dfrac{30.9857 + 68.213}{2(23.27 - 11.98)}$
$\approx \dfrac{30 + 70}{2(20 - 10)}$
$\approx \dfrac{100}{20}$
≈ 5

(e) $\dfrac{47.6 \times 0.18}{4.75}$
$\approx \dfrac{50 \times 0.2}{5}$
$\approx \dfrac{10}{5}$
≈ 2

Note that multiplying 50 by 0.2 is the same as finding $\dfrac{1}{5}$ of 50

(f) $\dfrac{98.75 \times 23.41}{0.506}$
$\approx \dfrac{100 \times 20}{0.5}$
$\approx \dfrac{2000}{0.5}$
≈ 4000

Note that dividing by 0.5 is the same as multiplying by 2

Sometimes a greater level of accuracy is required for your rounding.

Example 5

Estimate the value of the following: $\dfrac{620.4 - 489.6}{11.2 - 10.3}$

In the numerator, both numbers can be rounded to 1 significant figure, giving $600 - 500$

In the denominator, both numbers would round to 10 to 1 significant figure. Since it is not possible to divide by zero, we must use a greater level of accuracy when rounding these two numbers:

$\dfrac{620.4 - 489.6}{11.2 - 10.3}$
$\approx \dfrac{600 - 500}{11 - 10}$ by rounding both numbers in the denominator to 2 significant figures
$\approx \dfrac{100}{1} \approx 100$

Estimation is a useful skill for problem-solving questions.

Example 6

Pete is a hairdresser for ladies. Here is his price list.

Cut and blow dry	£47.50	Curling	£32.50
Wash and cut	£41.00	Colour	£39.50
Blow Dry:		Girls Under 12:	
Short	£28.00	Cut and blow dry	£37.50
Medium	£30.00	Wash and cut	£31.50
Long	£35.00		

On Saturday, Pete does 1 cut and blow dry and 2 wash and cuts for ladies. He also does three blow dries for ladies with short hair and 2 wash and cuts for girls under 12. Estimate how much money Pete makes on this day.

Total = 47.50 + 2 × 41 + 3 × 28 + 2 × 31.50
≈ 50 + 2 × 40 + 3 × 30 + 2 × 30   by rounding all numbers to 1 s.f. (significant figure)
≈ 50 + 80 + 90 + 60
≈ £280

You may be asked whether an estimate is an over-estimate or an under-estimate.

Example 7

(a) (i) Estimate the answer to this calculation: 160 × 8.78 + 48
   (ii) Is your answer to part (i) likely to be an over-estimate or an under-estimate of the true answer? Give a reason for your answer.
(b) (i) Estimate the answer to this calculation: 1492 + 1066 × 2024
   (ii) Is your answer to part (i) likely to be an over-estimate or an under-estimate of the true answer? Give a reason for your answer.

(a) (i)   160 × 8.78 + 48
         ≈ 200 × 9 + 50
         ≈ 1800 + 50
         ≈ 1850
   (ii) The answer in part (i) is an over-estimate of the true answer because all the numbers in the question were rounded up.
(b) (i)   1492 + 1066 × 2024
         ≈ 1000 + 1000 × 2000
         ≈ 1000 + 2 000 000
         ≈ 2 001 000
   (ii) The answer in part (i) is an under-estimate of the true answer because all the numbers in the question were rounded down.

Exercise 2D

1. Each of the following calculations is followed by three possible answers. By estimating the answer to the calculation, state which of the three answers is likely to be correct.

	Calculation	Options		
(a)	46.5 × 92.7	431.055	4310.55	43105.5
(b)	6392.12 ÷ 20.72	308.5	3085	30850
(c)	94 537.52 ÷ 327.8	2.884	28.84	288.4
(d)	96.7(321 + 240)	54248.7	542 487	5 424 870
(e)	$\dfrac{7.55 - 5.6}{0.48}$	40.625	4.0625	0.040625

2. Estimate the answer to each of the following by rounding each number to one significant figure.
   (a) 6.9 × 5.24   (b) 58.7 × 8.24   (c) 91.35 × 3.17   (d) 189 ÷ 21.65
   (e) 356.24 + 598.32 − 254.17   (f) 6.4 × (2.86)^2 (g) (0.48 × 17.2)^2

3. Estimate the answers to these calculations. Show each step of your working.
   (a) 37.3 × 4.48   (b) $\dfrac{52.7 \times 0.16}{2.19}$ (c) 4.2 × 24.92 − 39.4 ÷ 7.78
   (d) −6.94 − 21.98   (e) $\dfrac{1.597 \times 357.36}{0.48}$

4. Jamie has done some homework, which involves estimating the answer to some calculations. He has made a mistake in each part (a) and (b). Find and correct his mistakes.

(a) $\dfrac{612 \times 3.2}{27.94}$

$\approx \dfrac{600 \times 3}{30}$

$\approx \dfrac{1500}{30}$

$\approx \dfrac{150}{3}$

≈ 50

(b) $3.87(23.974 \times 6.979)$

$\approx 4(24 \times 7)$

≈ 882

5. Estimate the answers to the following calculations by rounding each number to one significant figure. In each case state whether your estimate is likely to be an over-estimate or an under-estimate.
   (a) $34.19 + 129.65$ (b) 28.8×86.7 (c) $0.053 + 0.0942$

6. Sharon is training for a running race. During one week she runs the following distances:
   5.65 km, 3.09 km, 8.87 km, 4.5 km
   Find an estimate for the total distance she runs this week.

7. A bin lorry collects rubbish from 1024 homes before it must be emptied. The rubbish in each bin has an average mass of 18.7 kg. Estimate the total mass of rubbish the bin lorry collects.

8. Jillian fills her car up with 57 litres of petrol, with each litre of petrol costing 79 p.
   (a) Estimate how much Jillian's petrol costs.
   (b) Is your answer to part (a) an over-estimate or an under-estimate? Explain your answer.

9. A walnut has a diameter of 3.825 cm. A hazelnut has a diameter of 0.842 cm. Roughly how many times bigger is the diameter of the walnut?

10. Caoimhe estimates the answers to the two calculations below by rounding each number to one significant figure. Which of her estimates will be closer to the true value? Explain your choice, but try to answer this question without estimating the answers yourself.
    Calculation 1: 5260×945
    Calculation 2: 5875×836

11. Estimate the answer to the following calculations, which are more challenging. Choose a suitable level of rounding for each number in the calculation. For example, you may decide, in some cases, that rounding a number to two or more significant figures, rather than one, makes the calculation easier, or that rounding to the nearest whole number is the best approach.
    (a) $\dfrac{32.4 - 28.7}{21.2 - 18.9}$ (b) $31.5 \div 0.813$ (c) $\dfrac{(3.71 + 6.32) \times 0.845}{0.548}$ (d) $\dfrac{67.1 \times 132.81}{2.12 - 1.88}$

2.4 Summary

In this chapter you have learnt how to:
- Round a number to a given number of decimal places.
- Round a number to a given number of significant figures.
- Estimate the answer for a calculation. Usually, you will do this by rounding each number in the calculation to one significant figure. However, there will be times when it is appropriate to round some or all of the numbers to a different level of accuracy.

Chapter 3
Percentages and Finance

3.1 Introduction

Percentages are used frequently in everyday life. For example, marks in school exams, the level of battery charge on your phone and reductions in prices during sales all use percentages. A percentage is a special way to write a fraction. Percent means out of 100, so a percentage can be expressed as a fraction with 100 as the denominator. For example, 35% means $\frac{35}{100}$

Percentages are used in many situations involving finance. For example:
- If a person invests money in a savings account they are given a certain percentage of interest, which depends on the amount invested and how long it is invested.
- If a person borrows money to buy something, a certain percentage of interest will be charged on the loan.
- Employees pay a certain percentage of their wage or salary in income tax and, in some cases, towards their pension.

Key words
- **Percentage**: Parts out of 100. For example, 60% means 60 parts out of 100, or $\frac{60}{100}$
- **Income tax**: A tax paid by most workers, calculated as a percentage of their income.
- **National Insurance**: Another tax paid by workers and employers, calculated as a percentage of a worker's income.
- **Pension**: A pension is a savings scheme, usually paid into by workers, then withdrawn and used during retirement.
- **Simple interest**: A method for calculating the interest earned on savings. Using this method, the amount of interest earned is the same each year.
- **Compound interest**: A method for calculating the interest earned on savings. Using this method, the amount of interest earned usually increases each year.
- **Hire purchase**: A method used to spread a large payment over a number of months or years.
- **Appreciation**: An item appreciates in value if its value goes up.
- **Depreciation**: An item depreciates in value if its value goes down.

Before you start you should know how to:
- Convert between fractions, percentages and decimals.

In this chapter you will learn how to:
- Calculate a percentage of a quantity.
- Perform a percentage increase or decrease.
- Write one number as a percentage of another.
- Calculate percentage change.
- Find the original quantity given the result of a percentage change.
- Work with income tax and national insurance.
- Work with VAT.
- Calculate simple and compound interest.
- Calculate appreciation and depreciation.
- Do calculations involving hire purchase and discounts.

Exercise 3A (Revision)

1. Convert each of these percentages to a simplified fraction and a decimal.
   - (a) 32%
   - (b) 6%
   - (c) 120%
   - (d) $3\frac{1}{2}$%
   - (e) 0.5%
2. Convert each of these decimals to a fraction and a percentage.
   - (a) 0.04
   - (b) 0.9
   - (c) 0.95
   - (a) 1.3
   - (b) 0.375

3. Convert each fraction to a percentage and a decimal.
   (a) $\frac{11}{50}$ (b) $\frac{2}{5}$ (c) $\frac{5}{8}$ (d) $\frac{19}{200}$ (e) $\frac{4}{9}$

4. Put each set of numbers in order from smallest to largest.
   (a) $35\%, \frac{1}{3}, 0.3$ (b) $0.7777, 77.7\%, \frac{77}{100}$ (c) $\frac{1}{5}, 0.5, 5\%$ (d) $133\%, 1.\dot{3}, \frac{13}{10}$

5. Out of all the pupils in a school, 31% of the pupils travel to school by bus, a proportion of 0.3 arrives by car and $\frac{1}{3}$ of the pupils walk. The rest take a train. Which way to travel to school is the most popular? Show your working.

3.2 Finding a Percentage of a Quantity With and Without a Calculator

You should know how to calculate a percentage of a quantity, both with and without a calculator.

Example 1

Calculate 45% of 600 (a) without a calculator (b) with a calculator.

(a) To find 45% of 600 without a calculator, find 10% and 5%:

10% of 600 means $\frac{1}{10}$ of 600, so we divide by 10:	10% of 600 = 60
Find 5% of 600 by halving 10% of 600:	5% of 600 = 30
Find four lots of 10%:	$4 \times 60 = 240$
Find 40% + 5%:	45% of 600 = 240 + 30 = 270

> **Note:** There are many ways to do this. For example, you could find 5% and then multiply by 9

(b) To find 45% of 600 with a calculator, remember that the word 'of' can be replaced with a multiplication. Type one of the following:

$45\% \times 600 =$ (On the Casio fx-83GT CW, the % sign is found under Catalog > Probability. On some older Casio models it can found using SHIFT-ANS.)

$0.45 \times 600 =$ (by changing 45% to a decimal)

$\frac{45}{100} \times 600 =$ (by changing 45% to a fraction)

All of these give the answer 270

Exercise 3B

1. Find, without a calculator:
   (a) 10% of 400 (b) 15% of 360 (c) 95% of 1600 (d) $\frac{1}{2}$% of 8000
   (e) 120% of 6 kg (f) 12% of 25 cm

2. Find, using a calculator:
   (a) 25% of 350 (b) 82% of 630 (c) 88% of 520 lb

3. Year 11 vote for a pupil to sit on the School Council. There are three pupils to choose from and everyone voted. Abi Appleby received 44% of the votes. 26% voted for Ben Bradley.
   (a) What percentage of pupils voted for Claire Cooper?
   (b) 50 pupils voted altogether. How many voted for the winning pupil?

4. On the first day of the new school year, the head teacher of Meadowlands School, Mrs Marshmallow, says 'At Meadowlands School the student population is now 105% of last year's number!' Last year there were 700 pupils in the school. How many pupils are there now?

3.3 Expressing One Quantity as a Percentage of Another

You should know how to express one quantity as a percentage of another.

Example 2

Express 36 as a percentage of 50 **(a)** without a calculator **(b)** with a calculator.

In both parts, first write 36 out of 50 as a fraction: $\frac{36}{50}$

(a) Find an equivalent fraction with a denominator of 100: $\frac{36}{50} = \frac{72}{100} = 72\%$

(b) Type $\frac{36}{50}$ into your calculator, press EXE, then press FORMAT. Move down to 'Decimal' using the down arrow key. On older Casio models, use $\boxed{S \leftrightarrow D}$ to change the fraction to a decimal.

Then multiply the decimal by 100% to change it to a percentage: $\frac{36}{50} = 0.72 = 72\%$

Exercise 3C

1. Copy and complete the following.
   (a) 17 is _____% of 100
   (b) 20 is _____% of 200
   (c) 26 is _____% of 104
   (d) 80 cm is _____% of 200 cm
   (e) 32 g is _____% of 40 g

2. (a) What percentage of 60 is 15?
   (b) What percentage is 60 of 90?
   (c) What percentage is 105 cm of 140 cm?
   (d) 1600 litres is what percentage of 1200 litres?

3. Each week Bethany saves £3 of the £15 she earns. What percentage of her earnings does she spend?

4. In 2012, 3.2 million people in the UK had a disease called diabetes. By 2020 that number had increased to 4 million.
   (a) What percentage of the 2020 figure is the 2012 figure?
   (b) What percentage of the 2012 figure is the 2020 figure?

5. In a church choir, there are 60 people. Twenty-one members of the choir are male.
   (a) What percentage of the choir is male?
   (b) What percentage of the choir is not male?

6. Fintan is running a marathon, which is 40 km long. He passes a sign saying: **18 km**
   (a) What percentage of the race has Fintan completed?
   (b) What percentage does he still have to complete?
   (c) Fintan's friend Alfie is 3 km behind him. What percentage of the race has Alfie completed?

3.4 Percentage Increase and Decrease

You may be asked to increase or decrease a quantity by a percentage of the original amount.

Example 3

Find the amount when £240 is **(a)** increased by 15% **(b)** decreased by 15%.

Two methods are shown.

Method 1: Finding 15%

There are many ways to find 15% of £240 without a calculator. One of them is:

10% of £240 is $\frac{1}{10} \times £240 = £24$

So: 5% is £12 (half of our 10%)

So: 15% is £24 + £12 = £36

(a) An increase of 15% gives £240 + £36 = £276
(b) A decrease of 15% gives £240 − £36 = £204

Method 2: Use a multiplier (calculator method)
(a) To increase by 15%, the multiplier is 1 + 0.15 = 1.15   £240 × 1.15 = £276
(b) To decrease by 15%, the multiplier is 1 − 0.15 = 0.85   £240 × 0.85 = £204

Alternatively, you may be asked to increase or decrease by a fraction of the original.

CHAPTER 3: PERCENTAGES AND FINANCE

Example 4

A rail fare of £28 is increased by $\frac{3}{7}$. Find the new fare.

The increase is $\frac{3}{7}$ of £28

First find $\frac{1}{7}$ of £28: $\frac{1}{7}$ of £28 = £4

So: $\frac{3}{7}$ of £28 = 3 × £4 = £12

The fare is increased by £12, so the new fare is £28 + £12 = £40

If you are told the price that was paid for an item and the percentage profit or percentage loss, you can calculate the selling price.

Example 5

(a) The cost price of a bedroom rug is £80
A carpet shop sells it at a profit of 12%. What is the selling price?

(b) A bookshop buys a maths textbook for £24. A more recent edition has been written by the same author, so the bookshop sells the book at a loss of 25%. What is the selling price?

(a) To find the selling price, the cost price is increased by 12%.
Using the multiplier method: Selling price = £80 × 1.12 = £89.60

(b) To find the selling price, the cost price is decreased by 25%.
Using the multiplier method: Selling price = £24 × 0.75 = £18

Exercise 3D

1. Work out the following.
   (a) Increase 40 by 10%   (b) Increase 40 by 25%   (c) Increase 39 by $33\frac{1}{3}$ %
   (d) Decrease 36 by 10%   (e) Decrease 41 by 100%

2. (a) Copy and complete the following table, finding 10%, 5% and 20% of each number in the left-hand column.

Number	10%	5%	20%
80	8	4	16
24	2.4		
360			
4200			

   (b) Using the answers in your table to help you, copy and complete the following two statements:
   10% of 80 is ____
   80 increased by 10% is 80 + ____ = ____
   (c) Using the information in your table, answer the following questions:
   (i) Increase 24 by 10%   (ii) Find 360 increased by 10%
   (iii) Increase 360 by 15%   (iv) Find 80 decreased by 15%
   (v) Find 4200 increased by 20%   (vi) Decrease 4200 by 15%

3. The length of a blue whale calf is 6 metres. As an adult it will be 250% longer. How long will the whale become?

4. Johnny writes articles for his local newspaper. In 2023 he wrote 108 articles. In 2024 his number of articles increased by $\frac{2}{9}$. How many articles did he write for the paper in 2024?

5. Between 2020 and 2024, the cost of a particular medicine fell by $\frac{2}{3}$.
If the medicine cost £54 for 100 tablets in 2020, how much does it cost for 100 tablets in 2024?

6. These items go up or down in value by the percentage shown. Find the new value of each item using a multiplier.
   (a) An old postage stamp £4, increased by 20%   (b) Scooter £45, down 10%
   (c) Antique pottery £130, increased by 65%   (d) Electric piano £540, down 25%

7. The government awards a 4% pay increase to all public sector workers. Work out the new salary for:
   (a) Carol, who earned £16 000
   (b) Cormac, who earned £18 000
   (c) Charlie, who earned £25 000

3.5 Calculating Percentage Change

Sometimes it is important to work out the percentage by which a quantity has increased or decreased. To do this we need to work out the change in the quantity (increase or decrease) and express it as a percentage of the original amount.

Example 6

Jim increases the number of vinyl records in his collection from 500 to 650
What is the percentage increase?

The increase is 650 − 500 = 150 records.

Write **the increase over the original amount** as a fraction: $\frac{150}{500}$

To convert to a **percentage increase**: $\frac{150}{500} = \frac{30}{100} = 30\%$

Example 7

Rockie the dog weighed 28.0 kg. His owner put him on a programme of long daily walks. After six months his weight is 22.4 kg. What percentage weight has Rockie lost?

Weight lost = 28.0 − 22.4 = 5.6 kg

Write **the decrease over the original amount** as a fraction: $\frac{5.6}{28.0} = \frac{56}{280} = \frac{8}{40} = \frac{1}{5} = 20\%$

The percentage decrease is 20%. Rockie has lost 20% of his weight.

Example 8

(a) A pair of trainers is bought for £60 and later sold for £102
    Find the percentage profit.
(b) A valuable painting is bought at auction for £19 200 and is later sold at a loss for £12 000
    What is the percentage loss?

(a) Calculate the profit made when the trainers are sold: £102 − £60 = £42
    Write the profit over the original amount (the buying price) as a fraction and convert this to a percentage: $\frac{42}{60} = \frac{7}{10} = 0.7 = 70\%$
    The trainers are sold at a profit of 70%.

(b) Calculate the loss made when the painting is sold: £19 200 − £12 000 = £7200
    Write the loss over the original amount (the buying price) as a fraction and convert this to a percentage: $\frac{7200}{19\,200} = \frac{3}{8} = 0.375 = 37.5\%$
    A percentage loss of 37.5% was made when the painting was sold.

Exercise 3E

1. Find the percentage increase or decrease that takes you from A to B in each case.

	(a)	(b)	(c)	(d)	(e)	(f)	(g)	(h)
A	100	10	100	150	20	64	660	200
B	105	11	96	100	28	16	495	199

2. Simon bought a car for £10 200 and later sold it for £9180. What percentage loss did Simon make?
3. Some gym equipment is purchased for £210 and later sold for £294. Find the percentage profit.

4. Look at the triangle shown on the right.
   (a) Sketch an enlarged shape with all the side lengths doubled.
   (b) What is the percentage increase in the side length?
   (c) Find the area of the original triangle. The formula for the area of a triangle is:
   Area = $\frac{1}{2}$ × base × perpendicular height
   (d) Find the area of the enlarged triangle.
   (e) What is the percentage increase in the area?

5. A bottle of fizzy drink contains 30 grams of sugar.
   (a) To make it a healthier drink, the factory reduces the sugar to 18 grams. What percentage reduction is this?
   (b) Sales of the drink go down and the company increases the sugar content back to 30 grams. What percentage increase is this?

6. A motor race is over 30 laps of a track that is 2 km long.
   (a) What distance is the race?
   (b) For the new season, the course organisers increase the length of the track by 10%. What is the new length of the track?
   (c) The organisers also increase the number of laps by 10%. How many laps do the cars have to travel now?
   (d) What is the new distance for the entire race?
   (e) What is the percentage increase in the length of the race?

7. You may use your calculator in this question. The table below gives the prices of four types of food in 2022 and 2023.
   (a) Copy and complete the table to show the percentage change for each item.

Food	2022 price	2023 price	Price increase	Increase or decrease	As a percentage
500 g sausages	£3.70	£4.40	£0.70	$\frac{70}{370}$ = 0.189	18.9%
400g jar of coffee	£4.10	£5.00			
1 kg potatoes	£1.50	£1.80			
12 eggs	£1.30	£1.50			

   (b) Phil's shopping list is shown on the right.
   What was the percentage increase in the cost of Phil's shopping between 2022 and 2023?

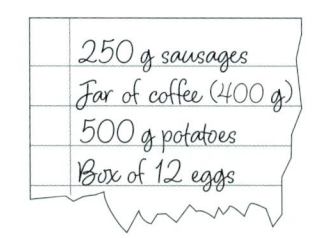

   250 g sausages
   Jar of coffee (400 g)
   500 g potatoes
   Box of 12 eggs

8. The table below shows the company profits at Andaman Productions for the years 2018 to 2023.

Year	2018	2019	2020	2021	2022	2023
Profit (£)	35 000		46 200	41 580	58 212	43 659
Percentage increase/decrease (compared with previous year's profit)		20% increase				

Using the values in the table, find:
(a) The company profit in 2019.
(b) The percentage decrease in profit in 2021, when compared with the 2020 figure.
(c) The percentage increase in profit in 2022, when compared with the 2021 figure.
(d) The percentage increase in profit in 2020, when compared with the 2019 figure.

3.6 Finding The Original Quantity

Sometimes you know the new cost or size of something and the percentage change. You may be asked to find the original cost or size. This is sometimes called a **reverse percentage**, because you are working backwards from the new cost or size to the original.

Example 9

A coat costs £45 in a sale. Its price has been reduced by 10%. Find the original price.

If the price has been reduced by 10%, the new price is 90% of the original price.

Method 1: Find 1% of the original price and then 100%

90% of the original price is £45

Divide by 90 to find 1% of the original price:

90% = £45
÷ 90
1% = £0.50

Multiply by 100 to find 100%, the original price:

× 100
100% = £50

Note: Remember, the original amount is always represented by 100%.

So the original price of the coat was £50

Method 2: Use a multiplier

The new price is 90% of the original price, so the original price has been multiplied by 0.9 to find the new price.

To reverse the process, **divide** by the multiplier to find the original price: $\frac{45}{0.9}$ = £50

So the original price of the coat was £50

Example 10

After the addition of an extra storey, a multi-storey car park is 12.5% bigger than last year.
It now has a total area of 14 400 m^2
(a) What was the area of the car park last year in square metres?
(b) What is the size in square metres of the new storey?

(a) If the size has increased by 12.5%, the new size is 112.5% of the original size.

Method 1: Find 1% of the original size and then 100%

112.5% of the original size is 14 400 m^2

Divide by 112.5 to find 1% of the original size:

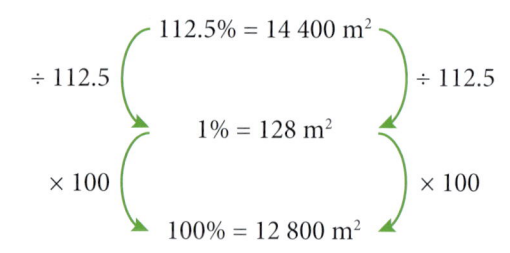

112.5% = 14 400 m^2
÷ 112.5
1% = 128 m^2

Multiply by 100 to find 100%, the original size:

× 100
100% = 12 800 m^2

The original size of the car park was 12 800 m^2

Method 2: Use a multiplier

The new size is 112.5% of the original size, so the original size has been multiplied by 1.125 to find the new size.

To reverse the process, **divide** by the multiplier to find the original size: $\frac{14\,400}{1.125}$ = 12 800 m^2

The original size of the car park was 12 800 m^2

(b) The size of the new storey is the difference between the car park's new size and old size.
14 400 − 12 800 = 1600   The size of the new storey is 1600 m^2

VAT is a tax added to the price of some goods in shops. Currently, VAT is charged at 20%.
Usually, prices in the shops already have VAT added.
We can use reverse percentages to find the price before the VAT was added.

Example 11

A school bag is priced at £24, including VAT at 20%. How much VAT was included in the price?

Since VAT is charged at 20%, the price of £24 is 120% of the price before VAT.
Here we use Method 2 (using a multiplier), but Method 1 is also possible.
The original price has been multiplied by 1.2 to find the new price.
To reverse the process, **divide** by the multiplier 1.2 to find the original price: $\frac{24}{1.2} = 20$
The price of the school bag was £20 before the VAT was added.
The VAT added was £24 − £20 = £4

Some questions involving percentage profit and loss may require reverse percentage calculations. In the following example, you are given the selling price and the percentage profit. Using reverse percentages, you can calculate the buying price.

Example 12

A violin is sold in a music shop for £55 at a percentage profit of 10%. What did the shop buy it for?

Using Method 2: if the shop's profit is 10%, the selling price is 110% of the buying price. In other words, the buying price is multiplied by 1.1 to give the selling price.
To reverse the process, divide by the multiplier 1.1 to find the buying price: $\frac{55}{1.1} = 50$
The violin was bought by the shop for £50

Exercise 3F

1. Find 100% if:
   (a) 120% is 9.6
   (b) 45% is 6.75 metres
   (c) 15% is 240 cm
   (d) 40% is 38.8
   (e) 25% is 19.25 kg
   (f) 60% is 53.4 litres
   (g) 75% is 8850 km
   (h) 95% is 798 cm
   (i) 25% is 48.75 miles
   (j) 80% is 20 000 cockroaches

2. The sale price of a pair of designer sunglasses is £96
   (a) The reduction in the sale is 20%. What percentage of the original price is the sale price?
   (b) Find the original price.

3. Mr Patterson sells his house at a 15% loss. The sale price is £340 000
   How much did Mr Patterson buy the house for?

4. Sale prices in a department store are 80% of the original price.
   Find the original price for each of these items.
   (a) Towels: sale price £24
   (b) Table lamp: sale price £70
   (c) Table and chairs: sale price £135
   (d) Cutlery set: sale price £225

5. A bottle of shower gel says it contains 30% extra free. The bottle contains 780 ml of shower gel. What volume of shower gel is in a bottle that does not have the extra 30%?

6. House prices in Belfast have risen by 2% in the last month.
   (a) Taking last month's value as 100%, what percentage is the current value?
   (b) A house in Belfast is sold for £367 200
        What was the value of the house one month ago?

7. Sale prices in a shoe shop are 70% of the original price. Find the original price for each of these items.
   (a) Trainers £63
   (b) Boots £84
   (c) Shoes £45.85

8. These prices include VAT at 20%, so they are 120% of the original prices. Find the price before VAT.
   (a) £36 (b) £180 (c) £1080 (d) £97.20

9. VAT (value added tax) is 20%. It is added to the original price (100%) of an item. The prices of the items shown include VAT. For each of the items (a) to (c), find:
   (i) the price excluding VAT, (ii) the amount of the VAT added.
   (a) Tablet £600 (b) Phone £180 (c) Printer £192

10. A restaurant buys a new cooker costing £480 including VAT at 20%.
    (a) What was the price of the oven excluding VAT?
    (b) How much was the VAT paid?

11. These prices include VAT at 20%. Find the price before VAT is added.
    (a) Printer ink £22.50 (b) Mobile phone bill £51.60
    (c) Burger and chips £9.12 (d) Hooded top £43.20

12. An ice sculpture begins to melt and loses 5% of its volume. If the sculpture now has a volume of 28.5 m^3
    (a) What was its original volume? (b) What volume of ice has melted?

3.7 Percentages in Finance

Tax, National Insurance and pensions

Money earned is called **gross pay**. Most people do not normally receive all their gross pay, as certain amounts of money are deducted from it. What they receive after the deductions is called their **net pay**, sometimes called **take-home pay**. The main deductions from wages and salaries are income tax, National Insurance and pension contributions.

Income tax

Income tax is used to finance government spending on things like schools, roads and hospitals. Most people have income tax deducted from their pay before they receive it.

A certain amount of each person's income is not taxed. This amount is called a **personal allowance**. A person's **taxable income** is their gross pay minus their personal allowance.

The income tax each person pays is a certain percentage of their taxable income. The income tax rate can vary. Some people have to pay a higher rate of tax if their taxable income is over a certain amount.

National Insurance

National Insurance is paid by both the employee and the employer. It helps pay for the National Health Service, statutory sick pay, unemployment benefits and the state pension.

The amount of National Insurance paid depends on a person's gross pay and whether they are in the state pension scheme. It is usually a certain percentage of total income.

Pensions

Some companies offer a **pension** or **superannuation** scheme that their employees can join. Employees can pay a percentage of their earnings into the pension scheme. When they retire, they get a pension through their employer as well as the state pension.

Example 13

Bill works for a manufacturing company. His weekly gross pay is £380
(a) There are 52 weeks in a year. Calculate Bill's gross annual salary.
(b) Bill has a personal allowance of £12 000 per year. What is his taxable income?
(c) Bill pays income tax at 20% on the remainder of his income. How much income tax does he pay per year?
(d) Bill pays National Insurance on his earnings above £12 584 at a rate of 13.25%. How much National Insurance does he pay per year?
(e) Bill also contributes 5% of his gross pay to a company pension scheme. How much does he pay into the pension scheme per year?
(f) Find Bill's total annual deductions.
(g) Calculate Bill's annual take-home pay.

(a) There are 52 weeks in a year. To calculate Bill's gross annual salary:
  52 × £380 = £19 760
(b) Bill's taxable income is the difference between his gross annual salary and his personal allowance:
  £19 760 − £12 000 = £7760
(c) To calculate the income tax payable: 20% of £7760 = 0.2 × £7760 = £1552
(d) Bill pays National Insurance on: £19 760 − £12 584 = £7176
  13.25% of £7176 = 0.1325 × £7176 = £950.82
(e) To calculate Bill's pension contribution: 5% of £19 760 = 0.05 × £19 760 = £988
(f) Bill's total deductions are: £1552 + £950.82 + £988 = £3490.82
(g) Bill's annual take-home pay is: £19 760 − £3490.82 = £16 269.18

Example 14

Bill's boss Seán earns £150 000 per year. Like Bill, Seán has a personal allowance of £12 000 per year.
He pays tax at 20% on his earnings between £12 000 and £50 000
He pays tax at 40% on his earnings above £50 000
(a) Calculate Seán's annual income tax bill.
(b) Seán also pays £8200 in National Insurance and pays £13 000 annually towards his pension. Calculate Seán's annual take-home pay.

On the first £12 000, Seán pays no income tax.

On the next £38 000, Seán pays income tax at 20%:
20% of £38 000 = 0.2 × £38 000 = £7600

On the remainder of his income, £100 000, Seán pays 40% income tax:
40% of £100 000 = £40 000

(a) Seán's total income tax bill is: £7600 + £40 000 = £47 600
(b) Seán's annual take-home pay is: £150 000 − £47 600 − £8200 − £13 000 = £81 200

Exercise 3G

1. Tim's gross wage is £280 per week. His take-home pay is £201.60
   (a) How much were the deductions?
   (b) What percentage of his gross wage were the deductions?
2. Áine earns £42 000 per year. She gets a tax-free allowance of £12 000 per year. She pays 15% tax on the first £15 000 of her taxable income and 20% tax on the rest. How much tax does Áine pay per year?
3. Jenna earns £50 000 and has a personal allowance of £12 500
   (a) What is Jenna's taxable income?
   (b) Jenna pays 20% tax on the first £17 500 of her taxable income. The remainder of her income is taxable at 40%. How much income tax does Jenna pay in total per year?

Simple interest

If you put money into a bank or building society account, you will be given **interest** every year. Two types of interest are covered in this chapter.
- With **simple interest**, the amount of interest given remains the same each year.
- The other type of interest calculation is **compound interest**, which is covered in the next section. Using this, the interest paid changes each year if the balance increases each year.

> **Note:** You will often see the words **per annum** (or **p.a.**) alongside an interest rate. This means 'per year'.

Example 15

Rhiannon puts £200 into a bank savings account. The bank account pays 6% simple interest per annum. Find out how much Rhiannon has in her savings account at the end of:
(a) 1 year   (b) 2 years   (c) 3 years

GCSE MATHEMATICS M3 AND M7

1% of £200 is $\frac{1}{100} \times £200 = £2$ 6% of £200 is $6 \times £2 = £12$

Using simple interest, the interest is £12 per year. Therefore:
(a) After 1 year Rhiannon has £200 + £12 = £212 in the account.
(b) After 2 years the balance is £224   (c) After 3 years it is £236

Compound interest

In the previous section you learnt about simple interest. Using this approach, the interest on savings remains the same every year. However, most banks and building societies use **compound interest**. Using this approach, the interest for the first year is calculated and added to the balance. At the end of the second year, the interest is calculated as a percentage of this **new** balance. And so on. Compound interest for the second year is usually greater than that calculated in the first year, because it is a percentage of a larger balance.

The following example shows two methods to calculate compound interest. The second method uses the compound interest formula:

$$A = P\left(\frac{100 + R}{100}\right)^n$$

where: A is the closing balance after n years
        P is the opening balance (or **principal**)
        R is the interest rate

Example 16

Carolyn saves for a new oven costing £400
She has been given £350
She sees the advertisement shown on the right.
Carolyn invests the money in the Platinum Saver Account for 4 years.

Platinum Saver Account
AER (Annual Equivalent Rate)
5.2% compound interest

(a) Will she have enough money in the savings account to be able to buy the new oven in 3 years' time? Show all your working.
(b) How much interest does Carolyn earn on her investment?

(a) Two methods are given to find the closing balance in Carolyn's account after 3 years.

Method 1
First year interest   = 5.2% of £350
                       = 0.052 × £350
                       = £18.20
Balance at end of first year = £350 + £18.20 = £368.20
Second year interest   = 5.2% of £368.20
                       = 0.052 × £368.20
                       = £19.15 (to the nearest penny)
Balance at end of second year = £368.20 + £19.15 = £387.35
Third year interest   = 5.2% of £387.35
                       = 0.052 × £387.35
                       = £20.14 (to the nearest penny)
Balance at end of third year = £387.35 + £20.14 = £407.49

Method 2
Use the compound interest formula:
$$A = P\left(\frac{100 + R}{100}\right)^n$$

$$A = £350 \left(\frac{100 + 5.2}{100}\right)^3 = £407.49$$

Carolyn will have more than £400, so she will have enough to buy the oven.

Note: Method 2 is the fastest method to find the closing balance, but it doesn't tell you the amount of interest earned each year.

(b) To calculate the interest, subtract Carolyn's initial investment from her final balance:
Total interest earned = £407.49 − £350 = £57.49

Alternatively, if Method 1 was used in part (a), the total interest can be calculated by adding up the

interest earned in each of the three years:
Total interest earned = £18.20 + £19.15 + £20.14 = £57.49

Appreciation and depreciation

The compound interest formula can be used to calculate **appreciation** and **depreciation** in the value of an item, as shown in the next example.

Example 17

(a) A house appreciates in value by 5% each year for 3 years. If it was purchased for £150 000 in 2020, how much is it worth in 2023?

(b) A government department buys a supercomputer for £2 000 000. Its value depreciates by 15% each year for 5 years. What is the supercomputer's value at the end of this time? Round the answer to the nearest £1000.

(a) To calculate appreciation and depreciation, we can use the compound interest formula

$$A = P\left(\frac{100 + R}{100}\right)^n$$

To find the value of the house after 3 years, use the compound interest formula with the initial value $P = 150\,000$, the rate $R = 5$ and the number of years $n = 3$

$$A = 150\,000 \left(\frac{100 + 5}{100}\right)^3$$

$A = £173\,644$ (to the nearest pound)

(b) For depreciation, use a negative value of R.
Since the supercomputer's value falls by 15% per year, $R = -15$

$$A = 2\,000\,000 \left(\frac{100 - 15}{100}\right)^5$$

$A = £887\,410.625 = £887\,000$ (to the nearest £1000)

Exercise 3H

1. Find the simple interest earned when the following amounts are invested.
   (a) £250 for 3 years at 2% per annum
   (b) £500 for 4 years at 4% per annum
   (c) £550 for 3 years at 3% per annum
   (d) £700 for 2 years at 1.5% per annum

2. Find the total amount in each of these simple interest accounts at the end of the time given.
   (a) £1000 for 2 years at 5% per annum
   (b) £350 for 3 years at 2% per annum
   (c) £1200 for 3 years at 1.75% per annum
   (d) £650 for 4 years at 3.5% per annum

3. Niall invests £350 in the Blackrock Building Society for one year. The building society pays simple interest of 2.5% per year. How much interest will Niall receive each year?

4. Sheena is saving up for her wedding in 3 years' time. She invests £4000 in a savings account that offers 4.9% simple interest. She hopes the balance on her account is at least £5000 at the time of her wedding. Does the balance on her account reach £5000? You must show all your working.

5. Find the total amount in each of these compound interest accounts at the end of the time given.
   (a) £300 for 3 years at 5% per annum
   (b) £1500 for 2 years at 2 percent per annum
   (c) £820 for 4 years at 3% per annum
   (d) £500 for 4 years at 5% per annum

6. Joanna saves £3500 for 4 years at 2% per annum **simple interest**. Her brother David saves the same amount for 4 years at 2% per annum **compound interest**. Who has more money at the end of the four years? Show all your working.

7. Niamh wants to put £10 000 into a savings account. She may need to withdraw the money at short notice for a large purchase. She sees an advertisement online comparing two savings accounts, both paying compound interest. The advertisement is shown overleaf.

	AER	Terms and conditions
Wise Owl Savings Account	8.88%	90 days' notice required for withdrawals
Bouncing Tiger Savings Account	5.54%	No notice required for withdrawals

(a) Which account should Niamh choose? Explain why.
(b) How much interest will she receive if she leaves the money in the account for one year?

8. Charlie invests £350 in a Monthly Target account for 3 years. The account pays an AER of 5.54% p.a. Will Charlie have enough money in the account to buy an electric scooter for £420 in 3 years' time? Show all your working and give a reason for your answer.

9. What will be the total amount of money if an initial investment of $5000 is made at a compound interest rate of 6% per annum for 5 years?

10. A classic car can be bought for £10 840. The value of the car will increase by 15% every year for the next three years. How much will the car be worth after three years? Give your answer to the nearest pound.

11. Lewis is planning to go on a cruise when he retires in 15 years' time. He has £5500 to invest. In 15 years' time he is hoping to have £7300 to spend on the cruise. He puts his money into a savings account paying 1.95% AER.
    (a) Does Lewis have enough money in the account when he retires? Show all your working.
    (b) What assumptions have you made?

Hire purchase

If a person buys an item, there may be different payment options. One option is that the buyer can pay the whole price immediately. This is sometimes called the **cash price**.

However, there may be an option to pay for the item over a period of time. In this option, the buyer pays a **deposit** immediately and then pays the rest of the cost (the **balance**) in instalments. We say they are paying by **credit**, or by **hire purchase**.

The difference between the cash price and the total price paid by high purchase is called a **credit charge**. Usually, paying by hire purchase is more expensive than paying in full at the time of purchase, but it has some advantages for the buyer as well, as the next example shows.

Example 18

Michael is buying an electric car. He is offered two payment options:
Option 1. He can pay the cash price of £19 000
Option 2. He can pay half of the cash price now. Then he must make 24 monthly payments of £400
(a) Which option is cheaper?
(b) What reasons may Michael have for choosing the more expensive option?

(a) Using Option 2, Michael would pay £9500 now, followed by 24 monthly payments of £400
9500 + 24 × 400 = £19 100
Under Option 2, Michael would pay £19 100 in total. So Option 1 is the cheaper.

(b) There are several reasons Michael may decide to choose Option 2, even though it is slightly more expensive:
- The **credit charge** is £100, which is not much compared with the total amount he is spending.
- He may not have the full amount of £19 000 now.
- He may have the full amount, but may decide that paying only £9500 now is better, as he can put the rest of the money into a savings account and earn interest.

Discounts

It is sometimes possible to get a **discount** on the price of some goods or services. This may happen as a reward for loyalty. Or a company may offer a discount to everyone for a period of time in order to attract customers. The discount may be a certain amount of money taken off the full price, or it may be a certain percentage taken off the full price.

Example 19

After 3 years of gym membership, Brigid is offered a 10% discount on her monthly payments. Her previous monthly payment was £22.50. How much does she pay now?

Calculate the discount: 10% of £22.50 = 0.1 × £22.50 = £2.25
This is the amount taken off the monthly payment.
So Brigid's new monthly payment is: £22.50 − £2.25 = £20.25

Exercise 3I

1. Pam wants to buy a new armchair. She can pay either:
   - £475 cash; or
   - £100 cash and then 12 monthly instalments of £35

   Pam decides to pay using the second option.
   (a) What is the deposit she must pay?
   (b) What is the balance after she has paid the deposit?
   (c) What is the total price Pam pays?
   (d) What is the credit charge?

2. Janet's vet's bill comes to £6050
   Her insurance covers £2500 of this bill.
   (a) How much does Janet have to pay herself?
   She has two payment options:
   - Option 1: Pay the full amount now.
   - Option 2: Pay £500 now and then another 7 monthly payments of £500
   (b) Assuming Janet chooses to pay using Option 2, what is the credit charge?

3. Jonny is a journalist. He reads an online news website called *All Ireland News*. The subscription costs him £8.50 per month. When he renews his subscription he is offered a loyalty discount of 10%. A different news website called *What's Going On?* has just started up. It charges £9 per month, but offers a discount of £1.20 per month to new subscribers. Find the monthly cost of each news website and state which news website is cheaper.

4. An electric bicycle is priced at £350
   It can be bought for a deposit of £70 followed by six instalments of £50
   (a) What is the price of buying the electric bicycle using the hire purchase option?
   (b) What is the difference between the cash price and the hire purchase price?

5. A wardrobe is on sale for a cash price of £380
   It can also be bought on hire purchase using a deposit of 25% of the cash price, followed by monthly instalments of £16 for two years.
   (a) What is the price of buying the wardrobe using the hire purchase option?
   (b) What is the difference between the cash price and the hire purchase price?

3.8 Summary

In this chapter, you have learnt about:
- Finding a percentage of a quantity with and without a calculator.
- Expressing one quantity as a percentage of another.
- Increasing or decreasing by a percentage, including salaries.
- Calculating a percentage increase or decrease, including percentage profit and loss.
- Applying successive percentage change.
- Money and percentages in the context of finance, including:
  - Salaries, income tax, National Insurance and pensions,
  - VAT,
  - Bank accounts: simple interest and compound interest,
  - Hire purchase,
  - Discounts.

Chapter 4
Bounds

4.1 Introduction

When we measure any quantity, we can only give our measurement correct to the degree of accuracy of the measuring device we are using. For example, when using a ruler marked in millimetres, you cannot accurately measure a length to a greater accuracy than the nearest millimetre, for example 68 mm.

Quantities such as length, mass and time are continuous quantities, i.e. they can take any value. So, for example, an object could have a length of 68.1428 mm.

A length measured as 68 mm to the nearest millimetre may in fact be anywhere between 67.5 mm and 68.5 mm. These two values are known as the **lower** and **upper bounds** for a measurement.

In a sense, finding the lower and upper bounds is the opposite of rounding. In the example given above, we are finding two lengths between which all lengths round to 68 mm.

Key words

- **Rounding**: Finding an approximation of a number or a quantity, for example rounding 68.1428 mm to the nearest millimetre gives 68 mm.
- **Lower and upper bounds**: Two numbers between which everything rounds to a given number or quantity. For example, for 68 mm (measured to the nearest mm), the lower and upper bounds are 67.5 mm and 68.5 mm. Anything between these values rounds to 68 mm.

Before you start you should know how to:

- Round to the nearest whole number.
- Round to a given number of decimal places.
- Round to a given number of significant figures.
- Estimate the value of a calculation by rounding each of the numbers involved.

In this chapter you will learn:

- About the continuous nature of measurements.
- How to calculate upper and lower bounds for measurements.

Exercise 4A (Revision)

1. Round these numbers to the nearest integer.
   (a) 20.95   (b) 3.125   (c) 99.75   (d) 49.45   (e) 0.489

2. Round these quantities to the given accuracy.
   (a) 4.2 kg to the nearest kilogram
   (b) 6.25 km to the nearest kilometre
   (c) 16.75 litres to the nearest litre
   (d) 3.47 s to the nearest second
   (e) 6.5596 km to the nearest **metre**

3. Round these numbers to the given accuracy.
   (a) 21.32 to 1 decimal place
   (b) 9.25 to 1 significant figure
   (c) 38.893 to 1 decimal place
   (d) 18.674 to 1 significant figure
   (e) 35.989 to 2 decimal places
   (f) 0.04527 to 2 decimal places
   (g) 0.03982 to 2 significant figures
   (h) 169 000 to 1 significant figure
   (i) 14 352 to 3 significant figures

4. Round these quantities to the given accuracy.
   (a) 6.78 km to 1 decimal place
   (b) 3.115 m to 2 decimal places
   (c) 7.145 s to 2 significant figures
   (d) 100.945 cm to 2 significant figures
   (e) 25.98 litres to 1 decimal place

5. Estimate the answer to each of the following calculations.
   (a) $\dfrac{3.14159 \times 9.81}{1.618}$
   (b) $2(4.962 + 5.188) - 9.18$
   (c) $\dfrac{(9.16 - 4.75)^2}{2.1}$
   (d) $\dfrac{8.75 \times 11.38}{31.2 - 29.7}$

4.2 Finding Upper and Lower Bounds for a Single Measurement

The lower and upper bounds are two numbers between which a measurement must lie. For example, a mass of 42 kg, measured to the nearest kilogram, must lie between 41.5 kg and 42.5 kg.

Whenever measurements are given, we must look at the degree of accuracy to determine the bounds.

If a mass is measured as 42.0 kg to one decimal place, it has been measured with an accuracy of 0.1 kg. So the mass must lie between 41.95 kg and 42.05 kg. Anything between these two values would round to 42.0 kg to one decimal place.

One method for finding the lower and upper bounds is to follow these steps:

- Halve the accuracy. For example, if a mass has been measured to the nearest kilogram, the accuracy is 1 kg. Halving this gives 0.5 kg.
- Subtract this from the measurement to find the lower bound; and add this to the measurement to find the upper bound. For the mass, the lower bound is 42 − 0.5 = 41.5 kg. The upper bound is 42 + 0.5 = 42.5 kg.

Example 1

Find the lower and upper bounds for:
(a) A measurement of 720 g measured to the nearest gram.
(b) A measurement of 720 g measured to the nearest ten grams.

(a) The accuracy is 1 gram. Halving this gives 0.5 g.
    The lower bound is 720 − 0.5 g = 719.5 g   The upper bound is 720 + 0.5 g = 720.5 g
(b) The accuracy is 10 grams. Halving this gives 5 g.
    The lower bound is 720 − 5 g = 715 g   The upper bound is 720 + 5 g = 725 g

Example 2

(a) A sprinter's time for a 100 m race is measured as 11.3 seconds to one decimal place. Find the lower and upper bounds for the sprinter's time.
(b) A feather's mass is given as 2.37 grams to three significant figures. Find the lower and upper bounds for the feather's mass.
(c) The size of a football crowd is reported as 15 600 to three significant figures. Find the lower and upper bounds for the crowd's size.

(a) If the measurement of 11.3 seconds is accurate to one decimal place, the accuracy is 0.1 s.
    Halving this gives 0.05 s
    The lower bound is 11.3 − 0.05 s = 11.25 s   The upper bound is 11.3 + 0.05 s = 11.35 s
(b) The feather's mass is given as 2.37 grams to three significant figures.
    The third significant figure (the 7) is in the hundredths column, so the mass has been measured to the nearest one hundredth of a gram. The accuracy is 0.01 grams.
    Halving this gives 0.005 g
    The lower bound is 2.37 − 0.005 g = 2.365 g   The upper bound is 2.37 + 0.005 g = 2.375 g
(c) The crowd's size is 15 600 to three significant figures.
    The third significant figure (the 6) is in the hundreds column, so the size of the crowd has been measured to the nearest hundred, i.e. the accuracy is 100
    Halving this gives 50
    The lower bound is 15 600 − 50 = 15 550   The upper bound is 15 600 + 50 = 15 650

Example 3

The distance d that a crow flies is 20.8 km, correct to one decimal place.
Copy and complete the statement: _____ $\leq d <$ _____

The distance is accurate to one decimal place, so the accuracy is 0.1 km
Halving the accuracy gives 0.05 km
The lower bound is 20.8 − 0.05 = 20.75 km
The upper bound is 20.8 + 0.05 = 20.85 km
So d must lie between 20.75 km and 20.85 km
Mathematically, this can be written 20.75 km $\leq d <$ 20.85 km

> **Note:** We use a $\leq$ symbol to the left of d because d could be equal to 20.75 km. But we use a $<$ symbol to the right of d because d could be any value up to, but not including, 20.85 km, since this would round up to 20.9 km.

Exercise 4B

1. Copy and complete the table below. The first row has been completed for you.

	Measurement	Accuracy	Lower bound	Upper bound
(a)	21 seconds	1 second	20.5 s	21.5 s
(b)	9.8 kg	1 decimal place		
(c)	2.75 cm	3 significant figures		
(d)	68 litres	Nearest litre		
(e)	2.568 grams	3 decimal places		
(f)	1050 miles	Nearest 10 miles		
(g)	30.175 inches	5 significant figures		
(h)	20 400 acres	3 significant figures		
(i)	6.08 m²	3 significant figures		
(j)	6.975 km	Nearest metre		

2. A number, when rounded to 1 decimal place, is 23.3
   Find the lower and upper bounds for this number.

3. At a medical, Michelle's height h is measured as 145 cm, correct to the nearest centimetre. Between which two values must her height lie? Copy and complete the line below.
   _____ $\leq h <$ _____

4. Sara would like to cover her window with curtains. The window measures 210 cm wide and 240 cm high, both measured to the nearest 10 cm. Find the lower and upper bound for both the width and height of the window.

5. A plot of land is being sold. It is described as having an area of 21 acres. This has been measured to the nearest acre. Find the smallest and largest possible area of this plot of land.

6. A laptop computer has a 15.0 inch screen, correct to one decimal place. Write down the lower and upper bounds for the size of the screen.

7. $Y = 660$ to the nearest 10 and $Z = 600$ to the nearest 100
   Sam says 'It is possible that Z is bigger than Y.'
   Is Sam correct? Explain your answer.

4.3 Lower and Upper Bounds for Combined Measurements

You may be asked to find the lower and upper bounds for a quantity that is calculated using two or more measurements. You should use the lower and upper bounds for each of the measurements used in the calculation. For the M3 module these calculations involve only addition and multiplication.

CHAPTER 4: BOUNDS

Example 4

A rectangle has a length of 10 cm and a width of 7 cm, both correct to the nearest centimetre.
(a) Calculate the range of values within which the area lies.
(b) Calculate the range of values within which the perimeter lies.

For both the length and the width, the measurements are to the nearest centimetre, so the accuracy is 1 cm. 1 cm ÷ 2 = 0.5 cm

For the length:
Lower bound: 10 − 0.5 = 9.5 cm   Upper bound: 10 + 0.5 = 10.5 cm
For the width:
Lower bound: 7 − 0.5 = 6.5 cm   Upper bound: 7 + 0.5 = 7.5 cm

(a) The minimum area is: minimum length × minimum width
                            = 9.5 × 6.5
                            = 61.75 cm^2

   The maximum area is: maximum length × maximum width
                            = 10.5 × 7.5
                            = 78.75 cm^2

   The area lies between 61.75 and 78.75 cm^2

(b) The minimum perimeter is: (2 × minimum length) + (2 × minimum width)
                                = (2 × 9.5) + (2 × 6.5)
                                = 32 cm

   The maximum perimeter is: (2 × maximum length) + (2 × maximum width)
                                = (2 × 10.5) + (2 × 7.5)
                                = 36 cm

   The perimeter lies between 32 and 36 cm

Example 5

A sprinter runs a race in 21 seconds to the nearest second. His speed was 9.5 m/s, correct to one decimal place. Find the range of possible values for the distance he runs.
You may use the formula: distance = speed × time

For the time taken, the accuracy is 1 second. Halving this gives 0.5 s
Lower bound: 21 − 0.5 = 20.5 s   Upper bound: 21 + 0.5 = 21.5 s

For the speed, the accuracy is 0.1 m/s. Halving this gives 0.05 m/s
Lower bound: 9.5 − 0.05 = 9.45 m/s   Upper bound: 9.5 + 0.05 = 9.55 m/s

For the distance use: speed × time

For the lower bound, use the lower bounds of speed and time: 20.5 × 9.45 = 193.725 m
For the upper bound, use the upper bounds of speed and time: 21.5 × 9.55 = 205.325 m
The sprinter runs between 193.725 m and 205.325 m. (It seems likely this was a 200 m race.)

Exercise 4C

1. Andrew is working out the smallest and largest possible perimeter of the rectangle shown on the right. The length and width of the rectangle have both been measured accurate to one decimal place. Copy and complete the following three tables:

   3.6 cm
   9.2 cm

Rectangle length	
Measurement: 9.2 cm	Accuracy: 0.1 cm
Lower Bound: 9.15 cm	Upper Bound: _____ cm

Rectangle width	
Measurement: 3.6 cm	Accuracy: 0.1 cm
Lower Bound: _____ cm	Upper Bound: _____ cm

39

Rectangle perimeter = 2l + 2w	
Lower Bound:	Upper Bound:
(2 × 9.15) + (2 × _____) = _____ cm	(2 × _____) + (2 × _____) = _____ cm

2. A swimming pool holds 540 m³ of water, correct to the nearest 10 m³.
   For every cubic metre of water, 2.0 grams of chlorine is added to the water, correct to the nearest one tenth of a gram. Find the lower and upper bounds for the total amount of chlorine added to the water.

3. The diagram on the right shows a scalene triangle. The lengths of the base and the perpendicular height are shown. These two measurements are both accurate to the nearest centimetre.
   Find the smallest and largest possible values for the area of the triangle.

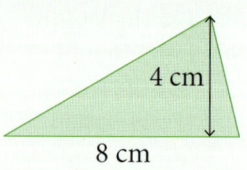

   4 cm
   8 cm

4. Sara measures her window as 210 cm wide and 240 cm high, both measured to the nearest 10 cm. Find the lower and upper bound for both the area and perimeter of the window.

5. A goods train travels across several states in the USA. It travels at an average speed of 23 mph to the nearest integer, taking 36 hours to the nearest hour. Find the lower and upper bound for the distance travelled by the train. You may use the formula: distance = speed × time

6. The density of a rock is measured as 2.6 grams per cm³, correct to two significant figures. Its volume is 600 cm³, correct to the nearest 10 cm³. Use the formula: mass = density × volume
   to find the range of possible values for the mass of this rock.

7. The area of a circle can be calculated using the formula $A = \pi r^2$, where r is the radius. A circle has a radius of 5 cm, correct to the nearest centimetre. Find the lower bound and upper bound for the circle's area. Give both answers to one decimal place.

8. Look at the quadrilateral on the right. Each side length has been measured accurate to the nearest tenth of a centimetre. Find the lower and upper bounds for the perimeter of the quadrilateral.

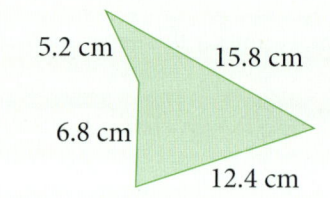

   5.2 cm
   15.8 cm
   6.8 cm
   12.4 cm

4.4 Lower and Upper Bounds for Expressions

You may be asked to find the lower and upper bounds for an expression that involves two or more variables. For these calculations, you should find the lower and upper bounds for each variable involved in the calculation. For the M3 module these calculations involve only addition and multiplication.

Example 6

The values of p, q and r are measured as 5.4, 4.9 and 10.0 respectively, each correct to one decimal place.
(a) Find the minimum possible value for (i) pq (ii) $p + r$
(b) Find the maximum possible value for (i) qr (ii) $q + r$

For p: lower bound 5.35, upper bound 5.45
For q: lower bound 4.85, upper bound 4.95
For r: lower bound 9.95, upper bound 10.05
(a) (i) Minimum value of pq is: 5.35 × 4.85 = 25.9475
    (ii) Minimum value of $p + r$ is: 5.35 + 9.95 = 15.3
(b) (i) Maximum value of qr is: 4.95 × 10.05 = 49.7475
    (ii) Maximum value of $q + r$ is: 4.95 + 10.05 = 15

Exercise 4D

1. The values of a, b and c are measured as 3.7, 5.8 and 13.1 respectively, each correct to one decimal place.
   (a) Find the minimum possible value for (i) ab (ii) $a + b$
   (b) Find the maximum possible value for (i) bc (ii) $b + c$

2. The value of *M* is given as 100 correct to the nearest 10
   (a) Find (i) the lower bound and (ii) the upper bound for M^2

   The value of *N* is given as 140 correct to the nearest 10
   (b) Find (i) the lower bound and (ii) the upper bound for $\sqrt{N}$, giving your answers to five significant figures.
   (c) Find (i) the lower bound and (ii) the upper bound for $M^2 + \sqrt{N}$, giving your answers to the nearest integer.
   (d) Find (i) the lower bound and (ii) the upper bound for $M^2\sqrt{N}$, giving your answers to three significant figures.

3. $x = 4.6$ and $y = 3.7$, both correct to one decimal place.
   (a) Find the lower and upper bounds for *x* and *y*.
   (b) Find the minimum value for each of the following.
       (i) $x + y$ (ii) xy (iii) $\sqrt{x}$ giving your answer to three significant figures
   (c) Find the maximum value for each of the following.
       (i) $x + y$ (ii) $2xy$ (iii) y^2

4. *P* and *Q* have been measured as 45 and 55 each to the nearest 5
   (a) Find (i) the lower bound and (ii) the upper bound for $P + Q$
   (b) Find (i) the lower bound and (ii) the upper bound for *PQ*

5. $x = 8.32$ correct to three significant figures. $y = 5.69$ correct to three significant figures.
   Find the least possible value of $x + y$

6. $Y = 200$ to the nearest 10 and $Z = 200$ to the nearest 100
   Find the lower and upper bounds for $Y + Z$ and *YZ*

7. The values of *A*, *B* and *C* are given below, each measured correct to two significant figures.
   $A = 7.1, B = 6.2, C = 9.3$
   (a) $D = A + B + C$
       Find the lower bound and upper bound for *D*.
   (b) $E = ABC$
       Find the lower bound and upper bound for *E*, giving these answers to the nearest whole number.

8. $A = 6$ to the nearest integer. $B = 4$ to the nearest integer. Copy and complete the table on the right.

	A	*B*	*A + B*	*AB*
Lower bound		5.5		
Upper bound				

9. The values *A*, *B* and *C* were measured as 55.1, 58.9 and 65.8 all correct to three significant figures.
   Copy and complete the following.
   (a) _____ ≤ $2A + 3B$ < _____
   (b) _____ ≤ $3B^2C$ < _____ rounding to three significant figures.
   (c) _____ ≤ $5AC^2$ < _____ rounding to three significant figures.
   (d) _____ ≤ $6\sqrt{A + C}$ < _____ rounding to three significant figures where appropriate.

4.5 Summary

In this chapter you have learnt about:
- The continuous nature of measurements.
- How the accuracy of a measurement depends on how that measurement was recorded. For example, a set of bathroom scales may give a mass to the nearest tenth of a kilogram.
- What is meant by lower and upper bounds for measurements and how to calculate them.
- How to calculate lower and upper bounds for expressions involving one or more measurement.

Progress Review
Chapters 1–4

This Progress Review covers:
- Chapter 1 – Numbers
- Chapter 2 – Approximation and Estimation
- Chapter 3 – Percentages and Finance
- Chapter 4 – Bounds

1. Write down the first 5 multiples of 8
2. Write down all the factors of 48
3. Write down all the prime numbers between 30 and 40
4. Stephen says 'Adding two prime numbers together never gives another prime number.' Ciara thinks he is wrong. Write down an example Ciara could use to show that Stephen is wrong.
5. (a) Copy the factor tree on the right, and fill in the empty circles.
   (b) Using your completed factor tree, write down the prime factorisation for 108
6. Write 800 as a product of prime factors using index notation.
7. (a) Find the lowest common multiple of 48 and 52
   (b) Find the highest common factor of 48 and 52
8. Along a straight stretch of road, there are streetlights every 400 metres. There are telegraph poles every 600 metres. At the start of the road there is both a streetlight and a telegraph pole. How far along the road is there again both a streetlight and a telegraph pole?
9. Round these numbers to the given level of accuracy.
   (a) 49.0056 to 3 decimal places
   (b) 1.979 to 3 significant figures
   (c) 24.455 to the nearest integer
   (d) 135.66 to the nearest ten
   (e) 16943 to 2 significant figures
   (f) 97.79 to 1 significant figure
   (g) 89.9901 to one decimal place
   (h) 135.6197° to the nearest tenth of a degree
   (i) 2548.9 cubic metres to the nearest hundred cubic metres
   (j) 79.5791 kg to the nearest gram
10. Find an estimate for the following calculations.
    (a) 74.57 + 35.479
    (b) $\dfrac{16.97 \times 31.907}{9.078 + 1.456}$
    (c) $\dfrac{731 - 109}{54.3 - 47.8}$
11. An estimate for each of these calculations is given. In each case, state whether it is an under-estimate or an over-estimate, giving a reason for your answer.
    (a) $2.3(6.1 + 4.4) \approx 2(6 + 4) \approx 20$
    (b) $6.9 + 2.9 + 0.48 \times 29 \approx 7 + 3 + 0.5 \times 30 \approx 7 + 3 + 15 \approx 25$
    (c) $100 - 31 - 42 \approx 100 - 30 - 40 \approx 30$
    (d) $\dfrac{60 + 30}{5.9 + 2.8} \approx \dfrac{60 + 30}{6 + 3} \approx \dfrac{90}{9} \approx 10$
12. Find the following. Give your answers to one decimal place where appropriate.
    (a) 22% of 600 mm
    (b) 5% of 240 cm^2
    (c) $7\frac{1}{2}$% of 200 kg
    (d) 27% of 90 tonnes
    (e) 42% of 610 km
    (f) 7% of 55 cm
    (g) 85% of 60 mm
    (h) 19% of 99 g

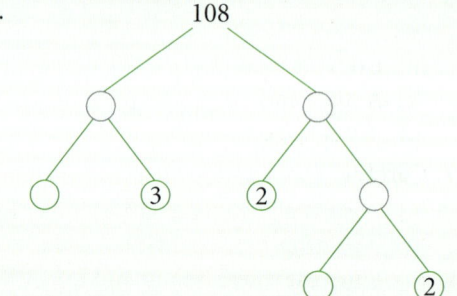

PROGRESS REVIEW: CHAPTERS 1–4

13. Cillian surveyed all the people in his school year about how they get to school. The results are shown in the pie chart on the right.
    (a) What percentage of pupils said they walk to school?
    (b) There are 120 pupils in Harry's year group.
        How many of them said:
        (i) Car
        (ii) Cycle
        (iii) Bus
        (iv) Walk?

14. What multiplier would you use to:
    (a) Increase by 18%?
    (b) Decrease by 20%?
    (c) Increase by 8%?
    (d) Decrease by 80%?
    (e) Increase by 300%?

15. Use a multiplier to find:
    (a) 60 kg increased by 25%
    (b) 120 cm increased by 40%
    (c) 220 metres increased by 22%
    (d) £290 decreased by 15%
    (e) 670 school lunches decreased by 60%

16. These items go up or down in value by the percentage shown. Find the new value of each item using a multiplier.
    (a) Exercise equipment £220, down 30%
    (b) A diamond ring £2200, increased by 124%
    (c) A vintage car £4800, increased by 26%

17. Caroline earns £22 000 per year. She pays 22% of her earnings in income tax and other deductions.
    (a) Calculate the total amount of the deductions.
    (b) What is Caroline's net income?

18. Find 100% if:
    (a) 50% is 16
    (b) 25% is 1350
    (c) 10% is 64
    (d) 70% is 3780 litres
    (e) 75% is 4.875 cm
    (f) 60% is 252 m
    (g) 90% is 3060
    (h) 45% is 1575 tonnes
    (i) 85% is 7055 years
    (j) 15% is 27

19. A carton of orange juice says it contains 20% extra free. The carton contains 396 ml of juice. What volume of juice is in a carton that does not have the extra 20%?

20. Find the original price of these items.
    (a) Jeans selling at £51.20 after a 20% reduction.
    (b) Shirt selling at £31.50 after a 30% reduction.

21. Ollie has just bought an office for his business. He needs to make the following purchases. The prices below include VAT at 20%. Find the price of each of these items before VAT was added.
    (a) A carpet £168
    (b) A wall painting £28.80
    (c) A computer printer £198
    (d) Table and chairs £307.20

22. In a building supplies shop, all prices include VAT at 20%. In each of parts (a) to (d) calculate (i) the price of these items before VAT is added (ii) how much VAT is added.
    (a) Shovel £93.60
    (b) Timber £17.64 per metre
    (c) 100 screws £7.20
    (d) Sand £57.60 per tonne

23. Mary stays at three different hotels on Friday, Saturday and Sunday. Her hotel bills are:

    Imperial Hotel: Friday £96
    Grand Hotel: Saturday £85
    Lagan View Rooms: Sunday £90

    (a) The hotel bills for Friday and Sunday nights include VAT at 20%. Find the cost of these two hotels before VAT is added.

(b) The hotel bill for Saturday night did not include VAT.
   (i) Which was the most expensive hotel Maria stayed at?
   (ii) Which was the cheapest?

24. Sharon's net income this year (her pay after deductions), is £18 240
The deductions are 24% of her gross pay. Calculate Sharon's **gross pay** (her pay before deductions).

25. Cars depreciate in value as they get older. By using a multiplier, find the value of each of these cars after one year's depreciation.
   (a) Ford Puma, originally £25 510, depreciation 10%
   (b) Kia Sportage, originally £28 350, depreciation 20%
   (c) Nissan Qashqai, originally £27 120, depreciation 15%

26. Lorcan puts £250 into a savings account paying 2% compound interest per year for 3 years. Calculate, giving your answers to the nearest penny:
   (a) The amount in his savings account at the end of the 3 years.
   (b) The interest Lorcan has earned during this time.

27. Natasha is thinking about putting some money into a savings account. An advert for the account is shown on the right. Natasha is trying to decide whether to put £450 or £500 into the account. She plans to keep her money in the account for 3 years.
   (a) Find the amount of interest she would receive over 3 years on both £450 and £500, giving your answers to the nearest penny.
   (b) What is the difference in the total amount of interest she would receive?

NBS Newtown Building Society
Active Savers Saving Account

- Balances below £500:
  2.5% compound interest per annum
- Balances £500 or above:
  2.75% compound interest per annum

28. (a) Jackson's weight increases from 45 kg to 51.75 kg between his 13th and 14th birthdays. What is the percentage increase in his weight during this year?
   (b) During the summer, Jackson's mum and dad both try to lose weight. Look at this table:

	Jackson's mum	Jackson's dad
Weight (start of summer)	72 kg	
Weight (end of summer)		72 kg

Jackson's mum and dad both lose 4% of their body weight. Copy and complete the table.

29. The concentration of carbon dioxide (CO_2) in the earth's atmosphere is measured in parts per million (ppm). The concentration of CO_2 in the atmosphere in 1980 was roughly 340 ppm. By 2020 this had increased by about 20%.
   (a) Using a multiplier, find the concentration of CO_2 in the atmosphere in 2020.
   (b) Since 2020, CO_2 levels have increased further. If the concentration of CO_2 in the atmosphere reaches 500 ppm, what will the percentage increase be, compared with:
      (i) The 2020 concentration?
      (ii) The 1980 concentration?
   Give both answers to one decimal place.
   (c) The 1980 level of 340 ppm was about 33% higher than what scientists call **pre-industrial levels**. Calculate the pre-industrial levels of CO_2 in the atmosphere, giving your answer to the nearest whole number.

30. There were 36 000 women and 30 000 men living in the town of Hamilton. The population increased over the next few years. The number of men increased by 10%. The number of women increased by only 4%.
   (a) How many men and how many women lived in Hamilton after this increase?
   (b) What was the total increase as a percentage of the original population?

31. Copy and complete the table by finding the lower and upper bound for each of the measured quantities.

	Measurement	Accuracy	Lower bound	Upper bound
(a)	53.0 seconds	0.1 second		
(b)	27.04 kg	2 decimal places		
(c)	9.5 cm	2 significant figures		
(d)	125 litres	Nearest 5 litres		
(e)	13.68 grams	2 decimal places		
(f)	50 miles	Nearest mile		
(g)	50 miles	Nearest ten miles		
(h)	420 hectares	2 significant figures		
(i)	61.8 m^2	3 significant figures		
(j)	9.375 kg	Nearest gram		

32. Mr Wales is very particular about the appearance of his garden. He likes to keep his grass to a height of 2.8 cm, correct to the nearest millimetre. Find the shortest and longest possible height of each blade of grass.

33. The rainfall in Katesbridge overnight was recorded as 23 mm, correct to the nearest millimetre. Find the lowest and highest possible rainfall in millimetres.

34. A train comprises an engine and 8 carriages. The engine is 12 metres long to the nearest metre. Each carriage has a length of 16 metres to the nearest metre. Find the lower bound and upper bound for the length of the train.

35. A child's drawing of a house is shown on the right. Each measurement shown is correct to the nearest centimetre. Find the lower bound and upper bound for the total area of the drawing.

36. In a track and field event, a javelin is thrown 64.4 metres to the nearest 10 cm. Find the lower bound and upper bound for the distance the javelin is thrown.

37. To calculate the perimeter of a semicircle, Johnny uses the formula:

$P = \pi r + 2r$ where r is the radius.

He draws a semicircle with a radius of 7 cm to the nearest centimetre. Find a lower bound and upper bound for the perimeter of Johnny's semicircle. Give your answers to one decimal place.

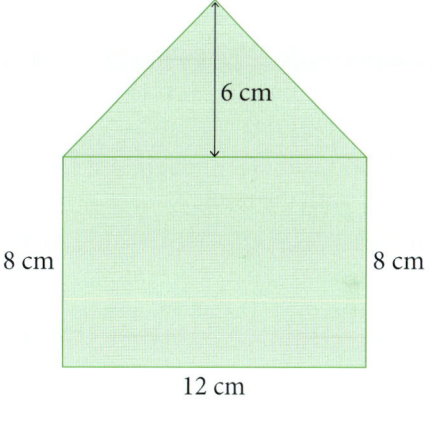

38. Sue draws a rectangle with a length of 12 cm and a width of 6 cm, both measured to the nearest centimetre. Seán draws a rectangle with a length of 8 cm and a width of 7 cm, both measured to the nearest centimetre.
  (a) Seán says to Sue 'Given the accuracy of our measurements, it's possible my rectangle has a bigger area than yours.' Is he correct? Show all your working.
  (b) Sue says to Seán: 'My rectangle definitely has a larger perimeter than yours.' Is she correct? Show all your working.

39. $R = 500$ to the nearest 10 and $S = 500$ to the nearest 100
Find the lower and upper bounds for each of the following.
  (a) R (b) S (c) $R + S$ (d) RS

40. $a = 5.3$ and $b = 4.7$, both measured accurately to one decimal place. Find the lower and upper bounds for each of the following, giving your answers to one decimal place where appropriate.
  (a) $a + b$ (b) ab (c) $\sqrt{ab}$ (d) $a^2 + b^2$

Chapter 5
Brackets

5.1 Introduction

In this chapter we will learn how to expand brackets in two situations. Firstly, when the brackets are multiplied by a variable. And secondly, when two sets of brackets are multiplied together.

It is important to understand what brackets mean when used in algebra. Imagine you are promised money from two people, Aoife and Bert. Aoife promises you A pounds while Bert promises you B pounds. You would receive $A + B$ pounds from them.

Now let's imagine both Aoife and Bert promise to double this money instead. How much will you now get? One way to work it out is to double the amount from each of them. Aoife now will give you $2A$ pounds. Bert will now give you $2B$ pounds. So, in total you receive $2A + 2B$ pounds.

But you can think of the total another way. Together they previously promised you $A + B$ pounds. They have now promised to double this. Multiplying $A + B$ by 2 you will get 2 times $A + B$ pounds $= 2(A + B)$.

This shows you that $2(A + B) = 2A + 2B$

Brackets are like this. The value on the outside of the bracket just multiplies every individual term inside.

The diagram on the right illustrates what a pair of brackets mean. Imagine working out the areas of the two identical rectangles shown.

Since the areas of the rectangles are clearly the same, we end up with the result:

$3 \times (4 + 6) = 3 \times 4 + 3 \times 6$

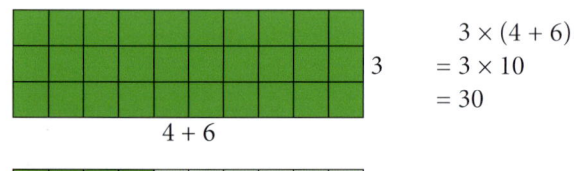

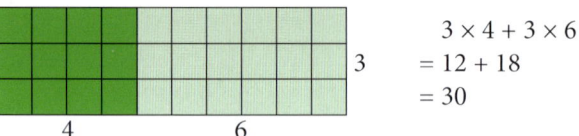

$3 \times (4 + 6)$
$= 3 \times 10$
$= 30$

$3 \times 4 + 3 \times 6$
$= 12 + 18$
$= 30$

> **Note:** Remember that when a value is placed outside a bracket, this value must be multiplied onto **each** term in the bracket.

Key words
- **Constant:** A constant is a number, for example 5 or –2.5

Before you start you should:
- Recall how to simplify and manipulate algebraic expressions by collecting like terms.

In this chapter you will learn:
- Multiply a bracket by a term involving a variable part such as $3x(a + b)$
- Multiply two brackets, each with two or more terms, using a variety of methods.

Example 1

Expand the brackets: **(a)** $4(x + 3)$ **(b)** $2(x - 6)$ **(c)** $-7(x - 5)$

(a) Multiply each term inside the brackets by 4: $4(x + 3) = 4x + 12$
(b) Multiply each term inside the brackets by 2: $2(x - 6) = 2x - 12$
(c) Multiply each term inside the brackets by –7: $-7(x - 5) = -7x + 35$
    Note that $-7 \times -5 = 35$

Exercise 5A (Revision)

Expand these brackets:

1. $2(3 + 8)$
2. $4(8 - 5)$
3. $9(7 + 3)$
4. $15(21 - 6)$
5. $-3(19 - 7)$
6. $12(34 - 18)$
7. $3(x + 5)$
8. $7(x + 4)$
9. $6(x + 5)$
10. $12(x - 6)$
11. $15(x - 3)$
12. $-3(6 - x)$
13. $2(3x - 8)$
14. $9(3x - 2)$
15. $11(8 - x)$
16. $7(4x - 30)$
17. $20(13 - 2x)$
18. $8(4 - 5x)$

5.2 Expanding Brackets Multiplied by Variables

The term outside the bracket does not have to be a number. It could be any term. To put it another way, you may be asked to multiply a bracket by a variable, or a combination of a number and a variable. The next example illustrates this.

Example 2

Expand the brackets:
(a) $3x(7 + 5x)$
(b) $4x(5y - 3)$
(c) $5x(7x - 2y)$
(d) $(8 - 5x)x^2$
(e) $-6x(4 - 5x)$

(a) $3x(7 + 5x) = 3x \times 7 + 3x \times 5x = 21x + 15x^2$ using $x \times x = x^2$
(b) $4x(5y - 3) = 4x \times 5y - 4x \times 3 = 20xy - 12x$
(c) $5x(7x - 2y) = 5x \times 7x - 5x \times 2y = 35x^2 - 10xy$ using $x \times x = x^2$
(d) $(8 - 5x)x^2 = 8 \times x^2 - 5x \times x^2 = 8x^2 - 5x^3$ using $x \times x^2 = x^3$
This example shows that it doesn't matter which side the bracket is on.
(e) $-6x(4 - 5x) = -6x \times 4 - (-6x) \times (5x) = -24x + 30x^2$ using $x \times x = x^2$

There may be more than two terms in the bracket. The following diagram shows what a bracket with three terms means. Imagine working out the areas of the two identical rectangles shown on the right.

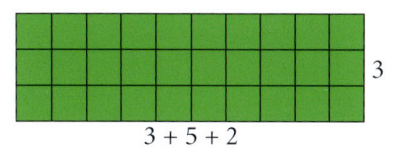

Since the areas of the rectangles are clearly the same, we end up with the result:

$3 \times (3 + 5 + 2) = 3 \times 3 + 3 \times 5 + 3 \times 2$

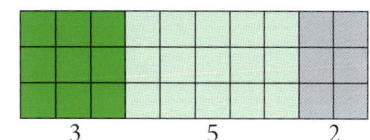

$3 \times (3 + 5 + 2)$
$= 3 \times 10$
$= 30$

$3 \times 3 + 3 \times 5 + 3 \times 2$
$= 9 + 15 + 6$
$= 30$

Terms, like 4 or $3x$ or $-4y$, can be multiplied across three or more terms inside brackets, as shown in the next example.

Example 3

Expand the brackets:
(a) $4(2 + x + y)$
(b) $3x(7 - 2x - 6y)$
(c) $-4y(12 - 2x - 5y)$

(a) $4(2 + x + y)$
     $= 4 \times 2 + 4 \times x + 4 \times y$
     $= 8 + 4x + 4y$

(b) $3x(7 - 2x - 6y)$
     $= 3x \times 7 - 3x \times 2x - 3x \times 6y$
     $= 21x - 6x^2 - 18xy$

(c) $-4y(12 - 2x - 5y)$
     $= -4y \times 12 - (-4y) \times (2x) - (-4y) \times (5y)$
     $= -48y + 8xy + 20y^2$

> **Note:** Don't forget that two minus signs multiply together to give a positive result: $-2 \times -2 = 4$

> **Note:** Don't forget that multiplying out brackets is often just the first part of a question. It may be possible to collect like terms afterwards.

Example 4

Expand the brackets and simplify.

(a) $3(2 - 5x) + 6(3x + 4)$

(b) $4(2x - 3y) - 6(3y - 5x)$

(a) $\quad 3(2 - 5x) + 6(3x + 4)$
$= 3 \times 2 - 3 \times 5x + 6 \times 3x + 6 \times 4$
$= 6 - 15x + 18x + 24$
$= 30 + 3x$

(b) $\quad 4(2x - 3y) - 6(3y - 5x)$
$= 4 \times 2x - 4 \times 3y - 6 \times 3y - 6 \times (-5x)$
$= 8x - 12y - 18y + 30x$
$= 38x - 30y$

Note: As always, don't forget that two negative numbers multiply to give a positive result!

Exercise 5B

1. Expand these brackets:
   (a) $2y(1 + x)$
   (b) $6x(2 + 7x)$
   (c) $3p(4 - 5y)$
   (d) $7y(8 - 5x)$
   (e) $8x(3y + 22p)$
   (f) $4q(2p - 5t)$
   (g) $3x(4x + 7y)$
   (h) $(3x - 8y)6y$
   (i) $-5x(3 + 4x)$
   (j) $5x(9p - 8x)$
   (k) $-2x(7 - 3x)$
   (l) $-2x(2x + 8q)$
   (m) $-4x(2x - 8)$
   (n) $-(3y - 8p)$
   (o) $-9p(2p + 3)$
   (p) $(5t - 3x)4x$
   (q) $-7p(3x - t)$
   (r) $4x(5y - 3x)$

2. Expand these brackets:
   (a) $3(2 - x + y)$
   (b) $6(2x + 3y - 8)$
   (c) $7(5 - 2x - 3y)$
   (d) $19(x - 2y + 2)$
   (e) $8(2 - 3x - 4y)$
   (f) $-9(2x - 4y - 6)$
   (g) $-4(2y - 3x - 7)$
   (h) $6t(2 + 3y - 6t)$
   (i) $-3(2x - 5y - 7)$
   (j) $7y(5y - 6x + y^2)$
   (k) $(3p - 2x + 7)5p$
   (l) $6(-3 - 2x - 9y)$
   (m) $2q(5 - q - 3q^2)$
   (n) $7x(2x - p + 5y - x^3)$

3. Expand and simplify:
   (a) $3(2 + 4x) + 5(x + 7)$
   (b) $5(3x - 6) + 3(2 - 4x)$
   (c) $7(2q - 5t) + 6(3t + 8q)$
   (d) $5(4x - 3w) - 3(4w - 3x)$
   (e) $6x(2 - 8y) - 7(3y + 4)$
   (f) $4(5t - 4x) - 5(4t + 2x)$
   (g) $6x(3x + 2y) - 7(3y - x^2)$
   (h) $4x(3y - 7t) + 8y(4t + 2x)$
   (i) $9(3 + 5xy) + 4y(2 - 8x)$
   (j) $6(5 - 3x + 2y) - 7(3x - 5y - 6)$
   (k) $p(4 - t) - t(2 - p) + 3(2p - 7t)$

5.3 Multiplying Out a Pair of Brackets

You may be asked to multiply a bracket by another bracket when each bracket contains more than one term. Examples of this situation are:

$(x + 2)(x + 3)$ $\quad\quad$ $(y - 3)(y + 4)$ $\quad\quad$ $(2x + 5)(3x + 4)$ $\quad\quad$ $(5x - 7)(2x - 9)$

When multiplying two brackets, such as in the above examples, **both** terms in the first bracket must multiply **both** terms in the second bracket. This can be seen by looking again at the example of a rectangle.

The area of the rectangle on the right is 48 units.

However, the length of each side can be written as two numbers added together, written inside brackets as shown. The result is still 48 units.

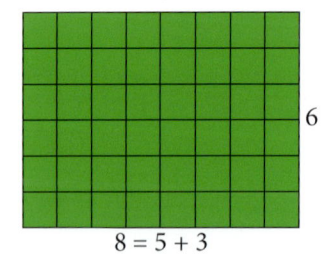

$6 = 4 + 2$
$8 = 5 + 3$

Area $= 8 \times 6$
$= (5 + 3) \times (4 + 2)$
$= 48$

You can also think about this as finding the area of four separate rectangles and adding the results together, as shown on the right. The result is still 48 units.

Area $= (5 \times 4) + (5 \times 2) + (3 \times 4) + (3 \times 2)$
$= 48$

CHAPTER 5: BRACKETS

This example shows how to multiply out brackets – you multiply every term in the first bracket by every term in the second bracket. There are three methods for doing this, as shown in the following examples.

Example 5

Multiply out $(x + 2)(y + 3)$

There are three different methods that you can use.

Method A

The most frequently used method to multiply a bracket of two terms by another bracket of two terms (and easiest to remember) is to use the word **FOIL**. Each letter stands for a particular pair of terms.

We are multiplying out:	$(x + 2)(y + 3)$		
F stands for multiplying the two first terms:	$(\boldsymbol{x} + 2)(\boldsymbol{y} + 3)$	giving	$x \times y = xy$
O stands for multiplying the two outer terms:	$(\boldsymbol{x} + 2)(y + \boldsymbol{3})$	giving	$x \times 3 = 3x$
I stands for multiplying the two inner terms:	$(x + \boldsymbol{2})(\boldsymbol{y} + 3)$	giving	$2 \times y = 2y$
L stands for multiplying the two last terms:	$(x + \boldsymbol{2})(y + \boldsymbol{3})$	giving	$2 \times 3 = 6$
Next, add the four FOIL terms together, giving:	$(x + 2)(y + 3)$		
	$= xy + 3x + 2y + 6$		

Method B

Multiply the whole of the second bracket separately by each term in the first bracket, and add the results together.

$(x + 2)(y + 3)$
$= x(y + 3) + 2(y + 3)$
$= xy + 3x + 2y + 6$

Method C

Arrange each bracket along the sides of a square.

	x	2
y		
3		

Then fill in the cells by multiplying the corresponding row and column.

	x	2
y	xy	$2y$
3	$3x$	6

Finally, add up the four cells in the body of the square:
$= xy + 3x + 2y + 6$

Example 6

Multiply out:
(a) $(x + 2)(x + 3)$ (b) $(y – 3)(y + 4)$ (c) $(2x + 5)(3x – 4)$ (d) $(5x – 7)(2x – 9)$

When the same letter appears in each bracket, the middle two terms can, and should, be combined.

(a) $(x + 2)(x + 3)$
$= x \times x + 3 \times x + 2 \times x + 2 \times 3$
$= x^2 + 5x + 6$

(b) $(y – 3)(y + 4)$
$= y \times y + 4 \times y – 3 \times y – 3 \times 4$
$= y^2 + y – 12$

Note: Be careful with minus signs when multiplying out.

(c) $(2x + 5)(3x – 4)$
$= 2x \times 3x – 4 \times 2x + 5 \times 3x + 5 \times(–4)$
$= 6x^2 + 7x – 20$

(d) $(5x – 7)(2x – 9)$
$= 5x \times 2x – 9 \times 5x – 7 \times 2x + 7 \times 9$
$= 10x^2 – 59x + 63$

Note: Again, watch out when multiplying negative signs, particularly the last term which ends up positive!

Methods B and C above allow you to multiply brackets with more than two terms, as shown in the following two examples.

Example 7

Multiply out $(x + 2)(y + 3 + t)$

Method B

$(x + 2)(y + 3 + t)$
$= x(y + 3 + t) + 2(y + 3 + t)$
$= xy + 3x + xt + 2y + 6 + 2t$ Notice that we end up with $2 \times 3 = 6$ terms

Method C

Arrange each bracket along the sides of a rectangle.
Arrange each bracket along the sides of a square, as before.
Then fill in the cells by multiplying the corresponding row and column.
Then add up the six cells in the body of the rectangle.

$(x + 2)(y + 3 + t)$
$= xy + 3x + xt + 2y + 6 + 2t$

	x	2
y	xy	$2y$
3	$3x$	6
t	xt	$2t$

Example 8

Multiply out:
(a) $(x + y)(3x + 2y + 1)$ **(b)** $(3p - 4q + 2)(p + 5q)$

(a) Using Method B:
$(x + y)(3x + 2y + 1)$
$= x(3x + 2y + 1) + y(3x + 2y + 1)$
$= x \times 3x + x \times 2y + x \times 1 + y \times 3x + y \times 2y + y \times 1$
$= 3x^2 + 2xy + x + 3xy + 2y^2 + y$
$= 3x^2 + 5xy + 2y^2 + x + y$

Note: We have added like terms to tidy up: $2xy + 3xy = 5xy$

(b) $(3p - 4q + 2)(p + 5q)$
$= (p + 5q)(3p - 4q + 2)$
$= p(3p - 4q + 2) + 5q(3p - 4q + 2)$
$= p \times 3p - p \times 4q + p \times 2 + 5q \times 3p - 5q \times 4q + 5q \times 2$
$= 3p^2 - 4pq + 2p + 15pq - 20q^2 + 10q$
$= 3p^2 + 2p + 11pq - 20q^2 + 10q$

Note: We have added like terms to tidy up: $-4pq + 15pq = 11pq$

Special cases

There are two special cases.
- When a bracket is squared, according to the rules of indices, it means multiplying the bracket by itself:
$(x + 4)^2$
$= (x + 4)(x + 4)$
$= x^2 + 4x + 4x + 16$
$= x^2 + 8x + 16$
- When two similar brackets just differ by the sign in the middle of each, this leads to the middle terms cancelling out when you multiply out the brackets:
$(x + 4)(x - 4)$
$= x^2 - 4x + 4x - 16$
$= x^2 - 16$ (Note that this result can be written as: $x^2 - 4^2$)

For this reason, the resulting expression is called the **difference of two squares**.

As always, if there are two or more pairs of brackets, you should seek to combine like terms.

Example 9

Multiply out:
(a) $(2x - 5)^2$ (b) $(1 - y)^2$ (c) $(3t - 4)(3t + 4)$ (d) $(7 - p)(7 + p)$

(a) $(2x - 5)^2$
$= (2x - 5)(2x - 5)$
$= 4x^2 - 10x - 10x + 25$
$= 4x^2 - 20x + 25$

Note: Remember that a minus in both brackets means that the last term ends up positive.

(b) $(1 - y)^2$
$= (1 - y)(1 - y)$
$= 1 - y - y + y^2$
$= 1 - 2y + y^2$

(c) $(3t - 4)(3t + 4)$
$= 9t^2 + 12t - 12t - 16$
$= 9t^2 - 16$

(d) $(7 - p)(7 + p)$
$= 49 + 7p - 7p - p^2$
$= 49 - p^2$

You may be given equations which contain more than one set of brackets to multiply out. The next example illustrates this. Don't forget to tidy up terms wherever possible.

Example 10

Multiply out:
(a) $(2x + 5)(1 + x) + (2 + x)^2$ (b) $(x + y)(x - y) - (2x + 3y)^2$ (c) $(x + y)^2 - (x - y)^2$

(a) $(2x + 5)(1 + x) + (2 + x)^2$
$= (2x + 5)(1 + x) + (2 + x)(2 + x)$
$= 2x + 2x^2 + 5 + 5x + 4 + 2x + 2x + x^2$
$= 3x^2 + 11x + 9$

(b) $(x + y)(x - y) - (2x + 3y)^2$
$= x^2 - xy + xy - y^2 - (2x + 3y)(2x + 3y)$
$= x^2 - xy + xy - y^2 - (4x^2 + 6xy + 6xy + 9y^2)$

Note: Keep brackets around the second expansion when it is being subtracted to get the signs correct.

$= x^2 - y^2 - 4x^2 - 6xy - 6xy - 9y^2$
$= -3x^2 - 12xy - 10y^2$

(c) $(x + y)^2 - (x - y)^2$
$= (x + y)(x + y) - (x - y)(x - y)$
$= x^2 + xy + xy + y^2 - (x^2 - xy - xy + y^2)$
$= x^2 + xy + xy + y^2 - x^2 + xy + xy - y^2$
$= 4xy$

Exercise 5C

1. Expand the following pairs of brackets. You may find it helpful to using all three of the methods explained above for the first few questions.
   (a) $(2x + 1)(x + 2)$ (b) $(q + 2p)(2q + 5p)$ (c) $(6r + t)(r + 6t)$ (d) $(3x + y)(4x + 7y)$
   (e) $(4p - q)(p + q)$ (f) $(2x + 7y)(3x - y)$ (g) $(2t + r)(r - 3t)$ (h) $(p - 3q)(4x - 3y)$
   (i) $(2x - 9y)(x - y)$ (j) $(x - y)(y - x)$ (k) $(p - 2q)(2p - q)$ (l) $(y - x)(q - p)$
   (m) $(a + b)(c - d)$ (n) $(a - b)(c - d)$ (o) $(2x + y)(y - 3t)$ (p) $(7p - q)(y + 2x)$
   (q) $(13x - 5y)(5x + 13y)$ (r) $(5x + 3y)(3x - 5y)$ (s) $(p + 2q)(2p + q)$ (t) $(3 - 5t)(9 - 6t)$

2. Recognise the special cases as you expand these pairs of brackets:
   (a) $(x + y)^2$ (b) $(p - q)^2$ (c) $(4x - 5y)^2$ (d) $(7t - 3y)^2$
   (e) $(p - q)(p + q)$ (f) $(t - 2r)(t + 2r)$ (g) $(3x + 4y)(3x - 4y)$ (h) $(8p - 3q)(8p + 3q)$
   (i) $(5y - 4x)^2$ (j) $(5y - 4x)(4x + 5y)$ (k) $(7x - 1)^2$ (l) $(6 - 5w)(5w + 6)$

3. Use either Method B or Method C to expand the following:
   (a) $(x + y)(1 + 2x - 2y)$
   (b) $(2x - 3y)(y + x + 5)$
   (c) $(3p + 2q)(3 - 2p - q)$
   (d) $(3x - y)(2 + 3x + 4y)$
   (e) $(a + b)(a + b + c)$
   (f) $(2t - 5r + 6)(t + r)$
   (g) $(p - q + 5)(2q + 7p)$
   (h) $(8x - 3y + 5)(3y - 8x)$

4. Expand the following expressions, writing them as simply as possible. Hint: This means combining like terms.
   (a) $(2x - y)^2 - (2x + y)^2$
   (b) $(x - y)(x + y) - (x + y)^2$
   (c) $(2p - 3q)^2 + (4p + q)^2$
   (d) $(4t - 3r)^2 + (4t + 3r)^2$
   (e) $(2x - 3t)(1 + x + t) - (1 - x)(3 + t)$
   (f) $(2x - 3t)(1 + x + t) + (3 - x)(2 + t)$
   (g) $q(3p + 2r - 5) + p(3q - 2r + 5)$
   (h) $(1 + x)(1 - y) + (1 + y)(1 - z) + (1 + z)(1 - x)$
   (i) $(3y - 2x)^2 + 4x(3y - 2x)$
   (j) $(x - y)^2 - 2(x - y)(x + y) + (x + y)^2$

5. A rectangle has length $(3x - 2)$ and breadth $(4 + 5x)$. Find an expression for its area, expanding the brackets.

6. A triangle has base $(2x - 6)$ and perpendicular height $(7x + 2)$. Find an expression for its area, expanding the brackets.

5.4 Summary

In this chapter you learned how to:
- Multiply a bracket by a term involving a variable part.
- Multiply two brackets, each with two or more terms, using a variety of methods.

Chapter 6
Factorisation

6.1 Introduction

You can factorise an expression by taking a **common factor** out of brackets. Factorising by taking a common factor out of brackets is the reverse of expanding brackets, which was discussed in the previous chapter.

Recall the example from the start of the previous chapter. If Aoife promises to give you $2A$ pounds and Bert promises to give you $2B$ pounds, then together they will give you:

$2A + 2B = 2(A + B)$ pounds.

The repeated 2 on the left-hand side above is called a **common factor** because 2 divides into both these terms. Instead of doubling both the A and B amounts separately, we can add them first and just double $A + B$.

The brackets show that the term outside the brackets must be multiplied on to every separate term inside the brackets. We can see this from the diagrams of the pair of rectangles pictured in the previous section. The area of the two rectangles considered separately is:

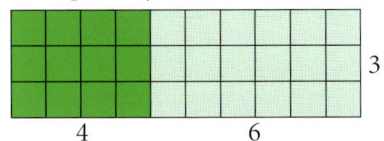

$3 \times 4 + 3 \times 6$
$= 12 + 18$
$= 30$

The area of the two rectangles joined together is the same:

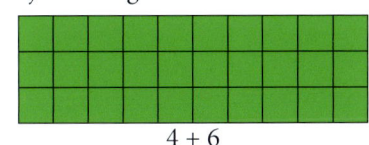

$3 \times (4 + 6)$
$= 3 \times 10$
$= 30$

Since the areas of the rectangles are clearly the same, we end up with the result: $3 \times 4 + 3 \times 6 = 3 \times (4 + 6)$

The common factor 3 has been taken out of the pair of terms and written in front of the brackets.

Key words
- **Common factor**: A constant or algebraic expression that can be taken outside brackets because it divides into every term in a list.
- **Difference of two squares**: a common term, involving algebra or numbers, of the form $x^2 - y^2$ or $p^2 - 5^2$

Before you start you should:
- Know how to expand brackets with a constant term outside the brackets.
- Know how to expand brackets with an algebraic term, for example x, outside the brackets.

In this chapter you will learn:
- How to factorise by taking out a common factor or factors.
- How to factorise using the difference of two squares.
- How to factorise expressions involving four terms.

Note: Factorising quadratic expressions of the form $x^2 + bx + c$ will be treated in the next chapter on quadratic expressions.

Example 1 (Revision)

Factorise by taking a constant common factor out of brackets.
(a) $3q + 12p$ (b) $2y - 4z$ (c) $49m + 7n$

(a) $3q + 12p = 3(q + 4p)$ where the common factor is 3
(b) $2y - 4z = 2(y - 2z)$ where the common factor is 2
(b) $49m + 7n = 7(7m + n)$ where the common factor is 7

53

Exercise 6A (Revision)

1. Expand the following brackets:
   (a) $3(x + 3y)$
   (b) $6(p + 2q)$
   (c) $4(2t - s)$

2. Take the common factor out of the following lists of terms:
   (a) $3x + 9y$
   (b) $6p + 12q$
   (c) $8t - 4s$
   (d) $15p + 25r$
   (e) $36t + 18y$
   (f) $48q - 144p$

6.2 Taking Common Factors Out From a List of Terms (Basic Factorisation)

Recall that factorisation is the reverse of expanding brackets. To put it another way, when asked to expand the bracket $3(x + 2)$ the answer is $3x + 6$, so when asked to factorise $3x + 6$ the answer is $3(x + 2)$

In this section we will take both numbers and variables out of a list of terms with common factors.

Individual variables may be common factors, or combinations of variables and numbers, as shown in the following examples.

Example 2

Factorise the following:
(a) $x + x^2$
(b) $xy - 4y$
(c) $12x - 4x^2$
(d) $7mn - 28mp$

(a) $x + x^2$ $= x(1 + x)$ where the common factor is x
(b) $xy - 4y$ $= y(x - 4)$ where the common factor is y
(c) $12x - 4x^2$ $= 4x(3 - x)$ where the common factor is $4x$
(d) $7mn - 28mp$ $= 7m(n - 4p)$ where the common factor is $7m$

It is also possible to have three, or more, terms.

Example 3

Factorise the following: (a) $3pq - 9p^2q + 6pq^2$ (b) $8xyz + 5wxy - x^2 y^2$

(a) $3pq - 9p^2q + 6pq^2$ $= 3pq(1 - 3p + 2q)$ where the common factor is $3pq$
(b) $8xyz + 5wxy - x^2 y^2$ $= xy(8z + 5w - xy)$ where the common factor is xy

Exercise 6B

1. Take the common factor(s) out of the following lists of terms:
   (a) $5x + 3xy$
   (b) $3q + 7pq$
   (c) $4uv + 2v$
   (d) $9y^2 + 7y$
   (e) $3t + 2t^2$
   (f) $x - x^2$
   (g) $3x + 21x^2$
   (h) $9pq - 3p$
   (i) $18y - 6xy$
   (j) $20p^2 - 15p^3$
   (k) $6xy - 4y^2$
   (l) $30xyz + 48yz$
   (m) $12px + 28x^2$
   (n) $72x^2 - 36$
   (o) $5qp - 15pt$
   (p) $84 - 24xy$
   (q) $y^3 - xy^2$
   (r) $4x^3y - 20xy$
   (s) $12xy - 20x^2y + 8xy^2$
   (t) $p^2qr - pq^2r + pqr^2$
   (u) $72xq - 36x^3 + 54q^2$
   (v) $4x - 2y + 6$
   (w) $40xyt - 36ytq$
   (x) $18p^4 + 15p^3 - 12p^2$
   (y) $38p - 19t + 57r$

6.3 The Difference of Two Squares

Consider the two expressions below. Do you notice what is in common in both examples after the brackets are multiplied out?

$(x + 3)(x - 3) = x^2 + 3x - 3x - 3^2$ $= x^2 - 3^2$ $= x^2 - 9$
$(4 - y)(4 + y) = 4^2 + 4y - 4y - y^2$ $= 4^2 - y^2$ $= 16 - y^2$

In both cases, when the brackets are multiplied out, the 'inner' and 'outer' terms cancel each other out leaving us with the pattern:

(first + last) × (first − last) = (first)² − (last)²

This is, of course, a difference of two squares, i.e. two terms squared and then subtracted.

Therefore, to factorise the difference of two squares, we simply write the above expression out backwards:
$$(\text{first})^2 - (\text{last})^2 = (\text{first} + \text{last}) \times (\text{first} - \text{last})$$

Example 4

Factorise the following using the difference of two squares:
(a) $x^2 - 9$ (b) $16 - y^2$ (c) $x^2 - y^2$ (d) $y^2 - 25z^2$

(a) $x^2 - 9 \quad = x^2 - 3^2 \quad = (x + 3)(x - 3) \quad$ by recognising that $9 = 3^2$
(b) $16 - y^2 \quad = 4^2 - y^2 \quad = (4 + y)(4 - y) \quad$ by recognising that $16 = 4^2$
(c) $x^2 - y^2 \quad = (x + y)(x - y)$
(d) $y^2 - 25z^2 \quad = y^2 - (5z)^2 \quad = (y + 5z)(y - 5z) \quad$ by recognising that $25 = 5^2$

> **Note:** You can check your answer is right by multiplying out the brackets again.

Sometimes there is a common factor in both terms that needs to be factorised out before using the difference of two squares pattern.

Example 5

Factorise the following using the difference of two squares:
(a) $2x^2 - 8$ (b) $xy^2 - 25x$ (c) $49x^2 - 9$ (d) $50ab^2 - 32a$ (e) $51^2 - 49^2$

(a) $2x^2 - 8 \quad = 2(x^2 - 4) \quad = 2(x^2 - 2^2) \quad = 2(x + 2)(x - 2)$
(b) $xy^2 - 25x \quad = x(y^2 - 25) \quad = x(y^2 - 5^2) \quad = x(y + 5)(y - 5)$
(c) $49x^2 - 9 \quad = (7x)^2 - 3^2 \quad = (7x + 3)(7x - 3)$
(d) $50ab^2 - 32a \quad = 2a(25b^2 - 16) \quad = 2a((5b)^2 - 4^2) \quad = 2a(5b + 4)(5b - 4)$
(e) $51^2 - 49^2 \quad = (51 - 49)(51 + 49) \quad = 2 \times 100 \quad = 200$

> **Note:** Remember, the best way to spot a difference of two squares is to look out for two square terms with a minus sign between them!

Exercise 6C

1. Which of the following quadratic expressions take the form of a difference of two squares?
   (a) $81 - t^2$ (b) $9 + y^2$ (c) $32x - 9$
   (d) $x^2 + 1$ (e) $x^2 - 7x$ (f) $q^2 - 1$

2. Factorise these expressions using the difference of two squares:
   (a) $d^2 - 4$ (b) $p^2 - q^2$ (c) $9 - x^2$ (d) $16 - x^2y^2$ (e) $81x^2t^2 - 144y^2$
   (f) $25q^2 - 16$ (g) $289 - 169q^2$ (h) $xp^2 - q^2x$ (i) $225 - 49t^2$ (j) $9 - 4x^2$
   (k) $20x^2 - 125$ (l) $2p^3 - 8p$ (m) $16 - 64p^2$ (n) $7a^2 - 28b^2$ (o) $a^3 - ab^2$
   (p) $81y^2 - 144p^2$ (q) $9p^3 - 100pr^2$ (r) $2y^2 - 98$

3. Evaluate $101^2 - 99^2$ without a calculator.

4. Evaluate $52^2 - 48^2$ without a calculator.

5. A right–angled triangle has hypotenuse and height as shown in the diagram on the right. Use Pythagoras' theorem and the difference of two squares to find the length of the base of the triangle.

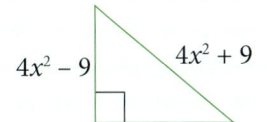

   (Triangle with height $4x^2 - 9$ and hypotenuse $4x^2 + 9$)

6.4 Factorising Expressions Involving Four Terms

In this section we are considering factorising four-term expressions that typically look like:
$$axy + bx + cy + d$$
This type of question can be answered by first picking any two terms with a common factor and factorising. Then factorise the remaining two terms, if possible. In a GCSE question pairs of terms can usually be factorised with a common bracket.

Example 6

Factorise: $6xy + 2x + 3y + 1$

First factorise the first pair of terms, if possible:	$6xy + 2x = 2x(3y + 1)$
Then factorise the second pair of terms, if possible. In this example they don't factorise, so we are left with:	$3y + 1$
Then write out the whole expression with our factorisations:	$2x(3y + 1) + (3y + 1)$
Notice that the bracket $(3y + 1)$ appears twice. So we can factorise using this bracket to get:	$(2x + 1)(3y + 1)$

> **Note:** We are treating $2x \times (3y + 1) + 1 \times (3y + 1)$ as two separate values each multiplied by the same bracket. In other words, $2x \times$ (bracket) $+ 1 \times$ (bracket) $= (2x + 1) \times$ (bracket).

Thus we can write: $\qquad 6xy + 2x + 3y + 1 = (2x + 1)(3y + 1)$

> **Note:** If we had factorised different pairs of terms, we would still obtain the same answer. Consider if we had started by factorising the first and third terms, as shown on the right:
>
> $6xy + 2x + 3y + 1$
> $= 3y(2x + 1) + (2x + 1)$
> $= (3y + 1)(2x + 1)$
> $= (2x + 1)(3y + 1)$

Example 7

Factorise: $xy - ax + by - ab$

First factorise the first pair of terms, if possible:	$xy - ax = x(y - a)$
Then factorise the second pair of terms, if possible:	$by - ab = b(y - a)$
Thus:	$xy - ax + by - ab$
	$= x(y - a) + b(y - a)$
	$= (x + b)(y - a)$

Exercise 6D

1. Factorise the following four-term expressions:
   - (a) $xy + x + y + 1$
   - (b) $pq + 3p + q + 3$
   - (c) $5st - s + 5t - 1$
   - (d) $12xy + 3x + 4y + 1$
   - (e) $pq - p - q + 1$
   - (f) $4st + s - 4t - 1$
   - (g) $xy + 2x + 3y + 6$
   - (h) $15mn - 5m - 6n + 2$
   - (i) $xy + qx + py + pq$
   - (j) $pq - cp - dq + cd$
   - (k) $xy + 3qx + 7py + 21pq$
   - (l) $21pq - 35tq + 6py - 10ty$
   - (m) $12xp + 21yp - 8xq - 14yq$
   - (n) $40py - 64pz - 25qy + 40qz$

2. By calculating the total area of the four rectangles on the right in two different ways, illustrate the factorisation of $xy + py + qx + pq$.

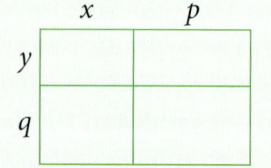

	x	p
y		
q		

6.5 Summary

In this chapter you have learned how to:
- Factorise by taking out a common factor or factors.
- Factorise using the difference of two squares.

Chapter 7
Algebraic Fractions

7.1 Introduction

Adding and subtracting algebraic fractions works in the same way as adding and subtracting fractions with ordinary numbers.

Key words
- **Numerator**: The top part of a fraction, for example the 4 in $\frac{4}{5}$
- **Denominator**: The bottom part of a fraction, for example the 5 in $\frac{4}{5}$
- **Improper fraction**: A fraction in which the numerator is greater than the denominator, for example $\frac{11}{7}$
- **Mixed number**: A value that has both a whole number part and a fraction part, for example $4\frac{5}{6}$

Before you start you should know how to:
- Convert mixed numbers into improper fractions and vice-versa.
- Show when fractions are equal.
- Add, subtract, multiply and divide basic fractions.
- Multiply out brackets.
- Factorise quadratic expressions.

In this chapter you will learn how to:
- Add or subtract algebraic fractions with constant denominators.
- Simplify algebraic fractions with linear or quadratic numerators and denominators.
- Multiply and divide algebraic fractions with linear or quadratic numerators and denominators.

Exercise 7A (Revision)

1. Write these fractions as improper fractions:
   (a) $3\frac{1}{2}$ (b) $4\frac{1}{3}$ (c) $1\frac{2}{3}$ (d) $3\frac{1}{3}$ (e) $5\frac{1}{4}$

2. Write these fractions as mixed numbers:
   (a) $\frac{5}{3}$ (b) $\frac{7}{3}$ (c) $\frac{9}{4}$ (d) $\frac{6}{5}$ (e) $\frac{8}{3}$

3. Each of the following fractions is equivalent to one of the other fractions given. Identify the four pairs of fractions which are equivalent.
   (a) $\frac{4}{6}$ (b) $\frac{12}{4}$ (c) $\frac{2}{12}$ (d) $\frac{12}{48}$ (e) $\frac{6}{24}$ (f) $\frac{12}{18}$ (g) $\frac{20}{120}$ (h) $\frac{27}{9}$

4. Evaluate the following without using a calculator:
   (a) $\frac{1}{2}+\frac{3}{4}$ (b) $\frac{2}{3}+\frac{1}{4}$ (c) $\frac{2}{5}+\frac{1}{6}$ (d) $\frac{1}{3}+\frac{5}{6}$ (e) $\frac{2}{5}+\frac{3}{7}$ (f) $\frac{3}{8}+\frac{3}{4}$ (g) $1\frac{1}{6}+\frac{2}{3}$
   (h) $3\frac{1}{4}+\frac{7}{8}$ (i) $5\frac{1}{2}+1\frac{1}{6}$ (j) $\frac{5}{8}+2\frac{2}{3}$ (k) $2\frac{2}{5}+3\frac{3}{4}$ (l) $6\frac{3}{5}+\frac{4}{7}$ (m) $\frac{3}{4}-\frac{1}{2}$ (n) $\frac{4}{7}-\frac{2}{5}$
   (o) $\frac{7}{8}-\frac{1}{3}$ (p) $\frac{6}{7}-\frac{2}{9}$ (q) $\frac{1}{8}-\frac{1}{12}$ (r) $\frac{5}{6}-\frac{3}{4}$ (s) $2\frac{1}{3}-\frac{2}{5}$ (t) $3\frac{2}{5}-1\frac{1}{6}$ (u) $4\frac{1}{4}-2\frac{2}{3}$
   (v) $\frac{21}{5}-2\frac{1}{6}$ (w) $8\frac{1}{2}-3\frac{5}{6}$ (x) $1\frac{11}{12}-\frac{5}{6}$

5. Multiply the following fractions:
   (a) $\frac{1}{3}\times\frac{1}{4}$ (b) $\frac{2}{5}\times\frac{3}{8}$ (c) $\frac{4}{9}\times\frac{5}{7}$ (d) $\frac{2}{5}\times\frac{3}{4}$ (e) $\frac{4}{5}\times\frac{1}{6}$ (f) $\frac{5}{6}\times\frac{8}{9}$ (g) $\frac{7}{15}\times\frac{5}{14}$
   (h) $\frac{3}{8}\times\frac{16}{30}$ (i) $\frac{1}{2}\times\frac{6}{7}$ (j) $\frac{24}{25}\times\frac{15}{36}$ (k) $\frac{2}{11}\times\frac{22}{39}$ (l) $\frac{7}{8}\times\frac{4}{21}$ (m) $\frac{3}{4}\times1\frac{1}{2}$ (n) $\frac{31}{2}\times\frac{5}{7}$
   (o) $2\frac{1}{5}\times\frac{10}{11}$ (p) $\frac{7}{9}\times1\frac{4}{5}$ (q) $4\frac{4}{5}\times3\frac{1}{8}$ (r) $1\frac{1}{4}\times10$

GCSE MATHEMATICS M3 AND M7

6. Perform the following divisions:
   (a) $\frac{4}{5} \div \frac{8}{15}$
   (b) $\frac{2}{3} \div \frac{5}{6}$
   (c) $\frac{1}{4} \div \frac{1}{2}$
   (d) $\frac{2}{7} \div \frac{4}{21}$
   (e) $\frac{4}{5} \div \frac{7}{10}$
   (f) $3 \div \frac{2}{3}$
   (g) $2\frac{1}{2} \div 4\frac{2}{3}$
   (h) $1\frac{1}{6} \div 2\frac{1}{2}$
   (i) $3\frac{3}{5} \div 2\frac{1}{4}$
   (j) $5\frac{2}{5} \div 4\frac{1}{2}$
   (k) $12\frac{1}{5} \div 20\frac{1}{2}$
   (l) $\frac{8}{9} \div 4$

7. Multiply out:
   (a) $3(x - 2)$
   (b) $(x - 1)(2x + 5)$
   (c) $(x + 2)(x - 5)$
   (d) $(x - 3)(x + 3)$

8. Factorise these quadratic expressions:
   (a) $x^2 + 4x + 3$
   (b) $x^2 + 5x - 6$
   (c) $x^2 + 10x + 24$
   (d) $x^2 - 5x + 6$

7.2 Adding and Subtracting Algebraic Fractions

The CCEA M3 specification requires you to add algebraic fractions like:

$\frac{2x - 1}{3}$ or $\frac{x + 7}{5}$ or $\frac{-3 - 4x}{6}$ or $\frac{2 - 5x}{15}$

In other words, fractions where only the numerator has an algebraic expression.

Adding two such fractions is performed in exactly the same way as adding two numeric fractions. The following examples show how to add algebraic fractions by illustrating each one alongside a similar numeric example.

Example 1

Calculate (a) $\frac{x - 4}{3} + \frac{2 + x}{4}$ and (b) $\frac{2}{3} + \frac{1}{4}$

First spot that the common denominator in both (a) and (b) must be $3 \times 4 = 12$

Next, make the denominator of every fraction equal to 12 by multiplying the first fraction by $\frac{4}{4}$ and the second fraction by $\frac{3}{3}$:

(a) $\frac{x - 4}{3} + \frac{2 + x}{4}$

$\frac{x - 4}{3} \times \frac{4}{4} + \frac{2 + x}{4} \times \frac{3}{3}$

$= \frac{4x - 16}{12} + \frac{6 + 3x}{12}$

(b) $\frac{2}{3} + \frac{1}{4}$

$\frac{2}{3} \times \frac{4}{4} + \frac{1}{4} \times \frac{3}{3}$

$= \frac{8}{12} + \frac{3}{12}$

Finish the question by adding the numerators to obtain:

$= \frac{4x - 16 + 6 + 3x}{12}$

$= \frac{7x - 10}{12}$

$= \frac{8 + 3}{12}$

$= \frac{11}{12}$

Example 2

Calculate (a) $\frac{x}{5} + \frac{2x - 3}{7}$ and (b) $\frac{1}{5} + \frac{2}{7}$

First spot that the common denominator in both (a) and (b) must be $5 \times 7 = 35$

Next, make the denominator of every fraction equal to 35 by multiplying the first fraction by $\frac{7}{7}$ and the second fraction by $\frac{5}{5}$:

(a) $\frac{x}{5} + \frac{2x - 3}{7}$

$\frac{x}{5} \times \frac{7}{7} + \frac{2x - 3}{7} \times \frac{5}{5}$

$= \frac{7x}{35} + \frac{10x - 15}{35}$

(b) $\frac{1}{5} + \frac{2}{7}$

$\frac{1}{5} \times \frac{7}{7} + \frac{2}{7} \times \frac{5}{5}$

$= \frac{7}{35} + \frac{10}{35}$

Adding the numerators:

$= \frac{7x + 10x - 15}{35}$

$= \frac{17x - 15}{35}$

$= \frac{7 + 10}{35}$

$= \frac{17}{35}$

Example 3

Calculate (a) $\frac{x + 2}{3} - \frac{3x}{4}$ and (b) $\frac{2}{3} - \frac{1}{4}$

First spot that the common denominator in both (a) and (b) must be $3 \times 4 = 12$

Next, make the denominator of every fraction equal to 12 by multiplying the first fraction by $\frac{4}{4}$ and the second fraction by $\frac{3}{3}$:

Subtract the numerators:

(a) $\dfrac{x+2}{3} - \dfrac{3x}{4}$

$\dfrac{x+2}{3} \times \dfrac{4}{4} - \dfrac{3x}{4} \times \dfrac{3}{3}$

$= \dfrac{4x+8}{12} - \dfrac{9x}{12}$

$= \dfrac{4x+8-9x}{12}$

$= \dfrac{8-5x}{12}$

(b) $\dfrac{2}{3} - \dfrac{1}{4}$

$\dfrac{2}{3} \times \dfrac{4}{4} - \dfrac{1}{4} \times \dfrac{3}{3}$

$= \dfrac{8}{12} - \dfrac{3}{12}$

$= \dfrac{8-3}{12}$

$= \dfrac{5}{12}$

Example 4

Calculate (a) $\dfrac{5x-7}{11} - \dfrac{3-2x}{7}$ and (b) $\dfrac{7}{11} - \dfrac{3}{7}$

First spot that the common denominator in both (a) and (b) must be $7 \times 11 = 77$

Next, make the denominator of every fraction equal to 77 by multiplying the first fraction by $\frac{7}{7}$ and the second fraction by $\frac{11}{11}$:

Subtract the numerators:

Note: Be careful to put brackets around the numerator in part (a) when you are taking it away, because it has more than one term.

(a) $\dfrac{5x-7}{11} - \dfrac{3-2x}{7}$

$\dfrac{5x-7}{11} \times \dfrac{7}{7} - \dfrac{3-2x}{7} \times \dfrac{11}{11}$

$= \dfrac{35x-49}{77} - \dfrac{33-22x}{77}$

$= \dfrac{35x-49-(33-22x)}{77}$

$= \dfrac{35x-49-33+22x}{77}$

$= \dfrac{57x-82}{77}$

(b) $\dfrac{7}{11} - \dfrac{3}{7}$

$\dfrac{7}{11} \times \dfrac{7}{7} - \dfrac{3}{7} \times \dfrac{11}{11}$

$= \dfrac{49}{77} - \dfrac{33}{77}$

$= \dfrac{49-33}{77}$

$= \dfrac{49-33}{77}$

$= \dfrac{16}{77}$

Exercise 7B

1. Add or subtract the following algebraic fractions:

   (a) $\dfrac{x+5}{2} + \dfrac{x-3}{4}$
   (b) $\dfrac{x-7}{5} + \dfrac{x-4}{3}$
   (c) $\dfrac{x+3}{2} + \dfrac{5-x}{4}$
   (d) $\dfrac{x-7}{6} + \dfrac{x+4}{5}$
   (e) $\dfrac{x+5}{8} + \dfrac{8-x}{2}$

   (f) $\dfrac{x+1}{3} + \dfrac{x-3}{2}$
   (g) $\dfrac{2x-1}{4} + \dfrac{2x+1}{5}$
   (h) $\dfrac{3x-2}{4} + \dfrac{8-x}{5}$
   (i) $\dfrac{2x-5}{3} + \dfrac{x+6}{8}$
   (j) $\dfrac{5x-4}{9} + \dfrac{4x+5}{3}$

   (k) $\dfrac{6x}{4} + \dfrac{1-x}{8}$
   (l) $\dfrac{5x+7}{9} + \dfrac{2-7x}{4}$
   (m) $\dfrac{2x-1}{3} - \dfrac{x+1}{4}$
   (n) $\dfrac{5x-3}{2} - \dfrac{5x}{3}$
   (o) $\dfrac{5x}{3} - \dfrac{2x}{7}$

   (p) $\dfrac{4-x}{10} - \dfrac{2+x}{12}$
   (q) $\dfrac{2x-7}{4} - \dfrac{x-6}{8}$
   (r) $\dfrac{9x}{10} - \dfrac{2}{3}$
   (s) $\dfrac{4x+9}{5} - \dfrac{x-5}{3}$
   (t) $\dfrac{2x}{7} - \dfrac{x}{9}$

   (u) $\dfrac{3x-1}{5} - \dfrac{x}{2}$
   (v) $\dfrac{5x+6}{4} - \dfrac{6}{7}$
   (w) $\dfrac{5x+7}{3} - \dfrac{x}{6}$
   (x) $\dfrac{12x-7}{4} - \dfrac{5x+2}{3}$

2. Rachel subtracts two algebraic fractions. Her working is shown on the right. Identify the error that she has made and work out the correct answer.

   $\dfrac{3x+2}{5} - \dfrac{1-3x}{4}$

   $= \dfrac{3x+2}{5} \times \dfrac{4}{4} - \dfrac{1-3x}{4} \times \dfrac{5}{5}$

   $= \dfrac{12x+8}{20} - \dfrac{5-15x}{20}$

   $= \dfrac{12x+8-5-15x}{20}$

   $= \dfrac{3-3x}{20}$

7.3 Cancelling Algebraic Fractions

We can simplify numeric fractions by cancellation. One factor from the product of factors in the numerator may be cancelled with an identical factor from the product of factors in the denominator. The following example illustrates this using numeric fractions, while Example 6 uses an algebraic fraction.

Example 5

Cancel (a) $\frac{3}{18}$ and (b) $\frac{20}{48}$

(a) Find a factor of both the numerator and denominator: $\frac{3}{18} = \frac{3 \times 1}{3 \times 6} = \frac{1}{6}$

(b) Find a factor of both the numerator and denominator: $\frac{20}{48} = \frac{4 \times 5}{4 \times 12} = \frac{5}{12}$

Example 6

Cancel (a) $\frac{3x}{18}$ and (b) $\frac{20}{48y}$

(a) Find a factor of both the numerator and denominator: $\frac{3x}{18} = \frac{3 \times x}{3 \times 6} = \frac{x}{6}$

(b) Find a factor of both the numerator and denominator: $\frac{20}{48y} = \frac{4 \times 5}{4 \times 12y} = \frac{5}{12y}$

Sometimes the numerator or denominator, or both, need first to be factorised before cancellation is possible.

Example 7

Cancel (a) $\frac{3x+6}{18}$ and (b) $\frac{7x-21}{2x-6}$

(a) $\frac{3x+6}{18} = \frac{3 \times (x+2)}{3 \times 6} = \frac{x+2}{6}$

(b) $\frac{7x-21}{2x-6} = \frac{7(x-3)}{2(x-3)} = \frac{7}{2}$

In part (b) of Example 7, the factors being cancelled involved a variable, x. The following examples show further examples of situations where this happens.

Example 8

Cancel (a) $\frac{3x^2}{18x}$ and (b) $\frac{2x^2y}{16xy}$ (c) $\frac{25x+15x^2}{30+18x}$ (d) $\frac{7p^2q}{14pq^2-35pq}$

(a) $\frac{3x^2}{18x} = \frac{3x \times x}{3x \times 6} = \frac{x}{6}$

(b) $\frac{2x^2y}{16xy} = \frac{2xy \times x}{2xy \times 8} = \frac{x}{8}$

(c) $\frac{25x+15x^2}{30+18x} = \frac{5x \times (5+3x)}{6 \times (5+3x)} = \frac{5x}{6}$

(d) $\frac{7p^2q}{14pq^2-35pq} = \frac{7pq \times p}{7pq \times (2q-5)} = \frac{p}{2q-5}$

If a quadratic expression is present, it may need to be factorised first.

Example 9

Cancel (a) $\frac{x^2+3x+2}{x+2}$ and (b) $\frac{x^2-3x+2}{x^2-4}$

(a) $\frac{x^2+3x+2}{x+2} = \frac{(x+1) \times (x+2)}{x+2} = \frac{x+1}{1} = x+1$

(b) $\frac{x^2-3x+2}{x^2-4} = \frac{(x-2) \times (x-1)}{(x-2) \times (x+2)} = \frac{x-1}{x+2}$

Exercise 7C

In each question, simplify the algebraic fractions, factorising and cancelling where necessary.

1. (a) $\frac{2x}{6}$ (b) $\frac{5y}{25}$ (c) $\frac{12p}{28}$ (d) $\frac{45pq}{5q}$ (e) $\frac{24x^2}{6x}$ (f) $\frac{36q}{72pq}$

2. (a) $\frac{2x-6}{x-3}$ (b) $\frac{8x+12y}{10x+15y}$ (c) $\frac{x-7}{5x-35}$ (d) $\frac{25x-5x^2}{10xy}$ (e) $\frac{7x^2y+14xy^2}{84xy}$ (f) $\frac{y-xy}{xy-y}$

3. (a) $\frac{x^2+4x+3}{x+3}$ (b) $\frac{x^2+8x+12}{x+2}$ (c) $\frac{x^2+3x+2}{x+1}$ (d) $\frac{x^2-x-2}{x+1}$ (e) $\frac{x^2-x-12}{x+3}$ (f) $\frac{x^2+6x+9}{x+3}$

4. (a) $\dfrac{x^2 + 4x + 4}{x^2 - 4}$ (b) $\dfrac{x^2 + 9x + 20}{x^2 + 8x + 16}$ (c) $\dfrac{x^2 - 2x - 15}{x^2 - 25}$ (d) $\dfrac{x^2 - 3x - 4}{x^2 - 8x + 16}$ (e) $\dfrac{x^2 + 2x + 1}{x^2 - 1}$ (f) $\dfrac{x^2 + 2x - 3}{x^2 - 2x + 1}$

7.4 Multiplying and Dividing Algebraic Fractions

Algebraic fractions are multiplied and divided in the same way as numeric fractions. With multiplication and division, there may be algebraic expressions in both the numerator and denominator. The first step is always to factorise every expression. The following example shows how to multiply algebraic fractions by illustrating each one alongside a similar numeric example.

Example 10

Calculate (a) $\dfrac{14x - 6}{9} \times \dfrac{15x + 3}{8}$ (b) $\dfrac{14}{9} \times \dfrac{15}{8}$

(a) $\dfrac{14x - 6}{9} \times \dfrac{15x + 3}{8}$

$= \dfrac{2(7x - 3)}{3 \times 3} \times \dfrac{3(5x + 1)}{2 \times 4}$ (factorising)

$= \dfrac{(7x - 3)}{3} \times \dfrac{(5x + 1)}{4}$ (cancelling 2s and 3s)

$= \dfrac{(7x - 3)(5x + 1)}{12}$

(b) $\dfrac{14}{9} \times \dfrac{15}{8}$

$= \dfrac{2 \times 7}{3 \times 3} \times \dfrac{3 \times 5}{2 \times 4}$ (factorising)

$= \dfrac{7}{3} \times \dfrac{5}{4}$ (cancelling 2s and 3s)

$= \dfrac{35}{12} = 2\dfrac{11}{12}$

Note: In the numeric example in part (b) we converted the answer to a mixed number. There is no need to convert an algebraic fraction in this way.

There may be algebraic expressions in the denominator as well as the numerator.

Example 11

Calculate (a) $\dfrac{5x - 30}{21x + 14} \times \dfrac{18x + 12}{x - 6}$ (b) $\dfrac{55}{21} \times \dfrac{18}{11}$

(a) $\dfrac{5x - 30}{21x + 14} \times \dfrac{18x + 12}{x - 6}$

$= \dfrac{5(x - 6)}{7(3x + 2)} \times \dfrac{6(3x + 2)}{x - 6}$ (factorising)

$= \dfrac{5}{7} \times \dfrac{6}{1}$ (cancelling both pairs of brackets)

$= \dfrac{30}{7}$

(b) $\dfrac{55}{21} \times \dfrac{18}{11}$

$= \dfrac{5 \times 11}{7 \times 3} \times \dfrac{6 \times 3}{1 \times 11}$ (factorising)

$= \dfrac{5}{7} \times \dfrac{6}{1}$ (cancelling 3s and 11s)

$= \dfrac{30}{7}$

Dividing algebraic fractions is carried out in a similar way. Turn the second fraction upside down and then multiply with it, as shown in the following example.

Example 12

Calculate (a) $\dfrac{3x - 12}{7} \div \dfrac{5x - 20}{14}$ (b) $\dfrac{3}{7} \div \dfrac{5}{14}$

(a) $\dfrac{3x - 12}{7} \div \dfrac{5x - 20}{14}$

$= \dfrac{3x - 12}{7} \times \dfrac{14}{5x - 20}$ (becomes a multiplication)

$= \dfrac{3(x - 4)}{1 \times 7} \times \dfrac{2 \times 7}{5(x - 4)}$ (factorising)

$= \dfrac{3}{1} \times \dfrac{2}{5}$ (cancelling twice)

$= \dfrac{6}{5} = 1\dfrac{1}{5}$

(b) $\dfrac{3}{7} \div \dfrac{5}{14}$

$= \dfrac{3}{7} \times \dfrac{14}{5}$

$= \dfrac{3}{7 \times 1} \times \dfrac{7 \times 2}{5}$ (factorising)

$= \dfrac{3}{1} \times \dfrac{2}{5}$ (cancelling twice)

$= \dfrac{6}{5} = 1\dfrac{1}{5}$

When quadratic expressions are involved, we first factorise them and then proceed as before, as shown in the following example.

Example 13

Calculate (a) $\dfrac{x^2 + 3x + 2}{x^2 + 4x + 3} \times \dfrac{x + 3}{x^2 + 6x + 8}$ (b) $\dfrac{x^2 - x - 6}{x^2 - 3x} \div \dfrac{x^2 - 3x - 10}{x^2 - 6x + 5}$ (c) $\dfrac{4x^2 - 49}{x^2 - x - 12} \div \dfrac{10x + 35}{x^2 + 6x + 9}$

(a) $\dfrac{x^2 + 3x + 2}{x^2 + 4x + 3} \times \dfrac{x + 3}{x^2 + 6x + 8}$

$= \dfrac{(x + 1)(x + 2)}{(x + 1)(x + 3)} \times \dfrac{x + 3}{(x + 2)(x + 4)}$ (factorising)

Cancel 3 pairs of brackets, giving:

$= \dfrac{1}{x + 4}$

(b) $\dfrac{x^2 - x - 6}{x^2 - 3x} \div \dfrac{x^2 - 3x - 10}{x^2 - 6x + 5}$

$= \dfrac{x^2 - x - 6}{x^2 - 3x} \times \dfrac{x^2 - 6x + 5}{x^2 - 3x - 10}$

$= \dfrac{(x - 3)(x + 2)}{x(x - 3)} \times \dfrac{(x - 5)(x - 1)}{(x - 5)(x + 2)}$ (factorising)

Cancel 3 pairs of brackets, giving:

$= \dfrac{x - 1}{x}$

(c) $\dfrac{4x^2 - 49}{x^2 - x - 12} \div \dfrac{10x + 35}{x^2 + 6x + 9}$

$= \dfrac{4x^2 - 49}{x^2 - x - 12} \times \dfrac{x^2 + 6x + 9}{10x + 35}$

$= \dfrac{(2x - 7)(2x + 7)}{(x - 4)(x + 3)} \times \dfrac{(x + 3)(x + 3)}{5(2x + 7)}$ (factorising, using the difference of two squares for $4x^2 - 49$)

Cancel 2 pairs of brackets, giving:

$= \dfrac{(2x - 7)(x + 3)}{5(x - 4)}$

Exercise 7D

1. Simplify the following:

   (a) $\dfrac{5x - 10}{2} \times \dfrac{8x + 2}{5}$ (b) $\dfrac{9x + 45}{7} \times \dfrac{35x + 42}{9}$ (c) $\dfrac{32x - 8}{5} \times \dfrac{20x - 15}{8}$ (d) $\dfrac{4x + 6}{3} \times \dfrac{15x + 12}{8}$

   (e) $\dfrac{4x + 6}{5} \div \dfrac{16x + 24}{8}$ (f) $\dfrac{3 - x}{6} \div \dfrac{12 - 4x}{7}$ (g) $\dfrac{x + 6}{5} \div \dfrac{9x + 54}{10}$ (h) $\dfrac{32x - 56}{5} \div \dfrac{28 - 16x}{4}$

   (i) $\dfrac{21x - 42}{9} \times \dfrac{3}{x - 2}$ (j) $\dfrac{8x - 24}{3} \times \dfrac{1}{x - 3}$ (k) $\dfrac{x + 4}{4} \times \dfrac{3}{5x + 20}$ (l) $\dfrac{x + 3}{7} \times \dfrac{14}{5x + 15}$

   (m) $\dfrac{x + 4}{x - 1} \times \dfrac{3x - 3}{2x + 8}$ (n) $\dfrac{8x + 12}{6x - 3} \times \dfrac{14x - 7}{10x + 15}$ (o) $\dfrac{x - 1}{2x + 8} \times \dfrac{3x + 12}{3 - 3x}$ (p) $\dfrac{2 - 3x}{1 - x} \times \dfrac{2x - 2}{2x - 3}$

2. Simplify the following:

   (a) $\dfrac{x^2 + 2x + 1}{x^2 + 3x + 2} \times \dfrac{x + 5}{x^2 + 6x + 5}$ (b) $\dfrac{x^2 + 5x + 6}{x^2 + 4x + 3} \times \dfrac{x^2 + 2x + 1}{x^2 + 3x + 2}$ (c) $\dfrac{x^2 + 3x}{x^2 + 4x + 4} \times \dfrac{x + 2}{6x + 9}$

   (d) $\dfrac{x^2 - 3x - 10}{x^2 + 4x + 4} \times \dfrac{x^2 + x - 2}{x^2 - 6x + 5}$ (e) $\dfrac{x^2 - 7x - 18}{x^2 - 10x + 9} \times \dfrac{x^2 - 2x + 1}{x^2 - 5x - 14}$ (f) $\dfrac{x^2 - 2x - 48}{x^2 - 10x + 16} \times \dfrac{x^2 - 4x + 4}{x^2 + 12x + 36}$

   (g) $\dfrac{x^2 + x - 6}{x^2 + 4x - 12} \div \dfrac{x^2 + 4x + 3}{x^2 + 7x + 6}$ (h) $\dfrac{x^2 + 8x + 15}{x^2 + 4x + 3} \div \dfrac{x^2 + 12x + 35}{x^2 + 8x + 7}$ (i) $\dfrac{x^2 - 10x + 16}{x^2 + x - 12} \div \dfrac{x^2 - 17x + 72}{x^2 - 12x + 27}$

3. Simplify the following:

   (a) $\dfrac{x^2 - 16}{x} \times \dfrac{x^2 + 5x}{x^2 + 9x + 20}$ (b) $\dfrac{49x^2 - 25}{7x^2 + 5x} \times \dfrac{30x^2}{14x - 10}$ (c) $\dfrac{x^2 + 5x - 24}{x^2 + 5x} \div \dfrac{x^2 - 9}{x^2 + 8x + 15}$

   (d) $\dfrac{x^2 + 4x}{x - 3} \div \dfrac{x^2 - 16}{x^2 - 6x + 9}$ (e) $\dfrac{x^2 - 5x - 66}{x^2 + 4x - 12} \div \dfrac{x^2 - 22x + 121}{x^2 + 7x - 18}$ (f) $\dfrac{x^2 + 15x}{x^2 - 9} \div \dfrac{x^2 + 18x + 45}{x^2 + 9x - 36}$

7.5 Summary

In this chapter you have learned how to:

- Add or subtract algebraic fractions with constant denominators.
- Simplify algebraic fractions with linear or quadratic numerators and denominators.
- Multiply and divide algebraic fractions with linear or quadratic numerators and denominators.

Chapter 8
Solving Equations and Identities

8.1 Introduction

In this chapter we will review the skills needed to solve different types of equation, for example, solving:

$$4x + 23 = 7x - 1 \qquad 3(x - 5) = 6 \qquad \frac{x}{6} - 7 = 3$$

We will then extend these skills to solve equations involving a number of fractional terms.

Key words
- **Expression**: An algebraic expression does not involve an equals sign.
  For example $2x$, $3y - 10$ and $5a^2$ are all expressions.
- **Equation**: An equation involves two expressions separated by an equals sign.
- **Linear equation**: A linear equation involves a variable, such as x, and numbers.
  It does not involve more complicated terms such as x^3

Before you start you should know how to:
- Add, subtract, multiply and divide basic fractions.
- Solve simple linear equations.
- Multiply out brackets.
- Factorise expressions involving like terms.

In this chapter you will learn how to:
- Solve simple linear equations.
- Solve linear equations with unknowns on both sides.
- Solve equations involving brackets.
- Solve equations with a fractional coefficient of x.
- Solve equations involving algebraic fractions.
- Set up equations and solve them.
- Work with and prove identities.

One method for solving equations is called 'balancing', as shown in the following example.

Example 1 (Revision)

Solve $5x - 6 = 24$

$$\begin{aligned} 5x - 6 &= 24 \\ 5x - 6 + 6 &= 24 + 6 \\ 5x &= 30 \\ \frac{5x}{5} &= \frac{30}{5} \\ x &= 6 \end{aligned}$$

The method of balancing involves making exactly the same change to both sides of the equation as in the 2nd and 4th lines above. So, if both sides of the equation were originally equal, then they remain equal.

As you get used to solving equations, you will probably omit the second and fourth lines. But when you are not completely sure what to do, it is safest to return to this approach.

Example 2 (Revision)

Solve $30 - 3x = 12$

Since the *x* term on the left is negative, adding 3*x* to each side will make the term involving *x* appear as positive on the right side:

$$30 - 3x + 3x = 12 + 3x$$
$$30 = 12 + 3x$$

Now isolate the term on its own on the right by taking 12 from each side:

$$30 - 12 = 12 - 12 + 3x$$
$$18 = 3x$$

Divide both sides by 3:

$$\frac{18}{3} = \frac{3x}{3}$$
$$x = 6$$

Note: In the rest of this chapter we will refer to the number part of a term as the **coefficient** of the term. For example, the coefficient of 3*x* is 3, the coefficient of 5*pq* is 5 and the coefficient of $4y^2$ is 4

Exercise 8A (Revision)

1. Solve the following equations:
   (a) $2x + 3 = 7$
   (b) $5x + 7 = 27$
   (c) $3x - 8 = 19$
   (d) $11x - 5 = 72$
   (e) $4x + 17 = 37$
   (f) $7x + 11 = 32$
   (g) $9x + 1 = 100$
   (h) $2x + 5 = 31$
   (i) $60 - 6y = 36$
   (j) $72 - 9q = 27$
   (k) $110 - 3p = 62$
   (l) $23 - 8x = 15$
   (m) $43 + 2q = 55$
   (n) $91 - 4w = 51$
   (o) $73 + 7x = 87$
   (p) $3 + 2y = 15$
   (q) $5 = 15 - 2p$
   (r) $34 = 69 - 5y$
   (s) $64 = 15 + 7p$
   (t) $97 = 23 + 4y$

2. Collect like terms in the following expressions:
   (a) $7x + 3x - x$
   (b) $3x - 6x + x$
   (c) $2x + 5y - 3x - 6y$
   (d) $x - y + x$
   (e) $2p - 4q + p - q$
   (f) $x - 2x + 3x - 4x$
   (g) $8y - x + 3z - y$
   (h) $2z - p + 3x + p - z$

3. Multiply out the following bracketed terms:
   (a) $3(2 + x)$
   (b) $21(x - 3)$
   (c) $8(-6x - 7)$
   (d) $-2(2 - 5x)$
   (e) $4(16x + 1)$
   (f) $(2 - x)7$
   (g) $16(4 - 5x)$
   (h) $7(9 - 8x + 3y)$

8.2 Linear Equations with Unknowns on Both Sides

This section covers equations in which the unknown appears on both sides of the equation.

Example 3

Solve $15x - 9 = 12x + 6$

Gather all the *x* terms on the side with the biggest *x* coefficient.
Do this by taking 12*x* from each side in this case:

$15x - 12x - 9 = 12x - 12x + 6$
$3x - 9 = 6$

Then add 9 to each side to gather the numbers on the opposite side to the *x* term:

$3x - 9 + 9 = 6 + 9$
$3x = 15$

Dividing both sides by 3 we obtain the answer:

$$\frac{3x}{3} = \frac{15}{3}$$
$$x = 5$$

Example 4

Solve $4x + 11 = 9x - 24$

Gather all the *x* terms on the side with the biggest *x* coefficient. Do this by taking 4*x* from each side:

$$4x - 4x + 11 = 9x - 4x - 24$$
$$11 = 5x - 24$$

Then add 24 to each side to gather the numbers on the opposite side to the x term:
$$11 + 24 = 5x - 24 + 24$$
$$35 = 5x$$
$$5x = 35$$

Dividing both sides by 5 we obtain the answer:
$$\frac{5x}{5} = \frac{35}{5}$$
$$x = 7$$

Not all answers are positive whole numbers, as shown in the next example.

Example 5

Solve $7x - 3 = 10x + 5$

Gather all the x terms on the side with the biggest x coefficient. Do this by taking $7x$ from each side:
$$7x - 7x - 3 = 10x - 7x + 5$$
$$-3 = 3x + 5$$

Then take 5 from each side to gather the numbers on the opposite side to the x term:
$$-3 - 5 = 3x + 5 - 5$$
$$-8 = 3x$$

Dividing both sides by 3 we obtain the answer:
$$\frac{-8}{3} = \frac{3x}{3}$$
$$\frac{-8}{3} = x \text{ or } x = \frac{-8}{3}$$

Exercise 8B

1. Solve the following equations:
   (a) $2x - 9 = x + 17$
   (b) $7x + 4 = 3x + 20$
   (c) $4x - 8 = 17 - x$
   (d) $16x - 5 = 23x + 16$
   (e) $27 - 5x = 3x - 5$
   (f) $7x - 9 = 24 + 5x$
   (g) $34 + 5x = 16x + 1$
   (h) $2x - 7 = x + 16$
   (i) $73 - 8x = 6x + 3$
   (j) $25 - 3x = 2x - 20$
   (k) $-12 - 4x = -3x - 19$
   (l) $52 + 4x = 7x + 76$
   (m) $27 - 3x = 22 + 7x$
   (n) $2 + 7x = 8x + 11$
   (o) $9 - 2x = 17 + 3x$
   (p) $12 - 7x = 9x + 4$
   (q) $18x = 24 - 5x$
   (r) $24 + 5x = 4 - 11x$
   (s) $15 + 9x = -21 - 7x$
   (t) $23 - 7x = 6x + 9$

8.3 Equations Involving Brackets

In these equations that have brackets, multiply out the brackets first. Then treat them as a usual linear equation.

Example 6

Solve $3(x + 5) = 36$

First multiply out the bracket:
$$3x + 15 = 36$$
$$3x + 15 - 15 = 36 - 15$$
$$3x = 21$$
$$\frac{3x}{3} = \frac{21}{3}$$
$$x = 7$$

Example 7

Solve $8(3 - 2x) = x - 61$

First multiply out the bracket:
$24 - 16x = x - 61$

Then rearrange to make the coefficient of x positive:
$24 - 16x + 16x = x - 61 + 16x$
$24 = 17x - 61$
$24 + 61 = 17x - 61 + 61$
$85 = 17x$
$\frac{85}{17} = \frac{17x}{17}$
$x = 5$

Be careful if the coefficient of the bracket is negative, as in the following example:

Example 8

Solve $37 - 2(3x - 10) = 3(2x - 5)$

First multiply out the brackets:
$37 - 6x + 20 = 6x - 15$

Then gather terms to make the coefficient of x positive:
$57 - 6x + 6x = 6x + 6x - 15$
$57 = 12x - 15$
$57 + 15 = 12x - 15 + 15$
$72 = 12x$
$\frac{72}{12} = \frac{12x}{12}$
$x = 6$

Exercise 8C

1. Solve the following equations:
   (a) $4(5 - 3x) = 8$
   (b) $6(11 - x) = 36$
   (c) $6x = 4(7 - 2x)$
   (d) $2(x + 3) = 8$
   (e) $8 - x = 2(x - 5)$
   (f) $7(x + 3) = 35$
   (g) $2(x + 4) = 5(x - 2)$
   (h) $7(2x + 5) = 63$
   (i) $12x - 2(3 - 5x) = 5$
   (j) $5(3x - 1) = 17 + 4x$
   (k) $4(3 + 2x) = 26x - 6$
   (l) $29x = 9(x + 2) + 2$
   (m) $23 - 2(1 - 5x) = 12x$
   (n) $7(3x + 7) = 11 + 2x$
   (o) $5(x - 9) = 9(5 - x) + 8$
   (p) $2(x - 7) = 3(x - 4)$
   (q) $13 + 4(3 - x) = x$
   (r) $2x = 6 - 7(x - 3)$
   (s) $3x + 4 = 8(x - 7)$
   (t) $8x - 7 = 3(2x + 5)$

8.4 Equations with a Fractional Coefficient of x

In this section we consider solving equations of the form $ax = b$ when a is a fraction.

Example 9

Solve $\frac{5x}{7} = 40$

This is an equation of the form $ax = b$ where $a = \frac{5}{7}$ and $b = 40$

First multiply both sides by the denominator of the fraction:
$\frac{7}{1} \times \frac{5x}{7} = \frac{7}{1} \times \frac{40}{1}$

Then cancelling to obtain:
$\frac{5x}{1} = \frac{7}{1} \times \frac{40}{1}$

Then divide both sides by 5:
$\frac{5x}{5} = \frac{7 \times 40}{5}$

Then cancelling to obtain:
$$\frac{x}{1} = \frac{7 \times 8}{5}$$
$$\frac{x}{1} = \frac{56}{1}$$

Giving:
$$x = 56$$

Example 10

Solve $\frac{4x}{9} + 3 = 17$

First, we must turn this into an equation of the form $ax = b$, by making sure that there is only an 'x term' on one side and a 'number term' on the other.

Take 3 from both sides of the equation:
$$\frac{4x}{9} + 3 - 3 = 17 - 3$$
$$\frac{4x}{9} = 14$$

Next multiply both sides by the denominator of the fraction:
$$\frac{9}{1} \times \frac{4x}{9} = \frac{9}{1} \times \frac{14}{1}$$

Then cancelling to obtain:
$$\frac{4x}{1} = \frac{9 \times 14}{1}$$

Then divide both sides of the equation by 4:
$$\frac{4x}{4} = \frac{9 \times 14}{4}$$

Then cancelling to obtain:
$$\frac{x}{1} = \frac{9 \times 7}{2}$$

Giving:
$$x = \frac{63}{2} \text{ or } 31\frac{1}{2}$$

Of course, solving an equation may involve a mixture of the approaches above, as shown in the next two examples.

Example 11

Solve $\frac{17x}{12} + 2 = x + 5$

Gather all the x terms on the side with the biggest x coefficient. Do this by taking x from each side.

$$\frac{17x}{12} - x + 2 = x - x + 5$$

$$\frac{5x}{12} + 2 = 5 \qquad \text{Using the fact that } \frac{17}{12} - 1 = \frac{17}{12} - \frac{12}{12} = \frac{5}{12}$$

Then take 2 from each side to gather the numbers on the opposite side to the x term:

$$\frac{5x}{12} + 2 - 2 = 5 - 2$$
$$\frac{5x}{12} = 3$$

Next multiply both sides by the denominator of the fraction:
$$\frac{12}{1} \times \frac{5x}{12} = \frac{12}{1} \times \frac{3}{1}$$

Then cancel to obtain:
$$\frac{5x}{1} = \frac{36}{1}$$

Then divide both sides by 5:
$$\frac{5x}{5} = \frac{36}{5}$$
$$x = \frac{36}{5} \text{ or } x = 7\frac{1}{5}$$

Example 12

Solve $\dfrac{3x}{4} = \dfrac{x}{3} - 25$

Gather all the x terms on the left-hand side. Do this by taking $\dfrac{x}{3}$ from each side:

$\dfrac{3x}{4} - \dfrac{x}{3} = \dfrac{x}{3} - \dfrac{x}{3} - 25$

$\dfrac{9x}{12} - \dfrac{4x}{12} = -25$ (Finding a common denominator on the left-hand side.)

$\dfrac{5x}{12} = -25$

To make the left-hand side equal to x, multiply both sides by $\dfrac{12}{5}$:

$\dfrac{12}{5} \times \dfrac{5x}{12} = \dfrac{12}{5} \times -\dfrac{25}{1}$

$x = 12 \times -5 = -60$

Exercise 8D

1. Find x in each of the following:

   (a) $\dfrac{3x}{7} = 18$
   (b) $\dfrac{4x}{9} = 44$
   (c) $\dfrac{5x}{7} = 35$
   (d) $\dfrac{4x}{13} = 28$
   (e) $\dfrac{2x}{3} = 7$

   (f) $\dfrac{8x}{11} = 3$
   (g) $\dfrac{7x}{13} = 21$
   (h) $\dfrac{33x}{26} = 77$
   (i) $\dfrac{x}{5} = 21$
   (j) $\dfrac{4x}{7} = 24$

   (k) $\dfrac{5x}{12} = 60$
   (l) $\dfrac{11x}{5} = 121$
   (m) $\dfrac{4x}{5} = \dfrac{8}{15}$
   (n) $\dfrac{25x}{9} = \dfrac{5}{8}$
   (o) $\dfrac{28x}{33} = \dfrac{56}{121}$

   (p) $\dfrac{3x}{8} = \dfrac{27}{8}$
   (q) $\dfrac{x}{5} + 6 = 4$
   (r) $\dfrac{7x}{3} - 5 = 9$
   (s) $\dfrac{4x}{9} + 4 = 13$
   (t) $3 - \dfrac{7x}{11} = 10$

   (u) $2 + \dfrac{5x}{8} = 6$
   (v) $\dfrac{4x}{9} + 1 = 13$
   (w) $5 - \dfrac{7x}{9} = 19$
   (x) $33 + \dfrac{4x}{5} = 5$

2. Solve the following equations:

   (a) $\dfrac{x}{3} = \dfrac{x}{2} + 3$
   (b) $\dfrac{3x}{5} - 7 = \dfrac{x}{5} + 3$
   (c) $5 - \dfrac{2x}{3} = \dfrac{x}{4} - 6$
   (d) $5 + \dfrac{x}{4} = \dfrac{x}{5} - 25$

   (e) $9 - \dfrac{x}{8} = \dfrac{x}{4} - 27$
   (f) $\dfrac{4x}{5} + \dfrac{7}{2} = \dfrac{x}{5} - \dfrac{5}{2}$
   (g) $\dfrac{5x}{2} - 3 = 4 + \dfrac{3x}{4}$
   (h) $11 + \dfrac{4x}{3} = \dfrac{8x}{3} + 3$

   (i) $20 - \dfrac{x}{5} = \dfrac{x}{2} + 6$
   (j) $3(\dfrac{x}{2} - 1) = \dfrac{3x}{4} + 3$
   (k) $\dfrac{5x}{8} - 9 = 2 + \dfrac{x}{6}$
   (l) $\dfrac{x}{12} + 3 = \dfrac{3x}{16} + 8$

   (m) $5 + \dfrac{x}{7} = \dfrac{3x}{8} + 18$
   (n) $\dfrac{x}{5} + 8 = 4(13 - \dfrac{x}{2})$
   (o) $\dfrac{4x}{9} = 10 + \dfrac{x}{6}$
   (p) $\dfrac{x}{4} - 7 = 3 + \dfrac{3x}{2}$

   (q) $5(23 - \dfrac{2x}{3}) = \dfrac{x}{2}$
   (r) $7(5 + \dfrac{3x}{5}) = 2 + \dfrac{x}{6}$
   (s) $3(\dfrac{x}{6} + 2) + \dfrac{x}{2} = 6$
   (t) $\dfrac{7x}{8} - 5 = 2 + \dfrac{3x}{5}$

8.5 Algebraic Fractions in Equations

In your examination, you may be asked to solve any of the previous types of equation. In this section we will consider equations which include **linear terms** in the numerators of algebraic fractions.

A linear term includes a variable, such as x, but does not involve powers of that variable such as x^2 or x^3.
Examples of linear terms are:

$2x - 5$ $6 - 7x$ $5x$ or $18x + 1$

So, examples of the type of algebraic fractions that we will look at in this chapter are:

$\dfrac{2x - 5}{7}$ $\dfrac{6 - 7x}{8}$ $-\dfrac{5x}{20}$ or $\dfrac{18x - 1}{4}$

The method is to write every term with the common denominator for the whole equation, as shown in the following examples.

CHAPTER 8: SOLVING EQUATIONS AND IDENTITIES

Example 13

Solve (a) $\dfrac{6x+2}{4} = \dfrac{5x-1}{2} - 2$ (b) $\dfrac{2x-1}{4} + \dfrac{5x+1}{8} = 1$

(a) The lowest common multiple of the two denominators, 4 and 2, is 4:
$$\dfrac{6x+2}{4} = \dfrac{5x-1}{2} \times \dfrac{2}{2} - 2 \times \dfrac{4}{4}$$
$$\dfrac{6x+2}{4} = \dfrac{10x-2}{4} - \dfrac{8}{4}$$

Now we can multiply every term by 4:
$6x + 2 = 10x - 2 - 8$ or
$6x + 2 = 10x - 10$

Then subtract $6x$ from each side:
$6x - 6x + 2 = 10x - 6x - 10$
$2 = 4x - 10$

Finally add 10 to each side:
$2 + 10 = 4x - 10 + 10$
$12 = 4x$

Dividing by 4 we finally obtain the answer:
$x = 3$

(b) The lowest common denominator of the two fractions is 8:
$$\dfrac{2x-1}{4} \times \dfrac{2}{2} + \dfrac{5x+1}{8} = 1 \times \dfrac{8}{8}$$
$$\dfrac{2(2x-1)}{8} + \dfrac{5x+1}{8} = \dfrac{8}{8}$$
$$\dfrac{4x-2}{8} + \dfrac{5x+1}{8} = \dfrac{8}{8}$$

Then we can multiply every term by 8:
$4x - 2 + 5x + 1 = 8$
$9x - 1 = 8$

Then add 1 to each side:
$9x - 1 + 1 = 8 + 1$
$9x = 9$
$x = 1$

Example 14

Solve (a) $\dfrac{x-1}{2} - \dfrac{x-2}{3} = 1$ (b) $\dfrac{x-1}{2} + \dfrac{x+1}{4} = 2$

(a) The lowest common multiple of the two denominators, 2 and 3, is 6:
$$\dfrac{x-1}{2} \times \dfrac{3}{3} - \dfrac{x-2}{3} \times \dfrac{2}{2} = 1 \times \dfrac{6}{6}$$
$$\dfrac{3(x-1)}{6} - \dfrac{2(x-2)}{6} = \dfrac{6}{6}$$
$$\dfrac{3x-3}{6} - \dfrac{2x-4}{6} = \dfrac{6}{6}$$

Then we can multiply every term by 6:
$3x - 3 - (2x - 4) = 6$

Multiply out the bracket:
$3x - 3 - 2x + 4 = 6$
$x + 1 = 6$

Then subtract 1 from each side:
$x + 1 - 1 = 6 - 1$
$x = 5$

(b) The lowest common multiple of the two denominators, 2 and 4, is 4:
$$\dfrac{x-1}{2} \times \dfrac{2}{2} + \dfrac{x+1}{4} = 2 \times \dfrac{4}{4}$$
$$\dfrac{2(x-1)}{4} + \dfrac{x+1}{4} = \dfrac{8}{4}$$
$$\dfrac{2x-2}{4} + \dfrac{x+1}{4} = \dfrac{8}{4}$$

Then we can multiply every term by 4:
$2x - 2 + x + 1 = 8$
$3x - 1 = 8$

Then add 1 to each side:
$3x - 1 + 1 = 8 + 1$
$3x = 9$

Finally, divide each side by 3:
$$\dfrac{3x}{3} = \dfrac{9}{3}$$
$x = 3$

Example 15

Solve $\dfrac{3x+1}{2} - 2 = \dfrac{4x-3}{3}$

The lowest common multiple of the two denominators, 2 and 3, is 6:
$$\dfrac{3x+1}{2} \times \dfrac{3}{3} - 2 \times \dfrac{6}{6} = \dfrac{4x-3}{3} \times \dfrac{2}{2}$$
$$\dfrac{3(3x+1)}{6} - \dfrac{12}{6} = \dfrac{2(4x-3)}{6}$$
$$\dfrac{9x+3}{6} - \dfrac{12}{6} = \dfrac{8x-6}{6}$$

Then we can multiply every term by 6:
$$9x + 3 - 12 = 8x - 6$$
$$9x - 9 = 8x - 6$$

Take $8x$ from each side:
$$9x - 8x - 9 = 8x - 8x - 6$$
$$x - 9 = -6$$

Add 9 to each side:
$$x - 9 + 9 = -6 + 9$$
$$x = 3$$

Exercise 8E

1. Identify the one mistake in the solution below and fix the solution from there on.
$$\frac{x-1}{2} - \frac{x-2}{4} = 1$$
$$\frac{2(x-1)}{4} - \frac{x-2}{4} = \frac{4}{4}$$
$$2x - 2 - x - 2 = 4$$
$$x - 4 = 4$$
$$x = 8$$

2. Solve the following equations for x:

(a) $\frac{x-2}{4} + \frac{3x+2}{10} = 3$

(b) $\frac{x-1}{2} + \frac{x-2}{3} = 3$

(c) $\frac{x-1}{2} = \frac{x+1}{4}$

(d) $\frac{3x+1}{2} + 2 = \frac{8x-3}{3}$

(e) $\frac{2x+3}{4} + \frac{3x+1}{2} = \frac{13}{4}$

(f) $\frac{x-2}{2} + \frac{x+2}{3} = 3$

(g) $\frac{2x+1}{2} - \frac{3x-2}{5} = 1$

(h) $\frac{3x+1}{2} - 2 = \frac{4x-3}{3}$

(i) $\frac{x+5}{2} - \frac{x+2}{3} = 6$

(j) $\frac{2x-1}{2} + \frac{3x-2}{3} = \frac{5}{6}$

(k) $\frac{2x-1}{3} + \frac{8x-7}{33} = 4$

(l) $\frac{5x+3}{4} + 1 = \frac{13-2x}{5}$

(m) $\frac{5x+2}{2} - \frac{7x+3}{4} = \frac{1}{2}$

(n) $\frac{8(x+1)}{15} + \frac{2x-1}{3} = \frac{1}{2}$

(o) $\frac{2x+1}{4} - \frac{7-6x}{12} = -\frac{1}{3}$

(p) $\frac{8x-3}{6} + \frac{x-1}{2} = \frac{9}{2}$

(q) $\frac{3x-2}{2} = \frac{4x+1}{5} - 1$

(r) $\frac{2x-5}{4} - \frac{4-x}{2} = \frac{3}{4}$

(s) $\frac{6x+1}{7} - \frac{4(x-2)}{9} = \frac{13}{3}$

(t) $\frac{2(3x+2)}{6} + \frac{9x-19}{15} = -1$

8.6 Setting Equations up and Solving Them

Many situations you encounter in the real world can be understood with a mathematical model. There are always three steps in this approach:

- Step 1: Translate the situation into an equation, usually by choosing a variable (x) to represent the value of a particular quantity.
- Step 2: Solve the equation to find the value of x that satisfies the equation.
- Step 3: Interpret the result in the context of the real-world situation.

Example 16

An isosceles triangle has its two equal sides of length $(3x + 14)$ cm and $(5x + 8)$ cm. Find this length.

Since the sides are equal:
$$3x + 14 = 5x + 8$$
$$3x - 3x + 14 = 5x - 3x + 8$$
$$14 = 2x + 8$$
$$14 - 8 = 2x + 8 - 8$$
$$6 = 2x$$
$$x = 3$$

To find the length of an edge, substitute $x = 3$ into either $3x + 14$ or $5x + 8$

We find that the length of both of the equal sides is 23 cm.

Example 17

One walker covers $(2x - 3)$ miles at a speed of 4 miles per hour.
A second walker covers $(3x + 1)$ miles at a speed of 6 miles per hour.
If the total time taken by them is 1 hour 25 minutes, find their separate times.

The time taken by the first rambler is given by $\frac{\text{distance}}{\text{speed}} = \frac{2x - 3}{4}$ hours.

The time taken by the second ramble is given by $\frac{\text{distance}}{\text{speed}} = \frac{3x + 1}{6}$ hours.

The total time is one hour 25 minutes $= 1\frac{25}{60} = \frac{17}{12}$ hours.

Adding:
$$\frac{2x - 3}{4} + \frac{3x + 1}{6} = \frac{17}{12}$$

Taking a common denominator:
$$\frac{6x - 9}{12} + \frac{6x + 2}{12} = \frac{17}{12}$$
$$6x - 9 + 6x + 2 = 17$$
$$12x - 7 = 17$$
$$12x = 24$$
$$x = 2$$

The first walker took $\frac{2x - 3}{4} = \frac{(2 \times 2) - 3}{4} = \frac{1}{4}$ hour or 15 minutes.

The second walker took $\frac{3x + 1}{6} = \frac{(3 \times 2) + 1}{6} = \frac{7}{6}$ hour = one hour 10 minutes.

Exercise 8F

1. The isosceles triangle shown on the right has its two equal sides written in terms of x as $(4x + 8)$ cm and $(80 - 5x)$ cm. Find this length.

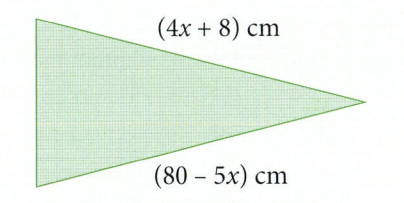

2. Four sides of a hexagon are $(7x - 13)$ cm long and the other two sides are $(3x - 4)$ cm long. The perimeter of the hexagon is 8 cm. Find x.

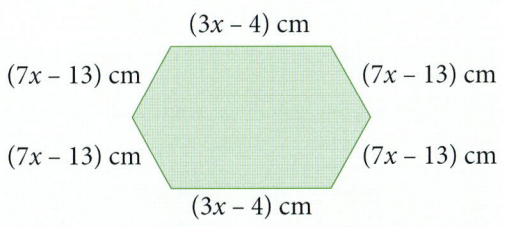

3. One walker covers $(2x + 1)$ miles at a speed of 3 miles per hour.
A second walker covers $(x + 1)$ miles at a speed of 2 miles per hour.
If the total time taken by them is 4 hours 20 minutes, find their separate times.

4. One walker covers $3x$ miles at a speed of 4 miles per hour.
A second walker covers $(x - 2)$ miles at a speed of 3 miles per hour.
If the total time taken by them is 2 hours 35 minutes, find their separate times.

5. One walker covers $(2x + 5)$ miles at a speed of 6 miles per hour.
A second walker covers $(12 - 3x)$ miles at a speed of 4 miles per hour.
If the total time taken by them is 3 hours 35 minutes, find which walker takes longer and by how much.

6. One travelling salesman makes £$(200x + 500)$ in 6 weeks. A second makes £$(300x - 500)$ in 4 weeks. They both make the same overall amount. How much does the second salesman make per week?

7. Two rectangles are shown on the right.
   One rectangle has a length of 5 cm and an area of $2(x + 3)$ cm².
   A second rectangle has a length of 3 cm and an area of $(x - 1)$ cm².
   If their combined widths total 6 cm, find the value of and their individual widths.

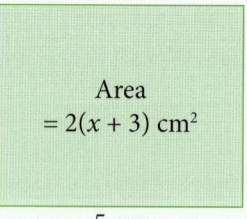

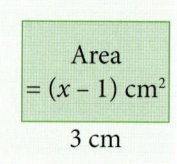

8. Two right-angled triangles have areas $(2x - 1)$ cm² and $(x + 3)$ cm² and heights 3 cm and 2 cm respectively. Find the length of the base of these triangles if they have equal base lengths.

9. One rectangle has a length of 6 cm and an area of $(x + 3)$ cm²
   A second rectangle has a length of 3 cm and an area of $(4x - 1)$ cm²
   If their combined widths total $4\frac{2}{3}$ cm, find their areas.

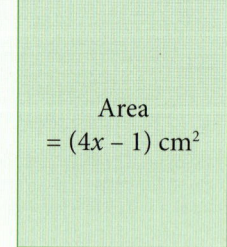

10. One cuboid has volume $(13x - 12)$ m³ and base area 6 m²
    A second cuboid has volume $(2x - 3)$ m³ and base area 3 m²
    If the first cuboid is taller and the difference in their heights is 2 m, find their respective heights.

8.7 Equations Versus Identities

- An **equation** is an algebraic statement that is *only* true when the variable takes a certain value. The value of x that makes an equation true is called its solution.
  For example, $2x = 6$ is only true when $x = 3$, but for every other possible value for x, the equation is not true. So if $x = 7$, then $2 \times 7 = 14$, not 6. Thus $2x = 6$ is an equation and $x = 3$ is its solution.

- An **identity** is a statement about a variable that is *always* true for *every* value of the variable.
  So $2(x + 3) = 2x + 6$ is an identity because both sides are always equal.
  Sometimes we write this as $2(x + 3) \equiv 2x + 6$ to emphasise that it is an identity.

Example 18

Prove that $x^2 + 8x + 15 = (x + 3)(x + 5)$ is an identity.

By multiplying out the right-hand side we get:
$(x + 3)(x + 5) = x^2 + 5x + 3x + 15 = x^2 + 8x + 15$
which is the same as the left-hand side, so this must be an identity.

The formal method to prove that an identity is true is to independently simplify the left-hand side and the right-hand side, showing that they both equal the same expression.

Example 19

Prove the identity $(x + k)^2 + (x - k)^2 \equiv 2(x^2 + k^2)$

The left-hand side $= (x + k)^2 + (x - k)^2$
$= (x + k)(x + k) + (x - k)(x - k)$
$= x^2 + kx + kx + k^2 + x^2 - kx - kx + k^2$
$= 2x^2 + 2k^2$

The right-hand side $= 2(x^2 + k^2) = 2x^2 + 2k^2$

As both sides equal the same thing, the identity is always true.

Example 20

In the identities below find the values of *a*, *b* and *c* where they occur:
(a) $3x^2 + ax + b \equiv cx^2 + 4x - 5$ (b) $(x + a)(x + b) \equiv x^2 + 4x + 3$ (c) $ax(1 - 3x) \equiv 4x - bx^2$

In general, we use the method that the terms in each power of *x* must be separately equal to each other. In addition, the constant terms on each side must be equal.

(a) Comparing the coefficients of x^2: $c = 3$
    Comparing the coefficients of *x*: $a = 4$
    Comparing the constant terms: $b = -5$

(b) First multiply out the left-hand side to get: $x^2 + (a + b)x + ab \equiv x^2 + 4x + 3$
    Comparing the coefficients of *x*: $a + b = 4$
    Comparing the constant terms: $ab = 3$
    So either: $a = 1$ and $b = 3$
    or: $a = 3$ and $b = 1$

(c) First multiply out the left-hand side to get: $ax - 3ax^2 \equiv 4x - bx^2$
    Comparing the coefficients of *x*: $a = 4$
    Comparing the coefficients of x^2: $-3a = -b$
    So: $b = 3 \times 4 = 12$

Exercise 8G

1. For each part (a)–(i), state whether it is an equation or an identity. If the question is an equation, find the value of *x*.
   (a) $5x + 6 = 9$
   (b) $3x - 9 = 3(x - 3)$
   (c) $3x = 2x + x$
   (d) $x^2 + 4x + 4 = 0$
   (e) $x^2 + 4x + 4 = (x + 2)^2$
   (f) $3x - 1 = 1 - 3x$
   (g) $2x - 12 = 2(x - 6)$
   (h) $3x = 2x$
   (i) $\dfrac{x}{2} = \dfrac{x}{4} + 3$

2. Find the value of *a*, *b*, *c* and *d* (where present) in the following identities:
   (a) $7x^2 + bx - 5 \equiv ax^2 - 6x + c$
   (b) $(x - 3)(x + 3) \equiv ax^2 + bx + c$
   (c) $x^2(x - 4) \equiv ax^3 + bx^2 + cx$
   (d) $(x - 2)(x + a) \equiv x^2 + bx - 8$
   (e) $ax^2 + bx - 24 \equiv (x + c)(2x - 3)$
   (f) $x(x^2 + a) + bx(3x - 4) \equiv cx^3 + 6x^2 - 13x + d$

3. Prove the following identities:
   (a) $(x + a)(x - a) \equiv x^2 - a^2$
   (b) $x^2(x - a) + x^2(x + a) \equiv 2x^3$
   (c) $(x + k)^2 - (x - k)^2 \equiv 4kx$
   (d) $x + (x - 1)x(x + 1) \equiv x^3$

8.8 Summary

In this chapter you have learnt how to:
- Solve simple linear equations.
- Solve linear equations with unknowns on both sides.
- Solve equations involving brackets.
- Solve equations with a fractional coefficient of *x*.
- Solve equations involving algebraic fractions.
- Set up equations and solve them.
- Work with and prove identities.

Chapter 9
Quadratic Expressions and Equations

9.1 Introduction

In this chapter we will consider how to factorise quadratic expressions, and how to set up and solve quadratic equations.

Key words
- A **quadratic expression** in the variable x has a term with x^2 as well as possible terms involving x and numbers. Examples of quadratic expressions are:

    $x^2 + 4x + 4$ $\qquad$ $x^2 - 9x + 20$ $\qquad$ $x^2 + 12x$ $\qquad$ $x^2 - 36$ $\qquad$ $7 - 3x - 8x^2$

Before you start you should:
- Make sure you have covered and understood the material on brackets in Chapter 5 and on factorisation in Chapter 6.

In this chapter you will learn how to:
- Factorise quadratic expressions of the form $x^2 + bx + c$
- Set up and solve quadratic equations using factorisation.
- Solve problems involving quadratic equations.

Exercise 9A (Revision)
Expand the following brackets:
1. $(x + 4)(x - 2)$
2. $(x - 8)(x - 5)$
3. $(2x + 5)(x - 7)$
4. $(5x - 9)(6 - 2x)$
5. $(2 + x)^2$
6. $(7x - 6)(7x + 6)$

9.2 Factorising a Quadratic Expression

Factorising a quadratic expression involves writing it as the product of two brackets having the form:
$(x + a)(x + b)$

We can explain how to factorise a quadratic expression by starting with the answer and working backwards. After factorising, let's assume we have the answer:
$(x + 2)(x + 3)$

Remember that this can be multiplied out to give
$x^2 + 3x + 2x + 6$
$= x^2 + 5x + 6$

Notice where the 5 and 6 come from compared to the original pair of brackets. The $5x$ comes from $3x + 2x$, so the coefficient of x is the **sum** of the original two numbers in the brackets. The 6 comes from the **product** of the 2 and the 3

This pattern works for every pair of brackets of the form $(x + a)(x + b)$ where a and b are any two numbers, positive or negative.

Example 1

Factorise $x^2 + 6x + 5$

We must find two numbers which add to give 6 and multiply to give 5	**Note:** You can check your answer by multiplying out the brackets to get $x^2 + 5x + x + 5 = x^2 + 6x + 5$, which is the same as the original problem, so our answer is correct.
We try 5 and 1 since $1 + 5 = 6$ while $1 \times 5 = 5$	
So, the factorised expression is $(x + 1)(x + 5)$	

CHAPTER 9: QUADRATIC EXPRESSIONS & EQUATIONS

Example 2

Factorise $x^2 + 7x + 12$

We must find two numbers which add to give 7 and multiply to give 12
We try 3 and 4 since $3 + 4 = 7$ while $3 \times 4 = 12$
So, the factorised expression is $(x + 3)(x + 4)$

Note: Again, you can check your answer by multiplying out the brackets to get $x^2 + 3x + 4x + 12 = x^2 + 7x + 12$, which is the same as the original problem, so our answer is correct.

Exercise 9B

Factorise the following quadratic expressions:

1. $x^2 + 3x + 2$
2. $x^2 + 2x + 1$
3. $x^2 + 6x + 8$
4. $x^2 + 6x + 5$
5. $x^2 + 8x + 7$
6. $x^2 + 8x + 15$
7. $x^2 + 8x + 12$
8. $x^2 + 14x + 13$
9. $x^2 + 14x + 48$
10. $x^2 + 14x + 49$
11. $x^2 + 11x + 18$
12. $x^2 + 11x + 24$
13. $x^2 + 11x + 30$
14. $x^2 + 12x + 20$
15. $x^2 + 12x + 27$
16. $x^2 + 12x + 35$
17. $x^2 + 12x + 36$
18. $x^2 + 15x + 50$
19. $x^2 + 15x + 54$
20. $x^2 + 15x + 56$
21. $x^2 + 16x + 39$
22. $x^2 + 16x + 48$
23. $x^2 + 16x + 63$
24. $x^2 + 16x + 64$
25. $x^2 + 18x + 32$
26. $x^2 + 18x + 45$
27. $x^2 + 18x + 77$
28. $x^2 + 18x + 72$
29. $x^2 + 18x + 80$
30. $x^2 + 19x + 18$
31. $x^2 + 20x + 96$
32. $x^2 + 20x + 75$
33. $x^2 + 20x + 51$
34. $x^2 + 20x + 64$
35. $x^2 + 20x + 91$
36. $x^2 + 20x + 100$
37. $x^2 + 24x + 140$
38. $x^2 + 27x + 140$
39. $x^2 + 30x + 200$
40. $x^2 + 32x + 231$

9.3 Factorising a Quadratic Expression of the Form $x^2 - px + q$

Sometimes you will need to factorise a quadratic expression with one minus sign and a positive product term. When minus signs appear in the quadratic expression, we must take care over the sum and product. As before, finding two numbers that multiply to give the product is the best starting point.

Example 3

Factorise $x^2 - 13x + 36$

We must find two numbers which add to give -13 and multiply to give 36
We try -4 and -9 since $-4 - 9 = -13$ while $(-4) \times (-9) = +36$
So, the factorised expression is $(x - 4)(x - 9)$

Note: Check your answer by multiplying out the brackets to get $x^2 - 4x - 9x + 36 = x^2 - 13x + 36$, which is the same as the original problem, so our answer is correct.

Example 4

Factorise $x^2 - 19x + 60$

We must find two numbers which add to give -19 and multiply to give 60
We try -4 and -15 as $-4 - 15 = -19$ while $(-4) \times (-15) = +60$
So, the factorised expression is $(x - 4)(x - 15)$

Exercise 9C

Factorise the following quadratic expressions:

1. $x^2 - 9x + 18$
2. $x^2 - 9x + 20$
3. $x^2 - 12x + 35$
4. $x^2 - 9x + 14$
5. $x^2 - 8x + 12$
6. $x^2 - 8x + 15$
7. $x^2 - 7x + 10$
8. $x^2 - 15x + 36$
9. $x^2 - 14x + 48$
10. $x^2 - 12x + 27$
11. $x^2 - 10x + 21$
12. $x^2 - 30x + 29$
13. $x^2 - 11x + 30$
14. $x^2 - 14x + 49$
15. $x^2 - 5x + 6$
16. $x^2 - 16x + 55$

17. $x^2 - 16x + 64$ 18. $x^2 - 16x + 63$ 19. $x^2 - 11x + 24$ 20. $x^2 - 23x + 60$

9.4 Factorising a Quadratic Expression of the Form $x^2 \pm px - q$

Sometimes you will need to factorise a quadratic expression with a negative product term. Consider what you know about the numbers a and b in the answer $(x + a)(x + b)$: one of a and b must be positive and the other negative to give a negative product term.

Example 5

Factorise $x^2 - x - 12$

We must find two integers which add to give -1 and multiply to give -12

The full list of pairs of integers that multiply to give -12 is:
1 and -12, -1 and 12, 2 and -6, -2 and 6, 3 and -4, -3 and 4

We also know that the sum of their values must be -1

All the possibilities are shown in the table on the right.

You can see from the table that the only pair that works is $+3$ and -4 since $+3 - 4 = -1$ while $(+3) \times (-4) = -12$

So, the factorised expression is $(x + 3)(x - 4)$

a	b	$a + b$
1	-12	-11
-1	12	11
2	-6	-4
-2	6	4
3	-4	-1
-3	4	1

Note: In practice we don't need to list out all the possibilities. We just keep guessing new pairs of values for a and b until we find one pair that works. We are showing the table here for illustration.

Example 6

Factorise $x^2 + 3x - 40$

We must find two numbers which add to give $+3$ and multiply to give -40

The full list of pairs of numbers that multiply to give -40 is:
1 and -40, -1 and 40, 2 and -20, -2 and 20, 4 and -10, -4 and 10, 5 and -8, -5 and 8

We also know that the sum of their values must be $+3$

All the possibilities are shown in the table on the right.

You can see from the table that the only pair that works is -5 and $+8$ since $-5 + 8 = +3$ while $(-5) \times (+8) = -40$

So, the factorised expression is $(x - 5)(x + 8)$

a	b	$a + b$
1	-40	-39
-1	40	39
2	-20	-18
-2	20	18
4	-10	-6
-4	10	6
5	-8	-3
-5	8	3

Exercise 9D

Factorise the following quadratic expressions:

1. $x^2 - 2x - 15$
2. $x^2 - 4x - 45$
3. $x^2 + 6x - 27$
4. $x^2 - x - 72$
5. $x^2 - 2x - 24$
6. $x^2 + x - 20$
7. $x^2 + 4x - 77$
8. $x^2 - 3x - 88$
9. $x^2 - 12x - 28$
10. $x^2 + 9x - 22$
11. $x^2 - 10x - 75$
12. $x^2 - 7x - 8$
13. $x^2 - 6x - 91$
14. $x^2 - 2x - 63$
15. $x^2 - x - 132$
16. $x^2 - 7x - 30$
17. $x^2 + x - 42$
18. $x^2 - 8x - 240$
19. $x^2 + 5x - 36$
20. $x^2 + 3x - 108$

9.5 Solving a Quadratic Equation

Solving a quadratic equation means finding the values of x that make the quadratic expression equal to zero. Let us begin by working out the value of a that is the solution to:

$5a = 0$

CHAPTER 9: QUADRATIC EXPRESSIONS & EQUATIONS

The answer is clearly $a = 0$

Now consider the values of *a* and *b* that make:

$a \times b = 0$

Either factor could be zero, so $a = 0$ or $b = 0$

These results give us a hint about how to solve a quadratic equation like:

$x^2 - 5x + 6 = 0$

First factorise the equation to get

$(x - 2)(x - 3) = 0$

Then realise that either $(x - 2) = 0$ or $(x - 3) = 0$

So, the solution to the quadratic equation is $x = 2$ or $x = 3$

Indeed, this can be checked by substituting $x = 2$ into $x^2 - 5x + 6$ to get $4 - 10 + 6 = 0$ and also by substituting $x = 3$ into $x^2 - 5x + 6$ to get $9 - 15 + 6 = 0$

The next three examples illustrate this method further.

Example 7

Solve the quadratic equation $x^2 + 7x + 12 = 0$

First factorise the quadratic expression to get $(x + 3)(x + 4)$

Note: Look back at Example 2 if you need reminded about how to do this.

Then realise that either $(x + 3) = 0$ or $(x + 4) = 0$

So, the solution to the quadratic equation is $x = -3$ or $x = -4$

Example 8

Solve the quadratic equation $x^2 + 3x - 40 = 0$

First factorise the quadratic equation to get $(x - 5)(x + 8)$

Note: Look back at Example 6 if you need reminded about how to do this.

Then realise that either $(x - 5) = 0$ or $(x + 8) = 0$

So, the solution to the quadratic equation is $x = 5$ or $x = -8$

Example 9

Solve the quadratic equation $x^2 - 4x + 4 = 0$

First factorise the quadratic equation to get $(x - 2)(x - 2) = 0$

This can also be written as $(x - 2)^2 = 0$

This means that the only solution is $(x - 2) = 0$, i.e $x = 2$

Note: When a quadratic equation takes the form of a bracket squared, there is only one solution.

Exercise 9E

Solve the following quadratic expressions:

1. $x^2 - 4x + 21 = 0$
2. $x^2 + 5x + 6 = 0$
3. $x^2 - x - 12 = 0$
4. $x^2 - 7x + 12 = 0$
5. $x^2 - 3x - 10 = 0$
6. $x^2 + 18x + 80 = 0$
7. $x^2 - 4x - 5 = 0$
8. $x^2 + 7x - 18 = 0$
9. $x^2 - 13x - 30 = 0$
10. $x^2 - 8x - 20 = 0$
11. $x^2 - 4x - 60 = 0$
12. $x^2 - x - 110 = 0$
13. $x^2 + 6x - 40 = 0$
14. $x^2 - 2x - 120 = 0$
15. $x^2 + 14x - 15 = 0$
16. $x^2 + 8x - 33 = 0$
17. $x^2 - 2x - 35 = 0$
18. $x^2 + 4x - 96 = 0$
19. $x^2 + 3x - 28 = 0$
20. $x^2 + 2x - 24 = 0$
21. $x^2 + 5x - 24 = 0$
22. $x^2 - x - 56 = 0$
23. $x^2 + 3x - 88 = 0$
24. $x^2 + x - 9900 = 0$

Sometimes you will see an equation which is a quadratic equation, but is not written in the form we have been considering above. We first need to use the rules of algebra to write the equation in the usual form before solving it.

Example 10

Solve the equation $x^2 = 4x - 4$

We begin by adding 4 to each side to get: $\qquad x^2 + 4 = 4x$
Then take $4x$ off each side to get: $\qquad x^2 - 4x + 4 = 0$
We can solve this in the usual way (see Example 9 above) to obtain $x = 2$

Example 11

Solve the equation $6 - x^2 = 5x$

We begin by adding x^2 to each side to get: $\qquad 6 = x^2 + 5x$
Then take 6 off each side to get: $\qquad x^2 + 5x - 6 = 0$
We seek two numbers that add to $+5$ and multiply to give -6
These are $+6$ and -1
Factorising this in the usual way we obtain $(x + 6)(x - 1) = 0$
So, the solutions are $x = -6$ and $x = 1$

Example 12

Solve the equation $x + \dfrac{9}{x} = 6$

At first sight this is not a quadratic equation.
But if we multiply all 3 terms by x we obtain: $\qquad x^2 + 9 = 6x$
Then taking $6x$ from each side we have: $\qquad x^2 - 6x + 9 = 0$
This factorises to give: $\qquad (x - 3)(x - 3) = 0$
which can also be written: $\qquad (x - 3)^2 = 0$
This has only one solution, $x = 3$

Exercise 9F

Solve the equations:

1. $x^2 - 10 = 3x$
2. $27 = x^2 + 6x$
3. $6 + x - x^2 = 0$
4. $x(x - 1) = 12$
5. $3x = x^2 - 40$
6. $18 = 7x + x^2$
7. $x + \dfrac{48}{x} = 14$
8. $x(x + 14) = -45$
9. $x + \dfrac{1}{x} = 2$
10. $x^2 + 20 = 9x$
11. $x - \dfrac{36}{x} = 9$
12. $x(4 - x) = 3$

9.6 Problems Involving Quadratic Equations

Many situations in the real world are described by quadratic equations. We first translate the situation into a quadratic equation. Then we solve the equation using the methods described in this chapter. Then we interpret the results in the context of the real-world situation, as shown in the following examples.

Example 13

Two positive whole numbers differ by 5 and their product is 84
Find the numbers.

We begin by introducing a variable to stand for one of the numbers.
Let's say the smallest one has the value x
Then the other one must be $x + 5$

Their product which is $x(x + 5)$ must then be equal to 84 so we can write:
$$x(x + 5) = 84$$
$$x^2 + 5x - 84 = 0$$

Next solve the equation. Factorising gives:
$$(x + 12)(x - 7) = 0$$

This has solutions $x = -12$ or $x = 7$

However, the initial problem stated that the numbers must be positive, so we must discount the solution $x = -12$

This means that the two numbers must be $x = 7$ and $x + 5 = 12$

Example 14

The area of a rectangular field is 1500 m² and its perimeter is 160 m. Find its dimensions.

This time we introduce the variable x to stand for the length of one of the two sides of the field. Then, considering the perimeter, we find that
$$2x + 2 \times \text{(the length of the other side)} = 160$$
So:
$$x + \text{(the length of the other side)} = 80$$
Thus:
$$\text{length of the other side} = 80 - x$$

Now, multiply both lengths together to find the area of the field: $x(80 - x) = 1500$
$$80x - x^2 = 1500$$
$$0 = x^2 - 80x + 1500$$

Factorising gives: $(x - 30)(x - 50) = 0$
This has solutions: $x = 30$ and $x = 50$

This means that the length of one of the sides of the field must be 30 m and the other side must be 50 m, with both sides adding up to 80 m.

So the dimensions of the field are 30 m × 50 m.

Exercise 9G

Solve the following problems by forming and solving a quadratic equation:

1. Two positive whole numbers differ by 9 and their product is 22   Find the numbers.
2. Two positive whole numbers differ by 3 and their product is 108   Find the numbers.
3. Two positive whole numbers differ by 7 and their product is 60   Find the numbers.
4. Two positive whole numbers differ by 12 and their product is 160   Find the numbers.
5. The area of a rectangular field is 5000 m² and its perimeter is 300 m. Find its dimensions.
6. The area of a rectangular page is 450 cm² and its perimeter is 90 cm. Find its dimensions.
7. The height of a triangle is 5 cm more than its base. Find the dimensions of the triangle if its area is 75 cm²
8. The base of a triangle is 3 cm more than its height. Find the dimensions of the triangle if its area is 35 cm²
9. The sum of a number and its reciprocal is 2
   Find the number.
10. The sum of a number and twice its reciprocal is 3
    Find both possible numbers.

Note: Remember that the reciprocal of a number is one divided by that number.

9.7 Summary

In this chapter you have learned how to:
- Factorise quadratic expressions of the form $x^2 + bx + c$
- Set up and solve quadratic equations using factorisation.
- Solve problems involving quadratic equations.

Chapter 10
Trial and Improvement

10.1 Introduction

Some maths equations are too hard to solve using normal algebra. But sometimes a possible solution can be guessed. Then the guess can be checked. If it turns out it's not quite good enough, then a slightly better guess can be tried. We can keep on improving the guess until it gives a close enough answer for our purposes.

Before you start you should know how to:
- Substitute values into an expression.
- Evaluate expressions on a calculator.

What you will learn

In this chapter you will learn how to:
- Solve an equation using trial and improvement

Exercise 10A (Revision)

1. Evaluate each of these expressions when $x = 2$
   (a) $4x - 5$ (b) x^2 (c) $x^2 - 2x$ (d) $x^2 + x + 1$
2. Evaluate each of these expressions when $p = 3$ and $q = -1$
   (a) $p \times q$ (b) $p + q^2$ (c) $p^2 + q^2$ (d) $p^3 - q^3$
3. Evaluate on a calculator:
   (a) $7 \times 4 - 5$ (b) $6^2 - 4$ (c) $3^2 + 4^2$ (d) $3^3 - 7 \times 2$

10.2 Solving by Trial and Improvement

The idea is a simple one: First, guess the answer. Then check to see if it is right. If not, make a better guess. You could use this method to solve the easiest equations.

When the equations can be solved another way, that usually takes less effort than trial and improvement. For example, if you were asked to find what number when multiplied by 4 gives 25 you *could* use trial and improvement:

- Let's guess 5 to start with: $4 \times 5 = 20$ which is too small.
- So, let's try 6: $4 \times 6 = 24$ which is closer but still too small.
- Let's try 7: $4 \times 7 = 28$ which is now too big.
- Try 6.5 (between 6 and 7): $4 \times 6.5 = 26$ which is only a little too big.
- We'll try 6.25: $4 \times 6.25 = 25$ which is the right value, so the answer is 6.25

But think about all this work. Maybe you were able to spot that 6.25 was the correct answer a lot sooner. The normal way to solve this problem is to divide 25 by 4
Immediately we see the answer is 6.25

So, for this question, trial and improvement takes a lot longer than simple division.

But there are other types of problem where trial and improvement is very useful. We know that $49 = 7^2$ so the square root of 49 is $\sqrt{49} = 7$

This is easy because 49 is a perfect square. But before calculators were invented, it was a hard task to find the square root of a number which is not a perfect square.

Try working out the square root of 40 without a calculator. It is very difficult. This is the type of problem which can be solved by trial and improvement.

Remember the basic idea is: Guess an answer → Check guess → Improve guess

It is a good idea to think about the first guess to save effort. Let's say we are trying to work out $\sqrt{40}$:
- As 40 is between $6^2 = 36$ and $7^2 = 49$ we can see that $\sqrt{40}$ lies between 6 and 7 so let's guess 6.5
- We check how good a guess this is by working out 6.5^2 which turns out to be $6.5 \times 6.5 = 42.25$
  As 42.25 is bigger than 40, our guess is too big.
- Next, we will try 6.4: 6.4^2 turns out to be $6.4 \times 6.4 = 40.96$, which is still too big.
- Next, we try 6.3: 6.3^2 turns out to be $6.3 \times 6.3 = 39.69$, which is now too small.
- So we have established that $\sqrt{40}$ lies between 6.3 and 6.4

Your calculator tells you that $\sqrt{40} = 6.324555320336759$ rounded to 15 decimal places. This illustrates that the square root of a number that is not a perfect square has unending decimals without a pattern. (These roots are called surds). We can never find them exactly. Every question will always state what number of decimal places you will need to find the root to.

The next example shows how to prove which value is closest to the exact answer.

Example 1

Find the square root of 20 accurate to one decimal place.

- We begin by finding two whole numbers: one above and one below the square root. As 20 is between 4^2 and 5^2 we can see that $\sqrt{20}$ lies between 4 and 5
- So a good first guess would be one between 4 and 5 so let us start with the guess 4.5 in the middle. $4.5^2 = 4.5 \times 4.5 = 20.25$, which is bigger than 20, so our guess is too big.
- Next we will try 4.4, a slightly lower number. $4.4^2 = 4.4 \times 4.4 = 19.36$, which is now too small.
- So we have established that $\sqrt{20}$ lies between 4.4 and 4.5: as we only have to find the answer to one decimal place, we need to pick one or the other. How do we tell which is closer? The answer is to work out 4.45^2, i.e. the number half way between our two guesses. $4.45^2 = 4.45 \times 4.45 = 19.8025$
- As this number is smaller than 20, it means that the actual square root of 20 must be between 4.45 and 4.5 Any number between 4.45 and 4.5 will round up to 4.5 to one decimal place.
- So, we have shown that $\sqrt{20} = 4.5$ to one decimal place.

Example 2

Find the square root of 30 to two decimal places.

- As 30 is between $5^2 = 25$ and $6^2 = 36$ we can see that $\sqrt{30}$ lies between 5 and 6
- So a good first guess is 5.5 and we calculate that $5.5^2 = 30.25$
- As this is too big, we then calculate 5.4^2 to find $5.4^2 = 29.16$
- We can see that 5.5^2 is closer to 30 than 5.4^2 so we try a number closer to 5.5 as our next guess – since in this question we have been asked to give the answer to two decimal places.
- Try 5.47 to get $5.47^2 = 29.9209$ but as this is too small, try $5.48^2 = 30.0304$
  So we see that the answer lies somewhere between 5.47 and 5.48
  Which one is closer to the true value?
- Work out the square of 5.475 which is their midpoint. $5.475^2 = 29.976$ (to 3 d.p.). This means that the actual answer lies between 5.475 and 5.48 Any number in this range will round to 5.48
- So, the square root of 30 is 5.48 to two decimal places.

Exercise 10B

1. Using trial and improvement find to one decimal place – and without using the square root function on your calculator – the square root of:
   (a) 60 (b) 10 (c) 90 (d) 72 (e) 111

2. Using trial and improvement, find to two decimal places – and without using the square root function on your calculator – the square root of:
   (a) 40 (b) 20 (c) 11

3. Using trial and improvement, find to one decimal place – and without using the cube root function on your calculator – the cube root of 10 (Hint: you need to solve $x^3 = 10$)

10.3 Solving Harder Equations by Trial and Improvement

In this section you will learn how to solve harder equations by trial and improvement. Structure your solution using a table, as in the following example.

Example 3

Find, accurate to one decimal place, the solution to the equation $x^2 + x = 35$

We begin by picking two whole numbers that are on either side of the solution.
We choose $x = 5$ and $x = 6$: $\quad 5^2 + 5 = 25 + 5 = 30 \quad$ and $\quad 6^2 + 6 = 42$
So, we can see that the value of x that makes $x^2 + x$ equal to 35 is between 5 and 6 so let's start with 5.5
We use a table layout:

Guess	Use the guess for x to evaluate the expression $x^2 + x$	Too small / big?
5.5	$5.5 \times 5.5 + 5.5 = 30.25 + 5.5 = 35.75$	too big
5.4	$5.4 \times 5.4 + 5.4 = 29.16 + 5.4 = 34.56$	too small

We have found two adjacent x values, 5.4 and 5.5, giving values for the expression on either side of 35
To decide which of 5.4 and 5.5 is closest to the exact solution, we evaluate the expression at the midpoint between them, namely 5.45:

5.45	$5.45 \times 5.45 + 5.45 = 29.7025 + 5.45 = 35.1525$	too big

As the value of the expression at 5.4 was too small and the value at 5.45 was too big, the actual solution must lie between 5.4 and 5.45
Every number in this range will round down to 5.4 to one decimal place.
Thus 5.4 is the value of x that is closest to the solution to one decimal place.

Exercise 10C

1. Using trial and improvement find, to one decimal place, the solution to the following equations:
   (a) $x^2 + x = 15$ starting with the value $x = 3$ (b) $x^2 + x = 7.5$ starting with the value $x = 2$
   (c) $x^2 - x = 35$ starting with the value $x = 6$ (d) $x^2 - x = 15$ starting with the value $x = 4$
   (e) $x^2 + 2x = 65$ starting with the value $x = 9$

2. Using trial and improvement find, to one decimal place, the solution to the following equations:
   (a) $x^3 + x = 15$ starting with the value $x = 2$ (b) $x^3 + x = 100$ starting with the value $x = 4$
   (c) $x^3 - 2x = 18$ starting with the value $x = 3$ (d) $x^3 + x^2 = 90$ starting with the value $x = 4$
   (e) $x^3 + 2x = 25$ starting with the value $x = 2$

3. Using trial and improvement find, to two decimal places, the solution to the following equations:
   (a) $x^2 + x = 17$ starting with the value $x = 3.6$ (b) $x^3 - x = 26$ starting with the value $x = 3$
   (c) $x^3 - x^2 = 200$ starting with the value $x = 6$ (d) $x^3 - 7x^2 + 50x = 220$ starting with the value $x = 5$
   (e) $3x^3 + 6x^2 - 45x = -100$ starting with the value $x = -6$

4. Using trial and improvement, find to three significant figures, the approximate solution to the following equations:
   (a) $x^3 = x^2 + 45$ (b) $x^3 + 7x = 24$ (c) $2x^3 - 25x = 37$ (d) $x^3 + 7x = 3$ (e) $x^3 - 10x = 29$

10.4 Rearrangement Problems Involving Trial and Improvement

The basic idea of trial and improvement can be applied to many more problems. Some involve guessing a good starting point to search for a solution. Often an equation needs to be rearranged into the pattern:

    Expression involving x = a value

as shown in the next example.

Example 4

A rectangle is $(x + 2)$ cm long and x cm wide. Its area is 10 cm^2
Find the value of x using trial and improvement accurate to one decimal place.

We begin by drawing a sketch of the rectangle, as shown on the right.
As we are told that the area is 10 cm² we can write: $x(x + 2) = 10$
Now rearrange this by multiplying out the brackets: $x^2 + 2x = 10$
Next, draw a table for our guesses:

Guess	Use the guess for x to evaluate the expression $x^2 + 2x$	Too small / big?
2	$2^2 + 2 \times 2 = 8$	Too small
3	$3^2 + 2 \times 3 = 15$	Too big
2.3	$2.3^2 + 2 \times 2.3 = 5.29 + 4.6 = 9.89$	Too small
2.4	$2.4^2 + 2 \times 2.4 = 5.76 + 4.8 = 10.56$	Too big
2.35	$2.35^2 + 2 \times 2.35 = 5.5225 + 4.7 = 10.2225$	Too big

Because 2.3 is too small and 2.35 is too big, the correct value must lie between 2.3 and 2.35
All values in this range will round down to 2.3 to one decimal place.
So the answer is 2.3 cm to one decimal place.

The following example demonstrates equations that should be rearranged before trial and improvement can be used.

Example 5

Rearrange the following into the form of an expression involving x = number:
(a) $x^2 - 15 = x - 4$ (b) $x^2 - x = 2 - 3x$ (c) $x(x - 5) + 5 = 8$

(a) $x^2 - 15 = x - 4$
$x^2 - 15 + 15 = x - 4 + 15$ adding 15 to each side
$x^2 - x = x - x + 11$ subtracting x from each side
$x^2 - x = 11$ which is of the form 'expression = number'

(b) $x^2 - x = 2 - 3x$
$x^2 - x + 3x = 2 - 3x + 3x$ adding $3x$ to each side
$x^2 + 2x = 2$ which is of the form 'expression = number'

(c) $x(x - 5) + 5 = 8$
$x^2 - 5x + 5 = 8$ multiplying out the bracket
$x^2 - 5x + 5 - 5 = 8 - 5$ subtracting 5 from each side
$x^2 - 5x = 3$ which is of the form 'expression = number'

Exercise 10D

1. A rectangle is x cm long and $(x - 1)$ cm wide. Its area is 15 cm² Find the value of x using trial and improvement, giving your answer accurate to one decimal place.
2. A cuboid has a square base, length of side x, and height 3 cm longer than any side of its base. Find the value of x, accurate to two decimal places, if the volume of the cuboid is 50 cm³
3. Solve the following expressions using trial and improvement, giving your answer accurate to one decimal place:
   (a) $x^2 - x = 12 - 7x$ (b) $x(x - 3) = 21 + 4x$ (c) $x(x^2 - 7x) = 23 - x$
   (d) $x(x^2 + 5x) = 121$ (e) $x(x^2 + 5x) + 4x = 6(x - 5)$ (f) $x(x^2 + 5) = 93 - 7x$

10.5 Summary

In this chapter you have learnt that:
- Nearly any equation may be solved using trial and improvement.
- This non-calculator method is useful when the solution to an equation involves multiple decimal places. The solution can be found to whatever accuracy is needed.
- The basic idea is: Guess an answer → Check guess → Improve guess

Progress Review
Chapters 5–10

This Progress Review covers:
- Chapter 5 – Brackets and Identities
- Chapter 6 – Factorisaton
- Chapter 7 – Algebraic Fractions
- Chapter 8 – Linear Equations
- Chapter 9 – Quadratic Equations
- Chapter 10 – Trial and Improvement

1. Expand the brackets:
   (a) $9(2 - 8p)$
   (b) $3x(2x - 7y)$
   (c) $12p(5 - 7q)$
   (d) $-7x(y - 2x)$

2. Expand the brackets:
   (a) $3(2x - 5y) - 4(x + y)$
   (b) $3x(y - 5) + 4y(3 - x)$
   (c) $3p(2q + 5p) - 4q(7p + 5q)$
   (d) $x(2x - 6) + 5(x - 3)$

3. Find the difference in areas of the two rectangles shown on the right.

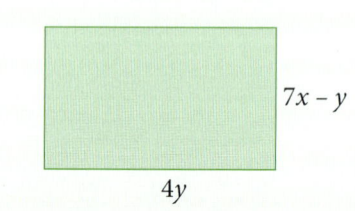

4. Expand the brackets:
   (a) $(x - 2)(x + 3)$
   (b) $(y - 3)(y - 7)$
   (c) $(3 - p)(8 + p)$
   (d) $(q + 12)(q - 2)$

5. Find the area of the rectangle shown on the right.

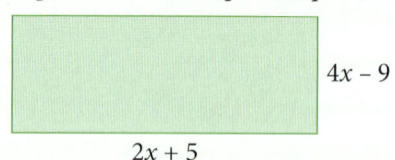

6. Expand the brackets:
   (a) $(2x + 7)(3 - 5x)$
   (b) $(1 - x)(2 - y)$
   (c) $(4p + 7q)(2q - 6p)$
   (d) $(x - 2y)(3y + 2x)$

7. Find the sum of the areas of the two rectangles shown on the right.

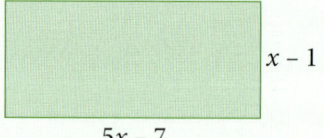

8. Simplify: $4(2 - x - y)(1 + x) - 5(y - 2)(x - 3)$

9. Factorise:
   (a) $5x + 4xy$
   (b) $12pq - 3q$
   (c) $24mn + 6n$
   (d) $-3 - 9xy$
   (e) $24p + 8pq$

10. Factorise:
    (a) $x^2 + 2x$
    (b) $3y - 6y^2$
    (c) $14p - 49p^2$
    (d) $3x + 12x^2 - x^3$
    (e) $16x^2 + 8x$

11. Factorise:
    (a) $p^2 - 36$
    (b) $5q^2 - 45$
    (c) $2y^2 - 98$
    (d) $3xy^2 - 27x^3$

12. Evaluate $151^2 - 149^2$ without using a calculator.

13. A right-angle triangle has largest side = $4y$ cm and another side = 8 cm. Find a factorised expression for the square of its third side. (Hint: You need to use Pythagoras' theorem.)

14. Factorise: $yz + y + z + 1$

15. Factorise:
  (a) $6pq + 3q + 2p + 1$
  (b) $6mt - 4ms + 3nt - 2ns$

16. Express as a single fraction:
  (a) $\dfrac{x}{5} + \dfrac{x}{6}$
  (b) $\dfrac{2x}{3} + \dfrac{x}{4}$
  (c) $\dfrac{3x}{7} + \dfrac{x}{6}$
  (d) $\dfrac{3x}{4} - \dfrac{2x}{5}$

17. Express as a single fraction:
  (a) $\dfrac{x-1}{3} + \dfrac{5x}{6}$
  (b) $\dfrac{2x+1}{4} - \dfrac{x}{3}$
  (c) $\dfrac{x-2}{3} + \dfrac{x+1}{4}$
  (d) $\dfrac{2-x}{3} - \dfrac{x-1}{4}$

18. Simplify by cancelling common factors:
  (a) $\dfrac{x+1}{3} \times \dfrac{3}{6x+6}$
  (b) $\dfrac{12x}{4x-2} \times \dfrac{8x-4}{3}$
  (c) $\dfrac{3x}{7x+21} \times \dfrac{5x+15}{6x}$
  (d) $\dfrac{3x-1}{x+3} \times \dfrac{2x+6}{6x-2}$

19. Simplify by cancelling common factors:
  (a) $\dfrac{x^2+2x+1}{4} \times \dfrac{5}{2x+2}$
  (b) $\dfrac{x^2-x-6}{4x} \times \dfrac{4x-4}{x^2-4x+3}$

20. Simplify by cancelling common factors:
  (a) $\dfrac{4x+2}{7} \div \dfrac{6x+3}{14}$
  (b) $\dfrac{12x-6}{3x} \div \dfrac{8x-4}{5}$
  (c) $\dfrac{1-5x}{x+2} \div \dfrac{10x-2}{9x+18}$
  (d) $\dfrac{x^2+8x+15}{3x+15} \div \dfrac{x^2+6x+9}{4x}$

21. Spot the error in the following working and correct it:

$\dfrac{3x+2}{5} - \dfrac{x-1}{3}$

$= \dfrac{3x+2}{5} \times \dfrac{3}{3} - \dfrac{x-1}{3} \times \dfrac{5}{5}$

$= \dfrac{9x+6}{15} - \dfrac{5x-5}{15}$

$= \dfrac{9x+6-5x-5}{15}$

$= \dfrac{4x+1}{15}$

22. Solve these equations:
  (a) $3x - 1 = x + 7$
  (b) $2x - 5 = 6x - 21$
  (c) $5x + 4 = 8 - 3x$

23. Solve these equations:
  (a) $5x - 7 = 19 - 8x$
  (b) $5x - 13 = 15 - 2x$
  (c) $2x - 27 = 33 - 13x$

24. Solve by first multiplying out the brackets:
  (a) $3(2 - 5x) = 4 - 3x$
  (b) $4(2x + 3) = 3(7 - x) - x$
  (c) $4(7x + 2) = 3(x + 1)$

25. Solve for x:
  (a) $\dfrac{3x}{4} - 3 = \dfrac{x}{3} + 2$
  (b) $\dfrac{x}{4} + 5 = \dfrac{x}{5} + 7$
  (c) $\dfrac{3x}{8} - 7 = \dfrac{x}{2} - 5$

26. Solve these equations, first multiplying out any brackets:
  (a) $3\left(\dfrac{x}{2} + 5\right) = \dfrac{x}{4} + 60$
  (b) $4x - 9 = 5\left(\dfrac{x}{3} + 8\right)$
  (c) $2\left(\dfrac{4x}{5} + 3\right) = 5x - 3\left(\dfrac{x}{3} + 2\right)$

27. Solve for x:
  (a) $\dfrac{5x+6}{7} + 2 = \dfrac{4x+2}{7} + 3$
  (b) $\dfrac{13-4x}{2} + 3 = \dfrac{3x+5}{2}$
  (c) $\dfrac{7x+3}{5} - 4 = \dfrac{6x-1}{7}$

28. Two rectangles are shown below. One has a width of 10 cm and an area of $4(x + 7)$ cm^2
The second rectangle has a width of 8 cm and an area of $(12x - 4)$ cm^2
Their lengths are equal. Find this length.

10 cm | Area = $4(x+7)$ cm^2 Area = $(12x-4)$ cm^2 | 8 cm

29. An isosceles triangle has two angles of size $(7x + 5)°$ and its third angle is $(11x – 5)°$
    Find the size of the angles in degrees.
30. Prove the identity $x(x + k) – x(x – k) \equiv 2xk$
31. Prove the identity $a(b – c) + b(c – a) + c(a – b) \equiv 0$
32. Find the values of a and b in the identity $(x – 4)(x – a) \equiv x^2 – 7x + b$
33. Factorise:
    (a) $x^2 + 12x + 32$
    (b) $x^2 + 10x + 24$
    (c) $x^2 + 13x + 40$
    (d) $x^2 + 4x + 4$
34. Factorise:
    (a) $x^2 – x – 12$
    (b) $x^2 – 5x – 24$
    (c) $x^2 – 14x + 40$
    (d) $x^2 + 6x – 16$
35. Solve the following quadratic equations:
    (a) $x^2 – 12x + 27 = 0$
    (b) $x^2 – 2x – 24 = 0$
    (c) $x^2 + 5x + 4 = 0$
    (d) $x^2 + 4x – 77 = 0$
36. Solve for x:
    (a) $x^2 – 22 = 9x$
    (b) $x^2 = 5x + 36$
    (c) $x^2 – 8 = 2x$
    (d) $x^2 = 15 – 2x$
37. The area of a rectangular field is 200 m² and its perimeter is 60 m. Find its dimensions.
38. Two positive numbers have a difference of 7 and they multiply to give 60
    By forming a quadratic equation, find the two numbers.
39. Using trial and improvement find, to one decimal place, the square roots of:
    (a) 15
    (b) 59
    (c) 41
    (d) 31
40. Find the solution to the equation: $x^2 + 4x = 70$
    which lies between 6 and 7, accurate to one decimal place.
41. Find the solution to the equation: $x^2 – 9x = 28$
    which lies between 11 and 12, accurate to one decimal place.
42. Using trial and improvement find, to two decimal places, the approximate solution to the following equation: $x^3 + x = 82$ starting with the value $x = 4.20$
43. Using trial and improvement find, to two decimal places, the approximate solution to the following equation: $x^3 – 3x = 136$ starting with the value $x = 5.30$
44. A rectangle is x cm long and $(2x + 7)$ cm wide. Its area is 35 cm²
    Find the value of x using trial and improvement giving your answer accurate to one decimal place.

Chapter 11
Straight Lines

11.1 Introduction

This chapter is about straight lines plotted on the coordinate grid.

The equation of a straight line is written in the form $y = mx + c$, where m represents the gradient of the line (its steepness) and c represents the y-intercept.

The gradient of a line is a measure of the amount that the line rises or falls for each unit it moves to the right. The y-intercept is the point at which the line crosses the y-axis.

Key words
- **Line segment**: A part of a straight line between two points on a graph.
- **Gradient**: The gradient is a measure of the steepness.
- **x-intercept**: Where a line passes through the x-axis.
- **y-intercept**: Where a line passes through the y-axis.
- **Midpoint**: The point at the centre of a line segment.
- **Parallel**: Parallel lines run in the same direction and never intersect.

Before you start you should know how to:
- Work with coordinates in all four quadrants.
- Substitute numbers into an expression.
- Solve linear equations.

In this chapter you will learn how to:
- Construct a table of values and draw a straight line graph from it.
- Take readings from a straight line graph.
- Find the midpoint of a line segment.
- Find the length of a line segment.
- Find the equation of a straight line.
- Find the gradient and intercepts of a line.
- How to interpret the gradient and y-intercept of a real-life linear graph in the context of the question.

Exercise 11A (Revision)

1. (a) Copy the diagram on the right.
   (b) Plot these points on the coordinate grid: A(1, 2), B(1, −3), C(−3, 0), D(−2, 3).
   (c) Shade or colour the quadrilateral ABCD.
   (d) Add these line segments to your diagram: AB, BC, CD, DA, AC and BD.
   (e) What shape have you drawn?

2. Find the value of the following expressions when $a = 2$ and $b = -3$.
   (a) $a + b + 2$
   (b) $2ab$
   (c) $2(a - b)$

3. Solve the following equations.
   (a) $x + 3 = 10$
   (b) $2 - y = 8$
   (c) $2x + 5 = 17$
   (d) $4 - 3y = -8$

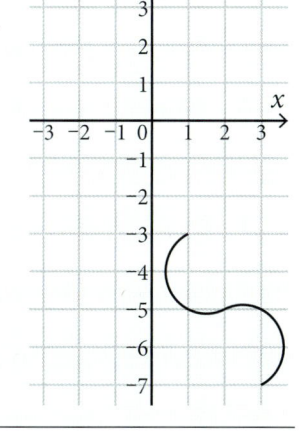

11.2 Drawing and Taking Readings from Straight Line Graphs

Drawing a straight line graph from a list of points

The first skill you need is being able to draw a straight line if you are given a list of points on it. This list could be a simple list of coordinate points.

Example 1

Plot the straight line passing through the points (1, 2), (2, 4), (3, 6), (4, 8) and (5, 10)

Note: In questions like this, the information may instead be given in the form of a table, for example:

x	1	2	3	4	5
y	2	4	6	8	10

Plot each point.

Join the points with a straight line, extending the line in both directions, as shown on the right.

Note: The points of every straight line are calculated from a formula. In this case, the formula is $y = 2x$

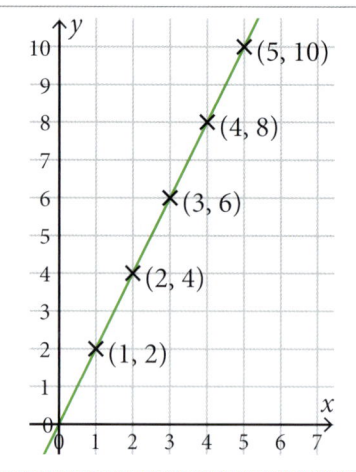

Drawing a straight line graph from the equation of the line

If you are given the equation of the line, you can use it to complete a table of x- and y-values. Using the x-values given, use the equation to find the corresponding y-values. Then, you can plot these points and draw a line through them.

Example 2

(a) Copy and complete the table of values on the right for the line $y = x - 2$
(b) Plot the line.

x	0	1	2	3	4	5
y	−2	−1		1		3

(a) To find the y-value corresponding to $x = 2$, substitute $x = 2$ into the equation of the line $y = x - 2$
So: $\quad y = 2 - 2 = 0$
Thus, when: $\quad x = 2, y = 0$

To find the y-value corresponding to $x = 4$, substitute $x = 4$ into the equation of the line $y = x - 2$
So: $\quad y = 4 - 2 = 2$
Thus, when: $\quad x = 4, y = 2$

Our completed table looks like:

x	0	1	2	3	4	5
y	−2	−1	0	1	2	3

(b) We can plot the line. The graph is shown on the right.

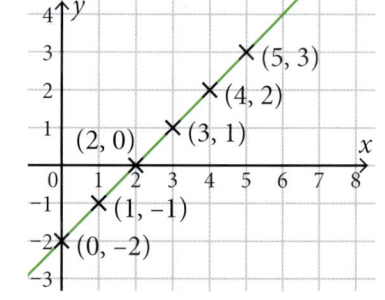

Horizontal and vertical lines

The line $y = 3$ is a horizontal line passing through 3 on the y-axis.
The line $x = 4$ is a vertical line passing through 4 on the x-axis.
The graph on the right shows these two lines.
The line $y = 3$ has this equation because every point on the line has a y-coordinate of 3
The line $x = 4$ has this equation because every point on the line has an x-coordinate of 4

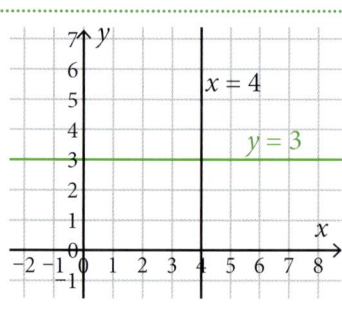

Taking readings from a straight line graph

You may be asked to take readings from a straight line graph. You may or may not be given the equation of the line.

Example 3

Use the graph on the right to find:
(a) The y-value corresponding to $x = 8$
(b) The x-value when $y = 6$

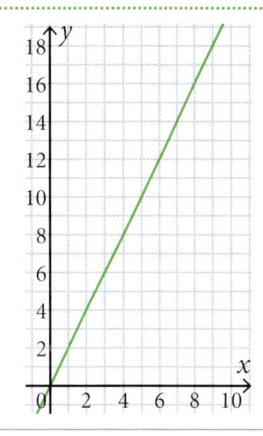

(a) Draw a vertical line (shown as a green dashed line) from $x = 8$ on the x-axis up to the graph. Then draw a horizontal line from the graph to the y-axis to find the corresponding y-value.

When $x = 8$, the y-value is 16

(b) Draw a horizontal line (shown as a black dashed line) from 6 on the y-axis to the graph. Then draw a vertical line from the graph to the x-axis.

When $y = 6$, the x-value is 3

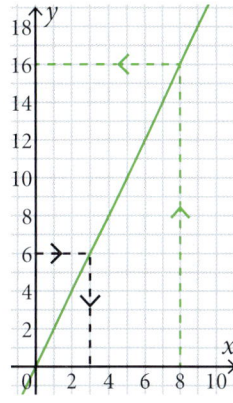

Checking whether a point lies on a line

To check whether a point lies on a straight line (or any curve), substitute the x- and y-coordinates of the point into both sides of the equation. If we find that the left-hand side equals the right-hand side, then the point lies on the line.

Example 4

Does the point $(4, -1)$ lie on the line $y = -x + 3$?

The equation of the line is $y = -x + 3$
Substitute in $x = 4$ and $y = -1$:

Left-hand side $= y$
$= -1$

Right-hand side $= -x + 3$
$= -4 + 3$
$= -1$

The left-hand side equals the right-hand side, so the point $(4, -1)$ does lie on the line.

Example 5

Determine whether the point $(5, -2)$ lies on the line $2x + 3y - 5 = 0$

The equation of the line is $2x + 3y - 5 = 0$
Substitute in $x = 5$ and $y = -2$:

Left-hand side $= 2x + 3y - 5$
$= (2 \times 5) + (3 \times -2) - 5$
$= -1$

Right-hand side $= 0$

The left-hand side does not equal the right-hand side, so the point $(5, -2)$ does not lie on the line.

Exercise 11B

1. Draw the graph of the line passing through the points given in the table on the right.

x	0	1	2	3	4	5
y	−1	1	3	5	7	9

2. Draw the graph of the line passing through the points given in the table on the right.

x	1	2	3	4	5	6
y	3	4	5	6	7	8

3. Copy and complete the table on the right and hence draw the graph of $y = 2x + 1$

x	−1	0	1	2	3	4
y	−1	1		5		9

4. Copy and complete the table on the right and hence draw the graph of $y = 3x - 1$

x	0	1	2	3
y		2		8

5. Copy and complete the table on the right and hence draw the graph of $y = 3x + 1$

x	−1	0	1	2
y		1		7

6. Copy and complete the table on the right and hence draw the graph of $y = x - 3$

x	0	1	2	3	4	5	6
y	−3		−1	0		2	

7. Copy and complete the table on the right and hence draw the graph of $y = 2 + 3x$

x	−1	0	1	2
y	−1		5	

8. Copy and complete the table on the right and hence draw the graph of $y = 2 - x$

x	0	1	2	3	4	5	6
y	2	1	0		−2	−3	

9. On the same graph, draw these lines.
   (a) $x = 3$ (b) $y = 1$ (c) $x = -5$ (d) $y = -3.5$

10. Write down the equation of each line shown on the graph below.

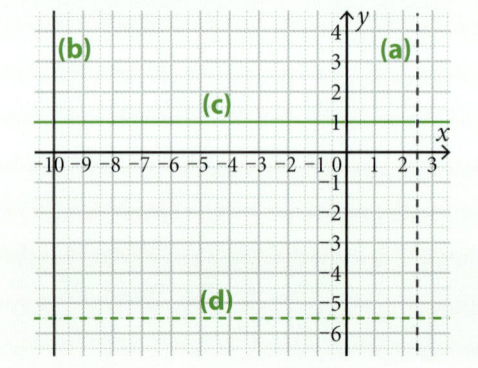

11. Look at the graph below.
    The equation of the straight line is $y = \frac{1}{2}x + 1$
    (a) Use the graph to find the y-value corresponding to an x-value of 8
    (b) Use the graph to find the x-value corresponding to a y-value of 6

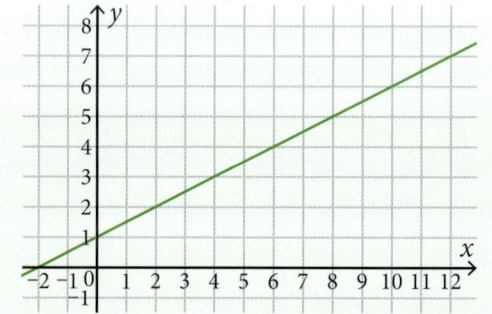

12. Look at the graph on the right.
    The equation of the straight line is $y = -2x + 4$
    (a) Use the graph to find the y-value corresponding to an x-value of 1
    (b) Use the graph to find the y-value corresponding to an x-value of −2
    (c) Use the graph to find the x-value corresponding to a y-value of 10
    (d) Use the graph to find the x-value corresponding to a y-value of 0

13. Does the point (7, 1) lie on the line $y = x + 6$?
    Show your working.

14. Determine whether the point (−6, −5) lies on the line $2x - 3y = 3$

15. The points $(p, -1)$ and $(2, q)$ both lie on the line $y = 3x - 4$
    Find the values of p and q.

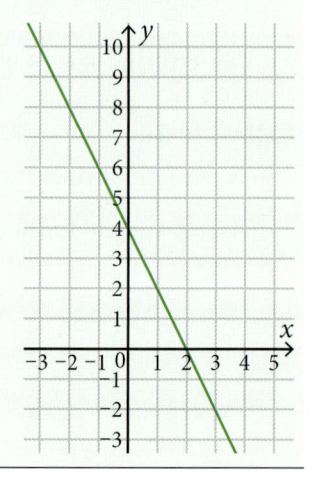

11.3 Midpoint and Length of a Line Segment

The midpoint of a line can be calculated by taking the average values of the coordinates of the end points of the line. This can be expressed by the formula:

Midpoint of line joining (x_1, y_1) and (x_2, y_2) is $\left(\dfrac{x_1 + x_2}{2}, \dfrac{y_1 + y_2}{2}\right)$

Example 6

Find the midpoint of the line joining the points:
(a) (1, 4) and (3, 8)
(b) (−3, 4) and (5, −3)

(a) The midpoint of the line joining (1, 4) and (3, 8) is the point with coordinates
= (average of the x-coordinates, average of the y-coordinates)
$= \left(\dfrac{1+3}{2}, \dfrac{4+8}{2}\right) = (2, 6)$

(b) The midpoint of the line joining (−3, 4) and (5, −3) is the point with coordinates
= (average of the x-coordinates, average of the y-coordinates)
$= \left(\dfrac{-3+5}{2}, \dfrac{4-3}{2}\right) = (1, 0.5)$

Example 7

The midpoint of the line joining (2, 5) and (a, b) is (6, 7). Find the end of the line, (a, b).

We use the fact that the midpoint coordinates are the average of the end points coordinates.

So: $\dfrac{2+a}{2} = 6$

Doubling both sides: $2 + a = 12$, giving $a = 10$

And: $\dfrac{5+b}{2} = 7$

Doubling both sides: $5 + b = 14$, giving $b = 9$

The missing end point is (10, 9)

To find the length of a line segment joining two points, it is best to plot these points and draw a right-angled triangle with the points at either end of the hypotenuse. Then the length of the line can be calculated using Pythagoras' Theorem.

Note: When finding the length of a line segment, you are not expected to give units, such as centimetres.

Example 8

Find the length of the line segment joining (1, 3) to (5, 6)

Method 1

Draw the line segment on the coordinate grid, as shown on the right.

Using Pythagoras' Theorem the hypotenuse (longest side) squared is given by the other two sides squared and added together:

Length$^2 = 3^2 + 4^2$
$= 9 + 16$
$= 25$

So, the length of the line is given by:
Length $= \sqrt{25}$
$= 5$

This method can be summarised by the formula:

Length of line joining (x_1, y_1) and (x_2, y_2) is: $\sqrt{(x_2 - x_1)^2 + (y_2 - y_1)^2}$

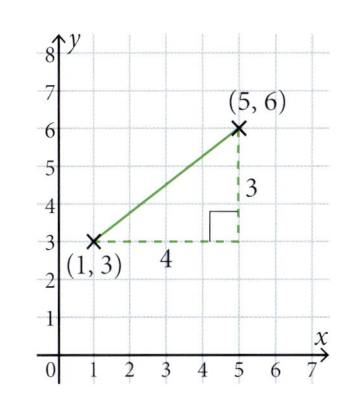

Note: This formula is simply a way of describing the method of using Pythagoras' Theorem. The example could be completed using the formula, as shown on the right.

Method 2

$(x_1, y_1) = (1, 3)$ and $(x_2, y_2) = (5, 6)$

Length of line segment $= \sqrt{(x_2 - x_1)^2 + (y_2 - y_1)^2}$
$= \sqrt{(5-1)^2 + (6-3)^2}$
$= \sqrt{4^2 + 3^2}$
$= \sqrt{16 + 9}$
$= \sqrt{25} = 5$

Exercise 11C

1. Find the midpoints of the lines joining the following pairs of points.
   (a) (2, 3) and (4, 8)
   (b) (3, 7) and (15, 9)
   (c) (5, −3) and (−6, 9)
   (d) (−7, −5) and (2, 5)
   (e) (3, 6) and (9, 6)
   (f) (7, 2) and (−8, −3)
   (g) (8, 2) and (6, 9)
   (h) (−5, −4) and (−5, 6)

2. Find the midpoint of the points A and B shown in the graph below.

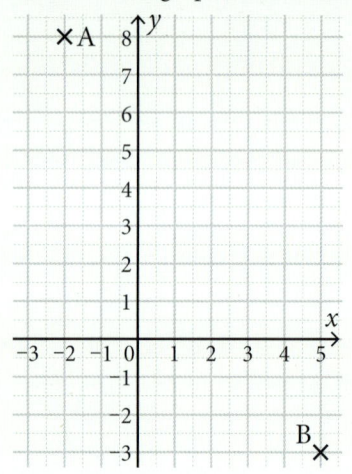

3. Find the midpoint of the points A and B shown in the graph below.

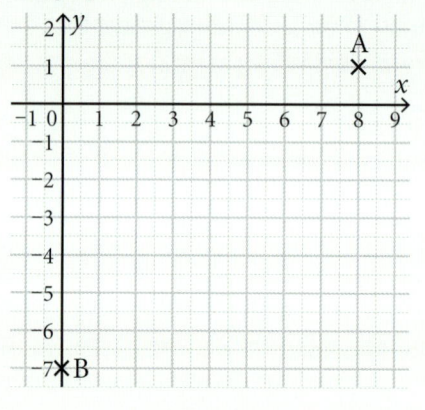

4. The midpoint of the line joining (4, 7) and (a, b) is (7, 7). Find the end of the line, (a, b).
5. The midpoint of the line joining (−5, 9) and (a, b) is (2, −3). Find the end of the line, (a, b).
6. The midpoint of the line joining (7, 2) and (a, b) is (1, 1). Find the end of the line, (a, b).
7. The midpoint of the line joining (2.3, 3.1) and (a, b) is (0.3, −4). Find the end of the line, (a, b).
8. Find the length of the line segment joining each of the following pairs of points.
   (a) (4, 7) and (4, 9)
   (b) (5, 3) and (1, 0)
   (c) (−8, −2) and (9, −2)
   (d) (−7, −2) and (−3, 1)
   (e) (1, 2) and (13, 7)
   (f) (−4, 1) and (4, 16)
   (g) (−3, −4) and (3, 4)
   (h) (−1, −2) and (20, 18)
   (i) (−3, −9) and (2, 3)
   (j) (−4, −5) and (2, 3)
   (k) (−12, 5) and (12, −2)

11.4 Gradients and Intercepts of Straight Line Graphs

The intercepts

The **y-intercept** of a straight line is the point at which the line meets the y-axis.
The **x-intercept** of a straight line is the point at which the line meets the x-axis.

Example 9

Write down the x- and y-intercepts of the line shown on the right.

The x-intercept is (7, 0)

The y-intercept is (0, 8)

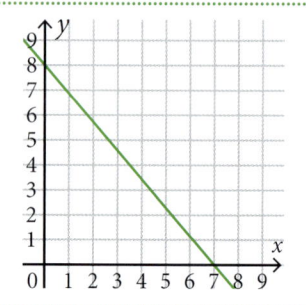

Given the equation of a line, you can work out the two intercepts:
- To find the *y*-intercept, set $x = 0$ in the equation of the line and solve to find *y*.
- To find the *x*-intercept, set $y = 0$ and solve to find *x*.

Example 10

Find the *y*-intercept and *x*-intercept of the line with the equation $y = 6x - 3$

Find the *y*-intercept by setting $x = 0$ in the equation of the line:
$y = 6(0) - 3$
$= -3$
So the *y*-intercept is the point $(0, -3)$
Find the *x*-intercept by setting $y = 0$ in the equation of the line:
$0 = 6x - 3$
$6x = 3$
$x = \frac{1}{2}$
So the *x*-intercept is the point $(\frac{1}{2}, 0)$

The gradient

The gradient of a line is a measure of its steepness.

We calculate the gradient from the coordinates of two points on the line.

If you are given a diagram, form a right-angled triangle between these points and use the formula:

Gradient = $\frac{\text{Rise}}{\text{Run}}$

Example 11

Find the gradient of the line shown in the diagram on the right.

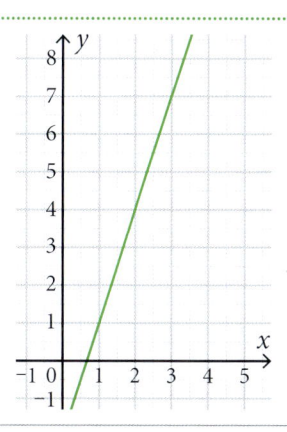

Find two points that lie on the line. In this case the points (1, 1) and (3, 7) have been used.

Moving from (1, 1) to (3, 7), we travel 2 units to the right and 6 units up, so the run is 2 and the rise is 6

> **Note:** When drawing the triangle, always start at the point furthest to the left and draw the 'run' first.

Gradient = $\frac{\text{Rise}}{\text{Run}} = \frac{6}{2} = 3$

> **Note:** We usually use the letter *m* for the gradient, so in this case $m = 3$

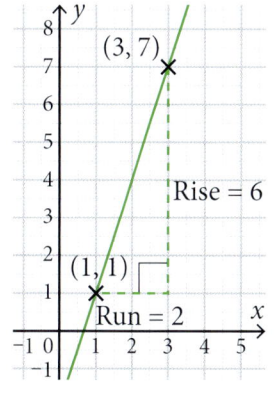

In the next example, the line slopes down the page from left to right. This means that the 'rise' is negative, resulting in a negative gradient.

GCSE MATHEMATICS M3 AND M7

Example 12

Find the gradient of the line shown in the diagram on the right.

Choose two points on the line.

In this case we choose (−5, 4) and (4, −5).

Add these to the diagram, as shown below, and draw a right-angled triangle between the points. Start at point (−5, 4) and draw the run first.

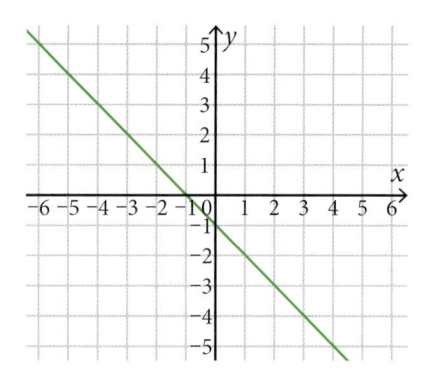

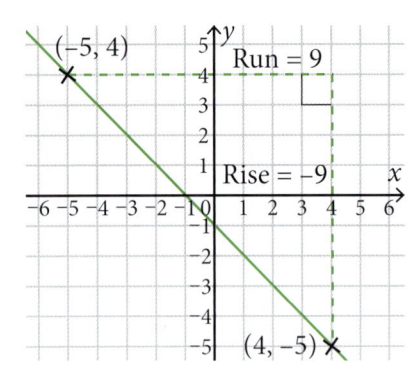

From the diagram on the left we can see that we have a negative value for the rise.

The run is 9 and the rise is −9

Find the gradient m using rise over run:

$$\text{Gradient} = \frac{\text{Rise}}{\text{Run}}$$
$$= \frac{-9}{9}$$
$$= -1$$

If no diagram has been given, it is easier to use a formula to calculate the gradient of a line.

The gradient m of the line joining the points (x_1, y_1) and (x_2, y_2) can be found using the formula:

$$m = \frac{y_2 - y_1}{x_2 - x_1}$$

Example 13

Find the gradient of the straight line that passes through the points (2, 8) and (6, −14).

Label the points given as (x_1, y_1) and (x_2, y_2).

$(x_1, y_1) = (2, 8)$

$(x_2, y_2) = (6, -14)$

The gradient of the line between these points is:

$$m = \frac{y_2 - y_1}{x_2 - x_1}$$
$$= \frac{-14 - 8}{6 - 2}$$
$$= \frac{-22}{4}$$
$$= -\frac{11}{2} \text{ or } -5.5$$

Key points about the gradient

It is important to remember the following key points about the gradient of a straight line:
- Lines sloping up the page (from bottom left to top right) have a positive gradient.
- Lines sloping down the page (from top left to bottom right) have a negative gradient.
- Horizontal lines have a zero gradient.
- For vertical lines, the gradient is not defined.
- Parallel lines have equal gradients.

Example 14

The diagram on the right shows five straight lines, labelled A to E.
(a) Which line has a positive gradient?
(b) Which two lines have a negative gradient?
(c) Which line has a gradient of zero?
(d) Which line's gradient is not defined?
(e) Which two lines have the same gradient?

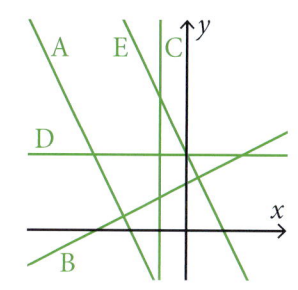

(a) Line B has a positive gradient, since it slopes upwards (from bottom left to top right).
(b) Lines A and E both have a negative gradient as they slope downwards (from top left to bottom right).
(c) Line D is horizontal, so this is the one that has a gradient of zero.
(d) Line C is vertical, so this is the line whose gradient is not defined.
(e) Lines A and E are parallel, so they have the same gradient.

The form $y = mx + c$

When the equation of a straight line is given in the form $y = mx + c$, for example $y = 3x + 1$, where m is the gradient and c is the y-coordinate of the y-intercept.

Example 15

The equation of a straight line is given by $y = 2x + 1$
(a) Write down the gradient and y-intercept of the line.
(b) Find the x-intercept of the line.
(c) Plot the line on a coordinate grid.

(a) The equation of the line is $y = 2x + 1$, which is in the form $y = mx + c$, where m is the gradient and c is the y-coordinate of the y-intercept. So, the gradient is 2 and the y-intercept is (0, 1).

(b) The x-intercept is the point at which the line passes through the x-axis. To find it, set $y = 0$ in the equation and solve for x:
$y = 2x + 1$
$0 = 2x + 1$
$-1 = 2x$
$x = -\frac{1}{2}$
So, the x-intercept is $(-\frac{1}{2}, 0)$

(c) To plot the line, choose two points on the line and plot them. For example, use the y-intercept and the x-intercept, (0, 1) and $(-\frac{1}{2}, 0)$.
Draw the straight line through these points, as shown on the right. Alternatively, you could complete a table of x and y-values, and plot several points from it. You learnt this method in section 11.2.

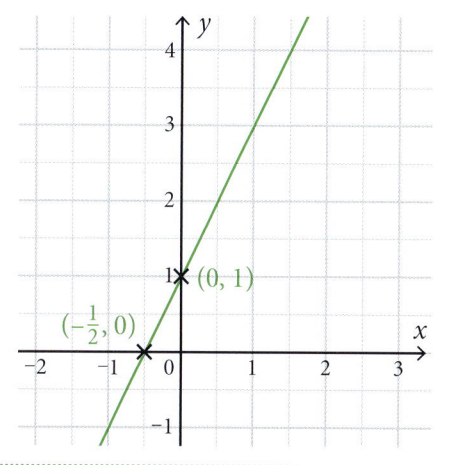

It may be necessary to rearrange the equation of the line to find the gradient and y-intercept.

Example 16

Find the gradient and y-intercept of the line $3x + 2y = 4$

Rearrange to obtain the form $y = mx + c$
Subtract $3x$ from both sides: $\quad 2y = -3x + 4$
Divide both sides by 2: $\quad y = -\frac{3}{2}x + 2$
In this form, it can be seen that the gradient is $-\frac{3}{2}$
The y-coordinate of the y-intercept is 2, so the point is (0, 2).

Note: It is more common to use fractions for the gradient and y-intercept of a straight line than decimal values. Top-heavy fractions are preferable to mixed numbers.

Exercise 11D

1. Find the gradients of the line segments joining the points:
   (a) (2, 4) and (8, 10)
   (b) (3, 5) and (7, 11)
   (c) (−3, −4) and (1, −3)
   (d) (−4, 6) and (2, −8)
   (e) (−3, −6) and (2, 4)

2. Find the gradients of the line segments shown.
   (a)
   (b)
   (c)
   (d)
   (e) h

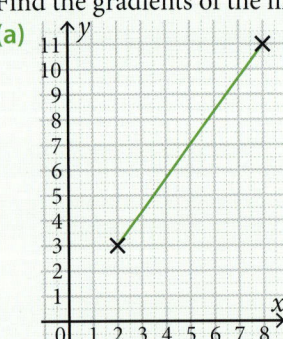

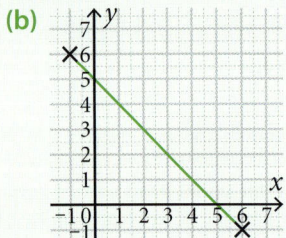

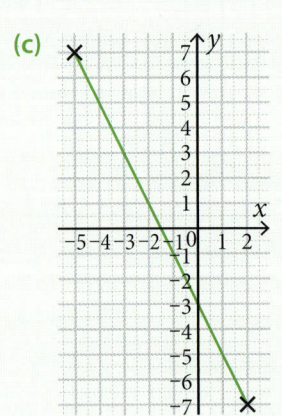

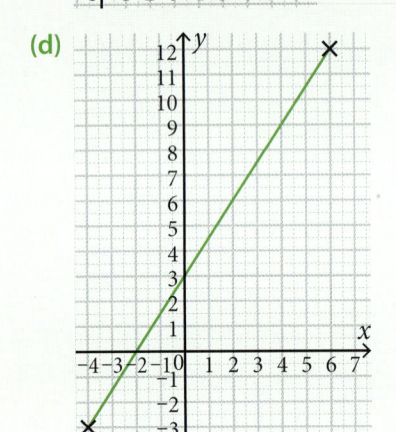

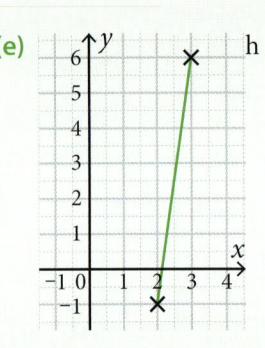

3. Using the equation of each of the following lines, find its:
   (i) gradient (ii) x-intercept; and (iii) y-intercept.
   (a) $y = x$
   (b) $y = 2x + 3$
   (c) $y = 2x$
   (d) $y = 3x - 4$
   (e) $y = -2x + 6$
   (f) $x + y = 10$
   (g) $2x - y = 6$
   (h) $x + 2y = 14$

4. Look at the following equations of straight lines:
   $y = 3x - 5$ $y = 3$ $y = 5 + 3x$ $x = 5$ $y = 3 - 5x$
   (a) State which two lines are parallel.
   (b) State which line has a zero gradient.
   (c) State which line has a negative gradient.

5. Look at the diagram on the right which shows five straight lines labelled A to E.
   (a) State which two lines have the same gradient.
   (b) State which line has a zero gradient.
   (c) State which line has a negative gradient.

6. The equation of a straight line is given by: $y = 3x - 2$
   (a) Write down the gradient and y-intercept of the line.
   (b) Find the x-intercept of the line.
   (c) Plot the line, using values of x from −1 to 2

7. Find the gradient, y-intercept and x-intercept of the straight line $2y - 7x = 42$

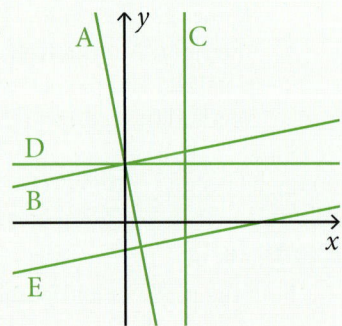

11.5 Finding the Equation of a Line

The equation of a line through two points

If you are given two points on a line, you can calculate the line's gradient as in the previous section.

Example 17

Find the equation of the line passing through the points (0, –4) and (4, 8).

Since no graph has been provided, use the formula to find the gradient:

$$m = \frac{y_2 - y_1}{x_2 - x_1} = \frac{8 - (-4)}{4 - 0} = \frac{12}{4} = 3$$

The *y*-intercept is given as (0, –4), so $c = -4$

The equation of the line is $y = 3x - 4$

If the *y*-intercept of the line is not given, it can be calculated, as shown in the next example.

Example 18

Find the equation of the straight line passing through the points A(6, 12) and B(2, 2).

Again, with no graph provided, use the formula for the gradient:

$$m = \frac{y_2 - y_1}{x_2 - x_1} = \frac{12 - 2}{6 - 2} = \frac{10}{4} = \frac{5}{2}$$

Note: We have used $(x_1, y_1) = (2, 2)$ and $(x_2, y_2) = (6, 12)$, to give positive numbers in both numerator and denominator of the fraction.

The value of *c* must be calculated, as follows. Write down the general equation of a straight line:

$y = mx + c$

Next, substitute the gradient *m* and the *x*- and *y*-coordinates of a point on the line. Here we choose (2, 2), but we could equally use (6, 12).

$2 = (\frac{5}{2})(2) + c$

$2 = 5 + c$

$c = -3$

So, the equation of the line is $y = \frac{5}{2}x - 3$

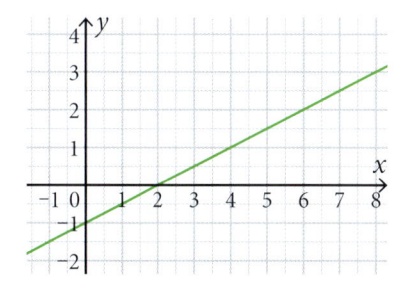

If you are given a diagram showing the straight line, you may find it easier to find the gradient using $\frac{\text{Rise}}{\text{Run}}$

Example 19

Find the equation of the straight line shown in the graph on the right.

Choose two points that lie on the line, for example (2, 0) and (8, 3). Then calculate the gradient.

Method 1

This can be done by choosing two points on the line, constructing a right-angled triangle and using:

Gradient = $\frac{\text{Rise}}{\text{Run}}$

The details here are left for the reader.

Method 2

Use the formula $m = \frac{y_2 - y_1}{x_2 - x_1} = \frac{3 - 0}{8 - 2} = \frac{3}{6} = \frac{1}{2}$

From the graph we can see that the *y*-intercept is –1

So, the equation of the line is $y = \frac{1}{2}x - 1$

The equation of a line through a point with a given gradient

In the following example, the gradient and *y*-intercept are both given.

GCSE MATHEMATICS M3 AND M7

Example 20

Find the equation of the straight line that has a gradient of −3 and passes through the point (0, 2).

The gradient and y-intercept are given, so we can write down the equation of the line:
$y = -3x + 2$ or $y = 2 - 3x$

If the y-intercept is not given, it can be calculated, as in the following example.

Example 21

The straight line L has a gradient of 4 and passes through the point (−7, 4).
(a) Find the y-intercept of L.
(b) Hence write down the equation of L.
(c) Find the x-intercept of L.

(a) We are given the gradient $m = 4$, but the y-intercept must be calculated. To find c, substitute m and the x- and y-coordinates of a point on the line into $y = mx + c$
$4 = (4)(-7) + c$
$4 = -28 + c$
$c = 32$
The y-intercept is the point (0, 32)

(b) The equation of the line is $y = 4x + 32$

(c) To find the x-intercept, set $y = 0$ in the equation of the line.
$0 = 4x + 32$
$4x = -32$
$x = -8$
The line passes through the x-axis at the point (−8, 0)

Exercise 11E

1. Find the equation of the straight line passing through the points A(3, 16) and B(2, 9).
2. Find the equation of each line, labelled A to D, in the diagram on the right. 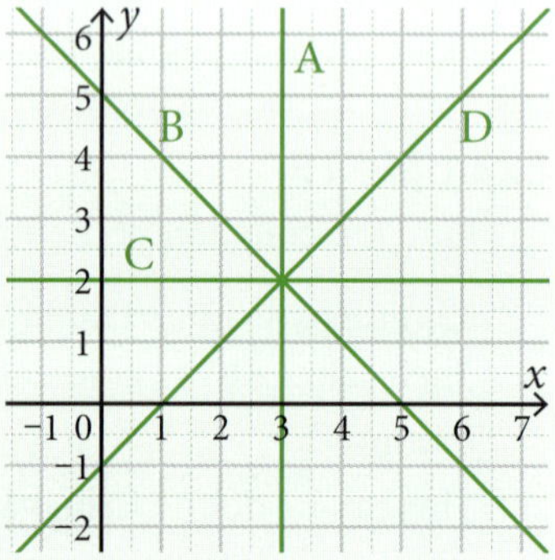
3. Find the equation of the straight line that has a gradient of 2 and passes through the point (0, −7).
4. Find the equation of the straight line that has a gradient of 2 and passes through the point (4, −1).
5. Find the equation of the horizontal line that passes through the point (6, 4).
6. Find the equation of the vertical line that passes through the point (−3, 7).
7. Find the equation of the line that passes through the point (−3, 2) with a gradient of −2
8. Find the equation of the line shown in the diagram on the right. 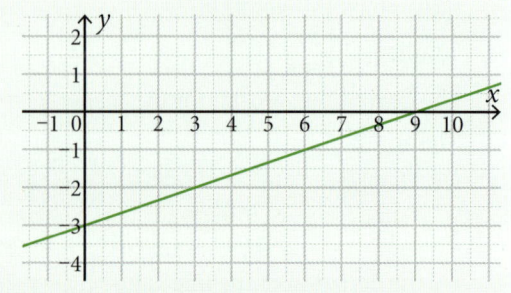
9. The straight line L has a gradient of 3 and passes through the point (−6, −3).
   (a) Find the y-intercept of L.
   (b) Hence write down the equation of L.
   (c) Find the x-intercept of L.
10. The lines L_1 and L_2 have gradients −1 and $\frac{1}{2}$ respectively. Both lines pass through the point A(−2, −3).
    (a) Sketch both lines on the same diagram, showing where each line passes through the coordinate axes.
    (b) Find the area of the triangle enclosed by the two lines and the y-axis.

11.6 Real-Life Linear Graphs

Earlier in this chapter you learnt how to calculate the gradient (the line's steepness) and the intercepts for a linear graph. For a real-life straight line graph you may be asked to interpret the gradient and y-intercept in the context of the question. Here are some examples of how linear graphs can be used to model real-world situations:

- To calculate the cost of hiring a car or a piece of equipment for a certain number of days.
- To calculate the total cost of a phone bill.
- To calculate the total cost of a taxi fare.

Example 22

The graph shows the cost C (£) of hiring a car for d days in Belfast.
(a) What is the cost of hiring a car for 2 days?
(b) If I pay £60, how many days have I hired a car for?
(c) What is the gradient of the line?
(d) Interpret the gradient of the line in the context of the question.
(e) What is the fixed cost of hiring a car, however many days it is hired for?

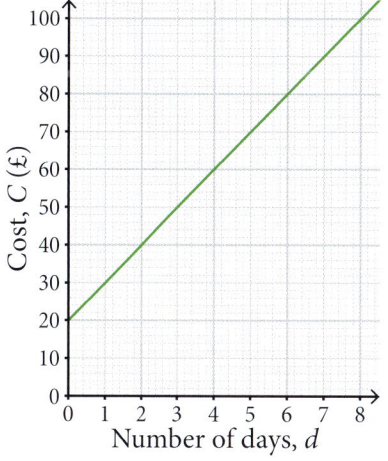

(a) Draw a vertical line from $d = 2$ on the horizontal axis to the graph (as shown on the right). Then draw a horizontal line from the graph to the vertical axis. Read off the corresponding C value.
$C = 40$, so the cost is £40

(b) Draw a horizontal line from $C = 60$ on the vertical axis to the graph (as shown on the right). Then draw a vertical line from the graph to the horizontal axis.
$d = 4$, so the car was hired for 4 days.

(c) Work out the rise and the run from the diagram, as shown on the right.
Gradient = $\frac{\text{Rise}}{\text{Run}} = \frac{60}{6} = 10$

(d) Every day the car is hired, the total cost increases by £10

(e) The fixed cost is the price that would be paid for hiring the car, but keeping it for 0 days. This can be read from the graph as the y-intercept.
So, in this case, the fixed cost is £20

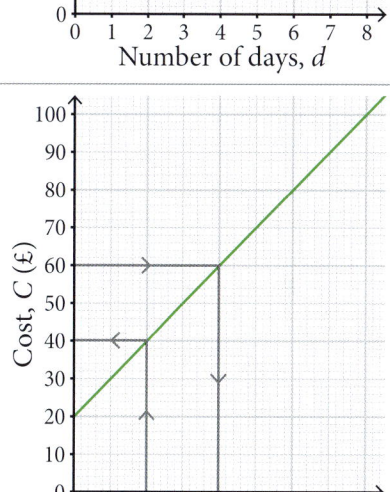

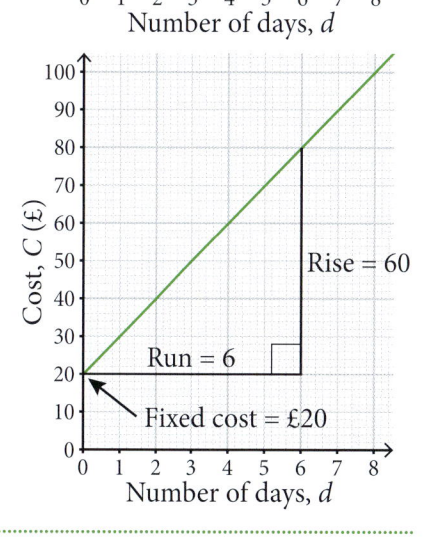

In the next example, the graph has a negative gradient.

Example 23

Linda is shopping for string and ribbon. She has £4 in total. String costs 20p per metre and ribbon costs 40p per metre. Linda must decide how much of each to buy. For example, if she buys no string she can afford 10 metres of ribbon. If she buys 4 metres of string for 80p, she has £3.20 left and she can afford 8 metres of ribbon.

(a) Copy and complete the table on the right to help her.
(b) Draw a straight-line graph showing the number of metres of string on the horizontal axis against the number of metres of ribbon on the vertical axis.
(c) Find the gradient of the graph.
(d) Interpret the gradient in the context of the question.

String (metres)	0	4	8		16	
Ribbon (metres)	10	8		4		0

(a)
String (metres)	0	4	8	12	16	20
Ribbon (metres)	10	8	6	4	2	0

(b) The graph is shown on the right.

(c) The gradient of the graph can be calculated using any two points on the line. The points (4, 8) and (12, 4) have been chosen.

Gradient = $\frac{\text{Rise}}{\text{Run}} = \frac{-4}{8} = -\frac{1}{2}$

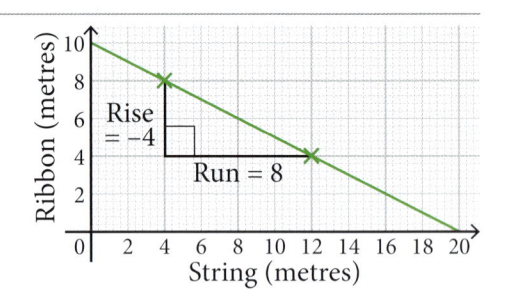

(d) The gradient of $-\frac{1}{2}$ means that for every extra metre of string Linda buys, she can afford to buy $\frac{1}{2}$ a metre **less** of ribbon.

Exercise 11F

1. The graph shows the cost (£C) of hiring a power tool for d days in Ballymena.
   (a) What is the cost of hiring the power tool for 5 days?
   (b) If Ben pays £90, how many days has he hired the power tool for?
   (c) What is the gradient of the line?
   (d) Interpret the gradient of the line in the context of the question.
   (e) What is the fixed cost of hiring the power tool?

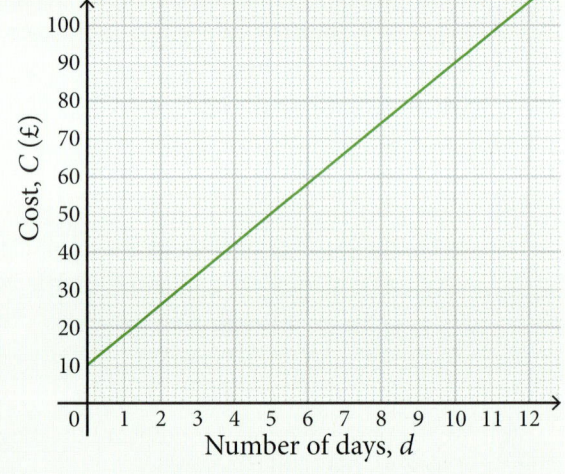

2. The graph on the right shows the fare £F for a taxi journey of m miles.
   (a) What is the cost of a taxi journey of 5 miles?
   (b) If Izzy's fare is £10.40, how long is her journey?
   (c) What is the gradient of the line?
   (d) Interpret the gradient of the line in the context of the question.
   (e) What is the fixed cost of a taxi journey?

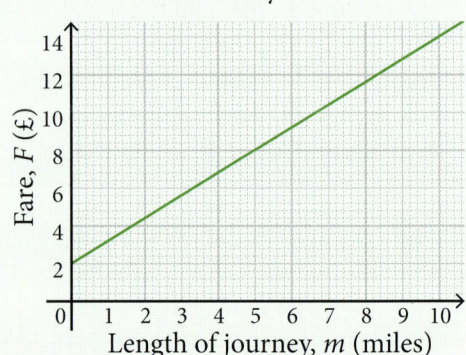

3. The graph shows the monthly cost £C of using a mobile phone for m minutes of calls.
   (a) Vicki makes 200 minutes of calls. What is her total bill this month?
   (b) If Siona pay £40, how many minutes has she used?
   (c) What is the gradient of the line?
   (d) Interpret the gradient of the line in the context of the question.
   (e) What is the monthly fixed cost of using this phone?
   (f) Write down a formula for C in terms of m.
   Hint: This formula is the equation of the straight line. You will use C in place of y and m in place of x.

4. A power company offers two tariffs. The graph shown can be used to calculate the total electricity bill £C from the number of units of electricity used, n, on each of these tariffs.
   (a) Calculate the gradient of both lines.
   (b) Interpret these gradients in the context of the question.
   (c) What is the standing charge on each tariff? (Hint: standing charge means the same as fixed cost.)
   (d) Dermot is on Tariff A and Carolyn is on Tariff B. They use the same number of units of electricity this month and the total cost was also the same. How many units did they each use and what was the cost?

5. This graph shows the depth, d metres of water in a reservoir t hours after a sluice has been opened to drain it.
   (a) What is the y-intercept of this graph?
   (b) What does the y-intercept represent?
   (c) Calculate the gradient of the line.
   (d) Interpret the gradient of the line in the context of the question.

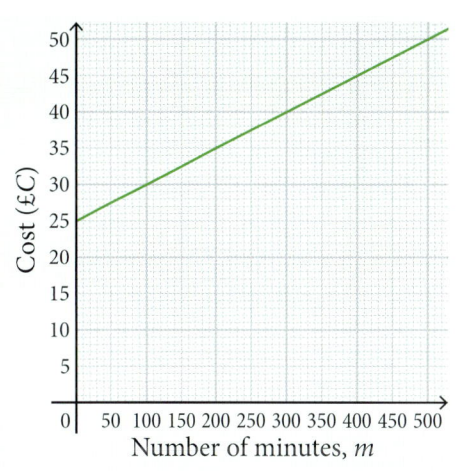

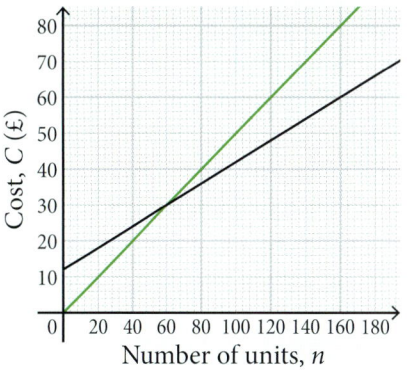

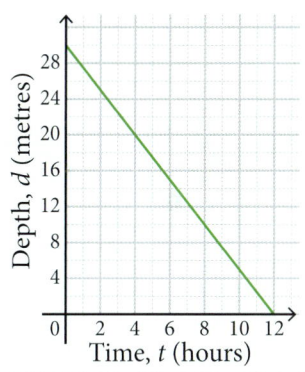

11.7 Summary

In this chapter you have learnt:
- How to graph and use linear equations.
- How to construct a table of values from the equation of the line and how to draw a straight line graph from it.
- How to take readings from a straight line graph.
- How to find the midpoint and the length of a line segment.
- How to find the gradient and the equation of a line between two points.
- How to find the equation of a line given its gradient and a point on the line.
- How to use the equation of a straight line to find the two intercepts.
- The following facts about the gradient of lines:
  - Parallel lines have the same gradient.
  - Horizontal lines have a gradient of zero.
  - For vertical lines, the gradient is undefined.
  - Lines with a positive gradient slope upwards (bottom left to top right).
  - Lines with a negative gradient slope downwards (top left to bottom right).
- How to calculate the **gradient** from a real-life linear graph and how to interpret the gradient.

Chapter 12
Angles

12.1 Introduction

This chapter covers angles:
- between straight lines,
- in triangles, and
- in quadrilaterals.

Key words

- **Interior angle**: An interior angle is an angle inside a shape, at one of its corners.

- **Exterior angle**: If one side of a shape is extended, an exterior angle is the angle between that extended side and the next side.

- **Corresponding angles**: Corresponding angles are equal. The letter F, or a reverse F, can be traced over corresponding angles. The 'arms' of the F must be parallel lines. Corresponding angles are sometimes called F-angles.

- **Alternate angles**: Alternate angles are equal. The letter Z, or a reverse Z, can be traced over alternate angles. The 'top' and 'bottom' of the Z must be parallel lines. Alternate angles are sometimes called Z-angles.

- **Vertically opposite angles**: Vertically opposite angles are equal. The letter X can be traced over vertically opposite angles.

- **Supplementary angles**: Supplementary angles add up to 180°
They may be between parallel lines or on a straight line.

Before you start you should

- Know the different types of triangle: scalene, isosceles and equilateral.
- Know that the three angles in a triangle add up to 180°
- Know that a quadrilateral is a four-sided shape.
- Know that the four angles in a quadrilateral add up to 360°

In this chapter you will learn how to:

- Find a missing interior or exterior angle in a triangle or quadrilateral.
- Find angles using alternate, corresponding, vertically opposite and supplementary angles.

CHAPTER 12: ANGLES

Exercise 12A (Revision)

1. Find the size of each missing angle in these diagrams.
   (a) 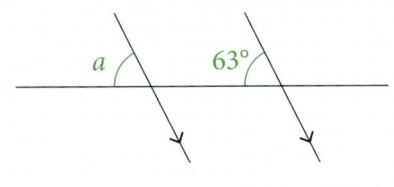
   (b)
   (c) 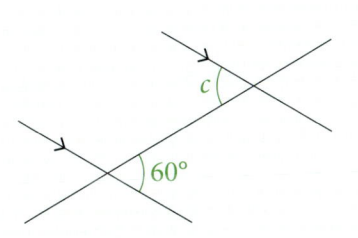
   (d)
   (e)

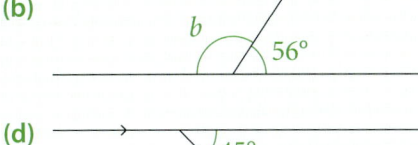

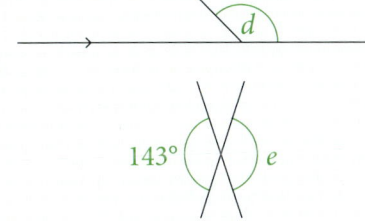

2. For each part (a) to (e):
   (i) Complete the first sentence above each diagram, using a word or phrase from this list:
   vertically opposite, alternate, corresponding, supplementary
   (ii) Complete the second sentence to state whether the angles in the question are equal, or add up to 180°
   The first one has been done for you.

   (a) (i) Angles a and b are
   <u>corresponding</u> angles.
   (ii) They <u>are equal</u>.

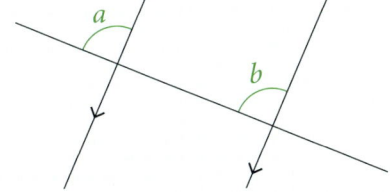

   (b) (i) Angles GHF and JHF are
   _____.
   (ii) They _____.

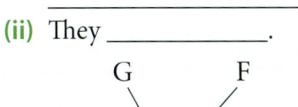

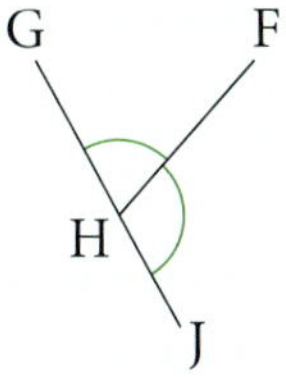

   (c) (i) Angles c and d are
   _____.
   (ii) They _____.

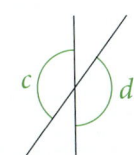

   (d) (i) Angles e and f are
   _____.
   (ii) They _____.

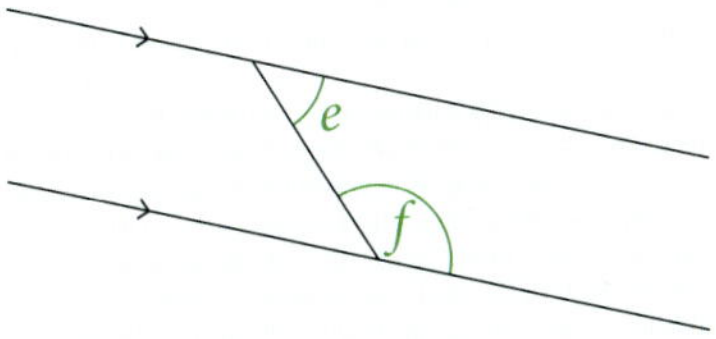

   (e) (i) Angles RVU and QUV are _____.
   (ii) They _____.

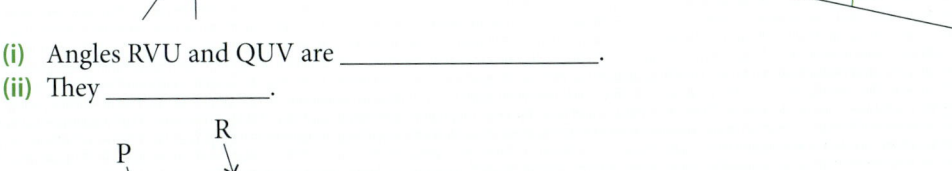

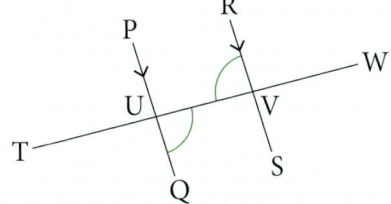

12.2 Angles in Triangles and Quadrilaterals

You may be asked to find missing interior and exterior angles in triangles and quadrilaterals. Use these important key results:
- The three angles in a triangle add up to 180°
- The four angles in a quadrilateral add up to 360°
- Angles on a straight line add up to 180°

GCSE MATHEMATICS M3 AND M7

Example 1

The diagram on the right shows a triangle ABC. The side CA has been extended to point D. Find the size of the exterior angle marked x.

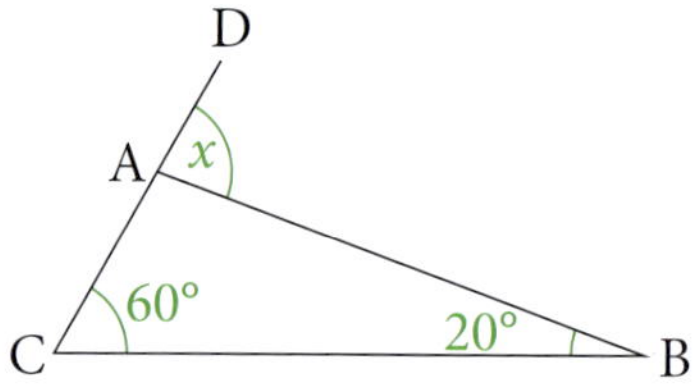

Begin by finding the third angle in the triangle, CAB.
The angles in the triangle add up to 180°, so:

CAB + 20 + 60 = 180
CAB + 80 = 180
CAB = 180 − 80
= 100°

Angle CAB and the angle marked x add up to 180°
x + 100 = 180
x = 180 − 100
= 80°

Example 2

The diagram shows a quadrilateral ABCD. The sides BC and AD are extended and meet at point E. Find the sizes of the angles marked w, x, y and z.

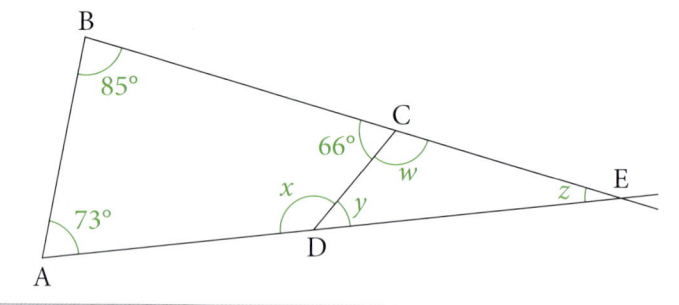

The angle marked w is on a straight line with the angle marked 66°, so these must add up to 180°
w + 66 = 180
w = 180 − 66 = 114°

The four angles inside quadrilateral ABCD must add up to 360°
x + 73 + 85 + 66 = 360
x + 224 = 360
x = 360 − 224 = 136°

The angle marked y is on a straight line with the angle marked x, so these must add up to 180°
y + x = 180
y + 136 = 180
y = 180 − 136 = 44°

The three angles inside triangle CDE must add up to 180°
z + w + y = 180
z + 114 + 44 = 180
z + 158 = 180
z = 180 − 158 = 22°

Exercise 12B

1. In the diagram on the right, ABC is a triangle. ABD is a straight line.
   (a) What type of triangle is ABC?
   (b) Find the angle marked x.
   (c) Find the angle marked y.

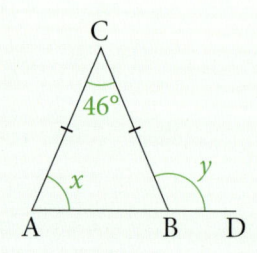

2. Work out the size of the angle marked x in the diagram below.

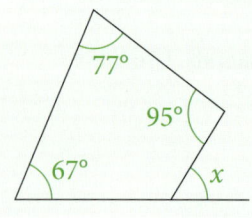

3. Find the size of the angle marked x in the diagram below.

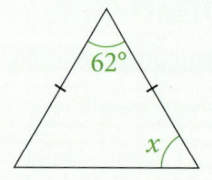

4. Work out the size of the angle marked x in the diagram below.

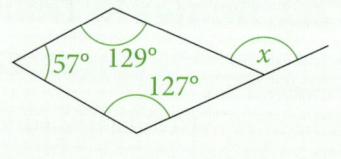

CHAPTER 12: ANGLES

5. Work out the size of the angle marked *y* in the diagram below.

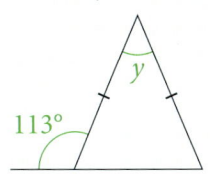

6. Find the sizes of the three angles A, B and C in the diagram below.

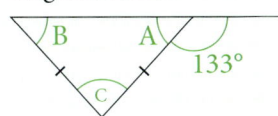

7. Find the size of the angle marked *x* in the diagram below.

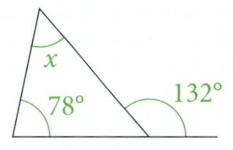

8. Triangle ABC, shown on the right, is isosceles. AC = BC
   (a) Find the size of the angle marked *x*.
   (b) Hence decide if the lines AC and DE are parallel. Explain your answer.

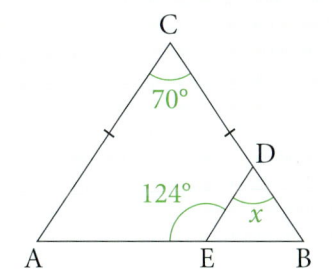

9. The diagram on the right shows a quadrilateral PQRS. The sides PS and QR are extended and meet at point T. Find the sizes of the angles marked *a*, *b*, *c* and *d*.

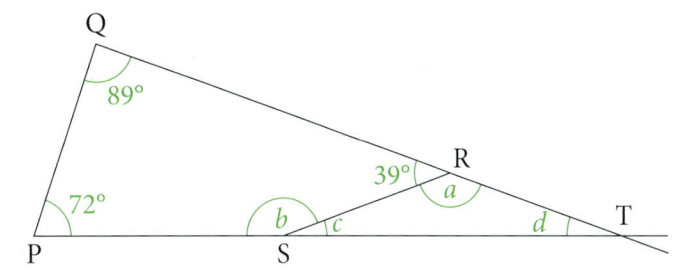

12.3 Angles and Lines

You may be asked to find angles in diagrams involving straight lines. Often parallel lines are involved. To do this you may need to use alternate angles, corresponding angles, vertically opposite angles and supplementary angles.

Example 3

The diagram on the right shows a pair of parallel lines and a line segment that intersects them.

Find the size of the angle marked *y*.

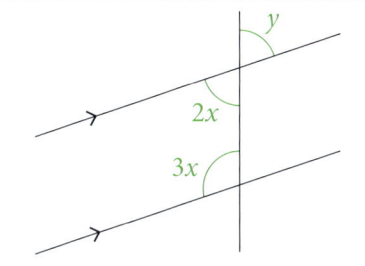

The angles marked $2x$ and $3x$ are supplementary angles between parallel lines. They add up to 180°.

$$2x + 3x = 180$$
$$5x = 180$$
$$x = \frac{180}{5} = 36°$$

The size of the angle marked $2x = 2 \times 36 = 72°$

The angle marked *y* is vertically opposite the angle marked $2x$, so they are equal in size.

$$y = 72°$$

Sometimes more than one step is involved to find an unknown angle.

Example 4

Find the angle marked *x* in the diagram on the right.

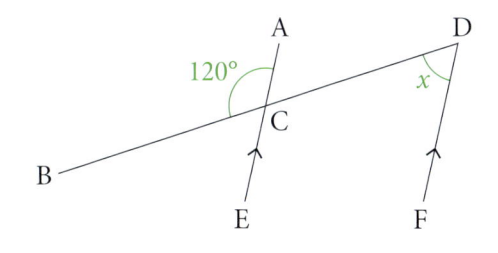

Step 1. Angle ECD is vertically opposite the angle marked 120°, so ECD = 120°

Step 2. Angle ECD and the angle marked x are supplementary angles between parallel lines, so they add up to 180°: $x + 120 = 180$
$x = 180 - 120$
$x = 60°$

In the following example, we must find the angles inside the triangle before then finding x.

Example 5

The diagram on the right shows parallel lines AC and EG and two line segments AF and BD. Angle DFG is 110° and angle BDF is 140°.

(a) Find all three angles in triangle ABD.
(b) What type of triangle is ABD?
(c) Find the size of angle DBC, marked x.

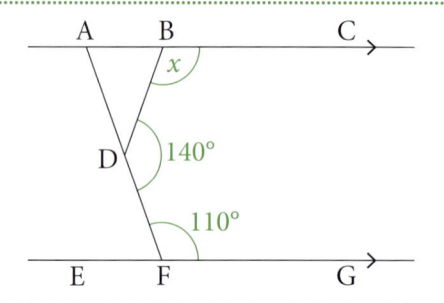

(a) Angle ADB is 40° (since it is on a straight line with the angle marked 140°)

Angle DAB is 70° (since angle DAB and the angle marked 110° are supplementary angles between the parallel lines).

Since the three angles in triangle ABD must add up to 180°:
Angle ABD = 180 − (70 + 40) = 70°

(b) Triangle ABD is isosceles, since it has two equal angles, DAB and ABD.

(c) Angles ABD and DBC are supplementary angles on a straight line, so add up to 180°
$70 + x = 180$
$x = 180 - 70$
$x = 110°$

You may have to add extra lines to the diagram to help you calculate the size of missing angles, as in the next example.

Example 6

In the diagram on the right, AB and DE are parallel lines. Find the size of the reflex angle marked p.

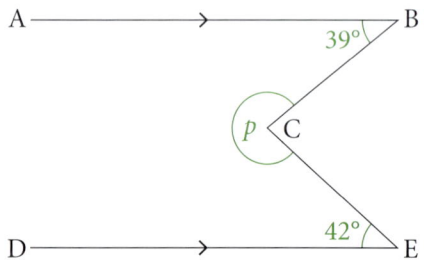

In the diagram on the right, a third parallel line has been added, ending at point C. In this way the angle p is split into two angles, labelled q and r.

The angles marked 39° and q are supplementary angles between parallel lines.
$q + 39 = 180$
$q = 180 - 39 = 141°$

The angles marked 42° and r are also supplementary angles between parallel lines.
$r + 42 = 180$
$r = 180 - 42 = 138°$

To find p, add q and r:
$p = q + r$
$p = 141 + 138 = 279°$

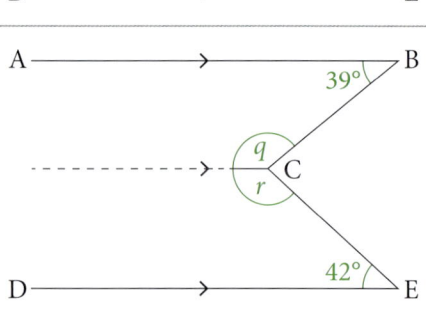

Note: There is more than one way to find p; other solutions are possible.

Exercise 12C

1. The diagram on the right shows two horizontal, parallel lines, AB and CD. The lines EF and EG intersect both parallel lines. Write down the size of the angles marked x and y.

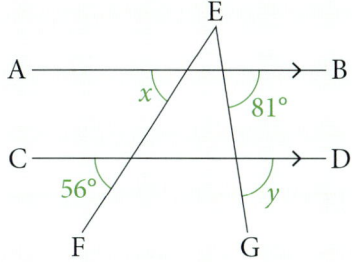

2. Look at the diagram on the right.
   (a) By forming and solving an equation, find the value of a.
   (b) What is the size of the larger of the two angles?

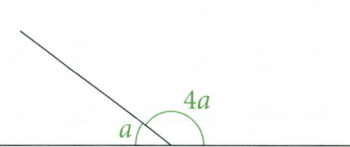

3. (a) Find the value of c in the diagram below. 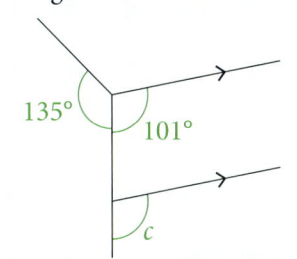
   (b) Find the values of d and e in the diagram below. 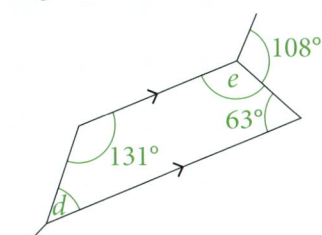
   (c) Find the value of f in the diagram below.

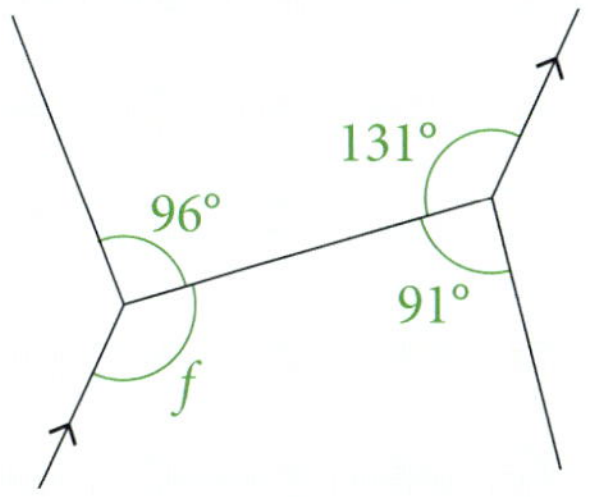

4. In the diagram below, AC and EG are parallel lines. Calculate the size of the angle marked x.

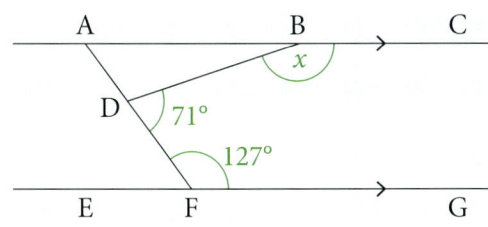

5. In the diagram below, PR and SV are parallel lines. Angle RQU = 48° and angle QTS = 96°. What type of triangle is QTU? Explain each step of your answer fully.

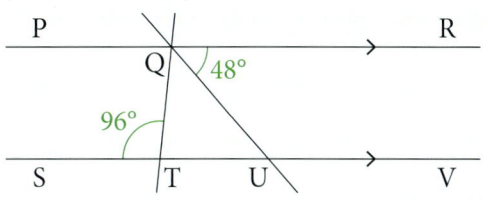

6. The diagram below shows three line segments and four angles. By forming an equation in x and solving it, find the sizes of all four angles.

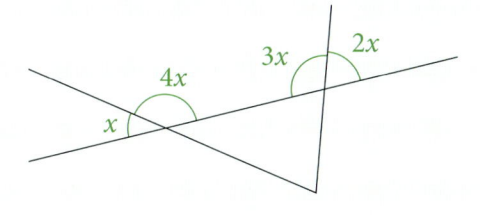

7. The diagram below shows two pairs of parallel lines and a fifth horizontal line. Find the sizes of the angles marked a, b and c.

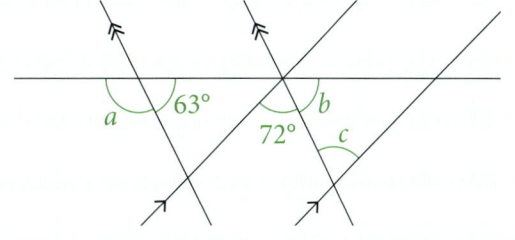

8. Look at the diagram on the below. There are two pairs of parallel lines. Find the size of:
   (a) Angle a (b) Angle b (c) Angle c

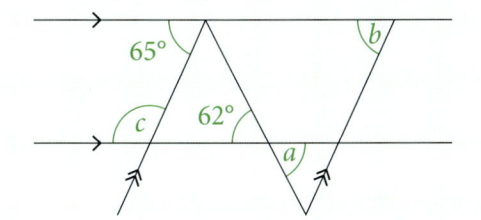

9. Find x in the diagram below. Explain each step of your reasoning.

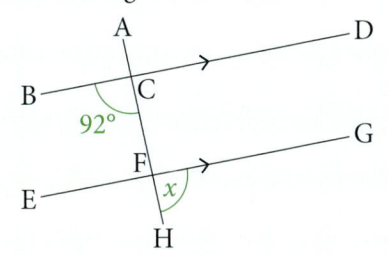

10. The diagram below shows a rectangle. Work out the sizes of the angles marked x, y and z.

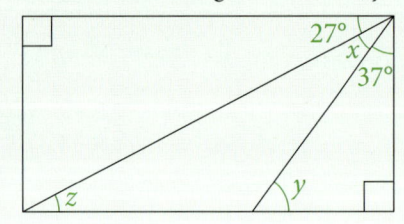

11. In the diagram below, CA is parallel to DB. Angle CAD = 97° and angle BDE = 55°
Find the size of the angle marked x. Show all the steps of your working.

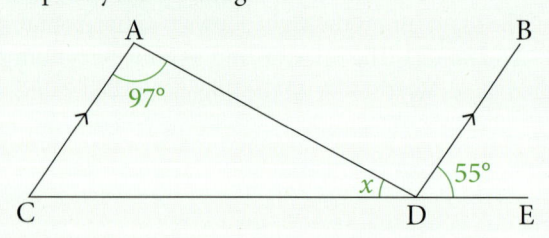

12. In the diagram below, lines DE and AB are parallel.
(a) Find the size of the angle marked m.
(b) Find the size of the angle marked n.

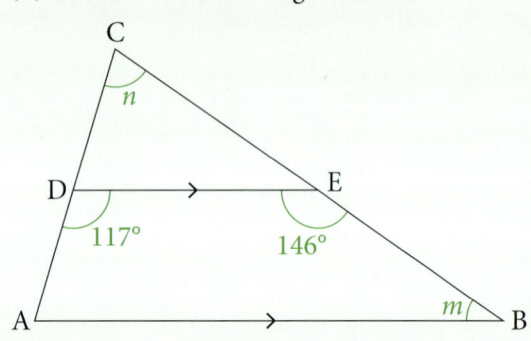

13. Find the size of the angle marked c in the diagram below.

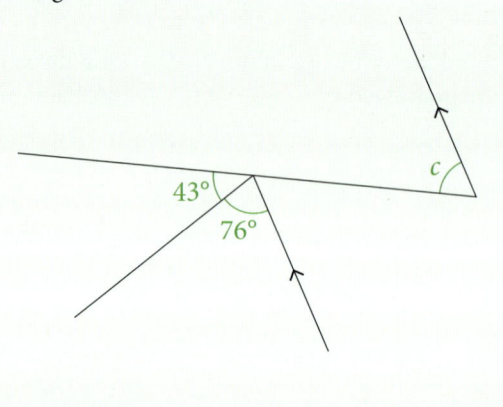

14. Find the size of the angle marked x in the diagram below.
Hint: You should copy the diagram and add any extra lines needed.

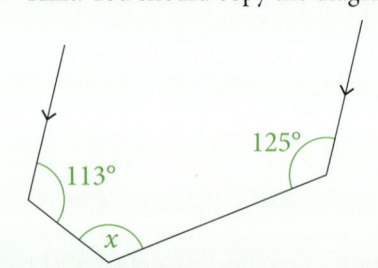

12.4 Summary

In this chapter you have learnt:
- That an interior angle is an angle inside a shape, at one of its corners.
- That if one side of a shape is extended, an exterior angle is the angle between that extended side and the next side.
- How to find missing angles in triangles and quadrilaterals.
- How to recognise pairs of corresponding, alternate and vertically opposite angles. In these cases the two angles are equal.
- How to recognise supplementary angles on a straight line or between parallel lines. Supplementary angles add up to 180°
- That many questions involve using a combination of these skills.
- That in some questions, extra lines must be added to the diagram – for example to split an angle up, so that each part can be found separately.

Chapter 13
Circles

13.1 Introduction

This chapter covers the important names for the different parts of a circle and related lines and areas. You will also learn how to use the properties of these geometrical features.

Key words
- The following key words: **Circumference**, **Radius**, **Diameter**, **Chord**, **Tangent**, **Arc**, **Sector** and **Segment** are defined visually in section 13.2 of this chapter.

Before you start you should
- Know the words **circumference**, **radius** and **diameter** and their meanings.

In this chapter you will learn:
- The names of various parts of a circle and related lines and areas.
- About the properties of these lines and areas.

Exercise 13A (Revision)

1. Copy the diagram on the right. Fill in each of the three boxes with one of the words from this list:
   - Radius
   - Diameter
   - Circumference

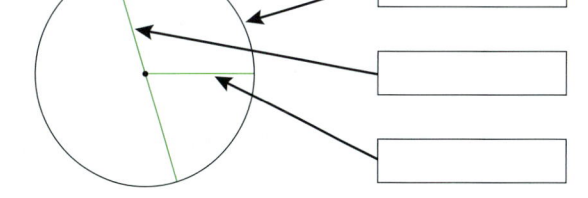

2. A circle has a diameter of 6 cm. What is the circle's radius?

3. (a) The radius of the centre circle on a football pitch is 9.15 m. What is the circle's diameter?
   (b) The football club's groundsman uses 1.5 litres of paint to mark out the radius of the centre circle. During the football season he paints the diameter 6 times. How many litres of paint does he use for this line throughout the season?

13.2 Parts of a Circle

You need to know the following definitions relating to the parts of a circle. You may be asked about the properties of these circle parts.

Circumference
The circumference is the special name for the perimeter of the circle.

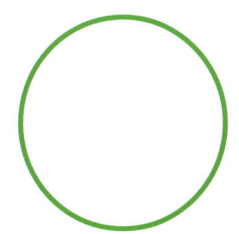

Radius
A radius is any line from the centre to the circumference.

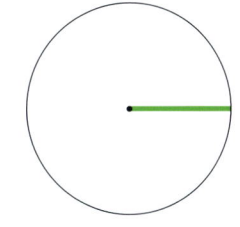

Chord
A chord is a straight line connecting any two points on the circumference.

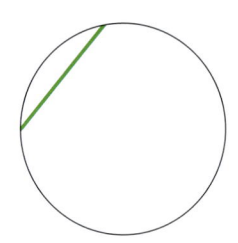

Diameter
The **diameter** is the distance across the circle, through the centre. A diameter is the longest chord that can be drawn.

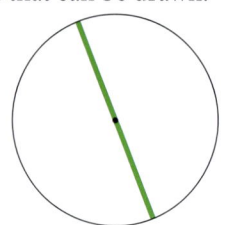

Arc
An arc is a part of the circumference.

A minor arc is less than half of the circumference.

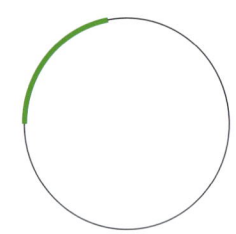

A major arc is more than half of the circumference.

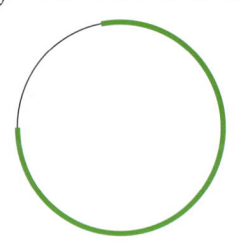

Sector
A sector is the area between two radii and the circumference.

A minor sector is less than half of the area of the circle.

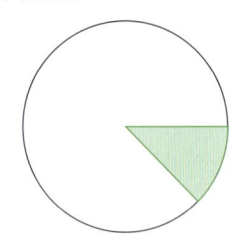

A major sector is more than half of the area of the circle.

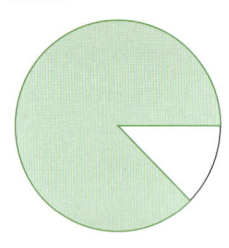

Segment
A segment is the area between a chord and the circumference.

A minor segment is less than half of the area.

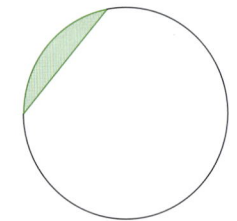

A major segment is more than half of the area.

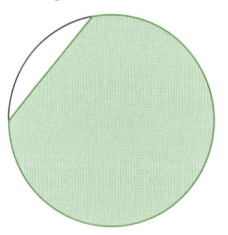

Tangent
A tangent is a straight line that touches the circle once.

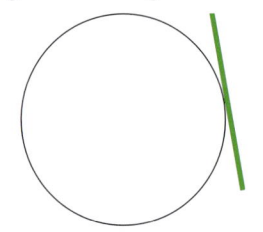

CHAPTER 13: CIRCLES

Example 1

Lynda draws a circle, a tangent to the circle and a chord.
(a) How many points of intersection are there between the circle and the tangent?
(b) How many points of intersection are there between the circle and the chord?

(a) The tangent touches a circle at one point, so there is one point of intersection.
(b) A chord connects two points on the circumference, so there are two points of intersection.

Exercise 13B

1. Write down the meaning of the following parts of a circle:
   (a) Chord   (b) Diameter   (c) Arc   (d) Sector

2. (a) Draw a circle using a pair of compasses.
   (b) Mark two points on the circumference of the circle. Label them A and B.
   (c) Using a different colour, show an arc of the circle between the points A and B.
   (d) Draw a chord of the circle between the points A and B.
   (e) Shade the minor segment of the circle between the arc and the chord of the circle.

3. What is the special name for the longest possible chord of a circle?

4. Jenny says 'The longest possible arc of a circle is the circumference.' Is she right? Explain your answer.

5. Pete says he can draw a tangent that passes through a circle in two places. Is he right?

6. Draw a circle, centre O, and label it C. Add two radii **that meet at 90°** and label them R_1 and R_2.
   (a) Shade the **minor sector** between R_1, R_2 and the circumference of C.
       What fraction of the circle have you shaded?
   (b) Shade the **major sector** between R_1, R_2 and the circumference of C.
       What fraction of the circle have you shaded?

7. Draw a circle. On the same diagram mark a point P that lies outside the circle. How many tangents to the circle can you draw from point P? Mark these tangents on your diagram.

8. (a) Draw a circle with a pair of compasses. Label the centre of the circle O.
   (b) On the circumference of the circle, mark two points A and B.
   (c) Draw the two radii OA and OB.
   (d) Draw the chord AB.
   (e) Shade the **minor sector** OAB.
   (f) Using a different colour, shade the **minor segment** between chord AB and the circumference of the circle.
   (g) Using your diagram to help you, copy and complete this sentence. The missing word is the name of a shape: You can find the area of a sector by adding the area of a segment and the area of a _____ .

13.3 Summary

In this chapter you have learnt:
- The names of various lines and areas relating to circles: **circumference**, **radius**, **diameter**, **chord**, **tangent**, **arc**, **sector** and **segment**.
- That the **circumference** is the special name for the perimeter of a circle, the distance around it.
- That the **diameter** is the distance across the circle, through the centre.
- That the **radius** is the distance from the centre to the circumference. It is half of the diameter.
- That a **chord** is a straight line that connects two points on the circumference.
- That a **tangent** is a straight line that touches the circle at one point.
- That an **arc** is a part of the circumference.
- That a **sector** is the area enclosed between two radii and the circumference.
- That a **segment** is the area enclosed between a chord and the circumference.

Chapter 14
Pythagoras' Theorem and Trigonometry

14.1 Introduction

This chapter is about using Pythagoras' Theorem and trigonometry in right-angled triangles.

If you know two of the sides of a right-angled triangle, Pythagoras' Theorem allows you to find the third side.

If you know one of the sides of a right-angled triangle and one of the acute angles, you can use trigonometry to find either of the remaining sides. If you know two of the sides, you can find either of the acute angles.

> **Note:** Pythagoras of Samos was an ancient Greek mathematician, who lived around 580 – 500 BC. He discovered the relationship between the three sides in a right-angled triangle. Pythagoras' Theorem is still named after him, over 2500 years later.

Key words
- **Hypotenuse**: The hypotenuse is the longest side in a right-angled triangle. It is always opposite the right angle.
- **Sine, cosine and tangent**: The three trigonometric functions you will use to find unknown sides or angles in a right-angled triangle.

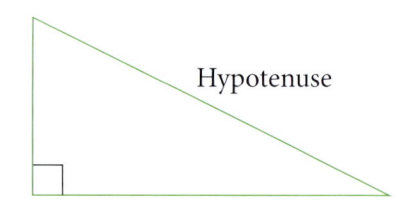

Before you start you should know how to:
- Find squares and square roots.
- Find approximations to numbers by rounding.
- Solve simple equations.

In this chapter you will learn how to:
- Use Pythagoras' Theorem in a right-angled triangle to:
  - Find the length of the hypotenuse in a right-angled triangle, given the other two sides.
  - Find the length of one of the shorter sides, given the other two sides.
- Use trigonometry in a right-angled triangle to:
  - Find the length of a side, given a side and an acute angle.
  - Find the size of an acute angle given two of the sides.

Exercise 14A (Revision)

1. Calculate the following.
   (a) 3^2 (b) 7^2 (c) 5^2 (d) $\sqrt{64}$ (e) $\sqrt{16}$ (f) $\sqrt{100}$

2. Round the following numbers to 1 decimal place.
   (a) 5.14 (b) 6.293 (c) 10.982 (d) 2.05 (e) 100.091

3. Solve the following equations. Round your answers to 1 decimal place where appropriate.
   (a) $5x + 2 = 32$ (b) $2y - 7 = 23$ (c) $a^2 = 9$ (d) $b^2 - 6 = 58$ (e) $c^2 + 20 = 100$

14.2 Pythagoras' Theorem

Pythagoras' Theorem states that $a^2 + b^2 = c^2$, where a and b are the lengths of the two shorter sides in the right-angled triangle and c is the length of the hypotenuse.

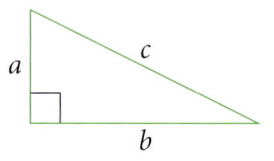

CHAPTER 14: PYTHAGORAS AND TRIGONOMETRY

To visualise Pythagoras' Theorem, try the following activity:

1. On squared paper, draw a right-angled triangle with a base of 4 cm and a height of 3 cm. Mark the right angle on your diagram. Carefully measure the length of the hypotenuse. What is its length? (Clue: it should be a whole number.)

2. On each of the three sides of your triangle, draw a square, as shown in the diagram on the right. Find the area of each square.

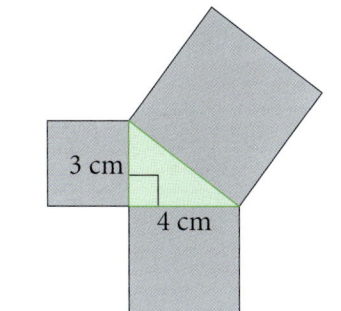

   Hint: Remember that the area of a square is base × height

3. Add together the two smaller areas. What do you notice?

Finding the length of the hypotenuse

Example 1

Find the length of the hypotenuse, marked x, in these right-angled triangles. Give your answers to one decimal place where appropriate.

(a) 8 cm, 6 cm, x

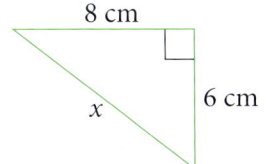

(b) x, 5.8 m, 3.2 m

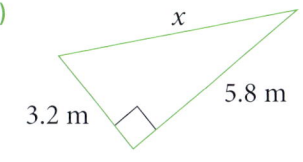

In each part, use Pythagoras' Theorem, $c^2 = a^2 + b^2$, where c is the length of the hypotenuse; a and b are the two shorter sides.

(a) $x^2 = 6^2 + 8^2$
$x^2 = 36 + 64$
$x^2 = 100$
$x = \sqrt{100}$
$x = 10$ cm

(b) $x^2 = 5.8^2 + 3.2^2$
$x^2 = 33.64 + 10.24$
$x^2 = 43.88$
$x = \sqrt{43.88}$
$x = 6.6$ m (1 d.p.)

In some questions you may not be given a diagram. You should draw your own diagram to help you answer the question.

Example 2

ABC is a right-angled triangle. AB = 15 cm, BC = 20 cm and angle ABC is 90°
Calculate the length of AC.

First draw the triangle, as shown. Then, by Pythagoras' Theorem:
$x^2 = 15^2 + 20^2$
$x^2 = 225 + 400$
$x^2 = 625$
$x = \sqrt{625} = 25$ cm

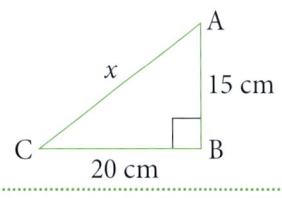

Exercise 14B

1. Find the length of the hypotenuse, marked x, for each of these right-angled triangles. Round your answer to 1 decimal place where appropriate.

   (a) x, 5 cm, 12 cm

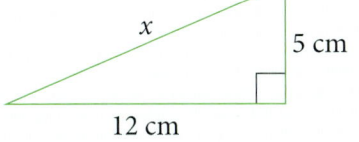

   (b) 3 cm, 4 cm, x

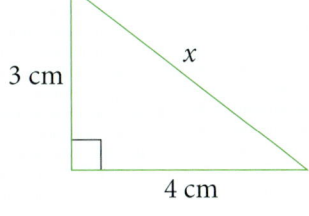

   (c) x, 6 km, 9 km

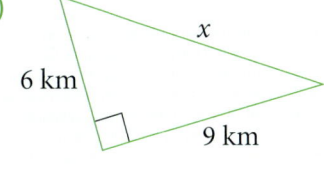

GCSE MATHEMATICS M3 AND M7

(d) triangle with sides 6.7 m, 5 m, and x (hypotenuse)

(e) triangle with 5.1 mm, x, 1.7 mm

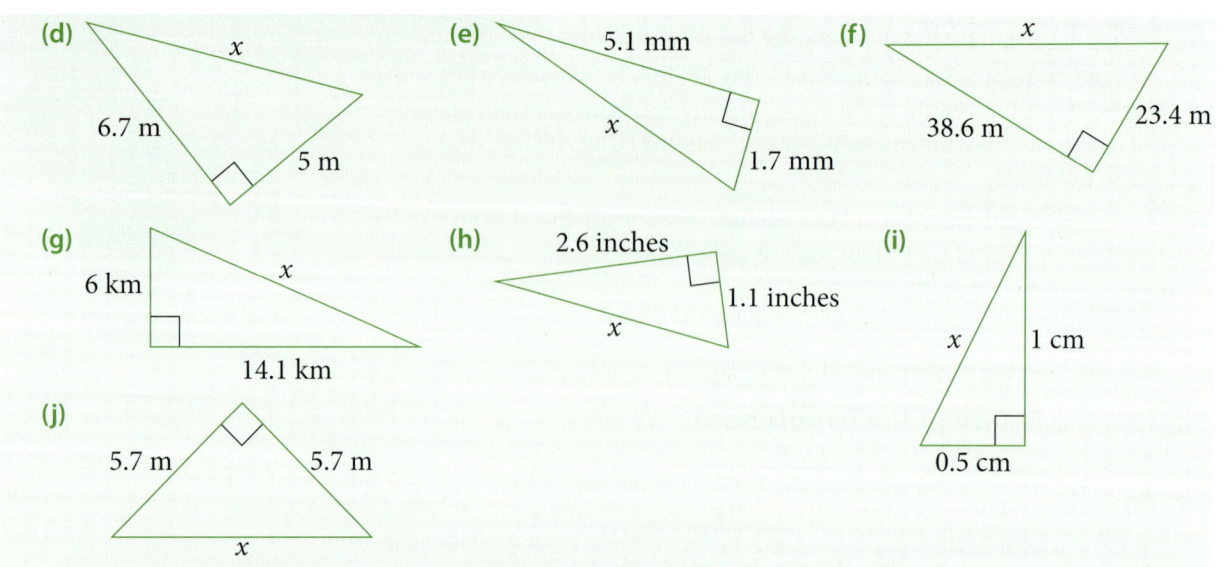

(f) triangle with x, 38.6 m, 23.4 m

(g) triangle with 6 km, x, 14.1 km

(h) triangle with 2.6 inches, x, 1.1 inches

(i) triangle with x, 1 cm, 0.5 cm

(j) triangle with 5.7 m, 5.7 m, x

2. A right-angled triangle has one side measuring 2.5 cm and another side measuring 6 cm. What is the length of the hypotenuse?

3. A right-angled triangle has hypotenuse of length z cm. The two shorter sides are x cm and y cm. Which of the following is a correct statement?
   (a) $x + y = z$ (b) $x^2 + y^2 = z^2$ (c) $x^2 + z^2 = y^2$ (d) $y^2 + z^2 = x^2$

4. What is the length of the hypotenuse of a right-angled triangle whose shorter sides are 9 and 12 units?

5. A triangle has sides of length 6, 8, and x cm, from smallest to largest. If the triangle is a right-angled triangle, what is the value of x?

6. The two shorter sides of a right-angled triangle are of length 10 and 24 mm. What is the length of the hypotenuse?

7. ABC is a right-angled triangle. AB = 7 cm, BC = 8 cm and angle ABC is 90°
   Calculate the length of AC to one decimal place.

8. PQR is a right-angled triangle, in which PQ = 15 cm, QR = 11 cm and angle PQR is 90°
   Calculate the length of PR to one decimal place.

9. The diagram on the right shows a square with a side length of 10 cm. Find the length of the diagonal of the square, marked x in the diagram.

Finding the length of one of the shorter sides

If you know the length of the hypotenuse and one of the shorter sides, you can find the length of the third side. Pythagoras' Theorem states that $a^2 + b^2 = c^2$, where c is the length of the hypotenuse; a and b are the two shorter sides. Re-arranging gives $b^2 = c^2 - a^2$ Use this form of Pythagoras' Theorem when finding one of the shorter sides.

Example 3

Find the length of the side marked x in these right-angled triangles. Give your answers to one decimal place where appropriate.

(a) triangle with 6.2 m, 3.7 m, x

(b) triangle with x, 15.2 km, 25.4 km

(a) $b^2 = c^2 - a^2$
$x^2 = 6.2^2 - 3.7^2$
$x^2 = 24.75$
$x = \sqrt{24.75}$
$x = 4.97...$
$x = 5.0$ m (1 d.p.)

(b) $b^2 = c^2 - a^2$
$x^2 = 25.4^2 - 15.2^2$
$x^2 = 414.12$
$x = \sqrt{414.12}$
$x = 20.349...$
$x = 20.3$ km (1 d.p.)

Exercise 14C

1. Find the length of the side marked x in each of these right-angled triangles. Round your answers to one decimal place where appropriate.

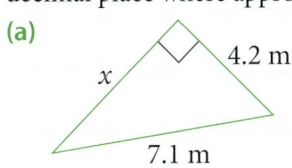

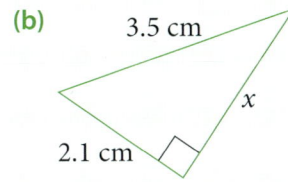

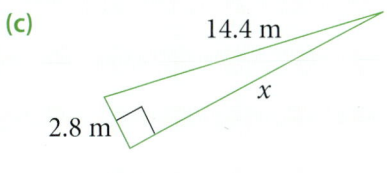

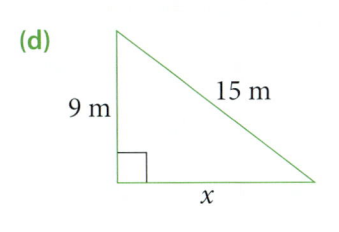

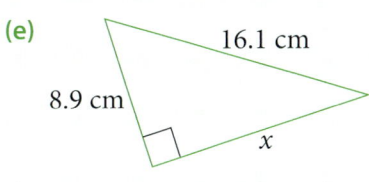

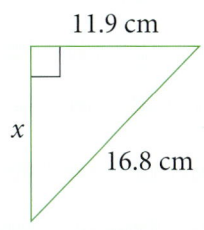

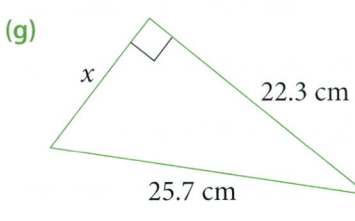

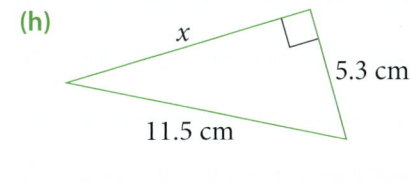

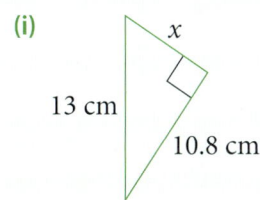

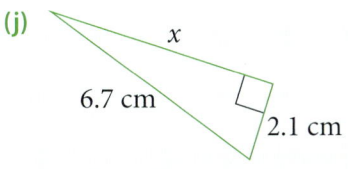

   (a) 4.2 m, 7.1 m, x

   (b) 3.5 cm, 2.1 cm, x

   (c) 14.4 m, 2.8 m, x

   (d) 9 m, 15 m, x

   (e) 8.9 cm, 16.1 cm, x

   (f) 11.9 cm, 16.8 cm, x

   (g) 22.3 cm, 25.7 cm, x

   (h) 11.5 cm, 5.3 cm, x

   (i) 13 cm, 10.8 cm, x

   (j) 6.7 cm, 2.1 cm, x

2. The hypotenuse of a right-angled triangle measures 10 units, and one of the other sides measures 6 units. What is the length of the remaining side?

3. In a right-angled triangle, one of the shorter sides is 5 cm long and the hypotenuse is 13 cm long. What is the length of the third side?

4. If a right-angled triangle has a hypotenuse of 15 cm and one side of length 12 cm, what is the length of the third side?

5. In a right-angled triangle, one of the shorter sides measures 7 units, and the hypotenuse measures 25 units. What is the length of the other shorter side?

6. A right-angled triangle has a hypotenuse of length 17 cm. One of the shorter sides is 8 cm long. What is the length of the third side?

Problem solving using Pythagoras' Theorem

You will be asked to solve problems using Pythagoras' Theorem.

Example 4

A vertical phone mast is supported by a cable attached to horizontal ground 5.5 m from the base of the mast. The mast is 16 m tall. Find the length of the cable.

First draw a diagram, as shown on the right.

The cable is the hypotenuse of the triangle. Label it x.

Then, by Pythagoras' Theorem:

$x^2 = 16^2 + 5.5^2$

$x^2 = 286.25$

$x = \sqrt{286.25}$

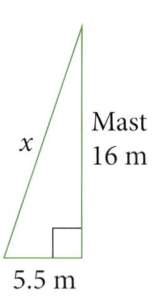

$x = 16.9$ m (3 s.f.)

Exercise 14D

1. A flagpole stands 10 metres tall. A rope is tied to the top of the pole and secured to the ground 8 metres away. What is the length of the rope?

2. The flag shown is in the shape of a right-angled triangle. The two shorter sides are 60 cm and 90 cm in length. Work out, correct to the nearest cm, the **perimeter** of the flag.

3. A ladder is leaning against a wall, as shown in the diagram on the right. The base of the ladder is 3 feet away from the wall, and the ladder itself measures 10 feet, as shown in the diagram. How high up the wall does the ladder reach?

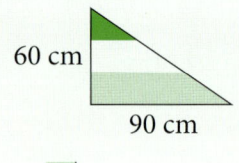

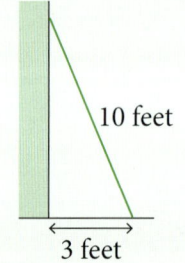

4. The field shown in the diagram has a length of 200 m and a width of 150 m. Elsie and Reuben want to get from corner A to corner C. Reuben wants to walk in a straight line AC across the field. Elsie wants to walk around the edge of the field, firstly from A to B, then from B to C. Whose route is longer and by how much?

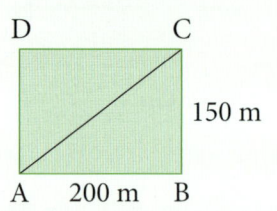

5. An air ambulance flies north 7 km from a hospital. It then turns and flies east for 6 km. Finally, it flies directly back to the hospital in a straight line. How far does the air ambulance fly on its return journey?

6. (a) With a ruler, measure the width and height of one page in your exercise book.
   (b) Using Pythagoras' Theorem, calculate the length of the diagonal across the page.
   (c) Measure the diagonal.
   (d) How close are your answers to parts (b) and (c)?

7. The diagram on the right shows the side view of a roller coaster ride in a funfair.
   (a) Find the height h of the tower AC.
   (b) Find the length x of the track DC.

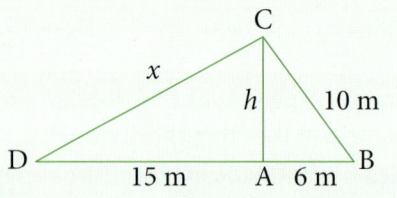

8. A ship sails due north 6 km from port A to port B. It then sails due east 6.16 km to port C, as shown in the diagram on the right.
   (a) A straight road runs from port A to port C, as shown. Find the length of the road. Give your answer to two decimal places.
   (b) From port C the ship sails 3.14 km to port D. Given that angle ADC = 90°, find the distance the ship is now from port A. Give your answer to two decimal places.

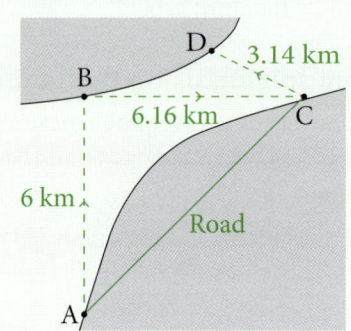

9. The diagram below shows the symmetrical front of a house with some of the dimensions shown. Calculate the length of DE, the slope of the roof marked *x*. Give your answer to 1 decimal place.

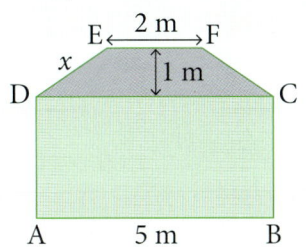

10. The diagram below shows ABCD, which is a square of side length 5 cm. The diagonal AC is shown. Calculate the difference between the lengths AC and AB. Give your answer to 1 decimal place.

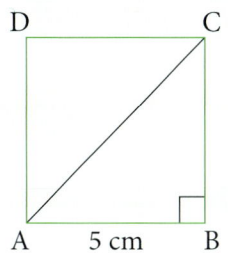

11. A ship sails from harbour H due north for 14 km. It then sails due east for 7 km to port P. Find the ship's distance from harbour H. Give your answer in kilometres to 1 decimal place.

12. Is it possible for a right-angled triangle to have sides of length 12 cm, 18 cm and 20 cm? Explain your answer.

14.3 Trigonometry

In this section you will use three trigonometric functions: sine, cosine and tangent, which are usually shortened to sin, cos and tan. You will need to remember three formulae:

$$\text{sine} = \frac{\text{opposite}}{\text{hypotenuse}} \qquad \text{cosine} = \frac{\text{adjacent}}{\text{hypotenuse}} \qquad \text{tangent} = \frac{\text{opposite}}{\text{adjacent}}$$

> **Note:** You may find it helpful to remember the mnemonic: SOH CAH TOA
> (**s**ine is **o**pposite over **h**ypotenuse; **c**osine is **a**djacent over **h**ypotenuse; **t**angent is **o**pposite over **a**djacent)

> **Note:** You will sometimes see the word 'trigonometry' abbreviated to 'trig'.

Finding a side length

You can use trigonometry to find the length of an unknown side in a right-angled triangle. The following examples show how to select and use the right formula from the three formulae above.

Example 5

Find the side length marked *x* in this triangle.

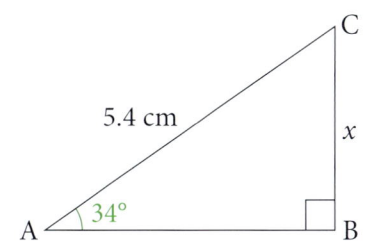

Step 1. Label the sides 'opposite', 'hypotenuse' and 'adjacent', as shown on the right.

The opposite side is always opposite the acute angle you know.

The hypotenuse is always the longest side of the triangle, opposite the right angle.

The adjacent is the third side. It lies next to the acute angle you know.

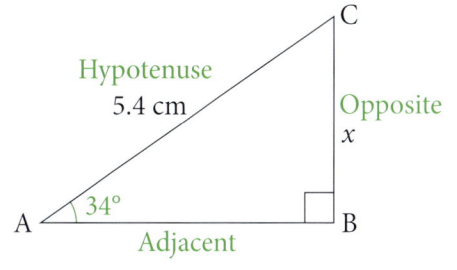

Step 2. Select the correct formula. In this case we use the formula involving opposite and hypotenuse. This is the formula for sine:

$$\text{sine} = \frac{\text{opposite}}{\text{hypotenuse}}$$

$$\sin 34 = \frac{x}{5.4}$$

> **Note:** Remember that a trig function always needs an angle. Writing the word sin, cos or tan on its own is meaningless.

Step 3. Re-arrange the equation to make x the subject. In this case there is just one step:
$$x = 5.4 \times \sin 34$$
Step 4. Use the calculator to find the value of x:
$$x = 3.0 \text{ cm}$$

In the next example, rearranging to make x the subject (step 3) is slightly harder, because x appears in the denominator of the fraction.

Example 6

Find the length of the hypotenuse, marked x, in this triangle.

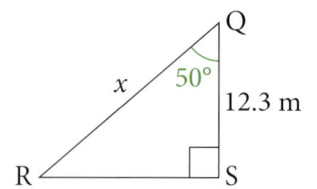

Step 1. Label the sides 'opposite', 'hypotenuse' and 'adjacent', as shown on the right.

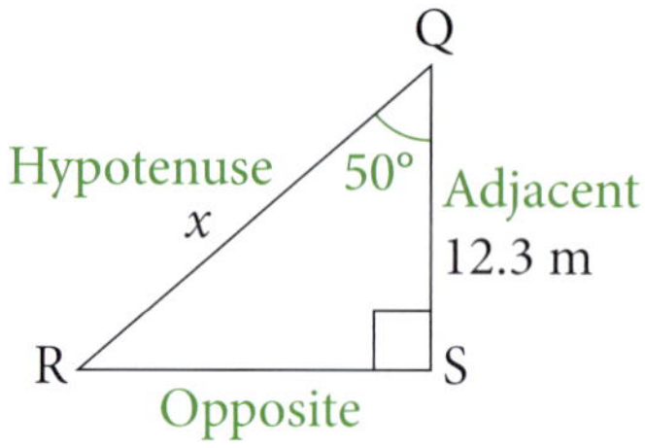

Step 2. Select the correct formula. In this case we use the formula for cosine, which involves adjacent and hypotenuse:
$$\cosine = \frac{\text{adjacent}}{\text{hypotenuse}}$$
$$\cos 50 = \frac{12.3}{x}$$

Step 3. Re-arrange the equation to make x the subject. Since x is in the denominator of the fraction, there are two steps:
$$x \cos 50 = 12.3$$
$$x = \frac{12.3}{\cos 50}$$

Step 4. Use the calculator to find the value of x:
$$x = 19.1 \text{ m}$$

Exercise 14E

1. Find the length of the side marked x in each of the following triangles. Give each answer to one decimal place.

   (a) 12.8 cm, 39°, find x (triangle ABC, right angle at A)

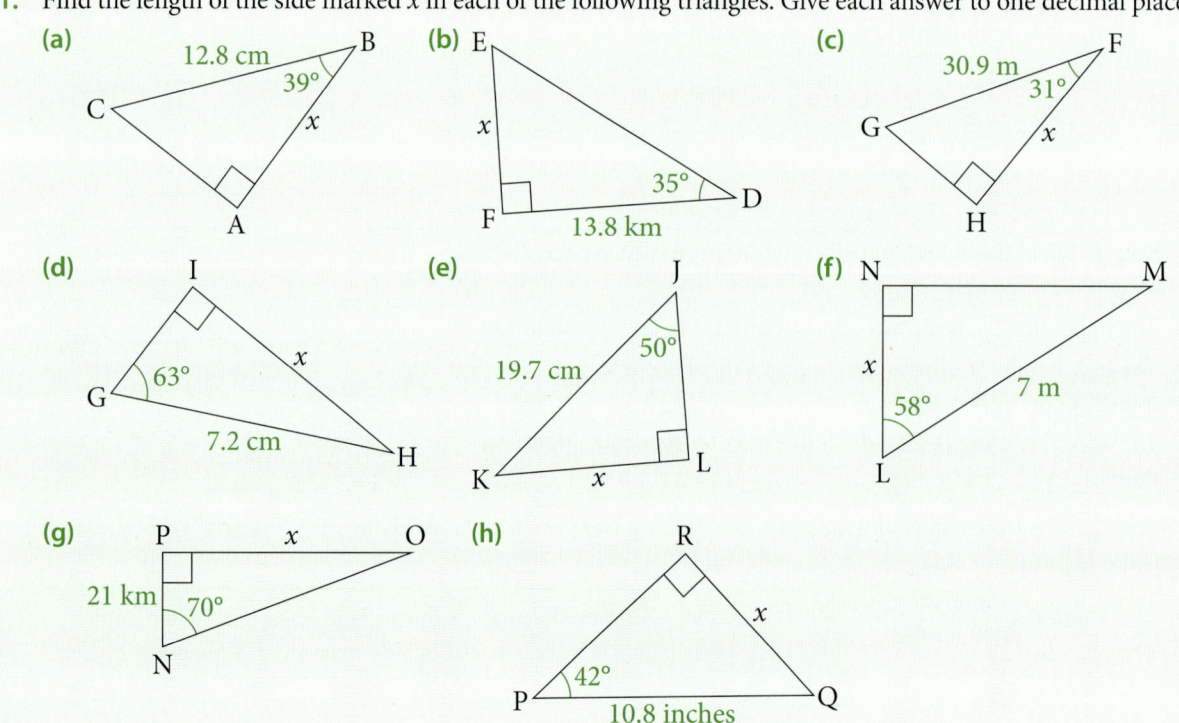

   (b) 13.8 km, 35°, find x (triangle EFD, right angle at F)

   (c) 30.9 m, 31°, find x (triangle GFH, right angle at H)

   (d) 7.2 cm, 63°, find x (triangle GIH, right angle at I)

   (e) 19.7 cm, 50°, find x (triangle JKL, right angle at L)

   (f) 7 m, 58°, find x (triangle NML, right angle at N)

   (g) 21 km, 70°, find x (triangle PNO, right angle at N)

   (h) 10.8 inches, 42°, find x (triangle PRQ, right angle at R)

CHAPTER 14: PYTHAGORAS AND TRIGONOMETRY

2. Find the length of the side marked *x* in each of the following triangles. Give each answer to one decimal place.

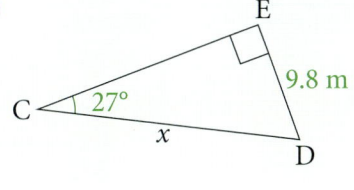

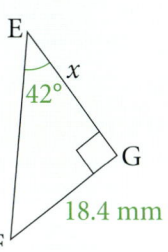

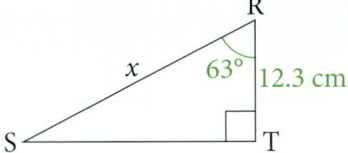

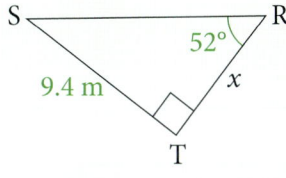

(a) Triangle ABC with angle 54° at A, right angle at C, BC = 6.7 mm, AC = *x*

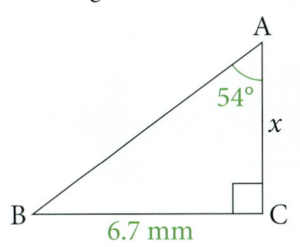

(b) Triangle CED with right angle at E, angle 27° at C, ED = 9.8 m, CD = *x*

(c) Triangle EFG with angle 42° at E, right angle at G, *x* = EG, FG = 18.4 mm

(d) Triangle SRT with angle *x* at S, angle 63° at R, RT = 12.3 cm, right angle at T

(e) Triangle SRT with angle 52° at S, right angle at T, ST = 9.4 m, TR = *x*

(f) Triangle TUS with TU = 12.5 cm, right angle at U, angle 62° at S, US = *x*

(g) Triangle VUT with right angle at V, VT = 30 cm, angle 54° at T, VU = *x*

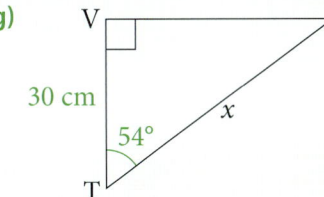

(h) Triangle WUV with right angle at U, WU = *x*, angle 74° at W, WV = 62.6 km

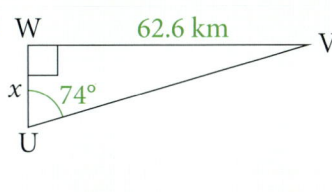

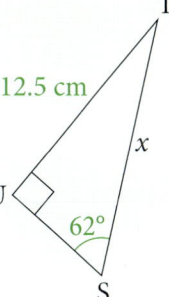

14.4 Finding an Angle

In this section we discuss finding the size of one of the acute angles in a right-angled triangle. When finding an angle using trigonometry, you will use one of the functions $\sin^{-1}$, $\cos^{-1}$ or $\tan^{-1}$.

Example 7

Find the size of the angle marked θ in the diagram on the right.

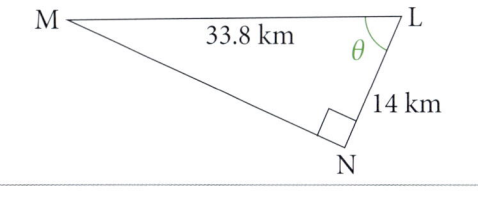

Triangle MLN with ML = 33.8 km, LN = 14 km, angle θ at L, right angle at N

Step 1. Considering the angle θ, label the sides opposite, hypotenuse and adjacent, as shown on the right.

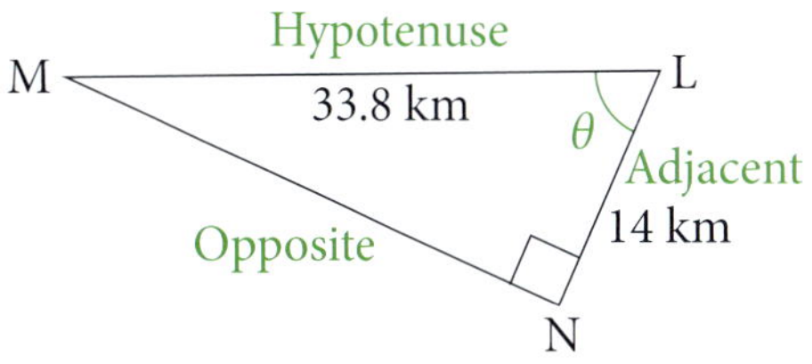

Hypotenuse = 33.8 km (ML), Adjacent = 14 km (LN), Opposite = MN

Step 2. Select the correct formula. Since we have the lengths of the adjacent side and the hypotenuse, the correct formula is the cosine formula:

$$\cos \theta = \frac{14}{33.8}$$

Step 3. Use the inverse cosine function $\cos^{-1}$ to find the angle:

$$\theta = \cos^{-1}\left(\frac{14}{33.8}\right) = 65.5° \text{ (1 d.p.)}$$

Note:
- On your calculator, press SHIFT-cos for the inverse cosine function.
- The inverse trig functions are only used to work out angles. You will never need them for side lengths.

119

Exercise 14F

1. Find the angle marked θ in each of the following diagrams.

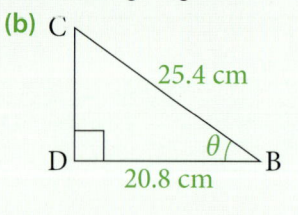

   (a) Triangle with B, C (right angle), A; CB = 31.3 km, CA = 18.4 km, angle θ at A.

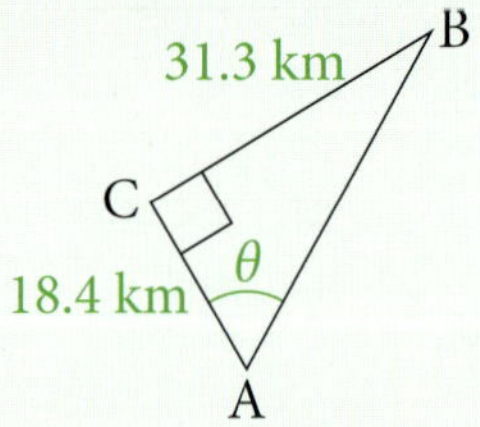

   (b) Triangle with C, D (right angle), B; CB = 25.4 cm, DB = 20.8 cm, angle θ at B.

   (c) Triangle with C, 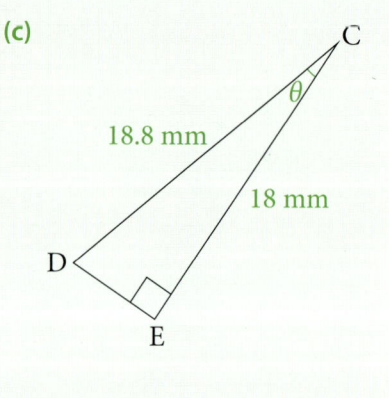 D, 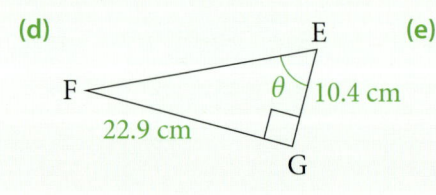 E (right angle); DC = 18.8 mm, DE = 18 mm, angle θ at C.

   (d) Triangle with F, E, G (right angle); EG = 10.4 cm, FG = 22.9 cm, angle θ at E.

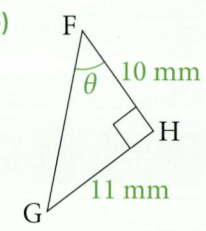

   (e) Triangle with F, H (right angle), G; FH = 10 mm, GH = 11 mm, angle θ at F.

   (f) Triangle with G, F, H (right angle); GF = 13.8 cm, GH = 11.2 cm, angle θ at F.

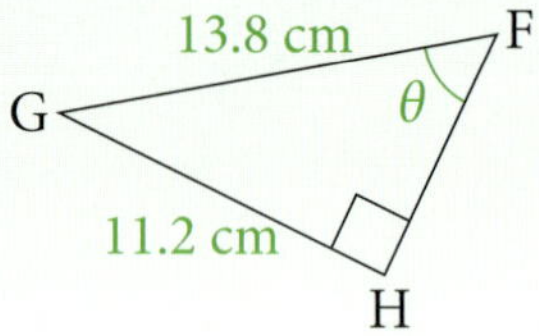

   (g) Triangle with L (right angle), K, J; LK = 10.2 feet, LJ = 8 feet, angle θ at J.

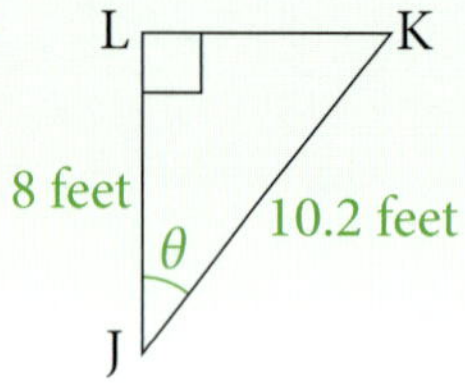

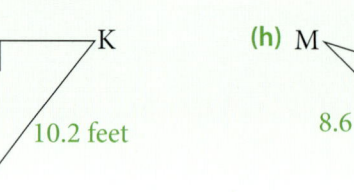

   (h) Triangle with M, L, N (right angle); ML = 9.7 cm, MN = 8.6 cm, angle θ at L.

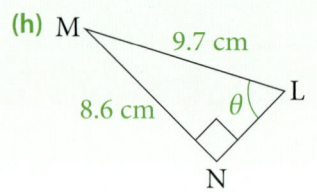

14.5 Problem Solving

In an exam setting, many of the questions are related to solving a problem set in context. These questions may require use of both Pythagoras' Theorem and trigonometry.

Example 8

Adam stands on horizontal ground, 50 m from the base of a tall vertical tower. From this point, the angle of elevation of the top of the tower is 42°
(a) Find the height of the tower to one decimal place.
(b) Beth is standing twice as far from the tower as Adam. Find the angle of elevation of the top of the tower from where she is standing. Give your answer to the nearest degree.

It is important to draw a diagram. Adam is shown at point A, 50 m from the base of the tower. The base and top of the tower have been marked as C and D respectively and its height is shown as h metres.

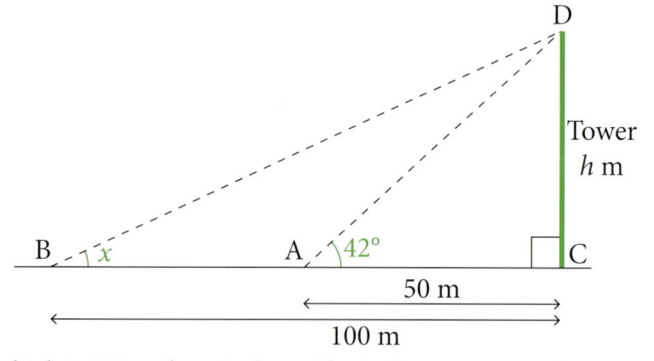

(a) Using triangle ACD:

$\tan 42 = \dfrac{h}{50}$

$h = 50 \tan 42 = 45.0$ m

(b) Beth is standing twice as far from the tower, which is 100 m from its base. She is shown at point B on the diagram. The angle of elevation of point D from B is labelled x.
Using triangle BCD:

$\tan x = \dfrac{45.0}{100} = 0.45$

$x = \tan^{-1}(0.45) = 24°$ (nearest degree)

CHAPTER 14: PYTHAGORAS AND TRIGONOMETRY

Exercise 14G

1. A boy is flying a kite with a string of length 6.7 m. The boy holds one end of the string 0.7 m above the ground. The kite's angle of elevation is 30° to the horizontal, as shown in the diagram. Calculate h, the height of the kite **above the ground**.

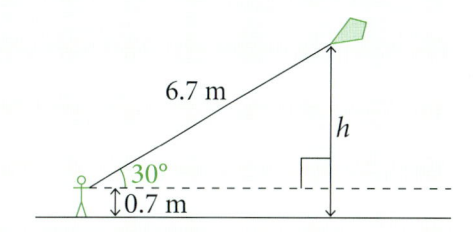

2. The longest side in a right-angled triangle is 14 cm. One of the shorter sides is 6 cm.
   (a) Calculate the perimeter of the triangle, giving your answer to 1 decimal place.
   (b) Find the angle between the 14 cm and 6 cm sides. Give your answer to 1 decimal place.

3. A swimmer dives into a pool at an angle of 16° below the horizontal, as shown in the diagram. The swimmer swims at 1 m/s for 5 seconds at this angle. Find the depth of the swimmer below the surface of the water after this time. Give your answer in metres to 2 decimal places.

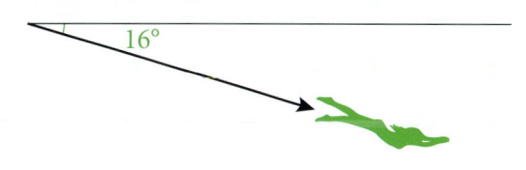

4. The diagram on the right shows a right-angled triangle. The two acute angles are marked, as well as the lengths of two of the sides. Find the **difference** in the size of the two acute angles.

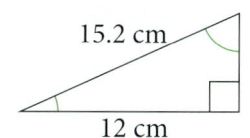

5. The diagram on the right shows a garden in the shape of a rectangle. Most of the area is grass, but the triangle shown is a flowerbed. Fencing is placed along the border between the grass and the flowerbed.
   (a) Find the length of the fencing used, giving your answer to 1 decimal place.
   (b) Find the sizes of the two acute angles in the flowerbed.

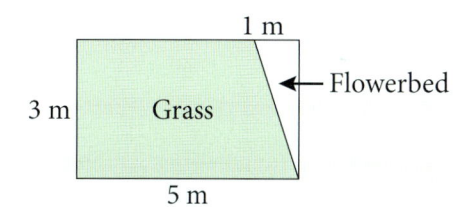

6. Rectangle A shown on the near right has a width of 12 cm.
   (a) Find the length of the diagonal shown in rectangle A.
   (b) Look at square B shown on the far right. Its diagonal is the same length as the diagonal of rectangle A in part (a). Find the length of one side of square B. Give your answer to 1 decimal place.

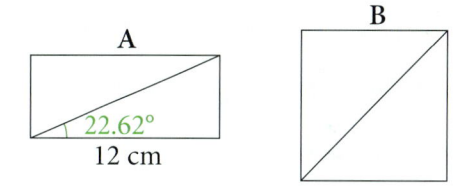

7. The towns Ardstrop, Ballybeg, Castlebridge and Porlock are shown on the map on the right as points A, B, C and P respectively. The straight road from Ardstrop to Castlebridge passes through Ballybeg. It runs perpendicular to the road from Castlebridge to Porlock. The distance from Ballybeg to Castlebridge is 3.16 km and the distance from Ballybeg to Porlock is 7.07 km, as shown. For the following, give your answers to 3 significant figures.
   (a) A person from Porlock walks directly to Castlebridge. How far does the person from Porlock walk?
   (b) A train travels from Ardstrop to Porlock along the straight track shown, measuring 8.56 km. Find the distance along the straight road from Ardstrop to Ballybeg.
   (c) Find the size of the angle at point A on the diagram, between the road to Castlebridge and the train line to Porlock.

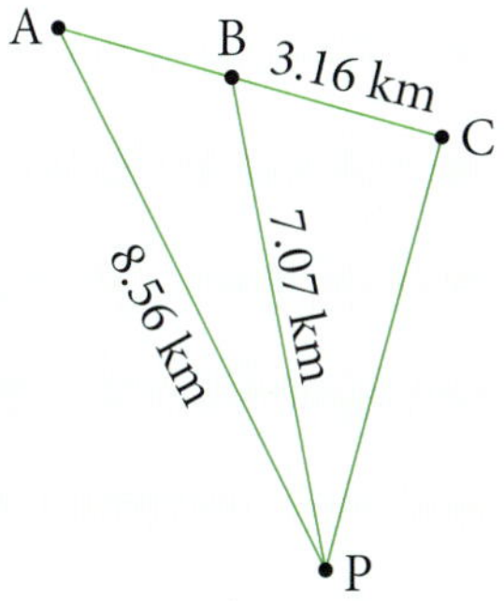

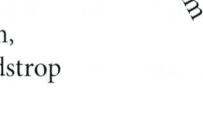

8. The diagram on the right shows a vertical cliff, which is 35 m high. From the top of the cliff, Pat can see two boats A and B on the sea, as shown on the diagram. From Pat's position, the angle of depression of boat A is 37.9° and the angle of depression of boat B is 30.3° Calculate the distance between the two boats, giving your answer to the nearest metre.

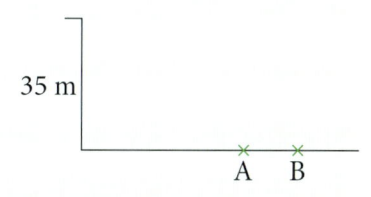

9. A zip wire is angled at 29° to the horizontal, as shown in the diagram on the right. The length of the wire is 59 m. The wire is tight so that it follows a straight line from the starting platform to the horizontal ground. Find the vertical height of the starting platform above the ground. Give your answer to 1 decimal place.

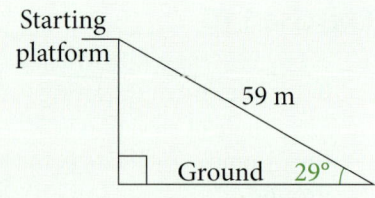

10. Calculate the perimeter of the isosceles triangle shown in the diagram below. Give your answer to 1 decimal place.

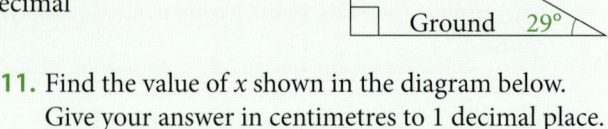

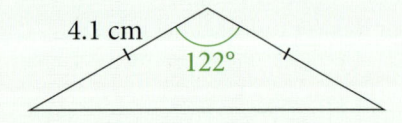

11. Find the value of x shown in the diagram below. Give your answer in centimetres to 1 decimal place.

12. Look at the trapezium in the diagram on the right. Giving your answers to 1 decimal place:
    (a) Calculate the perpendicular height of the trapezium, marked h in the diagram.
    (b) Find the size of the angle at A.

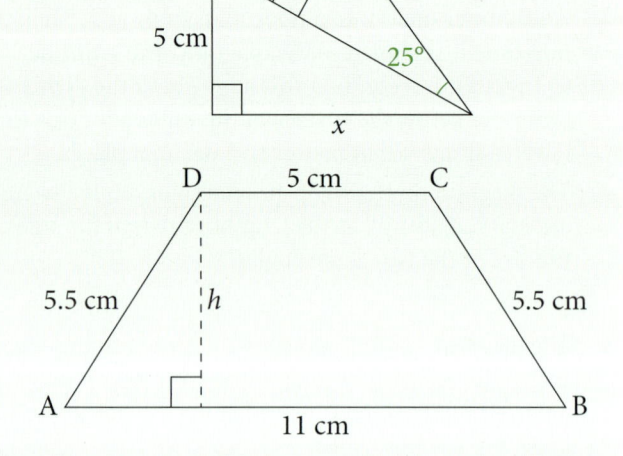

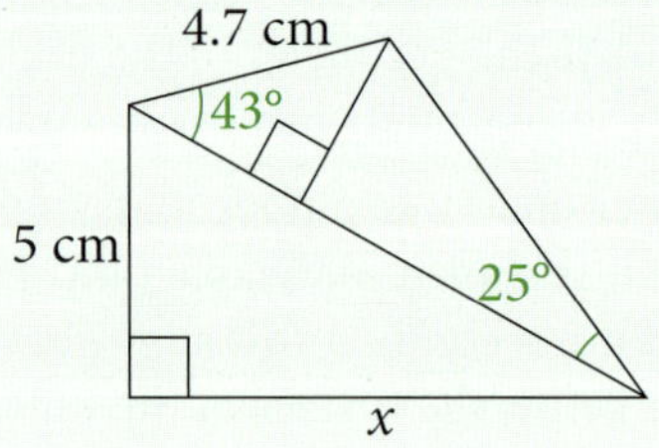

13. A large rectangular box is leaning against a vertical wall, as shown in the diagram on the right. The corner B rests on the horizontal floor, 2 metres from point O. Corner A rests against the wall, 4 metres above O.

    In each of parts (a) to (d) below, give all your answers to 1 decimal place.
    (a) Find the length of side AB of the box.
    (b) Find the angle between side AB and the floor.
    (c) Find the angle between the base of the box BC and the floor.
    (d) Given that point C is 1 metre above the floor, find the length of BC.

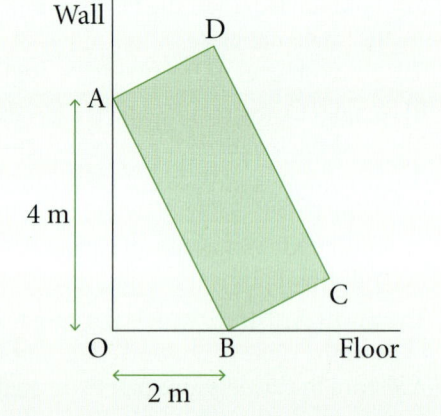

14. The diagram on the right shows a rhombus, with height and width 8 cm and 4 cm respectively.
    (a) Find the size of the largest angle in the rhombus. Give your answer to the nearest degree.
    (b) Find the length of each side of the rhombus, giving your answer in centimetres to 1 decimal place.

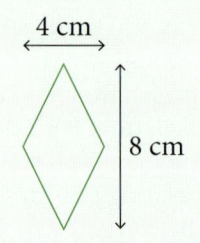

14.6 Summary

In this chapter you have learnt that:

- Pythagoras' Theorem states that $a^2 + b^2 = c^2$, where a and b are the lengths of the two shorter sides in the right-angled triangle and c is the length of the hypotenuse.
- In some questions you may not be given a diagram. You should draw your own diagram to help you answer the question.

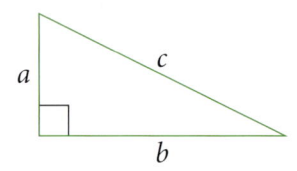

- In a right-angled triangle, Pythagoras' Theorem can be used to find the length of the **hypotenuse** if the two shorter sides are known.
- Pythagoras' Theorem can also be used to find one of the shorter sides if the hypotenuse and the other shorter side are known.
- You can also use trigonometry in a right-angled triangle. If you are given a side length and the size of one of the angles, you can find the other side lengths.
- If you are given two of the sides, you can find either of the acute angles.
- The key formulae you use are:

$$\text{sine} = \frac{\text{opposite}}{\text{hypotenuse}} \qquad \text{cosine} = \frac{\text{adjacent}}{\text{hypotenuse}} \qquad \text{tangent} = \frac{\text{opposite}}{\text{adjacent}}$$

- When finding an angle, you use one of the functions $\sin^{-1}$, $\cos^{-1}$ or $\tan^{-1}$
- Many questions involving Pythagoras' Theorem and trigonometry are problem-solving questions.

Chapter 15
Compound Measure

15.1 Introduction

This chapter deals with quantities that are **compound measures**. Examples include speed and density.

Key words
- **Density**: The mass of an object divided by its volume.
- **Speed**: The distance travelled divided by the time taken.
- **Pressure**: The force applied per square metre.

Before you start you should know how to:
- Add, subtract, multiply and divide using fractions.
- Substitute numbers into formulae.

In this chapter you will learn how to:
- Find the correct formula for a compound measure.
- Use formulae for compound measure.

Substituting numbers into a formula is a skill you will use a lot in this chapter, as in the following example.

Example 1 (Revision)

Given that $m = \dfrac{F}{p}$, find m if $p = 16$ and $F = 4$

Substitute the values of F and p into the formula:

$m = \dfrac{F}{p} = \dfrac{4}{16} = \dfrac{1}{4}$

You may need to convert between different units of time.

Example 2 (Revision)

(a) Convert 3.5 minutes to minutes and seconds.
(b) Convert 2 hours 15 minutes to hours.

(a) 3.5 minutes = $3\frac{1}{2}$ minutes, or 3 minutes 30 seconds.

(b) 2 hours 15 minutes is $2\frac{1}{4}$ hours, or 2.25 hours.

> **Note:** Remember that 2.25 hours **does not mean** 2 hours 25 minutes!

Exercise 15A (Revision)

1. Substitute the numbers into the formulae.
   (a) $A = BC$ Find A if $B = 2$ and $C = 5$
   (b) $Q = RS$ Find Q if $R = -6$ and $S = 5$
   (c) $X = \dfrac{Z}{Y}$ Find X if $Z = 12$ and $Y = 3$
   (d) $c = de$ Find c if $d = 6$ and $e = 2.5$
   (e) $s = \dfrac{r}{t}$ Find s if $r = 15$ and $t = 1.5$
   (f) $f = gh$ Find f if $g = \frac{1}{4}$ and $h = 8$
   (g) $p = \dfrac{q}{n}$ Find p if $q = 8$ and $n = \frac{1}{4}$

2. Convert:
   (a) 2 hours into minutes
   (b) $1\frac{1}{2}$ minutes into seconds
   (c) $5\frac{1}{4}$ hours into hours and minutes
   (d) 5.75 minutes to minutes and seconds
   (e) Convert 1 hour 45 minutes to hours

15.2 Speed as a Compound Measure

A compound measure is a measure involving two quantities.

An example is speed, which involves distance and time:

$$\text{speed} = \frac{\text{distance}}{\text{time}}$$

If the distance is measured in kilometres and the time is in hours, then the unit for speed is kilometres per hour (km/h). If distance is measured in metres and the time is in seconds, then the unit for speed is metres per second (m/s). Other possible units for speed are miles per hour, centimetres per second, kilometres per minute, etc.

Example 3

Padraig leaves Belfast in a car at 08:00 and travels to his parents' house 216 km away. He arrives at 12:00 Find his average speed in (a) kilometres per hour (b) metres per second.

Padraig's journey takes from 8 o'clock to 12 o'clock so it lasts 4 hours.

(a) average speed = $\frac{\text{distance}}{\text{time}} = \frac{216 \text{ kilometres}}{4 \text{ hours}}$ = 54 kilometres per hour

(b) For an answer in metres per second, use a distance in metres and a time in seconds:
216 kilometres = 216 × 1000 = 216 000 metres
4 hours = 4 × 60 × 60 = 14 400 seconds
average speed = $\frac{\text{distance}}{\text{time}} = \frac{216\,000 \text{ metres}}{14\,400 \text{ seconds}}$ = 15 metres per second

Using a formula triangle

For all compound measures, you may find it helpful to use a formula triangle.

Using the formula speed = $\frac{\text{distance}}{\text{time}}$, begin by labelling the triangle from the bottom left, then the top, then the bottom right.

To use the triangle:

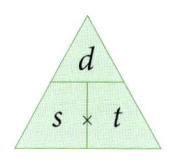

To find the **speed**, cover up s with your finger.

The formula is $s = \frac{d}{t}$

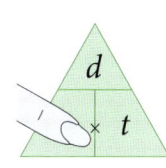

To find the **distance**, cover up d with your finger.

The formula is $d = s \times t$

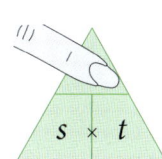

To find the **time**, cover up t with your finger.

The formula is $t = \frac{d}{s}$

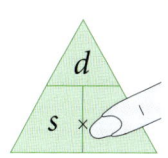

If you are not sure of the formula for a compound measure, you may be able to construct it using the units.

Example 4

Commonly used units for speed are metres per second (m/s) and miles per hour (mph). Which one of the following is the correct formula for speed?

speed = distance × time   speed = $\frac{\text{distance}}{\text{time}}$ speed = $\frac{\text{time}}{\text{distance}}$

Consider metres per second as the units for speed.

For the formula, we need a quantity measured in metres (distance) divided by a quantity measured in seconds (time).

So, the correct formula is: speed $= \dfrac{\text{distance}}{\text{time}}$

Example 5

You may use your calculator in this question.
(a) A lorry travels 169 miles in 3 hours 15 minutes on a motorway. What is the average speed of the lorry while it is on the motorway?
(b) The lorry leaves the motorway and travels a further 96 miles at a speed of 48 miles per hour. How long does it take for this part of the journey?
(c) For the final part of its journey, the lorry travels at an average speed of 44 miles per hour. It takes $1\tfrac{3}{4}$ hours to reach its final destination. What distance did the lorry travel for this third part of the journey?

(a) Convert 3 hours 15 minutes to a time in hours.

15 minutes is $\tfrac{1}{4}$ hour or 0.25 hours, so 3 h 15 mins = 3.25 hours.

Using the triangle and covering s you find that $s = \dfrac{d}{t}$

$s = \dfrac{169}{3.25} = 52$ miles per hour

(b) In this part of the question, you are finding the time, so cover t in the triangle.

$t = \dfrac{d}{s} = \dfrac{96}{48} = 2$ hours

(c) Find the distance d for the third part of the journey. Covering d in the triangle gives $d = s \times t$

$d = 44 \times 1.75 = 77$ miles

Exercise 15B

You may use your calculator in this exercise.

1. Alana drives a distance of 210 km in 3 hours. Work out Alana's average speed.
2. A baseball travels 50 metres to the edge of the ground at a speed of 12.5 m/s. How long does it take?
3. A leaf falls from a tree. It takes 10 seconds to reach the ground at an average speed of 0.8 m/s. How far does it fall?
4. A plane travels 180 kilometres from Belfast to Edinburgh at a speed of 240 km/h. How long does it take for the journey?
5. A rocket takes 10 minutes to reach orbit, travelling at an average speed of 1512 km/h. How far does it travel?
6. A cheetah runs 60 m in 5 seconds. Find its speed:
   (a) in metres per second,
   (b) in kilometres per hour.
7. Brianna swims for $1\tfrac{1}{2}$ hours at 1 km/h. How many lengths of the 25 metre pool does she swim?
8. Work out the average speed of each of these ships and boats in kilometres per hour.
   (a) A speedboat travelling 10 km in 20 minutes.
   (b) A cargo ship travelling 16 200 km from Taiwan to the Netherlands in 45 days.
   (c) A cruise ship covering 100 km in 8 hours.
9. Eoin's train leaves Belfast at 10:20 in the morning and arrives in Dublin at 12:40pm. The total distance for the journey is 105 miles. What is the train's average speed in miles per hour?
10. The sidewinder snake of North America is the fastest snake in the world. It can move at about 27 km/h. What is this speed in metres per second?
11. Freddie sees a fork of lightning and 4.5 seconds later he hears the thunder. How many metres away is the lightning? You may use the fact that sound travels 1 kilometre in 3 seconds. Note: Freddie sees the lightning almost instantly because light travels very, very quickly. The sound takes a longer time to reach him.
12. An Olympic athlete runs 100 m in 10 seconds. How fast, on average, was the athlete running:
    (a) in metres per second?
    (b) in kilometres per hour?
13. A worm moves 7 metres in 28 minutes. Work out the worm's average speed in
    (a) metres per minute
    (b) centimetres per second

14. **(a)** Peter is riding his motorbike at 40 miles per hour. Convert this speed into kilometres per hour, using the fact that 5 miles is approximately 8 kilometres.
  (b) Della is driving her car at 18 metres per second. Convert this speed into kilometres per hour.
  (c) Who is travelling faster and by how much?
  (d) Peter and Della are both travelling home from work. They both travel 12 km and both leave at the same time. What time elapses between Della and Peter getting home?

15.3 Density

Density as a compound measure

The formula for density is:

$$\text{density} = \frac{\text{mass}}{\text{volume}}$$

If the mass is measured in grams and the volume is in cm³, then the unit for density will be grams per cubic centimetre (g/cm³). If the mass is measured in kilograms and the volume is in m³, then the unit for density will be kilograms per cubic metre (kg/m³). Other possible units for density are kilograms per cm³, grams per m³, etc.

The example below demonstrates working with the formula for density.

Example 6

(a) A metal block has a volume of 1000 cm³ and a mass of 8.5 kg. Using the formula:

$$\text{density} = \frac{\text{mass}}{\text{volume}}, \text{ find the density of the block in:}$$

  (i) kilograms per cm³ **(ii)** grams per cm³

(b) A brick has a density of 1.8 grams per cubic centimetre. Its volume is 1470 cm³. Find the mass of the brick.

(a) (i) $\text{density} = \frac{\text{mass}}{\text{volume}}$

$= \frac{8.5}{1000}$

$= 0.0085$ kilograms per cm³

Note: To get an answer in kg/cm³ use a mass in kg and a volume in cm³

Note: This unit can also be written as kg/cm³

(ii) 8.5 kg = 8500 g

$\text{density} = \frac{\text{mass}}{\text{volume}}$

$= \frac{8500}{1000}$

$= 8.5$ grams per cm³

Note: To get an answer in g/cm³ use a mass in g and a volume in cm³

Note: This unit can also be written as g/cm³

(b) Using the formula:

$\text{density} = \frac{\text{mass}}{\text{volume}}$

$1.8 = \frac{\text{mass}}{1470}$

mass = 1.8 × 1470 = 2646 g = 2.646 kg

Using the formula density = $\frac{\text{mass}}{\text{volume}}$ you can construct the triangle, as shown on the right.

The following example demonstrates working with the triangle.

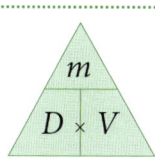

Example 7

(a) A plastic toy has a mass of 620 g and a volume of 3100 cm³. Calculate its density.

(b) A larger toy made from the same plastic material has a volume of 4280 cm³. Find its mass.

(a) When finding the density, cover up *D* in the triangle.
You divide the mass by the volume. So:

density = $\frac{620}{3100}$

= 0.2 g/cm³

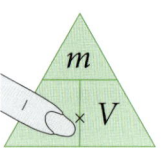

(b) Since the larger toy is made from the same plastic material, it also has a density of 0.2 g/cm³ It has a volume of 4280 cm³
To find the mass using the triangle, cover up *m*.
You multiply the density and the volume. So:
mass = 0.2 × 4280 = 856 g

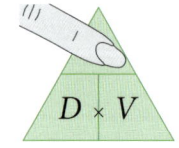

Exercise 15C

You may use your calculator in this exercise.

1. Calculate the density in g/cm³ for each of these items:
   (a) A piece of metal with a volume of 50 cm³ and a mass of 550 g
   (b) A litre of juice with a volume of 1000 cm³ and a mass of 1 kg
   (c) A block of plastic with a mass of 1.84 kg and a volume of 2000 cm³
   (d) A tonne of recycled cardboard with a volume of 1 400 000 cm³

2. The unit for density is grams per centimetre cubed (g/cm³).
   (a) Which one of the following is the correct formula for density?

   density = $\frac{mass}{volume}$ density = mass × volume   density = $\frac{volume}{mass}$

   (b) Which one of the following is the correct formula for mass?

   mass = $\frac{density}{volume}$ mass = density × volume   mass = $\frac{volume}{density}$

3. Find the density of these toast toppings in grams per cm³ where the volume of each jar is 500 cm³
   (a) Honey, weight 600 g.
   (b) Peanut butter, weight 750 g.
   (c) Plum jam, weight 454 g.

4. The mass of the block of oak shown is 120 g.
   (a) Calculate the density of the oak in g/cm³
   (b) Find the mass in kilograms of a cube of oak with a side length of 20 cm.

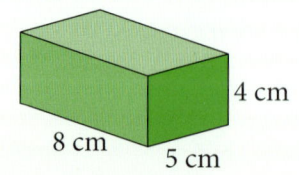

5. A small gold bar has a mass of 386 grams. Gold has a density of 19.3 g/cm³
   (a) Find the volume of the gold bar.
   (b) A fraudster makes a second bar out of a different metal. It has exactly the same size and shape, but its density is 15 g/cm³
   Find the mass of this metal bar.

6. Rubber has a density of 1.34 g/cm³
   (a) Find the mass of a small block of rubber with a volume of 2 cm³
   (b) Find the volume of a block of rubber with a mass of 2.68 kg.

7. Water has a density of 1 gram per cm³ or 1000 kg per m³. An object will float on water if its density is lower than the density of water. Work out the density of each of these objects. Then decide whether each object will float on water or sink.
   (a) An ice cube, mass 27.6 g, volume 30 cm³
   (b) An iceberg, mass 82 800 kg, volume 90 m³
   (c) A tree trunk, mass 7550 kg, volume 10 cubic metres
   (d) A rock, mass 4.5 kg, volume 2000 cm³

8. Human fat has a density of 0.9 grams per cm³
   The non-fat parts of the human body have an average density of 1.1 grams per cm³

(a) The table below shows the mass and volume of the fat and non-fat parts of Dylan's body. Copy the table and complete the four blank boxes.

	Mass (g)	Volume (cm³)
Fat – Density 0.9 g/cm³	13 248	
Non-fat – Density 1.1 g/cm³		40 320
Totals		

(b) Work out the overall density of Dylan's body, giving your answer in grams per cm³ to 2 decimal places.

9. A single pebble has a mass of 21 grams and a volume of 6 cm³. Kate fills a jar with similar pebbles and puts the jar on her scales. It weighs 2250 grams and has a volume of 1000 cm³
   (a) Work out the density of the single pebble in grams per cubic centimetre.
   (b) Find the density of the jar of pebbles in the same units.
   (c) Which object has the higher density: the single pebble or the jar of pebbles? Can you explain this?

10. (a) A small metal block A has a mass of 195 g and has volume 39 cm³
    Calculate the density of block A using the formula: density = $\frac{\text{mass}}{\text{volume}}$
    (b) A large metal block B has a mass of 36 000 kg and has volume 9 m³
        (i) Find the mass of block B in grams.
        (ii) There are 1 000 000 cm³ in 1 m³. Find the volume of block B in cubic centimetres.
        (iii) Calculate the density of block B in grams per cubic centimetre.
    (c) Which block is the most dense?

15.4 Pressure

Pressure as a compound measure

Pressure is a compound measure. It is calculated using the formula:

$$\text{pressure} = \frac{\text{force}}{\text{area}}$$

If the force is measured in newtons (N) and the area in square metres (m²), then pressure has the units N/m²

If the area is measured in square centimetres (cm²), then pressure has the units N/cm²

The formula triangle for pressure, force and area is shown on the right.

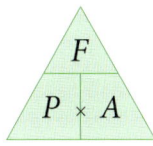

Example 8

A panda stands on the ground with its four paws. The panda's weight is 1650 N.
(a) If each paw has a surface area of 20 cm², find the pressure the panda exerts on the ground in N/cm²
(b) How many of the panda's paws are in contact with the ground if the pressure is 27.5 N/cm² ?

(a) The total area is 4 × 20 = 80 cm²
$$\text{pressure} = \frac{\text{force}}{\text{area}}$$
$$= \frac{1650}{80}$$
$$= 20.625 \text{ N/cm}^2$$

(b) Using the formula triangle:
$$\text{area} = \frac{\text{force}}{\text{pressure}}$$
$$= \frac{1650}{27.5}$$
$$= 60 \text{ cm}^2$$

Since each paw has an area of 20 cm², the panda has three paws in contact with the ground.

Be consistent with the units for area. If an area is measured in cm², a pressure in N/cm² must be used.

Example 9

The pressure in a bottle of fizzy drink is 24 000 N/m²
How much force is exerted on the bottle cap, which has an area of 3 cm²?

There are 10 000 cm² in 1 m² so:
24 000 N/m² ÷ 10 000 = 2.4 N/cm²

From the formula triangle:
$$F = PA$$
$$F = 2.4 \times 3$$
$$F = 7.2 \text{ N}$$

Exercise 15D

1. The units for pressure are newtons per square metre (N/m²). Which one of the following is the correct formula for pressure?

   pressure = force × area   pressure = $\frac{\text{area}}{\text{force}}$ pressure = $\frac{\text{force}}{\text{area}}$

2. A digger weighs 100 000 N. Its caterpillar tracks occupy an area of 5 m² on the horizontal ground. Calculate the pressure the digger exerts on this area.

3. Sophie pushes a table across her classroom. She exerts a pressure of 80 N/cm² over an area of 10 cm² Find the force she uses.

4. A man has a weight of 720 newtons. When the man stands on the ground, his weight is spread equally between his two feet. Each foot has an area of 0.025 m²
   What is the pressure the man exerts on the ground?

5. A cyclist applies both the front and rear brakes of her bike. For each brake there are two brake pads. Each pad applies a force of 90 newtons to the front wheel.
   (a) Find the total force acting against the wheels.
   (b) Given that each brake pad touches an area of 3 cm² on the wheel of the bike, find the pressure that each brake pad exerts.

6. A diagram of a symmetrical metal blade is shown on the right. The blade comprises a rectangle and three triangles, with the dimensions shown.
   (a) Calculate the total surface area of the blade.
   (b) The blade is used in a machine. The machine exerts a force of 500 N, which is distributed evenly across the surface area of the blade. Find the pressure applied, giving your answer in N/cm²
   (c) The blade breaks. It is replaced with a larger one so that the pressure is reduced. Find the area of the new blade if the force remains the same, but the pressure is reduced to 8 N/cm² ?

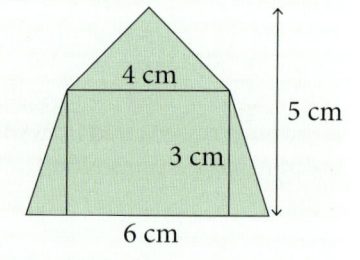

7. A swimming pool is 25 m long and 10 m wide. It is filled with water to a depth of 1.5 m.
   (a) Find the total volume of water in the pool, giving your answer in cubic metres (m³).
   (b) Find the surface area of the base of the pool.
   (c) One cubic metre of water has a weight of 10 000 N. Find the pressure that the water exerts on the base of the pool.
   (d) The pool designer says that the pressure on the base of the pool should not go above 19 000 N/m² Find the maximum depth of water that could be used in the pool.

15.5 Other Compound Measures

Heartbeats per minute

Heart rate is usually measured in beats per minute.

Example 10

You may use your calculator in this question.
Ginny's heart beats 29 times in 25 seconds. What is her average heart rate in beats per minute?

Method 1
First find the number of beats in one second = $\frac{29}{25}$
Then multiply by 60 to get the number of beats per minute:
$\frac{29}{25} \times 60 = 69.6$ beats per minute

Method 2

The unit for heart rate is beats per minute. This indicates that the formula is:

$$\text{heart rate} = \frac{\text{number of beats}}{\text{time in minutes}}$$

25 seconds is $\frac{25}{60}$ of a minute. Therefore:

$$\text{heart rate} = 29 \div \frac{25}{60} = 69.6 \text{ beats per minute}$$

Miles per gallon

The fuel economy of a car or other vehicle is often measured in miles per gallon (mpg).

Example 11

Seán's car travels 200 miles on 4.5 gallons of petrol. Calculate the fuel economy of Seán's car in miles per gallon.

The unit for fuel economy is miles per gallon. This indicates that the formula is:

$$\text{fuel economy} = \frac{\text{distance travelled}}{\text{fuel used}}$$

For Seán's car:

$$\text{Fuel economy} = \frac{200}{4.5} = 44.4 \text{ mpg (1 d.p.)}$$

Exercise 15E

1. Four members of a class do the same exercise routine. After the exercise, each person counts the number of times their heart beats over different periods of time. The results are in the table below.

	Jake	Ruari	Finn	Sue
Number of heartbeats	90	60	125	150
Time (seconds)	45	35	60	90

   Calculate each person's heart rate in beats per minute.

2. Phil does a survey into the fuel economy for some road vehicles.
   (a) Copy and complete the table to compare these vehicles.

	Smart Car	Large family car	Bus	Truck
Distance travelled (miles)	160	100		80
Amount of fuel used (gallons)	4		2	5
Miles per gallon		20	12	

   (b) Phil says that, looking at his results, the most environmentally friendly way to travel is by Smart Car, while the bus is the least environmentally friendly. Comment on his conclusions.

3. An old-fashioned record player has a turntable that can rotate at different speeds. The speed of rotation is measured in revolutions per minute (RPM).
   (a) Using the unit revolutions per minute for the speed of rotation, write down a formula for the speed of rotation.
   (b) For each of the following vinyl records, find the speed of rotation in revolutions per minute.
      (i) Madonna's *Material Girl*. The record lasts 4 minutes and spins 180 times.
      (ii) Bing Crosby's *I'm Dreaming of a White Christmas*. The record lasts for 3 minutes and spins 100 times.
      (iii) Frank Sinatra's *Close to you*. The record lasts 3 minutes 30 seconds and spins 273 times.

4. A measure of the quality of a printer is the number of dots per inch (DPI). The higher the dots per inch, the better the quality. Find the dots per inch for each of the following printers and state which model gives the best-quality prints.

	Dots	Width	DPI
Aramis 5 printer	650	1 inch	
Briskprint 500C printer	900	3 inches	
Cauldron 2000 printer	5000	2.5 inches	

15.6 Summary

In this chapter you have learnt about compound measures such as speed, density and heart rate.

You have also learnt that:

- If you know the unit for a compound measure you can write down the formula. For example, speed is measured in metres per second (m/s), so its formula is:

  $\text{speed} = \dfrac{\text{distance}}{\text{time}}$

- You can also construct a formula triangle, such as the one shown.
  To find the distance, cover up d with your finger. The formula is $d = s \times t$
  To find the time, cover up t with your finger. The formula is $t = \dfrac{d}{s}$

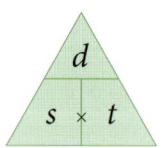

- Density is a compound measure. The density of an object is related to its mass and volume by the formula:

  $\text{density} = \dfrac{\text{mass}}{\text{volume}}$

- Pressure is also a compound measure and can be calculated using the formula:

  $\text{pressure} = \dfrac{\text{force}}{\text{area}}$

- For all compound measures, you can construct a formula triangle and use it in a similar way to the speed triangle.

Chapter 16
Perimeter, Area and Volume

16.1 Introduction

This chapter covers the surface area and volume of some three-dimensional shapes, such as prisms, cylinders, cones and spheres. Some composite shapes – which are made from two or more 3D shapes – are also considered.

The chapter also covers the length of an arc of a circle and the area of a sector.

A 3D object that has a constant cross-section and flat faces is called a **prism**. If there are curved faces, such as in a **cylinder**, the shape is not described as a prism.

Key words

- **Perimeter**: The distance around a shape.
- **Arc**: A part of the circumference of a circle.
- **Prism**: A 3D shape that has a constant cross-section. The example shown is a triangular prism.
- **Cone**: A 3D shape that has a circular base and a tapers to a single point.
- **Composite shape** or **compound shape**: Is made up of two or more other shapes. For example, a composite 3D shape may be a cone attached to a hemisphere.

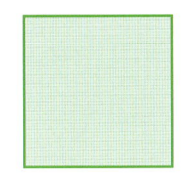

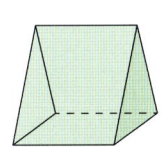

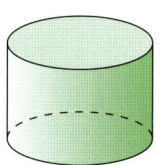

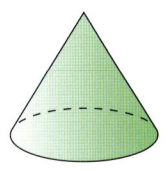

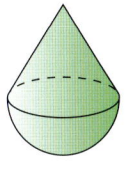

- **Circumference**: The distance around a circle.
- **Sector**: A part of the area of a circle, enclosed between two radii and an arc.
- **Cylinder**: A 3D shape that has a circle as a constant cross-section.
- **Sphere**: A 3D shape like a ball.

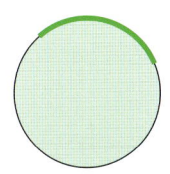

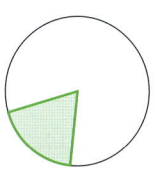

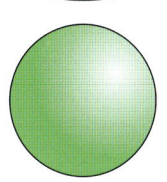

Before you start you should know:

- How to find the volume of a cube and a cuboid.
- How to find the perimeter of 2D shapes such as a square, rectangle and triangle.
- How to find the circumference of a circle.
- How to find the area of 2D shapes such as a circle, square, rectangle, triangle, parallelogram, rhombus, kite and trapezium.
- How to solve a simple equation.
- The names for different parts of a circle.
- How to use Pythagoras' Theorem.
- How to use trigonometry in a right-angled triangle.
- How to find square roots and cube roots.

GCSE MATHEMATICS M3 AND M7

In this chapter you will learn how to:
- Find the surface area and volume of a prism.
- Find the surface area and volume of a cylinder.
- Find the surface area and volume of a cone.
- Find the surface area and volume of a sphere.
- Find the surface area and volume of composite shapes.
- Find the arc length and area of a sector of a circle.

Example 1 (Revision)
The circle shown on the right has its centre at point O and a radius of 5 cm.
Calculate:
(a) the circumference of the circle,
(b) the area of the circle.

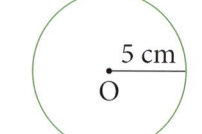

(a) Circumference = $2\pi r$
   = $2 \times 3.14 \times 5$
   = 31.4 cm

(b) Area = πr^2
   = 3.14×5^2
   = 78.5 cm²

Example 2 (Revision)
Find:
(a) the area, and
(b) the perimeter
of the compound shape ABCDEF shown on the right.

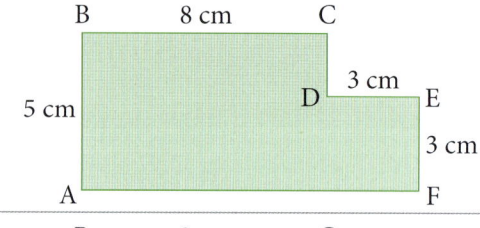

(a) To calculate the area, divide the shape into a rectangle and a square, as shown.

The area of the rectangle is $8 \times 5 = 40$ cm²

The area of the square is $3 \times 3 = 9$ cm²

The total area is $40 + 9 = 49$ cm²

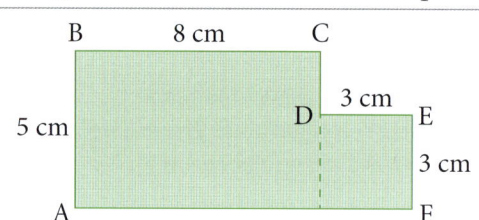

(b) To calculate the perimeter, calculate the missing side lengths CD and AF.
CD = 5 − 3 = 2 cm
AF = 8 + 3 = 11 cm
So the perimeter is $5 + 8 + 2 + 3 + 3 + 11 = 32$ cm

Example 3 (Revision)
Find the area of the trapezium shown on the right.
The measurements are all in centimetres.

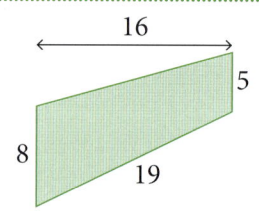

$A = \frac{1}{2} h (a + b)$

The two parallel sides are 5 cm and 8 cm, so use $a = 5$ and $b = 8$
The perpendicular distance between the parallel sides is 16 cm, so $h = 16$
The side length marked 19 cm is not needed in the calculation of the area.

$A = \frac{1}{2} \times 16 \times (5 + 8)$
   = 8×13
   = 104 cm²

Example 4 (Revision)
Find
(a) the area, and (b) the perimeter
of this semicircle.

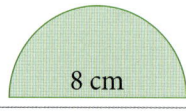

(a) The semicircle is half of a circle of radius 4 cm.
The formula for the area of a full circle is $A = \pi r^2$
So, the area of the semicircle is $\frac{1}{2} \times \pi \times 4^2 = 25.1$ cm² (1 d.p.)

(b) The circumference of a circle is $C = 2\pi r$
The perimeter of the semicircle is made up of half a circle and the diameter.
So, the perimeter is $\frac{1}{2} \times 2 \times \pi \times 4 + 8 = 20.6$ cm (1 d.p.)

Exercise 16A (Revision)

1. Find the circumference and area of a circle with a **diameter** of 8 cm.

2. Find the perimeter and area of each of these shapes. They are drawn on a 1 cm grid.

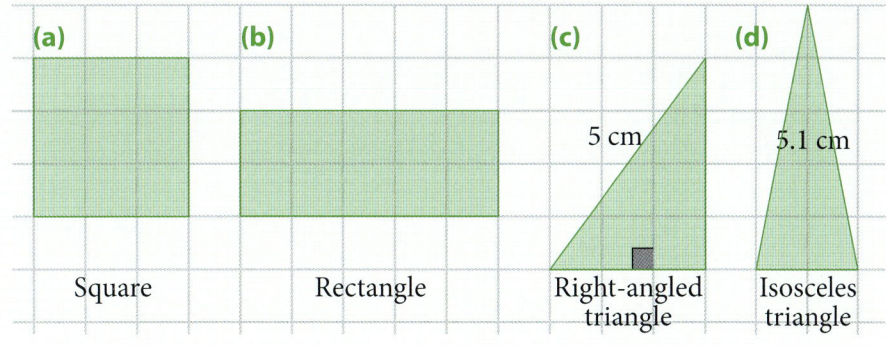

   (a) Square (b) Rectangle (c) Right-angled triangle (5 cm) (d) Isosceles triangle (5.1 cm)

3. A child's drawing of a sailing boat is shown in the diagram on the right. It has been drawn on 1 cm squared paper. Find the total area of the sail and the boat on the diagram.

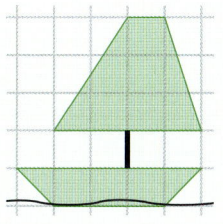

4. The card shown on the right is made up of a square ABCD with sides 6 cm long and four semicircles. Find
   (a) the area, and
   (b) the perimeter
   of the card, giving your answers to one decimal place.

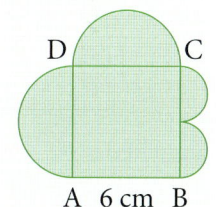

16.2 Surface Area and Volume

Surface area and volume of a prism

A **prism** is a 3D shape with a constant cross-section and no curved faces. There are many different types of prism. The two **solids** on the right are prisms.

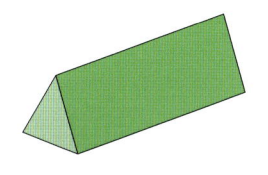

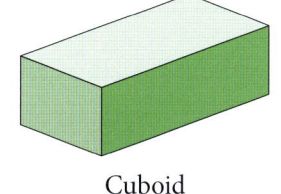

Note: **Solid** is another word for a 3D shape.

Cuboid   Triangular prism

The cross-section of the cuboid is a rectangle or a square all the way through the shape. The cross-section of the triangular prism is a triangle all the way through the shape. The cross-section of a prism can be any **polygon**. You may come across a hexagonal prism, an octagonal prism and other types of prism.

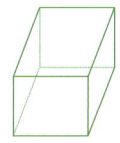

Cuboid   Pentagonal prism   Hexagonal prism

The volume of a prism is given by:

V = area of cross-section × length

or $\quad V = Al$

where: A is the area of the cross-section
l is the prism's length

Note: The cross-section can be any polygon. The method for finding its area will depend on what type of polygon it is. A polygon is a 2D shape with straight edges.

Example 5

Find
(a) the volume, and
(b) the total surface area
of the trapezoidal prism shown on the right.

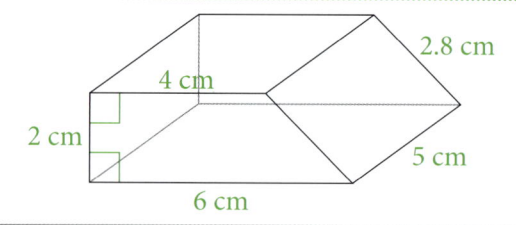

(a) The cross-section is in the shape of a trapezium. Its area can be found using the formula:
$A = \frac{1}{2}(a+b)h$
$= \frac{1}{2}(4+6) \times 2 = 10 \text{ cm}^2$

The volume of the prism can be found using the formula:
$V = Al$
$= 10 \times 5 = 50 \text{ cm}^3$

(b) The prism has 5 faces: a trapezium at each end and four rectangular faces.
The area of each trapezium has already been calculated as 10 cm².
Total surface area = $(2 \times 10) + (5 \times 4) + (5 \times 6) + (5 \times 2) + (5 \times 2.8) = 94 \text{ cm}^2$

The following example demonstrates that you do not necessarily need to know the type of prism in order to use the volume formula.

Example 6

Find the volume of a prism that has a cross-sectional area of 18 cm² and a length of 12 cm.

$V = Al = 18 \times 12 = 216 \text{ cm}^3$

If you are given the volume of a prism, you can find the cross-sectional area, as shown in the next example.

Example 7

Find the cross-sectional area of a prism with a length of 20 cm and a volume of 260 cm³

$V = Al$
$260 = A \times 20$
$A = \frac{260}{20} = 13 \text{ cm}^2$

The following example demonstrates how to find both the volume and surface area of a prism. To find the surface area, you need to add up the areas of all the faces.

Example 8

Find
(a) the volume, (b) the surface area
of the triangular prism shown on the right. All the length measurements shown are in centimetres.

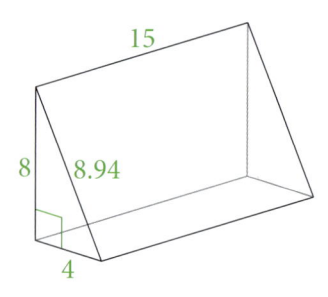

(a) The area of the triangular cross-section is given by:
$A = \frac{1}{2} \times \text{base} \times \text{perpendicular height}$
$= \frac{1}{2} \times 4 \times 8 = 16 \text{ cm}^2$

The volume of the prism is given by:
V = area of cross-section × length
$= 16 \times 15 = 240 \text{ cm}^3$

(b) The prism has 5 faces: a triangular face at each end and three rectangular faces.
Total surface area = $2(\frac{1}{2} \times 4 \times 8) + (15 \times 4) + (15 \times 8) + (15 \times 8.94) = 346.1 \text{ cm}^2$

CHAPTER 16: PERIMETER, AREA AND VOLUME

Exercise 16B

1. Find the volume of these triangular prisms. All lengths in the diagrams are measured in centimetres.
   (a) [prism with dimensions 20, 2, 6]
   (b) [prism with dimensions 8, 5, 10]

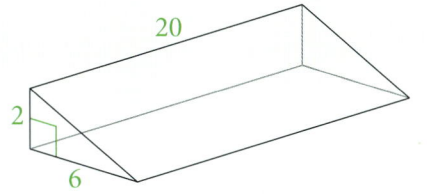

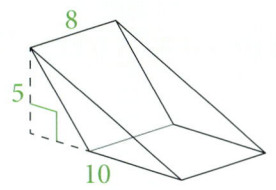

2. For each of parts (a) to (c), find (i) the volume and (ii) the surface area of these triangular prisms.
   (a) [prism with dimensions 4, 8.5, 10.4, 6]
   (b) [prism with dimensions 5, 7, 9.22, 6]
   (c) [prism with dimensions 10, 7.65, 5, 6.54, 10]

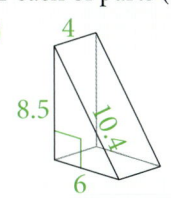

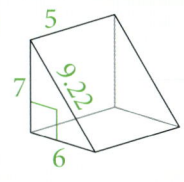

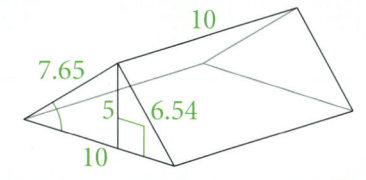

3. A packet of sweets is in the shape of a hexagonal prism. The hexagonal cross-section has an area of 3.5 cm² and the length of the packet is 12 cm. Find the volume of the packet.

4. Find
   (a) the volume, and
   (b) the total surface area of the prism shown on the right.

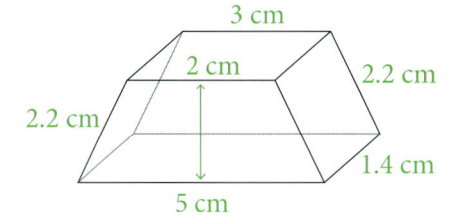

   [trapezoidal prism with dimensions: 3 cm, 2 cm, 2.2 cm, 2.2 cm, 1.4 cm, 5 cm]

5. Find the volume of a prism of cross-sectional area 23 cm² and length 16 cm.

6. Find the length of a prism with a volume of 225 cm³ and cross-sectional area of 10 cm².

7. A chocolate bar comes in a packet in the shape of a triangular prism. Find the area of the triangular cross-section if the volume of the packet is 100 cm³ and its length is 25 cm.

8. The diagram on the right shows a tank at a chemical factory. The height of the tank is 2 m and base of the tank has an area of 30 m².
   (a) What is the name of this 3D shape?
   (b) Find the volume of the tank in cubic metres.
   (c) The volume of the liquid in the tank is 51 m³. Find the height of the top of the liquid above the base.

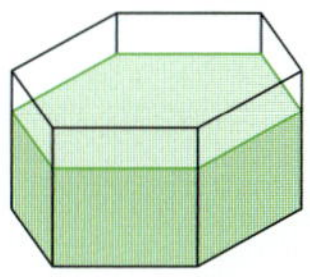

9. The first diagram below shows the cross-section of a wheelbarrow. The cross-section is in the shape of a trapezium with dimensions shown. The second diagram shows the three-dimensional view of the wheelbarrow. Its width is 32 cm.
   (a) What is the area of the wheelbarrow's cross-section?
   (b) The wheelbarrow is used to carry cement. What volume of cement can it carry?
   (c) How many times must the wheelbarrow be filled to move 2 cubic metres of cement? You may use the fact that 1 m³ = 1 000 000 cm³

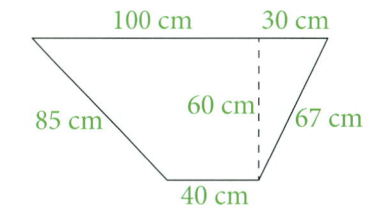

   [trapezium: 100 cm, 30 cm, 85 cm, 60 cm, 67 cm, 40 cm]
   [3D wheelbarrow: 32 cm, 130 cm, 40 cm]

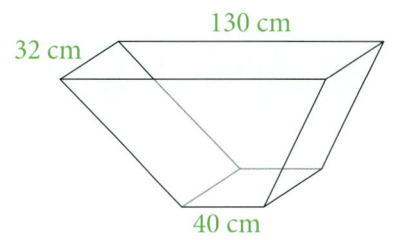

10. The polygon ABCDEF is shown in the first diagram on the right. It has been drawn on a centimetre grid.
    (a) Calculate the area of the polygon.
    (b) The prism shown in the second diagram above has polygon ABCDEF as its cross-section and a length of 5 cm. Find the volume of the prism.

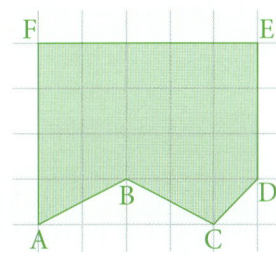

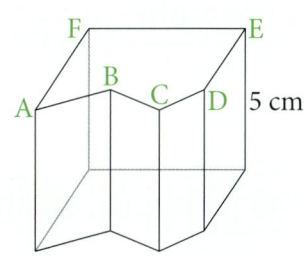

137

GCSE MATHEMATICS M3 AND M7

11. The diagram on the right shows the front of a shed. The shed is in the shape of a prism. The front and back both have the trapezium shape shown in the diagram. The length of the shed from front to back is 3.5 m. Find its volume.

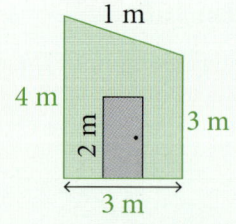

12. A prism is shown in the first diagram on the right. The cross-section is a regular hexagon. The hexagon can be divided into a rectangle and two identical isosceles triangles as shown in the second diagram above. The length of the prism is 20 cm.
    (a) Find the volume of this prism.
    (b) Find the total surface area.

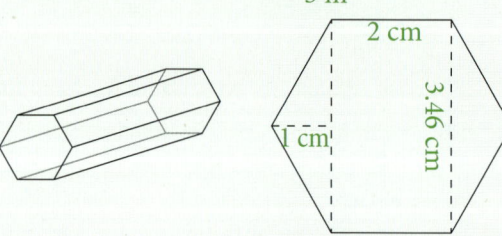

Surface area and volume of a cylinder

Two cylinders are shown on the right. A cylinder has a circle as its constant cross-section, but a cylinder is not classed as a prism because it has a curved face.

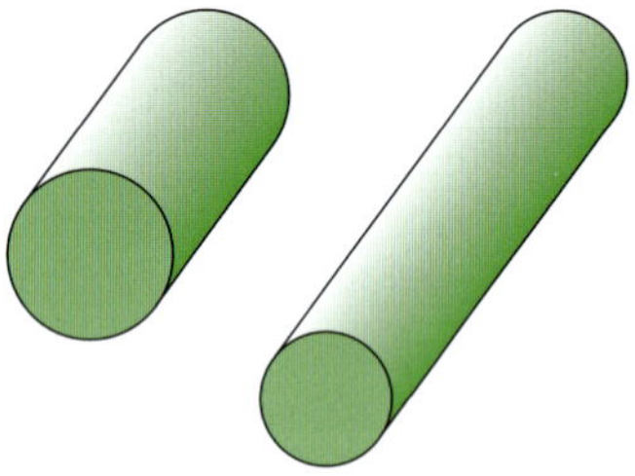

To find the volume of a cylinder, we can use the formula for the volume of a prism, because it has a constant cross-section:

V = area of cross-section × length

For a cylinder, the cross-section is a circle. The area of a circle is given by $A = \pi r^2$

So using h for the length or height of the cylinder, we have:

$V = \pi r^2 h$

In this section we also consider other 3D objects with a constant cross-section and a curved face.

Example 9

Find the volume of the piece of drainpipe shown in the diagram on the right. It has a diameter of 10 cm and a length of 40 cm.

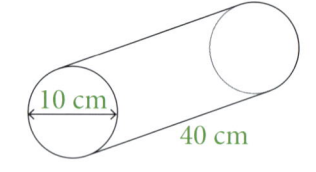

The diameter is 10 cm, so the radius is 5 cm.

Note: Read the question carefully. If the diameter is given, you must divide it by 2 to find the radius.

$V = \pi r^2 h$
$V = \pi \times 5^2 \times 40 = 3142$ cm³ (to the nearest whole number)

For any other solids with curved faces and a constant cross-section, the area of the cross-section will be given to you.

Example 10

Find the volume of the solid shown on the right. It has a constant cross-section of 130 cm² and a height of 7 cm.

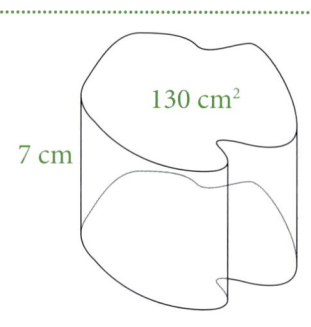

This is a solid with curved faces. However, since it has a constant cross-section, the volume can be calculated as with a prism:

V = area of cross-section × length
$V = 130 \times 7 = 910$ cm³

Exercise 16C

1. What is the volume of a cylinder with a radius of 2.5 feet and a height of 6 feet? Round your answer to a sensible level of accuracy.

CHAPTER 16: PERIMETER, AREA AND VOLUME

2. What is the volume of the cylinder shown on the right? It has a radius of 2 cm and a height of 6 cm. Round your answer to a sensible level of accuracy.

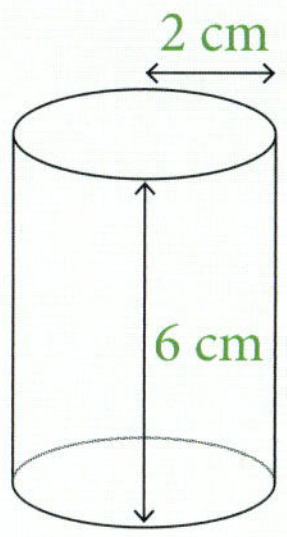

3. What is the volume of a cylinder with a radius of 3 inches and a height of 8 inches? Give your answer to the nearest cubic inch.

4. What is the volume of a cylinder with a diameter of 11 cm and a height of 15 cm? Give your answer to the nearest cubic centimetre.

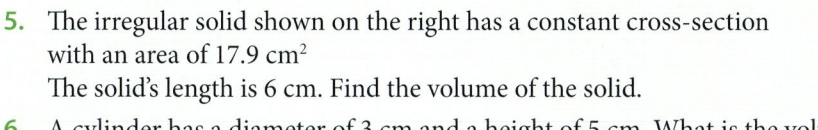

5. The irregular solid shown on the right has a constant cross-section with an area of 17.9 cm²
   The solid's length is 6 cm. Find the volume of the solid.

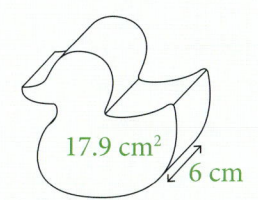

6. A cylinder has a diameter of 3 cm and a height of 5 cm. What is the volume of the cylinder? Give your answer to one decimal place.

7. A cylinder is filled with water. The cylinder has a radius of 4 cm and a height of 6 cm. How much water is in the cylinder? Give your answer to the nearest cubic centimetre.

8. The rainwater gutter shown has a constant cross-section in the shape of a semicircle. It has a length of one metre and a width of 20 cm. Find the volume of water that the gutter can hold, giving your answer in cubic centimetres.

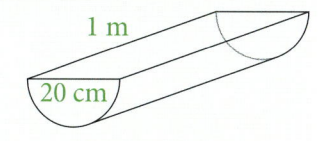

9. The measuring cylinder shown on the right has a diameter of 5 cm. The height of the water is 2 cm.
   (a) Find the volume of water in the cylinder in millilitres (ml).
   You may use the fact that 1 cm³ of water is equivalent to 1 ml.
   (b) The measuring cylinder has a maximum capacity of 225.8 ml. What is the height of the cylinder? Give your answer to one decimal place.

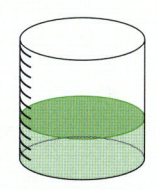

Surface area and volume of a cone

The diagram shows a cone. An ice cream cone is a real-world example of a cone. It has a circle at one end and tapers to a point.

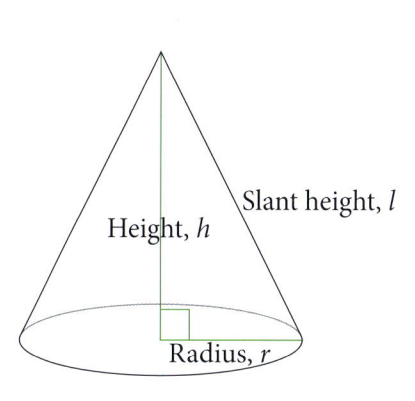

The **curved surface area** of a cone is given by the formula:
$$\text{curved surface area} = \pi r l$$
where: l is the slant height, shown in the diagram.

The **total surface area** includes the circle at the base. This is given by:
$$\text{total surface area} = \pi r l + \pi r^2$$

The volume, V, is given by:
$$V = \tfrac{1}{3}\pi r^2 h$$

Example 11

For the cone shown in the diagram on the right, calculate:
(a) the volume   (b) the curved surface area,
(c) the total surface area.

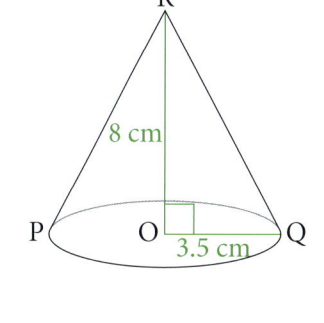

(a) The volume is given by
$$V = \tfrac{1}{3}\pi r^2 h = \tfrac{1}{3} \times \pi \times 3.5^2 \times 8 = 102.6 \text{ cm}^3$$

(b) To find the curved surface area we need the slant height l. Draw triangle OQR (shown on the right) and use Pythagoras' Theorem to find the slant height l, which is the hypotenuse of the triangle: 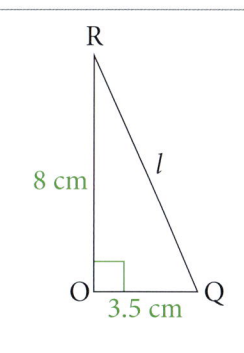
$$l^2 = 3.5^2 + 8^2 = 76.25$$
$$l = \sqrt{76.25} = 8.73 \text{ cm}$$

So the curved surface area = $\pi r l$
$$= \pi \times 3.5 \times 8.73 = 96.0 \text{ cm}^2$$

(c) Total surface area = $\pi r l + \pi r^2$
$$= 96.0 + \pi \times 3.5^2 = 134.5 \text{ cm}^2$$

The following example requires trigonometry to calculate the radius and height of the cone.

Example 12

For the cone shown in the diagram on the right, calculate:
(a) the curved surface area,
(b) the total surface area,
(c) the volume.

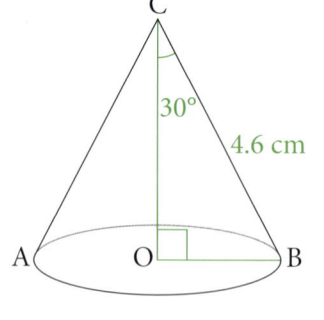

(a) Curved surface area = $\pi r l$

The slant height l is given as 4.6 cm, but the radius r must be calculated. Draw the right-angled triangle COB (shown below right) and use trigonometry to find r.

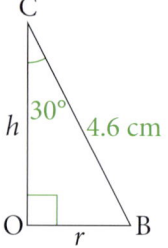

$$\sin 30 = \frac{r}{4.6}$$

$r = 4.6 \sin 30 = 2.3$ cm

So curved surface area = $\pi \times 2.3 \times 4.6$
$$= 33.2 \text{ cm}^2 \text{ (1 d.p.)}$$

(b) Total surface area = $\pi r l + \pi r^2$
$$= 33.2 + \pi \times 2.3^2$$
$$= 49.9 \text{ cm}^2$$

(c) For the volume calculation h is needed. Again, use trigonometry with triangle COB:

$$\cos 30 = \frac{h}{4.6}$$

$h = 4.6 \cos 30$
$= 3.98$ cm

Volume, $V = \frac{1}{3}\pi r^2 h$
$$= \frac{1}{3} \times \pi \times 2.3^2 \times 3.98$$
$$= 22.1 \text{ cm}^3$$

In the following two examples the volume or the curved surface area of the cone is given. From either of these you can calculate the height or the radius.

Example 13

A traffic cone has a volume of 50 000 cm³ and a base radius of 25 cm.
Find the height of the traffic cone.

$$V = \frac{1}{3}\pi r^2 h$$
$$50\,000 = \frac{1}{3} \times \pi \times 25^2 \times h$$
$$50\,000 = 654.498 \times h$$
$$h = \frac{50\,000}{654.498} = 76.4 \text{ cm (1 d.p.)}$$

Example 14

The cone shown has a curved surface area of 900 cm² and a slant height of 17.9 cm.
(a) Find the base radius of the cone.
(b) Find the cone's volume.

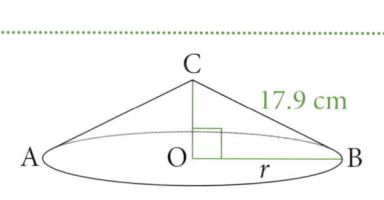

(a) Curved surface area = $\pi r l$
$900 = \pi \times r \times 17.9$
$900 = 56.235 \times r$
$$r = \frac{900}{56.235} = 16.0 \text{ cm (1 d.p.)}$$

(b) To calculate the volume, we need the height, which can be found using Pythagoras' Theorem.
$h^2 = 17.9^2 - 16^2$
$\quad = 64.41$
$h = \sqrt{64.41} = 8.026$
$V = \frac{1}{3}\pi r^2 h = \frac{1}{3} \times \pi \times 16^2 \times 8.026 = 2151.6 \text{ cm}^3$ (1 d.p.)

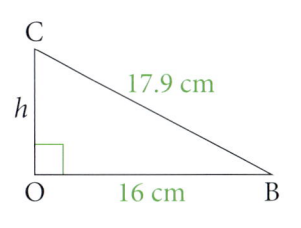

Exercise 16D

1. For each cone (a) to (f), calculate:
   (i) the curved surface area **(ii)** the total surface area **(iii)** the volume.

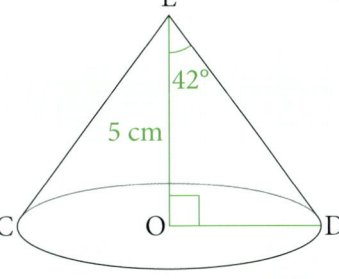

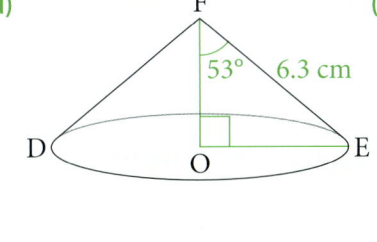

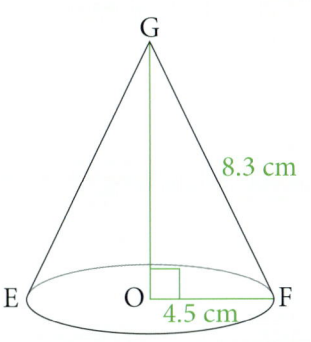

   (a) [Cone with apex D, base centre O, B on base, angle 27°, radius 3 cm]
   (b) [Cone with apex E, base centre O, C and D on base, angle 42°, height 5 cm]
   (c) [Cone with apex E, base centre O, C and D on base, height 3.9 cm, radius 2.6 cm]
   (d) [Cone with apex F, base centre O, D and E on base, angle 53°, slant 6.3 cm]
   (e) [Cone with apex G, base centre O, E and F on base, slant 8.3 cm, radius 4.5 cm]
   (f) [Cone with apex H, base centre O, F and G on base, height 6 cm, slant 6.2 cm]

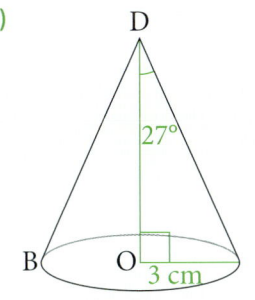

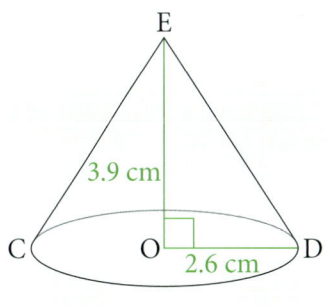

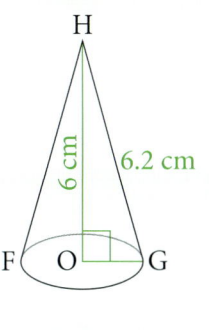

2. A cone has a volume of 8000 cm³ and a base radius of 25 cm. Find the height of the cone.

3. An ice cream cone has a volume of 100 cm³ and a perpendicular height of 10 cm. Find the radius of its circular end.

4. The cone shown on the right has a curved surface area of 405.3 cm² and a slant height of 12.9 cm.
   (a) Find the base radius of the cone.
   (b) Find the cone's volume.

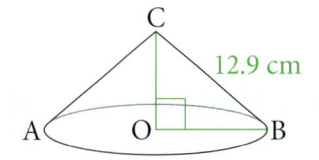

Surface area and volume of a sphere

The diagram shows a sphere. An everyday example of a sphere is a football. The radius r is measured from the centre to the surface of the sphere.

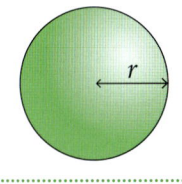

The formula for the volume of a sphere is: $\quad V = \frac{4}{3}\pi r^3$

The formula for the curved surface area (CSA) is: $\quad \text{CSA} = 4\pi r^2$

Example 15

Find **(a)** the volume and **(b)** the surface area of a sphere that has a **diameter** of 20 cm.

The diameter of the sphere is 20 cm, so the radius is 10 cm.

(a) $V = \frac{4}{3}\pi r^3$
$\quad = \frac{4}{3} \times \pi \times 10^3$
$\quad = 4188.8 \text{ cm}^3$ (1 d.p.)

(b) $\text{CSA} = 4\pi r^2$
$\quad = 4 \times \pi \times 10^2$
$\quad = 1256.6 \text{ cm}^2$ (1 d.p.)

GCSE MATHEMATICS M3 AND M7

Example 16

The diagram shows a hemispherical bowl with a closed circular lid. The **total** surface area is 150 cm³. Find the volume of the bowl.

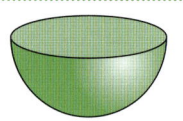

Firstly, we can use the total surface area to find the radius. The curved surface area of a sphere is given by $4\pi r^2$, so the curved surface area of a hemisphere is given by $2\pi r^2$

In addition, this solid has a circular lid with area πr^2, so, the total surface area is $2\pi r^2 + \pi r^2$ or $3\pi r^2$

$3\pi r^2 = 150$
$9.245 r^2 = 150$
$r^2 = \dfrac{150}{9.245}$
$= 15.915$
$r = \sqrt{15.915} = 3.99$ cm

Now we can use the volume formula:

$V = \dfrac{4}{3}\pi r^3$

$= \dfrac{4}{3} \times \pi \times 3.99^3$

$= 266.1$ cm³ (1 d.p.)

Exercise 16E

1. Find the volume of a sphere that has a radius of 8 cm, giving your answer to the nearest cubic centimetre.
2. Find the volume of a football that has a diameter of 22 cm. Give your answer to the nearest cubic centimetre.
3. The paperweight shown in the diagram on the right is the shape of a solid hemisphere. It has a base radius of 5.6 cm. 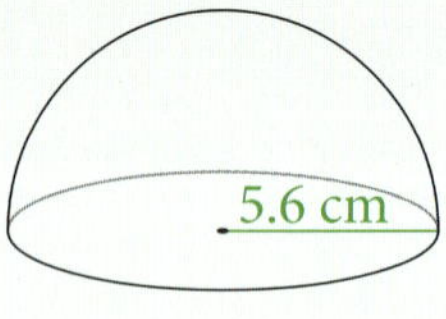
   (a) Find the volume of the paperweight. Give your answer in cubic centimetres to 1 decimal place.
   (b) Find the total surface area of the paperweight. Give your answer in square centimetres to 1 decimal place.
4. A sphere has a radius of 5 cm. It is cut into 8 equal pieces and one of those pieces is shown in the diagram on the right. 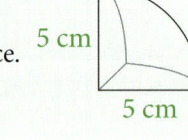
   (a) Find the volume of this solid. Give your answer in cubic centimetres to 1 decimal place.
   (b) Find the total surface area of the solid. Give your answer in square centimetres to 1 decimal place.
5. Find the radius of a hemisphere that has a volume of 60 cm³
   Give your answer in centimetres to 2 decimal places.
6. The solid hemisphere shown in the diagram on the right has a diameter of 20 cm.
   Three friends calculate the solid's total surface area to the nearest square centimetre. 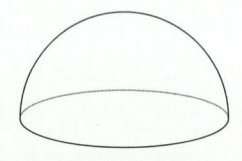
   • Ali calculates its total surface area as 628 cm²
   • Bella calculates its total surface area as 942 cm²
   • Charlie calculates its total surface area as 1257 cm²
   Who has calculated the total surface area correctly? Show all your working.

16.3 Arcs and Sectors

You will be asked to find the length of an arc, the area of a sector of a circle and perimeters of shapes that include parts of circles.

If r is the radius of the circle and x is the angle at the centre, the length of an arc can be calculated using the formula:

Arc length $= \dfrac{x}{360} \times 2\pi r$

The area of a sector can be calculated using the formula:

Sector area $= \dfrac{x}{360} \times \pi r^2$

CHAPTER 16: PERIMETER, AREA AND VOLUME

Example 17

The diagram on the right shows an arc AB of a circle of radius 5 cm. The arc subtends an angle of 74° at the centre of the circle.
(a) Calculate the length of the arc.
(b) Calculate the area of the sector AOB.

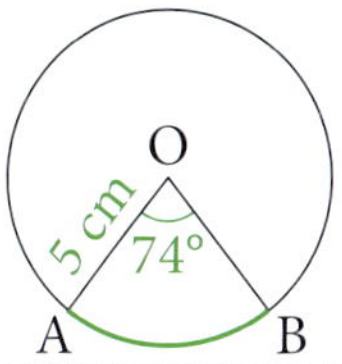

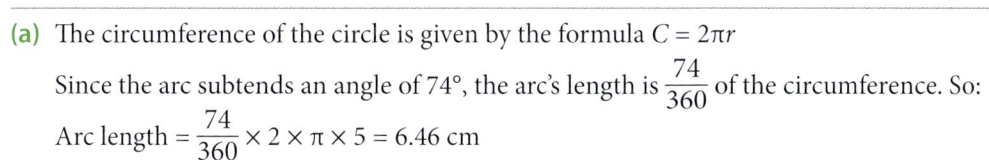

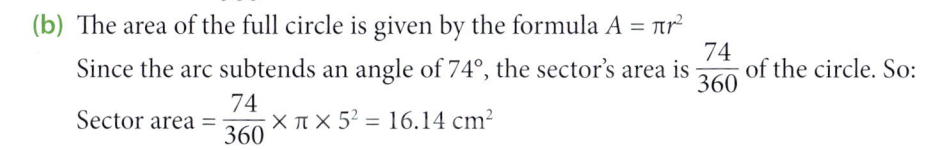

(a) The circumference of the circle is given by the formula $C = 2\pi r$

Since the arc subtends an angle of 74°, the arc's length is $\frac{74}{360}$ of the circumference. So:

Arc length $= \frac{74}{360} \times 2 \times \pi \times 5 = 6.46$ cm

(b) The area of the full circle is given by the formula $A = \pi r^2$

Since the arc subtends an angle of 74°, the sector's area is $\frac{74}{360}$ of the circle. So:

Sector area $= \frac{74}{360} \times \pi \times 5^2 = 16.14$ cm^2

Given the length of an arc you can work out the radius of the circle, or the angle that the arc subtends.

Example 18

The circle shown in the diagram on the right has a radius of 4 cm. The arc AB shown has a length of 2.79 cm and subtends an angle of $x°$ at the centre of the circle. Find x.

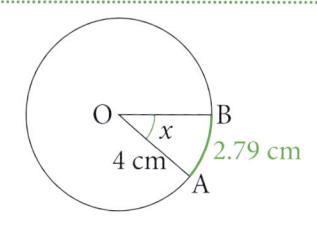

Since we are given the arc length, use the arc length formula.

Arc length $= \frac{x}{360} \times 2\pi r$

$2.79 = \frac{x}{360} \times 2 \times \pi \times 4$

On the right-hand side, calculate $2 \times \pi \times 4$ and divide by 360 on the calculator:

$2.79 = 0.0698x$

$x = \frac{2.79}{0.0698}$

$= 40.0°$ (1 d.p.)

Given the area of a sector, you can work out the radius of the circle, or the angle at the centre.

Example 19

Look at the circle in the diagram on the right. The sector shown subtends an angle of 120° at the centre of the circle and has an area of 9.42 cm^2
(a) Find r, the radius of the circle.
(b) Find the **perimeter** of the sector.

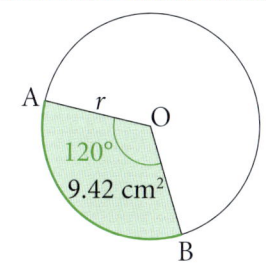

(a) Since we are given the sector area, use the sector area formula.

Sector area $= \frac{x}{360} \times \pi r^2$

$9.42 = \frac{120}{360} \times \pi \times r^2$

Calculate $\frac{120}{360} \times \pi$ on the calculator:

$9.42 = 1.047r^2$

$r^2 = \frac{9.42}{1.047}$ 9.0 (1 d.p.)

$r = \sqrt{9.0} = 3.0$ cm (1 d.p.)

(b) Now use the radius to calculate the arc length.

Arc length $= \frac{x}{360} \times 2\pi r$

$= \frac{120}{360} \times 2 \times \pi \times 3.0$

$= 6.3$ cm (1 d.p.)

The perimeter is the distance all the way around the sector. We must add the arc length and two radii:

Perimeter $= 6.3 + 3.0 + 3.0 = 12.3$ cm

Exercise 16F

1. In each part (a) to (d) below, find **(i)** the length of the arc AB and **(ii)** the area of the sector AOB, giving your answers to 1 decimal place.

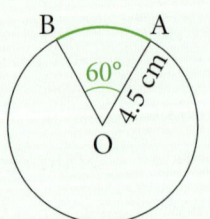

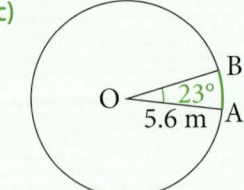

   (a) Circle with radius 5 cm, angle 140°

   (b) Circle with radius 4.5 cm, angle 60°

   (c) Circle with radius 5.6 m, angle 23°

   (d) Circle with radius 3.2 m, angle 82.6°

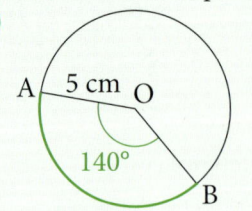

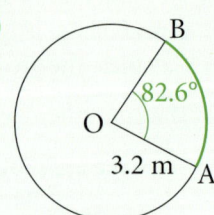

2. Each diagram shows a sector of a circle. In each part, find **(i)** the area and **(ii)** the **perimeter** of each sector.

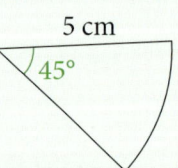

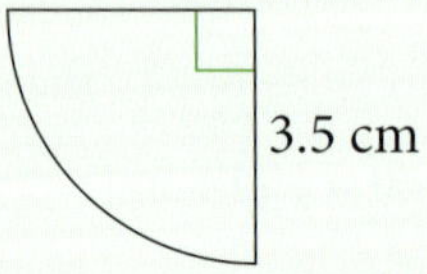

   (a) Semicircle, diameter 4 m

   (b) Sector, radius 5 cm, angle 45°

   (c) Quarter sector, radius 3.5 cm

   (d) Sector, radius 2.5 cm, angle 60° (major sector shown)

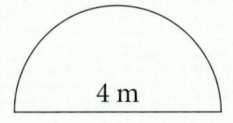

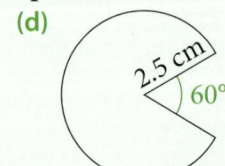

3. The circle shown in the diagram on the right has a radius of 8 m. The arc AB shown has a length of 6.98 m and subtends an angle of $x°$ at the centre of the circle. Find x, giving your answer to 1 decimal place.

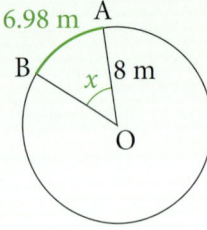

4. The arc of the circle shown in the diagram on the right has a length of 8.2 cm and it subtends an angle of 70° at the centre of the circle.
   (a) Find the radius of the circle.
   (b) Find the area of the sector AOB.

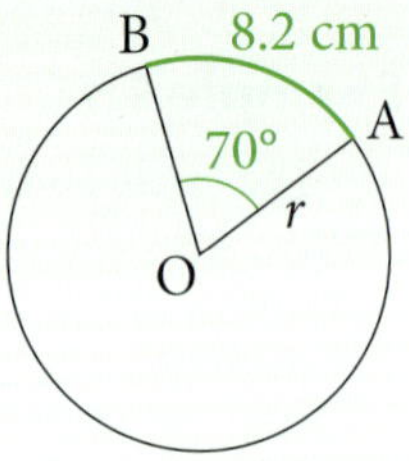

5. Look at the circle in the diagram on the right. The sector shown subtends an angle of 80° at the centre of the circle and has an area of 6.28 cm².
   (a) Find r, the radius of the circle.
   (b) Find the arc length AB.

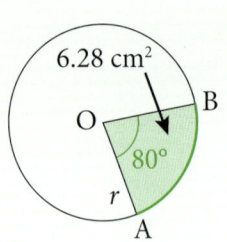

6. The circle shown in the diagram on the right has a radius of 11 m. The sector shown subtends an angle of $x°$ at the centre of the circle and has an area of 26.4 m².
   (a) Find x, giving your answer in degrees to 1 decimal place.
   (b) Find the length of arc AB, giving your answer in metres to 1 decimal place.

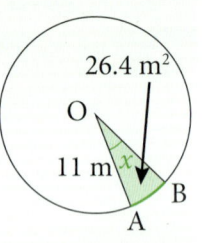

16.4 Mixed Questions

This section includes questions that require understanding from more than one of the sections above.

Example 20

A cylinder and sphere have the same volume and the same radius. The cylinder has a height of 12 cm.
(a) Find the radius of both objects.   (b) Find the volume of both objects.

(a) Use $h = 12$ for the height of the cylinder. Use V and r for the volume and radius of each object.

For the cylinder: $V = \pi r^2 h$

For the sphere: $V = \frac{4}{3}\pi r^3$

Since these two volumes are equal: $\pi r^2 h = \frac{4}{3}\pi r^3$

Divide both sides by π and by r^2: $h = \frac{4}{3}r$

Since $h = 12$: $\frac{4}{3}r = 12$

$r = 12 \div \frac{4}{3} = 9$ cm

(b) For the volume of the cylinder: $V = \pi r^2 h$

$V = \pi \times 9^2 \times 12$

$= 3053.6$ cm³ (1 d.p.)

Note: For part (b) of the question we could instead find the volume of the sphere. This should give the same answer, since the two volumes are equal.

Exercise 16G

1. The cylinder and cone shown on the right have the same radius r cm and the same volume. The height of the cone is 12 cm. Find the height h of the cylinder.

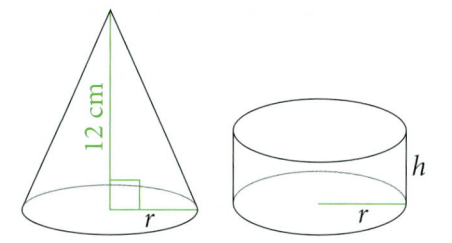

2. The diagram shows a weight used for fitness training. It is made up of two spheres connected by a cylinder. The cylinder has a radius of 10 cm and a length of 30 cm. The spheres have a radius of 20 cm. Find the volume of the compound shape, giving your answer to the nearest cubic centimetre. You can assume there is no overlap between the cylinder and the spheres.

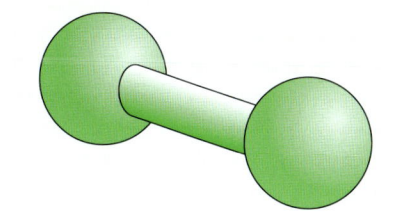

3. The cone and hemisphere shown on the right have the same radius r cm and the same total surface area. If the height of the cone is 8 cm, find the value of r, giving your answer in centimetres to 2 decimal places.

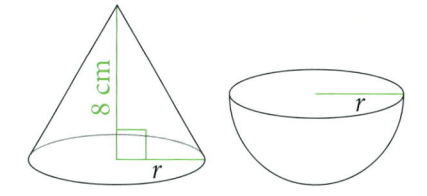

4. The diagram shows a cone, a cylinder and a sphere. Each solid has a radius of r cm and a height of h cm.
   (a) Find the volume of each solid in terms of π and r only.
   (b) Put the three solids in order from smallest to largest volume.

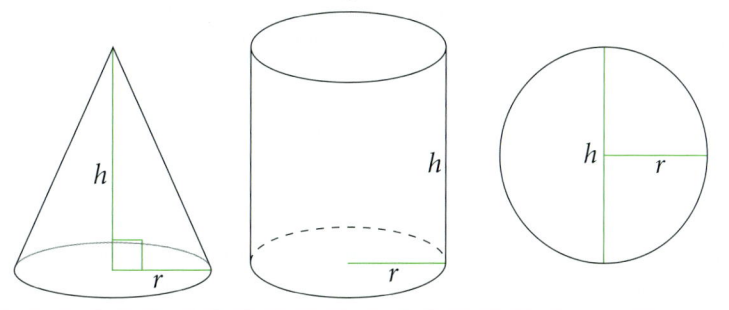

16.5 Summary

In this chapter you have learnt that:
- A prism is a three-dimensional shape with flat faces and a constant cross-section.
- The cross-section can be any type of polygon, for example a triangle, pentagon or hexagon.
- The volume of a prism can be found using the formula:
   V = area of cross-section × length
- The same formula can be used for three-dimensional shapes that have a constant cross-section and curved faces. For example, a cylinder is not a prism because it has curved faces.
- Since a cylinder has a constant cross-section, its volume is given by:
   V = area of cross section × length = $\pi r^2 h$

   where area of cross section = $A = \pi r^2$ and h is the length or height of the cylinder.

Progress Review
Chapters 11–16

This Progress Review covers:
- Chapter 11 – Straight Lines
- Chapter 12 – Angles
- Chapter 13 – Circles
- Chapter 14 – Pythagoras' Theorem and Trigonometry
- Chapter 15 – Compound Measure
- Chapter 16 – Perimeter, Area and Volume

1. (a) Copy the table on the right. Use the equation $y = -3x + 4$ to complete the table.
   (b) Hence draw the graph of the straight line $y = -3x + 4$

y	−1	0	1	2	3
x		4		−2	

2. For this question, use a graph with an x-axis from −2 to 5 and a y-axis from −4 to 3
   (a) Draw the line $x = 2$
   (b) On the same graph, draw the line $y = -1$
   (c) Write down the coordinates of the point of intersection of these lines.

3. The line shown on the graph on the right has the equation $y = \frac{1}{2}x + 1$
   Copy the graph.
   (a) Add construction lines to the graph to find the value of y when $x = 6$
   (b) Add construction lines to the graph to find the value of x when $y = 3$
   (c) Where on this straight-line graph does $x = y$? Write down the coordinates of the point.

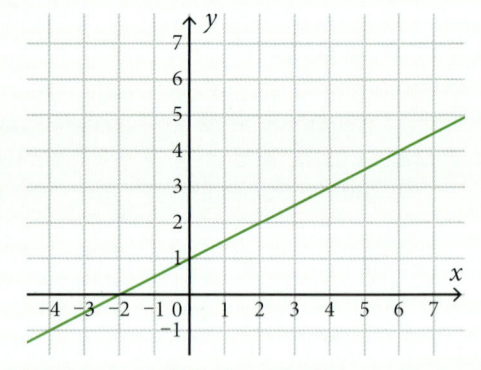

4. Determine whether the point (6, −4) lies on the straight line $y = -5x + 16$

5. A line segment connects the points A(9, 2) and B(3, −6). Find:
   (a) the gradient of the line segment,
   (b) the midpoint of the line segment,
   (c) the length of the line segment.

6. A line segment joins the points A and B. Point A has the coordinates (2, −3) and the midpoint has the coordinates (5, 15). Find the coordinates of point B.

7. The equation of a straight line is given by $y = 2x - 2$
   (a) Write down the gradient and y-intercept of the line.
   (b) Find the x-intercept of the line.
   (c) Plot the line, using values of x from −1 to 3

8. The straight line L has a gradient of 3 and passes through the point (−1, −3).
   (a) Find the y-intercept of L.
   (b) Hence write down the equation of L.
   (c) Find the x-intercept of L.

9. Find the size of the angles x and y in the diagram below.

10. Find the size of the angle x in the diagram below.

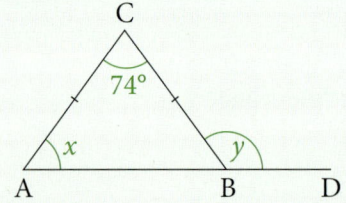

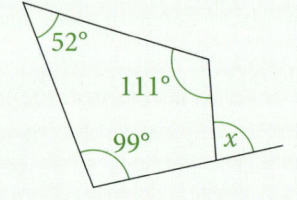

PROGRESS REVIEW: CHAPTERS 11–16

11. Find the sizes of angles a, b, c and d in the diagram below.

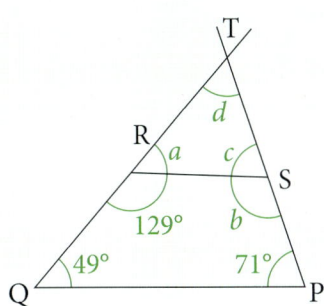

12. (a) Look at the diagram below. By forming an equation, find the value of b.
(b) Hence find the sizes of the two angles marked.

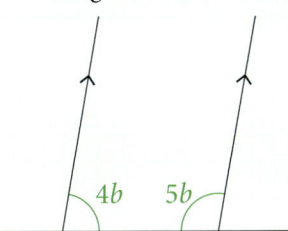

13. Find y in the diagram below. Explain each step of your reasoning.

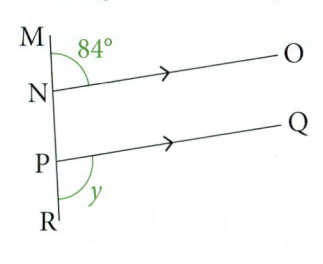

14. Given that ADB is a straight line in the diagram below, find the size of the angle marked b.

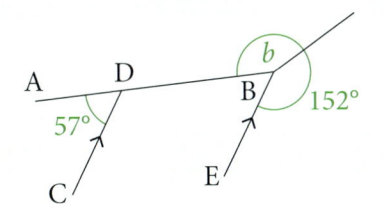

15. In the diagram below, AGB and CGD are straight lines. The lines AF and GE are parallel. Find the size of the angle marked a.

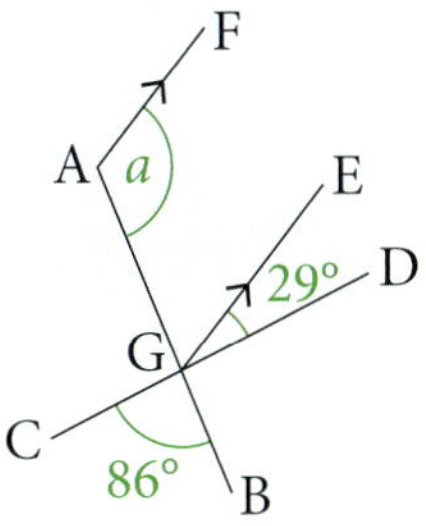

16. Find the size of the angle marked d in the diagram below. Hint: You should copy the diagram and add any extra lines needed.

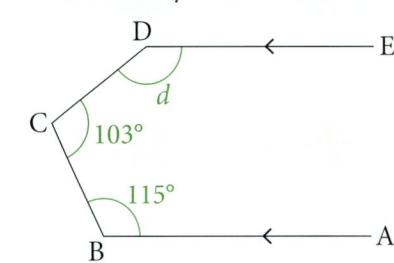

17. Copy this diagram. Fill in the boxes with the correct words from this list:
- Circumference
- Radius
- Diameter
- Chord
- Sector
- Arc

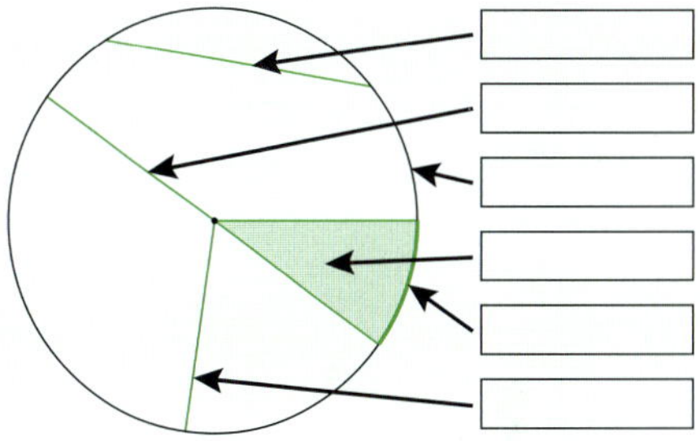

18. Sinéad draws a circle and marks a radius and a diameter of the circle on her diagram. The two lines meet at one point. Where is this point?

19. The diagram shows two overlapping circles.
The centre of Circle 1 is point A.
The centre of Circle 2 is point B.

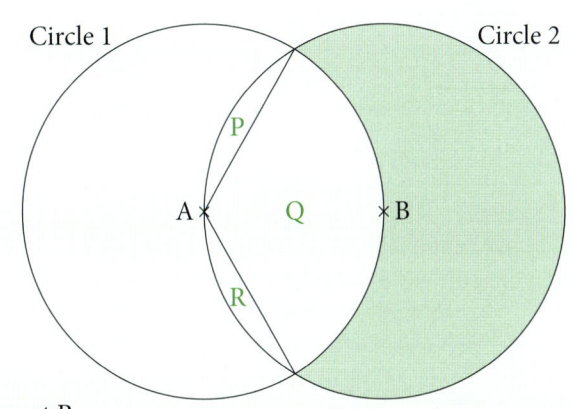

(a) Fill in the blanks in these sentences:

Area P is a _____ of Circle ___.

Area Q is a _____ of Circle ___.

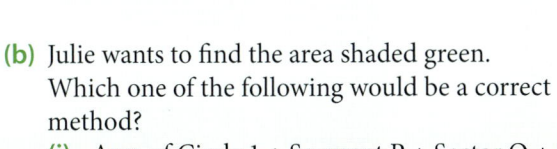

(b) Julie wants to find the area shaded green. Which one of the following would be a correct method?
(i) Area of Circle 1 + Segment P + Sector Q + Segment R
(ii) Area of Circle 1 + Segment P − Sector Q + Segment R
(iii) Area of Circle 2 + Segment P + Sector Q + Segment R
(iv) Area of Circle 2 − Segment P − Sector Q − Segment R
(v) Area of Circle 2 + Segment P − Sector Q + Segment R

147

20. The diagram shows a major sector OPQR of a circle. The sector has an angle of 300° at its centre. What type of triangle is OPR? Explain your answer.

21. Look at the diagram on the right. It shows four circles and a line segment AB. Point A is the centre of Circle 1.

    Fill in the blanks in these sentences. Use each of the numbers 1 to 4 once.

    (a) Line AB is a chord of Circle ___
    (b) Line AB is a diameter of Circle ___
    (c) Line AB is a radius of Circle ___
    (d) Line AB is a tangent to Circle ___

22. Find the length of the hypotenuse, marked x, in the triangle on the right.

23. Look at the right-angled triangle on the right.
    (a) Find the length of the side marked x.
    (b) Find the size of both acute angles.

24. Find the side lengths marked x and y in the triangles shown on the right. Give your answers to 1 decimal place.

25. Jake and Kieran are running together at the same speed, with Jake 5 metres ahead of Kieran. They approach a 90° corner and Jake turns left first.
    (a) When Jake has run 1 metre past the corner, Kieran still has 4 metres to run before he reaches the corner, as shown in the diagram. Find the distance x between the two boys at this time.
    (b) Find the distance between the two boys when Jake has run 2 metres past the corner.
    (c) Find the distance between the two boys when they are both 2.5 metres from the corner.

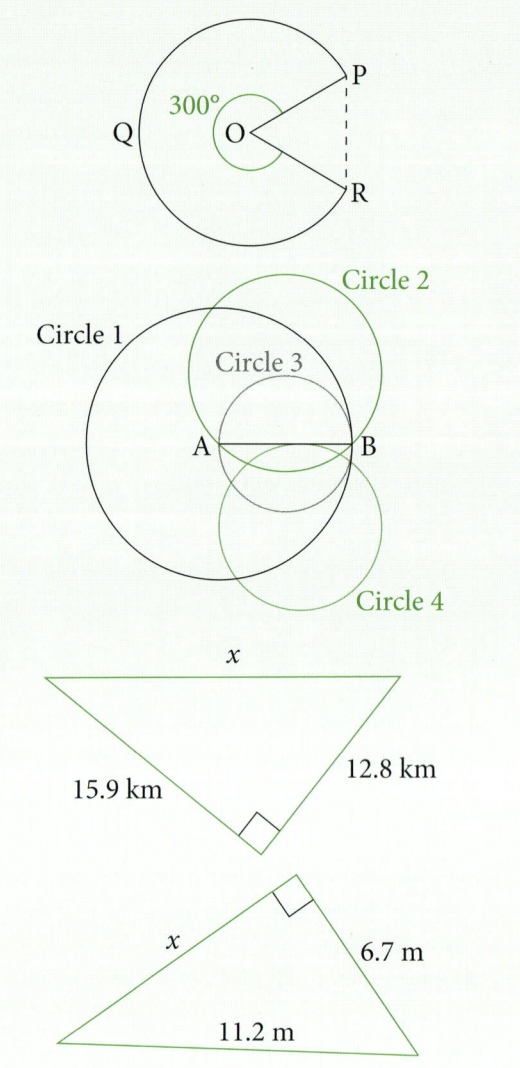

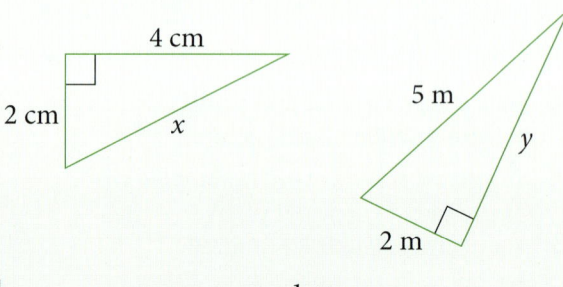

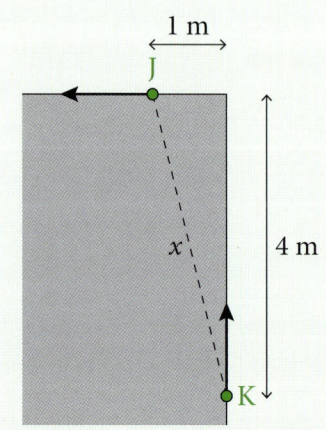

PROGRESS REVIEW: CHAPTERS 11–16

26. Look at the kite on the right.
    (a) Find the side length marked x.
    (b) Find the side length marked y.
    (c) Find the area of the kite.

27. A car travels north 5 km along a straight road. It then turns east and travels 10 km. How far is the car from its starting position? Give your answer to 3 significant figures.

28. The diagram shows a triangular plot of land. Find the lengths marked x and y.

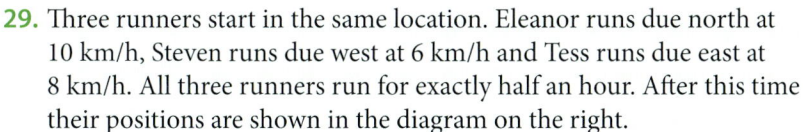

29. Three runners start in the same location. Eleanor runs due north at 10 km/h, Steven runs due west at 6 km/h and Tess runs due east at 8 km/h. All three runners run for exactly half an hour. After this time their positions are shown in the diagram on the right.
    (a) Find the distance between Eleanor and Steven after this time.
    (b) The positions of the three runners form a triangle. Find the sizes of the three angles in the triangle.

30. The diagram shows two tall buildings in Belfast. Building A is 60 metres tall and building B is 45 metres tall. The angle of depression from the top of Building A to the top of Building B is 27°
    Find the horizontal distance between the two buildings, giving your answer in metres to 1 decimal place.

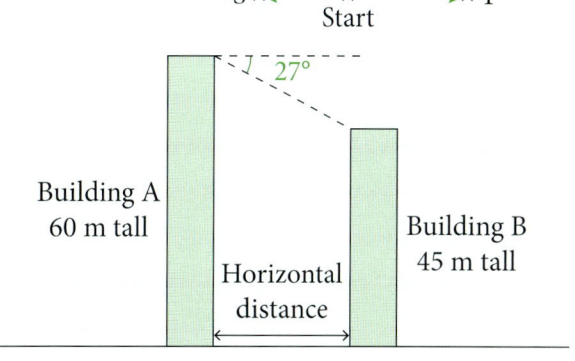

31. A car travels 70 miles from Belfast to Derry/Londonderry at an average speed of 52.5 mph. How long does it take for the journey? Give your answer in hours and minutes.

32. Lee, Kay and Siobhan are travelling home for a family gathering.
    (a) Lee travels 60 km in 50 minutes. Find his speed.
    (b) Kay travels 100 km at an average speed of 120 km per hour. Find the time she takes.
    (c) Siobhan travels for three quarters of an hour at an average speed of 60 km per hour. Find her distance travelled.
    (d) If Lee, Kay and Siobhan all start their journeys at 9:00 am, who arrives at the family gathering first and at what time do they arrive?

33. A bin lorry picks up 20 tonnes of waste, having a volume of 9000 m^3
    What is the density, in kg per m^3, of this waste? You may use the fact that 1 tonne = 1000 kg.

34. Typically, seawater has a density of 1.03 kg per litre. However, the seawater in the Dead Sea has a density of 1.24 kg per litre.
    (a) What volume of ordinary seawater has a mass of 5 kg?
    (b) What volume of seawater from the Dead Sea has a mass of 5 kg?

35. A car parks on a manhole cover with an area of 0.2 m^2
    The car exerts a force of 20 000 newtons on the cover.
    The manhole cover is designed to withstand a pressure of 250 000 N per m^2
    Does the manhole cover break? Show all your working.

36. A person's mass on a chair exerts a force of 600 newtons and a pressure of 2000 N/m^2
    Find the area of contact between the person and the chair.

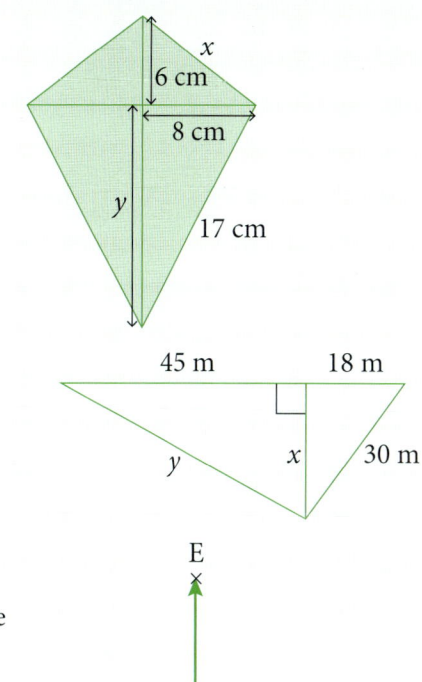

GCSE MATHEMATICS M3 AND M7

37. A doctor measures her patient's heart rate as 120 beats per minute after exercise. If the doctor counted 30 beats, how long did she take for her observation? Give your answer in seconds.

38. Two cars fill up at the same petrol station at the same time. Car A has a tank that holds 18 gallons, while car B holds 15 gallons. Car A can travel 45 miles per gallon, while car B can travel 50 miles per gallon. If both cars travel along the same road until they run out of fuel, which one travels further? Show all your working.

39. Find **(a)** the circumference and **(b)** the area of a circle with a diameter of 20 cm. Give both answers correct to the nearest whole number.

40. Look at the diagram of the cone on the right. It has a slant height of 13.7 cm. Find:
    (a) the base radius,
    (b) the height of the cone,
    (c) the volume of the cone,
    (d) the total surface area.
    Give all your answers to 1 d.p.

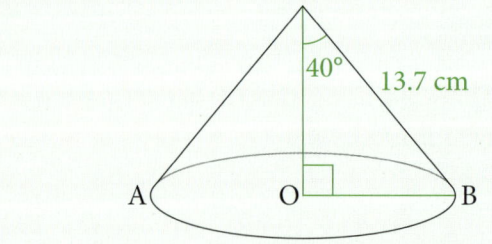

41. The diagram shows a triangular prism.
    (a) Calculate the total surface area of the prism, giving your answer to the nearest square centimetre.
    (b) Calculate the volume of the prism.

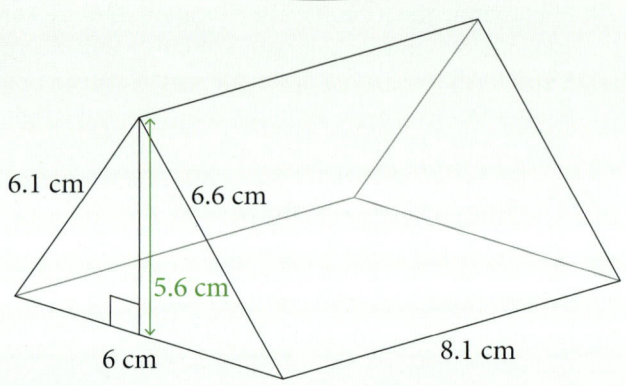

42. The Giant's Causeway is made up of hundreds of hexagonal prisms like the one shown on the right. The regular hexagons at each end have a side length of 10 cm and an area of 260 cm²
    The height of the prism is 70 cm.
    (a) Calculate the total surface area of the prism, giving your answer to the nearest square centimetre.
    (b) Calculate the volume of the prism.

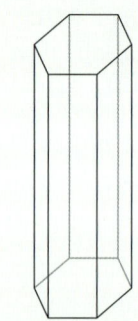

43. Find **(a)** the volume and **(b)** the surface area of a sphere of radius 2 m.

44. The diagram on the right shows a sector of a circle subtending an angle of 120° at the circle's centre.
    The sector has an area of 16.76 cm²
    (a) Calculate the radius r of the circle.
    (b) Find the length of the arc AB.

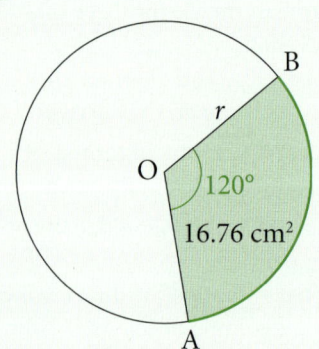

45. A cylinder has a radius of 5 cm and a height of 10 cm.
    (a) What is the volume of the cylinder? Give your answer to the nearest cubic centimetre.
    (b) By how many times does the volume of the cylinder increase if its height is tripled?
    (c) If the radius of the original cylinder is doubled, how does its volume change?

Chapter 17
Collecting Data

17.1 Introduction

This chapter is about collecting data using a **questionnaire**. You will learn about the **data handling cycle**. As well as learning how to design a suitable questionnaire, you will learn how to design a **recording sheet** to record your data.

Key words
- **Data handling cycle**: A process that should be followed for any data collection task.
- **Questionnaire**: A form, given to each person taking the survey, containing spaces for responses.
- **Recording sheet**: A form used by you to record the data on the questionnaire forms.

Before you start you should:
- Know how to use tally marks.
- Know the rules of inequalities.

In this chapter you will learn how to:
- Use the data handling cycle.
- Design and use a recording sheet.
- Design and use a questionnaire.

Exercise 17A (Revision)

1. Copy and complete the tally chart below, using the following ages of people in a waiting room:
   61, 28, 18, 52, 13, 43, 58, 15, 20, 58, 52, 53, 38, 30, 21

Age	Tally	Frequency
$10 \leq a < 20$		
$20 \leq a < 30$		
$30 \leq a < 40$		
$40 \leq a < 50$		
$50 \leq a < 60$		
$60 \leq a < 70$		

2. For each of (a) to (h), state whether an x-value of 5 lies within the range given. The first one has been done for you.

(a)	$x > 3$	Yes
(b)	$x \geq 0$	
(c)	$x > 5$	
(d)	$x \geq 5$	
(e)	$x \geq -5$	
(f)	$x < 5$	
(g)	$x \leq 5$	
(h)	$-1 \leq x < 6$	

17.2 The Data Handling Cycle

To carry out an investigation or survey, you should follow the **data handling cycle**, which consists of the following steps:
- Planning
- Collecting the data
- Processing
- Representation
- Interpreting and Analysing
- Repeat

We will consider each step in turn.

Planning

When planning your investigation, you first need a **hypothesis**, which is a statement that you are testing. A good hypothesis is something that can be investigated by collecting data, for example 'Girls eat more slowly than boys'. An example of a poor hypothesis is 'Girls are better than boys' since this statement is too vague and cannot be investigated by collecting data.

You then need to plan how you are going to carry out the investigation. You should be clear what data is required, how much of it you need to collect, and how it is to be collected. This may involve conducting a survey to collect data. A survey can be carried out by asking people to fill in a **questionnaire**. If questionnaires are being used, you will need to know:

- how many to print;
- how they are going to be distributed;
- how and when they are going to be collected.

In some cases, it may be possible to collect data using a spoken survey, rather than using a written questionnaire.

There are other ways to collect data. For example, a hypothesis such as 'Girls eat more slowly than boys' could be investigated by timing a group of girls and boys in the school canteen. A hypothesis such as 'Pupils with birthdays in September do better in their GCSE exams' could be investigated by collecting data on the school's exam results.

You may have problems accessing some kinds of data. If your survey requires data from school databases (for example, exam results) you will need to speak to teachers about gaining access to the correct data. In some cases, it may not be possible to access all the data you would like (for example, for privacy reasons) and you will have to modify your plans.

Section 17.3 in this chapter goes into more detail about the format of a questionnaire.

Collecting the data

If your planning has been done well, then collecting the data should run smoothly. If questionnaires are being used, they should be distributed and collected at the agreed times.

There may still be problems – for example questionnaires can be lost or forgotten and you will have to decide whether to give extra time in such cases, or whether to start the processing step without this missing data.

If the survey requires access to school databases, you will need to arrange to meet the right teacher at the right time, and be prepared to receive the data in electronic form.

Processing

If you used a questionnaire, you will now have a lot of pieces of paper to process. You will need to prepare a **recording sheet** to summarise your results. Section 17.4 in this chapter goes into more detail about the format of the recording sheet.

If you have got information from a database, you may need to filter the data, leaving only the data that is relevant to your investigation. For example, you may only need to look at the girls' exam results, or only those pupils with September birthdays.

You may also need to do some calculations with your data. You may find it useful to calculate an average (the mean, median or mode) and the range of your data. For example, you may decide it is useful to calculate the mean English exam score for Year 12 boys.

Chapter 18 goes into more detail about calculating averages and the range.

Representation

In the representation stage, you must decide how best to present your results. You can choose to use a mixture of methods, for example tables, bar charts, pie charts, scatter graphs and so forth. All graphs and tables should be accompanied by some text to help explain them.

Interpreting and Analysing

You will also need to interpret your results to draw conclusion from the tables and charts. In addition, you should analyse your study by asking yourself questions such as 'How accurate was the data?' and 'How valid are my conclusions?'

Reflection

It is good to reflect upon your investigation, asking yourself 'How could I have done things better?' Here are examples of some possible reflections:

- I should have printed and distributed more questionnaires.
- I should have given more time for the questionnaires to be completed.
- I should have included question Y on the questionnaire instead of question X.
- I should have looked at data over five years rather than one year.

Repeat

Following your reflections you may decide to repeat the investigation with some changes. Here are examples of some changes you might make:

- The hypothesis could be modified.
- A different question could be included on the survey.
- The sampling method could be changed.

Example 1

Ciaran would like to investigate this hypothesis:
Girls work harder than boys.
(a) Write down some different datasets that Ciaran could collect for his investigation.
(b) How could Ciaran collect these datasets?
(c) What limitations are there?

(a) Ciaran could collect data on the number of hours of homework pupils do each night. In addition, he may want to ask teachers for their opinions.
(b) The data on the number of hours of homework could be collected using a questionnaire. If teachers are willing to share their opinions, these could be given verbally or in electronic form.
(c) Pupils may not be entirely accurate or truthful on a questionnaire. Teacher responses may be subjective, i.e. just opinions. Teachers may not be able to give information about each individual pupil, but may be able to summarise about classes, etc.

Exercise 17B

1. Gemma is investigating this hypothesis: **The most popular pet is a cat.**
   (a) Write down two datasets that Gemma should collect to help with the investigation.
   (b) How should Gemma collect the data?

2. Shelley wants to investigate the hypothesis: **White is the most popular colour of car.**
   (a) Describe a dataset Shelley could collect to help her with this investigation.
   (b) How could she collect the relevant data?
   (c) What type of chart could Shelley produce to represent her findings?

3. Molly has decided to investigate this hypothesis: **Food prices have gone up.**
   (a) Write down some different datasets Molly could collect for her investigation.
   (b) How could Molly collect these datasets?
   (c) What limitations are there?
   (d) In her reflection step, state one improvement could Molly make to her investigation.

4. Matt is investigating the hypothesis: **Cheaper supermarkets sell poor quality food.**
   His teacher suggests that this may be difficult to investigate because of difficulties measuring the quality of food.
   (a) How could Matt change the hypothesis?
   (b) Using the new hypothesis, what data should he collect?

5. Cheryl is investigating the hypothesis: **Larger families have more arguments.**
   (a) Cheryl carries out her investigation using a questionnaire, asking the pupils in her class to record the number of arguments that take place in their families over one week. She produces the following graph. State something that appears to be wrong with the graph.

GCSE MATHEMATICS M3 AND M7

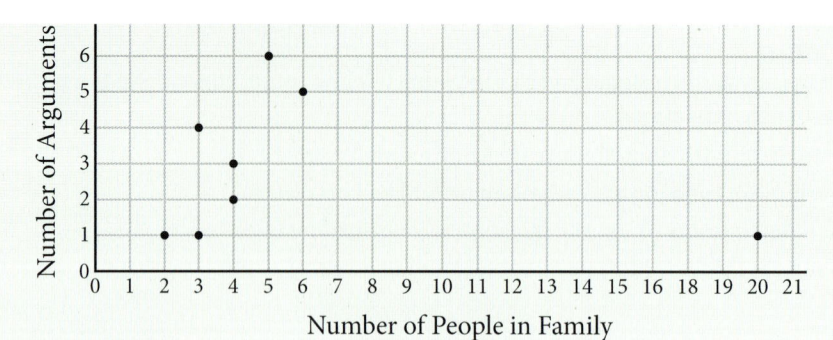

[Scatter graph: Number of Arguments (y-axis, 0-6) vs Number of People in Family (x-axis, 0-21). Points approximately at (2,1), (3,1), (3,4), (4,2), (4,3), (5,6), (6,5), (20,1).]

(b) Do you think it was a good idea for Cheryl to investigate this hypothesis using a questionnaire? Give one reason **for** using this approach and one reason **against**.

6. Tim works for the council. He wishes to investigate the hypothesis: **The increase in the number of out-of-town supermarkets has led to the closure of shops in town centres.**
   (a) Write down two sets of data that Tim could collect to investigate this.
   (b) How could Tim collect this data?

17.3 The Questionnaire

A questionnaire should have the following features:
- It should have more than one question.
- The questions should be as short and precise as possible.
- The questions should be fair, not biased.
- The questions should be closed rather than open. A closed question can be answered with a single, simple response such as 'Yes' or '38'. An open question may need several sentences to answer, for example a description of the water cycle.
- The response options should be balanced.
- The response options should be clearly and uniquely defined.
- There should be no gaps in the response options.

Example 2

(a) What is wrong with this question on a questionnaire about pets?
   Dogs are much friendlier than cats – don't you agree?
   Yes ☐   No ☐
(b) Design a better question.

(a) This is a biased question in favour of dogs.
(b) A better question would be:
   Which of these animals makes for the friendliest pet?
   Dog ☐   Cat ☐   Rabbit ☐   Goldfish ☐   Hamster ☐   Other ☐ (please state) _____

Example 3

Which of these questions is a **closed question**, which one an **open question**?
- Question 1: Is there a fish special on the menu tonight?
- Question 2: Why did World War II happen?
- Question 3: Which internet browser do you prefer?

- Question 1 is a closed question. The answer is either yes or no.
- Question 2 is an open question. It could be answered using more than one sentence, perhaps a paragraph, an essay, or even a whole book! It is not suitable for a questionnaire.
- Question 3 is a closed question. It can be answered with a single, simple response, for example 'Edge' or 'Chrome'.

CHAPTER 17: COLLECTING DATA

Example 4

What is wrong with the set of response boxes in this questionnaire question?

I like pop music. Tick one box to indicate your level of agreement.

Disagree ☐   **Neither agree nor disagree** ☐   **Agree** ☐   **Strongly agree** ☐

The response boxes are not balanced. There should be a 'Strongly disagree' option, giving five options, with 'Neither agree nor disagree' in the middle.

Example 5

Max has written a questionnaire on sport. However, in each question, the response options are poorly chosen. In each case, state why and how it could be improved.

Question 1. What is your favourite sport?

Football ☐   Rugby ☐   Hockey ☐   Gaelic Football ☐   Netball ☐

Question 2. How many different sports do you play regularly?

1 ☐   2 ☐   3 ☐   4 ☐   More than 4 ☐

Question 3. How long in total each week do you spend playing sport?

0–1 hours ☐   1–2 hours ☐   2–3 hours ☐   3–4 hours ☐   More than 4 hours ☐

For Question 1 the response options should include 'Other'. For example, somebody may prefer cricket or swimming.

For Question 2 there should be an option '0', as some people may not play any sports regularly.

For Question 3, it is unclear which box should be ticked for exactly 2 hours. The question could be rewritten in this way:

Question 3. How long in total each week do you spend playing sport?
Tick one box for the time t hours.

$0 \leq t < 1$ ☐ $1 \leq t < 2$ ☐ $2 \leq t < 3$ ☐ $3 \leq t < 4$ ☐ $t \geq 4$ ☐

Exercise 17C

1. Closed questions are suitable for a questionnaire, whereas open questions are not. State whether each of these questions is an open question or a closed question.
   (a) Have you ever made a cake?
   (b) Do you prefer to eat at home or go out?
   (c) Are there any members of your family working for the emergency services (fire, police, ambulance)?
   (d) What makes the leaves change colour in the Autumn?
   (e) If your family has a car, who is the main driver?
   (f) How do you know when it is the right time to get married?
   (g) Why is it important to save money for retirement?
   (h) Would you prefer fried or baked chicken for dinner, or neither?
   (i) What is your favourite kind of pie?
   (j) How did you and your best friend meet?
   (k) What is it like to raise children as a single parent?
   (l) Where do you go to school?
   (m) How could you prepare to buy your first house?

2. State one thing that is wrong with the response options for this question.

   How tall are you? Tick one box for your height h in centimetres.

   $h < 120$ ☐ $120 \leq h \leq 140$ ☐ $140 \leq h \leq 160$ ☐ $h > 160$ ☐

3. State one thing that is wrong with the response options for this question.

   How many pieces of fruit did you eat yesterday?

   0 ☐   1 ☐   2 ☐   4 ☐   More than 4 ☐

4. The following question appeared on a questionnaire about mathematics tutors. State one thing that is wrong with the response options.

   How satisfied are you with your tutor's standard of teaching?

   Completely dissatisfied ☐   Somewhat dissatisfied ☐   Neither satisfied nor dissatisfied ☐   Satisfied ☐

5. These questions appeared on a questionnaire about attitudes to schoolwork and homework. For each question, state something that is wrong with either the question or the response options.

   (a) Don't you think students are much lazier than they used to be?

   Yes ☐   No ☐   Don't know ☐

   (b) What is your favourite school subject?

   Maths ☐   English ☐   Science ☐   Music ☐
   PE ☐   Art ☐   Geography ☐   History ☐

   (c) Why is it important to do homework?

   (d) How much time do you spend doing homework each evening?

   0–1 hours ☐   1–2 hours ☐   2–3 hours ☐   3–4 hours ☐   More than 4 hours ☐

   (e) Read the statement below and tick one box to show how much you agree.

   I get a lot of satisfaction doing my homework.

   Strongly disagree ☐   Disagree ☐   Neither agree nor disagree ☐   Agree ☐

6. The following two questions appeared on a questionnaire about pupils' journeys to school. For each question, state one thing that is wrong with the response options.

   (a) How do you get to school?

   Car ☐   Bus ☐   Walk ☐   Cycle ☐

   (b) How long does your journey take? Tick one box for the time t.

   $t < 10$ minutes ☐
   10 minutes $\leq t < 30$ minutes ☐
   30 minutes $\leq t < 1$ hour ☐
   1 hour 30 minutes $\leq t < 2$ hours ☐
   $t > 2$ hours ☐

17.4 The Recording Sheet

Once you have collected your questionnaires, a single recording sheet can then be used to summarise the data. Frequently, a tally mark is used to record the response to each question. After tallying each question, the totals can be calculated.

Example 6

Shona distributes the following questionnaire to the 30 school pupils in her class.

Question 1: How old are you?

11–12 ☐   13–14 ☐   15–16 ☐   17–18 ☐   Other ☐ (please state) _____

Question 2: What is your favourite school subject?

Maths ☐   English ☐   Science ☐   Music ☐   PE ☐
Art ☐   Geography ☐   History ☐   Other ☐ (please state) _____

Design a recording sheet that could be used to record and summarise the information from the 30 questionnaires.

One possible recording sheet is shown on the next page. Tally marks have been added and the totals calculated.

Question 1: How old are you?

11–12	||||	4
13–14	⊦⊦⊦⊦ |	6
15–16	⊦⊦⊦⊦ ⊦⊦⊦⊦ ||	12
17–18	⊦⊦⊦⊦ ||	7
19	|	1

One pupil stated 'Other' and gave 19 as their age.

Question 2: What is your favourite school subject?

Maths	||	2
English	|||	3
Science	⊦⊦⊦⊦	5
Music	|||	3
PE	⊦⊦⊦⊦ ||	7
Art	|||	3
Geography	||||	4
History	||	2
French	|	1

One person stated 'Other' and gave French as their favourite subject.

Inferring properties of the population

When you carry out a survey, you may choose to hand out a questionnaire to a **sample** of the entire **population**. For example, if you want to investigate the mean height of the pupils in Year 12, you may decide only to distribute questionnaires to the 30 Year 12 pupils in your class. In this case, all the Year 12 pupils are the 'population'; and the 30 pupils in your class are the 'sample'.

If the mean height of your sample is 162 cm, you may conclude that the mean height of the population is also 162 cm. This may or may not be accurate. For example, if there are a lot of taller people in one of the other classes, the true mean might be 165 cm.

It is also important to understand that different samples may provide different results. For example, suppose only 20 pupils in your class completed the questionnaires, and you calculate a mean of 162 cm. If you had carried out the survey the following week, all 30 pupils may have been able to complete the questionnaires, and the mean height calculated might have been lower or higher. Larger samples generally give more reliable results.

Despite these limitations, using a sample to infer properties of the population is an important statistical technique. This is particularly true when it would be too time-consuming or expensive to carry out a survey of the entire population. Nevertheless, you should be aware of the limitations discussed here.

Example 7

Shona asked 30 pupils for their favourite school subject. Three said English, while two said Maths. Shona concludes that English is the favourite subject of more pupils in her school than Maths. Comment on her conclusion.

Shona is basing her conclusion on a fairly small sample of size 30. Since the numbers preferring English and Maths were fairly close (3 and 2), she should be careful about drawing conclusions about the whole school population. She should probably repeat her survey, and, if possible, use a larger sample size, for example her entire year group. Better still, she could ask some pupils from each year group in the school. Larger samples generally give more reliable results and the sample should be representative of the population.

Exercise 17D

1. The following questionnaire has been used to collect data on the colours of cars.

   What colour is your family car? You may choose more than one option if your family has two or more cars.

   Black ☐   White ☐   Silver/grey ☐
   Red ☐   Blue ☐   Other ☐ (please state) _____

   (a) State one thing that is wrong with the design of the questionnaire.
   (b) Design a recording sheet to record and summarise the information on the questionnaires.

2. Monty carries out a survey on the types and ages of pets. He prints questionnaires and hands them out to the pupils in his class. Nathan's returned questionnaire is shown overleaf.

Do you have any pets?

Yes ✓   No ☐

If you have pets, please complete the table on the right.

Pet	Number	Ages
Dog		
Cat	2	5, 6
Rabbit		
Goldfish		
Hamster / gerbil		
Horse		
Tortoise	1	45

Monty records all the data from the questionnaires on the recording sheet shown below.

Type of pet

Pet	Tally	Total
Dog	|||||	
Cat	|||||		
Rabbit	||	
Goldfish	|||	
Hamster / gerbil	|||	
Horse			
Tortoise			
No pets	||||| ||||	

Age of pet

Age (years)	Tally	Total
< 1	||||| |||	
1 – 2	||||| ||||	
3 – 5	||||| ||	
6 – 10	||||	
11 – 20			
> 20			

(a) Copy and complete the tables, by filling out the 'Total' column in each case.
(b) What is the most popular pet in the class?
(c) What are the two least popular pets?
(d) What is the type and age of the oldest pet in the class?
(e) State something that is wrong with Monty's questionnaire.

3. The recording sheet on the right records and summarises the medals won by Team GB and Team Ireland during the 2024 Olympic Games.
   (a) How many silver medals were won by Team GB?
   (b) How many bronze medals were won by Team Ireland?
   (c) Calculate the total number of medals won by Team GB.

	Gold		Silver		Bronze	
Team GB	||||| ||||| ||||	14	||||| ||||| ||||| ||		||||| ||||| ||||| ||||| ||||	29
Team Ireland	||||	4		0	|||	

4. A questionnaire is designed to ask people in a hotel restaurant for (1) their ages and (2) what they were choosing as their main course from the menu shown.
   (a) Design the questionnaire.
   (b) Design a recording sheet to record this information.

> **MENU**
> Pasta
> Fish
> Chicken
> Beef
> Pork
> Vegetarian Lasagne
> *Children's meals are available for all of the above*

5. Dan is carrying out a survey into the speed of internet connections on his road, Ashvale Drive. There are 100 houses on Ashvale Drive, but Dan decides to deliver questionnaires to the houses with numbers 5, 10, 15, 20, etc, up to 100.
   (a) To how many houses does Dan deliver questionnaires?
   (b) Half of the questionnaires are returned. Dan records the internet connection speeds on a recording sheet. What is the size of his sample?

(c) From the recording sheet, Dan calculates the mean speed as 95 Mbps. He concludes that the mean internet connection speed on Ashvale Drive is 95 Mbps. Describe briefly two limitations of Dan's method and conclusion.

6. Eoin is interested in how often the people of Northern Ireland order takeaway meals. He carries out a survey by handing out questionnaires to 28 people. Eoin's results show that 14 of these people have takeaway meals at least once a week. Eoin says 'I conclude that half the population of Northern Ireland gets a takeaway meal at least once a week'. Comment on Eoin's conclusion.

17.5 Summary

In this chapter you have learnt that:

- For any investigation, data should be collected. The data handling cycle should be followed. A hypothesis should be formed. The following steps should be followed:
  - Planning a survey.
  - Collecting data.
  - Processing the data.
  - Representing the data.
  - Interpreting and analysing your results.
  - Reflecting on your survey.
  - If necessary, repeating the survey with some modifications.
- A questionnaire is often used and distributed to a sample of people. It should have the following features:
  - It should have more than one question.
  - The questions should be as short and precise as possible.
  - The questions should be fair, not biased.
  - The questions should be closed rather than open.
  - The response options should be balanced.
  - The response options should be clearly and uniquely defined.
  - There should be no gaps in the response options.
- A single recording sheet should be used to record and summarise the data from the set of questionnaires. Tally marks are frequently used to record the data for each question. After tallying each question, the totals can be calculated.
- When carrying out a survey, you should be aware that drawing conclusions about the population based on your sample has limitations.
- Different samples may provide different results and that larger samples generally give more reliable results.

Chapter 18
Statistical Averages and Spread

18.1 Introduction

Three types of statistical averages are covered in GCSE mathematics:
- The **mean**, which is calculated as the sum of all the values divided by the number of values.
- The **mode**, which is the value that occurs most frequently.
- The **median**, which is the middle value when the values are listed in order from smallest to largest. If there are two values in the middle, then the median is halfway between these two values.

You have also learnt about the **range**, which measures the spread of the values.

In this chapter you will learn about the **interquartile range**, which is another measure of spread.

You will also learn about frequency tables and grouped frequency tables, and calculating or estimating statistical averages from them.

Key words
- **Mean**: The sum of all the values divided by the number of values.
- **Median**: The middle value when the values are listed in order from smallest to largest.
- **Mode**: The value that occurs most frequently.
- **Range**: The difference between the highest value and the lowest value.
- **Lower quartile**: The value that lies 25% of the way through the list.
- **Upper quartile**: The value that lies 75% of the way through the list.
- **Interquartile range**: The difference between the upper quartile and the lower quartile.

Before you start you should know how to:
- Calculate the mean from a list of values.
- Calculate the mode from a list of values.
- Calculate the median from a list of values.

In this chapter you will learn how to:
- Calculate quartiles and the interquartile range from a list of values.
- Calculate the mean from a frequency table, and identify the mode and the median.
- Estimate the mean from a grouped frequency distribution, and identify the modal class and the median class.
- Use an average and a measure of spread to compare two distributions.

Example 1 (Revision)

These were the test scores for a small group of interview candidates: 18, 17, 9, 8, 13, 17, 10, 12, 20, 16
(a) Find (i) the mean (ii) the median and (iii) the mode of these scores.
(b) Calculate the range of the scores.

(a) (i) To find the mean, add the values and divide by the number of values, 10:

$$\text{Mean} = \frac{18 + 17 + 9 + 8 + 13 + 17 + 10 + 12 + 20 + 16}{10} = 14$$

(ii) To find the median, put the test scores in order: 8, 9, 10, 12, 13, 16, 17, 17, 18, 20
Since there is an even number of scores, there are two in the middle: 13 and 16
The median is the number halfway between 13 and 16, which is 14.5
This number is found by calculating the mean of the middle two numbers:

$$\text{Median} = \frac{13 + 16}{2} = 14.5$$

(iii) The mode is the number that occurs most frequently, 17

> **Note:** If more than one value occurs most frequently, you can say there is more than one mode, or you can say there is no mode. For example, in the list 5, 6, 6, 7, 8, 8 you could say 6 and 8 are both modes, or there is no mode.

(b) The range is the highest score minus the lowest:
Range = 20 − 8 = 12

Exercise 18A – Revision

1. Five yachts are moored in a harbour with these lengths: 16 m, 12 m, 18 m, 17 m, 18 m
   (a) Calculate the mean length.
   (b) Find the median.
   (c) Find the mode.
   (d) Find the range of the lengths.

18.2 Measures of Spread

It is important to remember that the range is a measure of spread. It is not an average like the mean, median and mode.

If Class A's test scores have a range of 45, while Class B's scores have a range of 15, it shows that Class A's scores were more variable than Class B's. Class A's scores have a bigger spread.

In this section you will learn how to calculate a different measure of spread: the interquartile range.

The quartiles

You already know that the median lies halfway through the data when the values are listed in order.

In other words, the **median** is the value that lies 50% of the way through the list.

The **lower quartile** is the value that lies 25% of the way through the list.

The **upper quartile** is the value that lies 75% of the way through the list.

The position of the median item is given by the formula $\frac{n+1}{2}$ where n is the number of values in the list.

The position of the lower quartile is given by the formula $\frac{n+1}{4}$

The position of the upper quartile is given by the formula $3\left(\frac{n+1}{4}\right)$

> **Note:** You may see the variable Q_2 used for the median, while Q_1 and Q_3 are used for the lower and upper quartiles respectively.

The interquartile range

The interquartile range is the difference between the upper quartile and the lower quartile.

The following example demonstrates two methods for finding the quartiles.

Example 2

Ten carrots had these lengths in centimetres: 6, 9, 8.5, 11, 14, 7, 9.5, 12.5, 8, 9
(a) Find (i) the median Q_2, (ii) the lower quartile Q_1 and upper quartile Q_3 for this set of data.
(b) Find the interquartile range.

(a) (i) To find the median, list the values in order from smallest to largest:
6, 7, 8, 8.5, 9, 9, 9.5, 11, 12.5, 14

Since there is an even number of values, there are two values in the middle: 9 and 9
The median Q_2 is the number halfway between them (or the mean of them), which is 9

(ii) Two methods are given below to find the lower and upper quartiles.
Method 1
When working with a list of values, the lower quartile is the value that lies halfway through the first

half of the list. The upper quartile is the value that lies halfway through the second half of the list.
The first half of the ordered values is: 6, 7, 8, 8.5, 9
The lower quartile is the value that lies halfway through this list. The lower quartile Q_1 is 8
The second half of the ordered values is: 9, 9.5, 11, 12.5, 14
The upper quartile is the value that lies halfway through this list, so the upper quartile Q_3 is 11

Method 2
The position of the lower quartile Q_1 is $\frac{10+1}{4}$
This is closer to 3 than 2, so look for the third value in the ordered list, which is 8
The position of the upper quartile Q_3 is $3\left(\frac{10+1}{4}\right)$
This is closer to 8 than 9, so look for the eighth value in the ordered list, which is 11

(b) The interquartile range is the difference between the upper quartile Q_3 and the lower quartile Q_1
Interquartile range = $Q_3 - Q_1$ = 11 − 8 = 3

Exercise 18B

1. Find the median and interquartile range for the following sets of data. Hint: Remember to order the values from smallest to largest.
   (a) 13, 18, 15, 11, 6, 9, 23, 8, 17, 6
   (b) 6, 7, 9, 10, 11, 13, 14
   (c) 30, 35, 45, 50, 55, 65, 70, 80, 82, 85, 90
   (d) 15, 10, 12, 17, 9, 20, 16
   (e) 1, 5, 7, 8, 11, 14, 17, 17, 25

2. In a hospital 7 newborn babies have the following weights: 2.5 kg, 3.1 kg, 3.4 kg, 3.5 kg, 3.5 kg, 4 kg, 4.1 kg
   Find (a) the median weight; and (b) the interquartile range of the weights of the babies.

18.3 Frequency Tables

A frequency table summarises a list of values. The frequency is the number of times a value occurs.

Example 3

Peter does a survey into the number of cars owned by families. He asks 12 pupils in his class. The results are:
1, 2, 2, 1, 3, 1, 2, 1, 2, 1, 0, 1
Summarise this dataset using a frequency table.

For these families, the number of cars lies between 0 and 3
- One pupil said their family had no cars.
- Six said one car.
- Four said two cars.
- One said three cars.

This information can be summarised in the frequency table shown on the right.

Number of cars	Frequency
0	1
1	6
2	4
3	1

Finding averages from a frequency table

You may be asked to calculate the mean from a frequency table, and to identify the mode and median.

The mean
To calculate the mean from a frequency table, follow these steps:
- Label the columns x and f, where x is the quantity whose mean we are calculating and f is the frequency.
- Add an extra column to the table and label it fx. Calculate each value in the fx column by multiplying the x value and the f value.
- Add up the f column and the fx column. Place the totals at the bottoms of the columns. We use the Greek letter Σ for the sum, so we are calculating Σf and Σfx.
- Calculate the mean using the formula: Mean = $\frac{\Sigma fx}{\Sigma f}$

CHAPTER 18: STATISTICAL AVERAGES & SPREAD

Example 4

Using the data from Peter's survey in Example 3, calculate the mean.

Copy the table, label the columns x and f, and calculate the fx column.

The totals Σf and Σfx have also been calculated.

Note that Σf is the total number of pupils in the survey, also referred to as n.

Number of cars, x	Frequency, f	fx
0	1	0
1	6	6
2	4	8
3	1	3
	$\Sigma f = 12$	$\Sigma fx = 17$

To calculate the mean, use the formula: Mean $= \dfrac{\Sigma fx}{\Sigma f} = \dfrac{17}{12} = 1.42$ (to 2 decimal places)

Note: The calculated mean looks like a sensible value as it lies between the smallest and largest number of cars, 0 and 3. If this were not the case, there must be an error, and we should check the working.

The median

To find the median value from a frequency table use **cumulative frequencies**. The cumulative frequency for a row is the sum of all the frequencies up to and including that row.

Calculate the position of the median item using the formula $\dfrac{n+1}{2}$ and use the cumulative frequencies to identify the correct row of the table, as in the next example.

The mode

The mode is the value that has the highest frequency.

Finding the range from a frequency table

The range is difference between the highest and lowest values. These can be read from the top and bottom rows of the table, as shown in the following example.

Example 5

A survey was carried out to find the median number of pets owned by each child in class 8C1. The results of the survey are shown in the table on the right.

(a) Copy the table. Add a cumulative frequency column to the table and complete it.
(b) Find the position of the median value.
(c) Find the median value.
(d) Find the modal value.
(e) Find the range.

Number of pets	Frequency
0	3
1	5
2	2
3	7
4	10
5	3
6	1

(a) The cumulative frequency column has been added to the table, as shown. The cumulative frequencies are the sum of all the frequencies up to that row. So, the first one is 3, the second one is $3 + 5 = 8$
For the third cumulative frequency, add on the third frequency of 2, giving 10
And so on.

Number of pets	Frequency	Cumulative frequency
0	3	3
1	5	8
2	2	10
3	7	17
4	10	27
5	3	30
6	1	31

(b) There are 31 children in the class, i.e. $n = 31$
The position of the median item is:
$\dfrac{n+1}{2} = \dfrac{31+1}{2} = 16$ i.e. the median is the 16th value.

(c) Using the cumulative frequency column, the 16th item is in the 3 pets row.
Median = 3

(d) The mode (or modal value) is the value with the highest frequency. So, the mode is 4

(e) The highest number of pets is 6 and the lowest is 0 so: Range = 6 − 0 = 6

Exercise 18C

1. For the following frequency distributions, copy the table. By adding appropriate columns, find:
   (i) The mean value of x.
   (ii) The median value of x.
   (iii) The modal value of x.
   (iv) The range for the x-values.

   (a)
x	f
1	2
2	4
3	1

   (b)
x	f
5	2
6	5
7	4
8	2
9	2
10	1

   (c)
x	f
1	20
2	10
4	10
10	8
20	2

   (d)
x	f
12	8
13	11
14	6
15	2
16	4
17	9

   (e)
x	f
18	7
19	8
20	6
21	4

2. In this question, the frequency distributions are given in rows rather than columns. Copy each table, converting it to columns as in question 1. Then add appropriate columns to find:
   (i) The mean value of x.
   (ii) The median value of x.
   (iii) The modal value of x.
   (iv) The range for the x-values.

   (a)
x	1.8	1.9	2.0	2.1	2.2	2.3
f	6	8	10	9	4	2

   (b)
x	2.10	2.11	2.12	2.13	2.14	2.15
f	5	6	12	13	7	1

   Note: In part (b), each value of x is given to two decimal places. Two decimal places is a suitable level of accuracy for the mean, median and mode.

3. The table on the right shows the results of Peter's survey into the number of cars per family. Find the median number of cars.

Number of cars	Frequency
0	1
1	6
2	4
3	1

4. A survey is carried out into the purchases of 1000 customers visiting a popular online shop. The frequency table on the right shows the number of items purchased by these customers. Calculate:
   (a) the mean,
   (b) the median,
   (c) the mode,
   (d) the range
   for the number of items purchased. Give your answers to 1 decimal place where appropriate.

Number of items (x)	Number of customers (frequency f)
1	496
2	317
3	104
4	65
5	10
6	7
7	1

5. Sarah does a survey into the number of pets her friends have. The results are shown in the frequency table on the next page.

(a) How many people did Sarah survey?
(b) Find the total number of pets.
(c) Find the mean number of pets.
(d) Find the median.
(e) There are two modes for this distribution. What are they?
(f) What is the range in the number of pets?

Number of pets (x)	Number of people (frequency f)
0	6
1	6
2	4
3	2
4	1
5	1

6. Simon is investigating the average number of Smarties in a tube. He opens fifty tubes and counts the number of Smarties in each tube. The results are shown in the frequency table below.

Number of Smarties x	29	30	31	32	33	34	35
Frequency f	1	3	7	28	7	3	1

(a) Calculate the mean number of Smarties in a tube.
(b) Calculate the median.
(c) What is the mode?
(d) What is the range of the number of Smarties?

7. The table on the right shows the ages of 10 staff members in a meeting.
   (a) Calculate the mean age of these staff members.
   (b) What is the median age?
   (c) What is the modal age?

Age	Number of staff members
38	2
39	2
40	3
41	2
42	1

8. On a cub camp, one of the activities is archery. The cubs fire arrows at the target shown. There are 50 cubs on the camp and they each fire one arrow. The scores are shown in the table on the right.
   (a) What is the modal score?
   (b) What is the median score?
   (c) Find the mean score.
   (d) What percentage of cubs scored **more than** the mean?

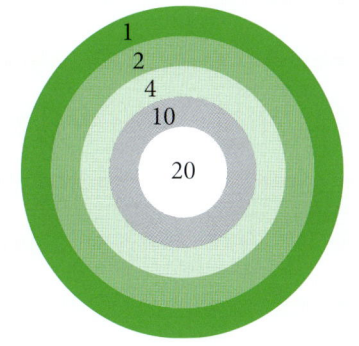

Score	Frequency
1	20
2	10
4	10
10	8
20	2

18.4 Grouped Frequency Tables

If there is a large amount of data, it is helpful to group the data. For example, if we measure 100 people's heights we may get 100 individual values such as 1.45 m, 1.455 m, 1.46 metres, etc. So instead we group them, for example with groups from 130 cm to 140 cm, 140 cm to 150 cm, and so on.

Mode and median

We cannot find the exact mode or median for a grouped frequency distribution but we can give the limits of the group or class in which the mode or median lies. These classes are called the **modal group** and the **median group**.

> **Note:** The modal group is sometimes called the modal class or the modal class interval. The median group is sometimes called the median class or the median class interval.

Finding these groups can be done in the same way as finding the mode and median in a frequency table.

Mean

When data are grouped, we cannot find the exact mean, but we can calculate an estimate of the mean by choosing the midpoint of each group or class to represent the group. We assume the midpoint to be the x-value.

GCSE MATHEMATICS M3 AND M7

Example 6

Look at the grouped frequency table on the right. It summarises the scores for a class of pupils in a science test.

(a) How many pupils are there in this class?
(b) State the modal class.
(c) Find the median class.
(d) Calculate an estimate for the mean.

Score	Frequency f
15 – 19	7
20 – 24	3
25 – 29	6
30 – 34	10
35 – 39	4

(a) Add up the frequency column:
$\Sigma f = 7 + 3 + 6 + 10 + 4 = 30$
There are 30 pupils in the class.

(b) The modal class is 30 – 34, since this is the class (or group) with the highest frequency.

(c) To find the median class, we add the cumulative frequency column to the table.

$n = 30$, so $\frac{n+1}{2} = 15.5$

The 15th and 16th pupils both lie in the 25 – 29 class (or group), so this is the median class.

Score	Frequency f	Cumulative frequency
15 – 19	7	7
20 – 24	3	10
25 – 29	6	16
30 – 34	10	26
35 – 39	4	30

(d) To estimate the mean, add two columns:
- first add the x column, which is calculated as the midpoint of each class;
- then add the fx column, which is calculated by multiplying f and x.

Score	Frequency f	Midpoint x	fx
15 – 19	7	17	119
20 – 24	3	22	66
25 – 29	6	27	162
30 – 34	10	32	320
35 – 39	4	37	148

Note: To find the midpoint, calculate the mean of the upper and lower limits of the group.

For example, in the 15 – 19 class, the midpoint = $\frac{15+19}{2} = 17$

Proceed as before: find Σfx and Σf, then calculate $\frac{\Sigma fx}{\Sigma f}$

$\Sigma fx = 119 + 66 + 162 + 320 + 148 = 815$
$\Sigma f = 30$
Estimate of mean = $\frac{\Sigma fx}{\Sigma f} = \frac{815}{30} = 27.2$ (1 d.p.)

Note: We cannot be sure that this is the true mean score, since we do not have the individual scores. By finding the midpoint for each group, we are assuming that all 7 pupils in the 15 – 19 group scored 17 in the test, all 3 pupils in the 20 – 24 class scored 22, etc.

Example 7

The table on the right shows the times it took a group of adults to solve a puzzle.

(a) Calculate an estimate for the mean time.
(b) State the modal class.
(c) Which class interval contains the median?

Note: The question in part (c) is another way of asking for the median class.

Time t (seconds)	Frequency f
$0 \leq t < 10$	18
$10 \leq t < 20$	26
$20 \leq t < 30$	35
$30 \leq t < 40$	32
$40 \leq t < 50$	12

CHAPTER 18: STATISTICAL AVERAGES & SPREAD

> **Note:** $10 \leq t < 20$ means times between 10 and 20 seconds. Somebody taking exactly 10 minutes is included in this group, but not somebody taking 20 minutes. They would be included in the $20 \leq t < 30$ group.

Add the x column (the midpoint) and the fx column to the table.

(a) Find Σfx and Σf, then calculate $\dfrac{\Sigma fx}{\Sigma f}$

$\Sigma fx = 90 + 390 + 875 + 1120 + 540$
$ = 3015$

$\Sigma f = 18 + 26 + 35 + 32 + 12 = 123$

Estimate of mean $= \dfrac{\Sigma fx}{\Sigma f} = \dfrac{3015}{123} = 24.5$ seconds (1 d.p.)

Time t (seconds)	Frequency f	x	fx
$0 \leq t < 10$	18	5	90
$10 \leq t < 20$	26	15	390
$20 \leq t < 30$	35	25	875
$30 \leq t < 40$	32	35	1120
$40 \leq t < 50$	12	45	540

(b) The modal class is $20 \leq t < 30$ seconds, since this is the class with the highest frequency.

(c) To find the median class, add the cumulative frequency column.

$n = 123$, giving $\dfrac{n+1}{2} = 62$

The median group is the group containing the 62nd item, which is $20 \leq t < 30$ seconds.

Time t (seconds)	Frequency f	Cumulative frequency
$0 \leq t < 10$	18	18
$10 \leq t < 20$	26	44
$20 \leq t < 30$	35	79
$30 \leq t < 40$	32	111
$40 \leq t < 50$	12	123

Exercise 18D

1. A botanist is studying the number of petals on daisies. She carries out a survey of 50 daisies and the results are shown in the frequency table below.

Number of petals	10 – 14	15 – 19	20 – 24	25 – 29	30 – 34	35 – 39
Frequency	1	6	20	18	4	1

   (a) Find an estimate for the mean number of petals for this group of 50 daisies.
   (b) State the modal class.
   (c) Find the median class.

2. Mr Smyth sets his maths class some homework and asks them each to record the time they take to do it. The results are shown in the table on the right.
   (a) How many pupils are in Mr Smyth's maths class?
   (b) Write down the modal class for the time taken.
   (c) Which class interval contains the median?
   (d) Calculate an estimate for the mean time taken. Give your answer to one decimal place.

Time t (minutes)	Frequency f
$0 \leq t < 10$	2
$10 \leq t < 20$	8
$20 \leq t < 30$	14
$30 \leq t < 40$	6
$40 \leq t < 50$	1

3. Séan carried out a survey into the heights of pupils in class 11F. His results are shown in the table on the right.
   (a) Find an estimate for the mean height of the pupils in class 11F, giving your answer in centimetres to 1 decimal place.
   (b) State the modal class.
   (c) Find the median class.

Height h (cm)	Number of pupils
$140 \leq h < 150$	7
$150 \leq h < 160$	8
$160 \leq h < 170$	9
$170 \leq h < 180$	3
$180 \leq h < 190$	1

4. The table on the right shows the times patients spent in the waiting room of a hospital department one day.
   (a) How many patients visited this hospital department on this day?
   (b) Calculate an estimate for the mean time the patients waited. Give your answer in minutes to 1 decimal place.
   (c) What is the modal class?
   (d) Find the median class.

Time t (minutes)	Number of patients
$10 \leq t < 15$	10
$15 \leq t < 20$	8
$20 \leq t < 25$	4
$25 \leq t < 30$	2
$30 \leq t < 35$	1

5. The weights of 20 newborn babies are recorded in pounds (lb) and are shown in the grouped frequency table on the right.
   (a) Estimate the mean weight of these 20 babies.
   (b) What is the median class interval?

Weight w (lb)	Frequency f (number of babies)
$6 \leq w < 7$	3
$7 \leq w < 8$	6
$8 \leq w < 9$	5
$9 \leq w < 10$	6

6. The speeds of 300 cars were recorded on a section of road in county Fermanagh. The table shows the distribution of these speeds.
   (a) The speed limit in this area is 40 mph. What percentage of cars were travelling faster than the speed limit?
   (b) Find an estimate for the mean speed of these 300 cars.
   (c) What was the modal group for these speeds?
   (d) What was the median group?

Speed s (mph)	Frequency f
$0 \leq s < 10$	15
$10 \leq s < 20$	34
$20 \leq s < 30$	86
$30 \leq s < 40$	135
$40 \leq s < 50$	27
$50 \leq s < 60$	3

18.5 Harder Problems Involving Statistical Averages and Spread

This section gives examples of harder problems involving the three statistical averages and spread.

Working backwards

The following example shows how, if you are given the mean, you can calculate a missing frequency in a frequency table.

Example 8

Three boys and four girls travel to a summer camp. The distances travelled by the boys are:
4 miles, 8 miles, 6 miles. For the four girls, the mean distance travelled is 7 miles.
(a) Find the mean distance travelled by the boys.
(b) Find the total distance travelled by the girls.
(c) Find the mean distance travelled by all the children, giving your answer to 1 decimal place.

(a) To calculate the mean for the boys: $\frac{4 + 8 + 6}{3} = 6$ miles
(b) There are 4 girls and they travel a mean distance of 7 miles. Their total distance travelled is:
$4 \times 7 = 28$ miles
(c) The total distance travelled by the boys is $4 + 8 + 6 = 18$ miles.
The total distance travelled by all the children is $18 + 28 = 46$ miles.
The mean distance travelled is $\frac{46}{7} = 6.6$ miles (1 d.p.)

Example 9

Some children look for crabs in rock pools. The frequency table on the right shows how many crabs were found. Each child found either 2 crabs or c crabs.
The mean number of crabs found was 3
Find the value of c. Show all your working.

Number of crabs x	Number of children f
2	6
c	3

> **Note:** When you see the wording 'show all your working', it means you won't be allowed to solve the problem using trial and improvement.

The mean is 3 and the total number of children is 9
Therefore, the total number of crabs found is 27
$$12 + 3c = 27$$
$$3c = 15$$
$$c = 5$$

Choosing an appropriate statistical average

You may be asked which of the three statistical averages may be the most appropriate to use in a particular context.

Example 10

Which would be the most appropriate statistical average to use for each of the following?
(a) The fat content in 8 different microwave meals: 5%, 5%, 6%, 8%, 10%, 15%, 15%, 15%
(b) The cost of these five plane tickets: £140, £95, £165, £130, £3500
(c) The colours of ten cars passing through a checkpoint:
grey, black, white, black, red, white, blue, blue, red, silver, white, yellow

(a) The mean may be a good average to use. The mode, which is 15%, would not be appropriate since 15% is the highest value, hence as an average it would be misleading. The mode is not a good choice of average if it is the highest or lowest value in a dataset.
(b) The median may be appropriate here. The mean should not be used, since the one very high value of £3500 would distort the mean. There is no mode.
(c) The mode, which is white. It is not possible to find the median or the mean with non-numerical data.

Comparing datasets

You may be asked to compare two datasets. You will usually perform two comparisons:
- Compare one of the averages: the mean, median or mode.
- Compare the spread within the two datasets by comparing the ranges or the interquartile ranges.

Example 11

In Orchard Way there are 10 houses. The annual incomes for these households are:
£20 000, £25 000, £32 500, £26 500, £36 000, £300 000, £19 750, £37 000, £42 000, £26 500

(a) Find (i) the mean (ii) the median (iii) the mode and (iv) the range for these income figures.
(b) A researcher is comparing the incomes for these households with those of a nearby street, Manor Way. The averages for Manor Way are shown in the table on the right.
 (i) Choose a suitable average (mean, median or mode) to compare the average household income for these two streets.
 (ii) Compare the spread in the income figures for the two streets.

Manor Way Household Incomes	
Mean	£40 100
Median	£27 500
Mode	£26 000
Range	£75 000

(a) (i) To find the mean from a list of 10 numbers, add them up and divide by 10:
$$\text{Mean} = \frac{20\,000 + 25\,000 + 32\,500 + 26\,500 + 36\,000 + 300\,000 + 19\,750 + 37\,000 + 42\,000 + 26\,500}{10}$$
$$= £56\,525$$

(ii) To find the median, begin by putting the numbers in order from smallest to largest.
£19 750, £20 000, £25 000, £26 500, £26 500, £32 500, £36 000, £37 000, £42 000, £300 000

There are 10 values, so $n = 10$ and $\frac{n+1}{2} = 5.5$

So, we look for the 5th and 6th values in the ordered list, which are £26 500 and £32 500

Find the mean of these two: $\text{Median} = \frac{26\,500 + 32\,500}{2} = £29\,500$

(iii) The mode is £26 500
This is the only figure that appears more than once in the list.

(iv) The range is the difference between the highest and lowest values.
Range = 300 000 − 19 750 = £280 250

(b) (i) When comparing income data, the median is usually the best average to use, since the mean is distorted by very large values. In the case of the Orchard Way data, the mean of £56 525 is not representative of the data, since it is larger than all the values except one.

Comparing the median values, the median in Orchard Way, £29 500 is higher than the median in Manor Way, £27 500

We conclude that the average household income in Orchard Way is higher.

(ii) When comparing the spread, compare the ranges of £280 250 and £75 000
We conclude that there is a greater spread in the Orchard Way household incomes.

Example 12

Look at the data below summarising the scores for two teams that took part in a basketball competition. Compare the statistics for Team A and Team B.

Team A		
Lower quartile	Median score	Upper quartile
25	51	79

Team B		
Lower quartile	Median score	Upper quartile
41	51	63

The median score is the same for each team, telling us that, on average, the two teams were evenly matched.
The interquartile range for Team A's scores is 79 − 25 = 54
The interquartile range for Team B's scores is 63 − 41 = 22
There is a larger interquartile range for Team A. This tells us that Team A's scores were more variable, or that Team B was more consistent.

Exercise 18E

1. Laura's percentage scores in her end of year exams are shown in the table below.

Maths	English Language	English Literature	Science	French	Geography	History	Religious Studies	Art
45	51	27	49	51	42	36	13	51

(a) Laura is not happy with her results. She is wondering how to tell her parents. Which average – the mean, median or mode – should she use when telling her parents her average score, to make her results sound as good as possible?
(b) If Laura chooses this average to make her results sound as good as possible, explain why it could be misleading.

2. In a clothing shop, men's trousers are sold in the following sizes:
28 inches, 30 inches, 32 inches, 34 inches, 36 inches and 38 inches
The manager works out the mean, median and mode size of all the pairs of trousers sold this year. He then places an order for new stock from the supplier. If he can only order one size, which average should he use: the mean, median or mode? Explain your answer.

3. Mr Byrne is setting a maths test for his class of 26 pupils. He wants half of the class to pass and half to fail. Which of the three averages (mean, median or mode) should Mr Byrne use as the pass mark for the test? Explain your answer.

4. Class 4G are going on a school trip. They vote to decide on which type of lunch to have. The results are shown in the table below.

Type of food	Pizza	Sandwich	Chips	Burger	Noodles
Number of votes	7	1	6	6	4

Which type of food should the class have for lunch? Explain your answer.

5. Twelve players took part in a football penalty shoot-out. Each player took 10 penalties. Nine players scored 3 penalties, while the remaining 3 scored x penalties. The frequency table on the right shows this information. Given that the mean number of penalties scored is 4, find the value of x.

Number of penalties scored	Number of players
3	9
x	3

6. Johnson's Pork Pies is a food company that employs 5 people. Their annual salaries are shown in the table below.

Name	Job title	Annual salary
Mr Johnson	Company director	£1 000 000
Mrs Fisher	Secretary	£20 000
Mr Palmer	Pork pie maker	£19 000
Mr Parker	Cleaner	£15 000
Miss Singleton	Trainee	£12 000

   (a) Find the mean salary of the 5 workers at Johnson's Pork Pies.
   (b) Find the median salary.
   (c) Mrs Fisher has been asked to put an advert in the paper for a new worker. She is to include the average salary of workers at the company. Mr Johnson suggests she uses the mean salary, as this will make the company seem a more attractive place to work. Do you think she should use the mean or the median? Explain your answer.

7. Pádraig carries out a survey into the average shoe size of the boys in class 10H. These are the results:
   5, 7½, 6½, 6, 6, 5½, 7, 8, 4½, 7, 8, 6, 7½
   (a) What is the range in the shoe sizes of the boys?
   (b) Which average should Pádraig use if he wishes the result to be one of the shoe sizes in the list?

8. There are 4 houses with gardens in May Tree Close. Three gardens have areas of 25 m^2, 37 m^2 and 64 m^2 Find the area of the fourth garden if the mean of the four areas is 35.5 m^2

9. The 20 pupils in a maths set take a test. For the 8 boys the scores are:
   6 7 2 5 6 8 10 4
   (a) Find the mean score for the boys.

   For the 12 girls, the mean score is 8
   (b) For these 12 girls, what is their **total** score?
   (c) Find the mean score for the whole class.

10. Lucy records daily rainfall amounts in millimetres for the month of June. Her results are listed below.

0	2	1	0	3	5	1	2	2	3
1	0	0	0	2	1	0	0	2	4
3	0	0	0	0	1	5	3	2	0

   (a) Find the range in the daily rainfall.
   (b) Explain why the mode would not be a suitable choice of average.

11. Colin records the number of birds visiting his garden between 10 and 11 am every day for 15 days. His results are shown below.
    3 8 2 1 6 5 5 2 4 5 6 2 0 1 3
    Colin says 'The median is 2, because that's the number in the middle.' Explain why Colin is not correct.

12. Two environmental scientists, Clare and Callum, are studying the number of insects on a patch of land. The table on the right shows the number of insects found in their samples during one week.
    (a) Find the median number of insects for both Clare and Callum.
    (b) Find the interquartile range for both Clare and Callum.
    (c) Using your answers to parts (a) and (b), compare Clare's data with Callum's.

Day	Clare	Callum
Sunday	55	40
Monday	58	92
Tuesday	71	122
Wednesday	65	138
Thursday	70	86
Friday	165	52
Saturday	93	138

18.6 Summary

In this chapter you have learnt that:
- The mean, median and mode are the three types of average commonly used.
- The mean is the sum of all the values divided by the number of values.
- The median is the value in the middle of the list of ordered values.
- The mode is the value that occurs most frequently.
- The range is a measure of spread within a dataset. It is calculated as the difference between the highest and lowest values.
- The lower quartile and upper quartile are the values that are 25% and 75% of the way through the ordered list.
- The interquartile range is the difference between the upper quartile and lower quartile.
- To calculate the mean from a frequency table, add a column labelled fx, which is the x value multiplied by its frequency.
- In the case of grouped frequency tables, the x value is the midpoint of the group.
- The mean of the frequency table is given by $\frac{\Sigma fx}{\Sigma f}$
- To find the median, add a cumulative frequency column. Calculate $\frac{n+1}{2}$, where $n = \Sigma f$
  Then use the cumulative frequencies to find the median or median group.
- The mode is the value with the highest frequency.
- When comparing two datasets it is common to compare an average (either mean, median or mode) and to compare a measure of spread (either range or interquartile range).
- It is important to have a good understanding of statistical averages, the range and frequency tables, because a wide variety of questions can be asked.

Chapter 19
Statistical Diagrams

19.1 Introduction

In this chapter you will learn to construct and interpret a wide range of graphs and diagrams: stem and leaf diagrams, pie charts, two-way tables and Venn diagrams. Scatter graphs are considered in the next chapter.

Key words

- **Stem and leaf diagram**: Represents numerical data in a graphical way, to help visualise the data.
- **Pie chart**: A circular chart divided into sectors that is used to display data. The area of each sector is proportional to the frequency.
- **Two-way table**: A table whose rows and columns shows information categorised in two different ways.
- **Venn diagram**: A diagram comprising some circles and an enclosing rectangle. The rectangle represents all the data, while the circles represent individual categories.

Before you start you should:

- Recall the work covered in Chapter 18 on frequency tables.
- Understand how to calculate the number of items in a category, given certain information such as the total and the numbers in other categories.

In this chapter you will learn about:

- Stem and leaf diagrams.
- Bar charts.
- Pie charts.
- Line graphs.
- Two- and three-circle Venn diagrams.

Exercise 19A (Revision)

1. At a party, there are 25 balloons. Six balloons are red, three are blue, nine are green and the rest are orange. Work out:
   (a) how many balloons are orange,
   (b) how many balloons are red or orange,
   (c) how many balloons are not red,
   (d) the **percentage** of blue balloons.

2. Kaff Coffee is sold in packets. Deirdre measures the weights of 100 packets from her stock. The results are shown in the frequency table on the right.
   (a) Write down the modal class.
   (b) Find the median class.
   (c) Calculate an estimate for the mean weight of a packet of Kaff Coffee.

Weight w (g)	Frequency f
$240 \leq w < 245$	10
$245 \leq w < 250$	17
$250 \leq w < 255$	37
$255 \leq w < 260$	25
$260 \leq w < 265$	11

19.2 Stem and Leaf Diagrams

You need to know how to construct and read a stem and leaf diagram. The next example demonstrates the construction.

Example 1

15 people are playing in a 5-a-side football competition. Their ages are as shown on the right.
(a) Display these results in a stem and leaf diagram.

32	19	12	48	46
16	13	21	22	30
30	50	17	26	54

(b) What is the modal age of these players?
(c) What is the range in their ages?

(a)

```
1 | 2 3 6 7 9
2 | 1 2 6
3 | 0 0 2
4 | 6 8
5 | 0 4
```

Choose a suitable stem. In this case it makes sense to use the 'tens' part of the age. Then the 'leaf' section of the diagram contains the units part of each age.

In the 'leaf' section, the numbers in each row should be ordered from smallest to largest. Keep these values lined up in columns.

When you have put all the data in the diagram, make sure you have the correct number of items by counting them.

Key: 1 | 2 means 12 ← Always include a key.

(b) The modal age is 30
This is the age that appears most frequently.

(c) The oldest player is 54 and the youngest is 12
The range is 54 − 12 = 42

You may come across a back-to-back stem and leaf diagram, such as in the following example. Note that the lowest leaf values are closest to the stem on both sides of the diagram.

Example 2

The back-to-back stem and leaf diagram on the right shows the heights of 24 pupils in a class: 12 girls and 12 boys.

(a) What is the difference in the median heights for the girls and the boys?
(b) Give one advantage of displaying the data in a stem and leaf diagram.

```
       Girls        Boys
           4 | 12 |
       7 5 2 | 13 | 3 7
       6 5 3 | 14 | 1 3 5 7
       8 7 2 | 15 | 0 6 7
           2 | 16 | 2 7
             | 17 | 1
```

Key (boys): 13 | 3 means 133 cm
Key (girls): 4 | 12 means 124 cm

(a) There are 12 girls and 12 boys. We look for the 6th and 7th people for both the boys and the girls. Take the mean of the 6th and 7th heights.

For the girls: median = $\frac{143 + 145}{2}$ = 144 cm.   For the boys: median = $\frac{147 + 150}{2}$ = 148.5 cm.

The difference is 148.5 − 144 = 4.5 cm.

(b) A stem and leaf diagram allows you to visualise the distribution. For example, from the shape of this diagram it is clear that the boys are generally taller. The girls' heights occupy the first five rows of the table from 120 cm to 160 cm. The boys' heights are in the 130 cm to 170 cm rows.

Exercise 19B

1. Plot these values in a stem and leaf diagram. On the leaf side of the diagram, remember to keep the columns lined up.

    (a) 77   51   79   40   49
         64   60   28   43   58

    (b) 5.7 3.4 8.5 9.6 6
         5.4 3.5 8.1 7.5 4.4

    (c) 0.13 0.4   0.35 0.16 0.7
         0.39 0.23 0.6   0.34 0.1
         0.33 0.12 0.12 0.42 0.8

2. The stem and leaf diagram on the right shows the ages of the people in a TV talent show. Each person takes part on their own.
    (a) How many people take part in the talent show?
    (b) What is the modal age of the people taking part?
    (c) What is the median age of the people taking part?
    (d) What fraction of the competitors are under 20?

```
0 | 7 8
1 | 0 2 5 5 8 9
2 | 0 1 3 7 9
3 | 1 2 3 6 7
4 | 0 5 7 9
5 | 1 3
```

Key: 1 | 0 means 10

CHAPTER 19: STATISTICAL DIAGRAMS

3. The stem and leaf diagram on the right shows the ages of people in a hospital waiting room.
   (a) How many people are in the waiting room?
   (b) What is the modal age of the people waiting?
   (c) What is the median age of the people waiting?
   (d) If a person under 16 years of age is considered a child, what fraction of those waiting are children?

```
0 | 9
1 | 1 5
2 | 0 3 7
3 | 2 3 7
4 | 1 6 7 9
5 | 0 0 4 8
6 | 1 1 1 4 7
7 | 0 2 6 9
8 | 1 2 3 5
```
Key: 1 | 5 means 15

4. The back-to-back stem and leaf diagram on the right shows the cost of petrol in pence per litre in 22 petrol stations: eleven in Belfast and eleven in Omagh.
   (a) Compare the distribution of prices in Belfast with the distribution in Omagh.
   (b) State one advantage of displaying the data in a stem and leaf diagram.

Belfast		Omagh
9 8	12	
7 4 3	13	4 7
8 6 3	14	0 1 5 7
6 5 1	15	0 6 8
	16	2 3

Key (Belfast): 4 | 13 means £1.34
Key (Omagh): 13 | 4 means £1.34

5. There are 20 patients in a dentist's waiting room. They are asked how many minutes they have been waiting to see the doctor. Their answers are as shown on the right.
   (a) Display these results in a stem and leaf diagram.
   (b) Find the median time these patients have waited.

12	38	41	32	38
23	41	32	45	25
23	42	7	11	16
51	53	6	50	53

6. Pete asks his 15 colleagues how many hours they normally work in a week. The results are shown in the stem and leaf diagram on the right.
   (a) Find (i) The median number of hours worked. (ii) The mode. (iii) The range.
   (b) Pat joins Pete's company. She works 43 hours per week. Pete adds this value to the stem and leaf diagram.

```
0 | 9
1 | 1 3 5 8
2 | 2 3 5 7
3 | 1 5 5 8
4 | 0 2
```
Key: 1 | 3 means 13 hours

In each line below, choose the correct word or words to make the sentence true.

The mode will:	decrease	stay the same	increase
The median will:	decrease	stay the same	increase
The range will:	decrease	stay the same	increase

7. The stem and leaf diagram on the right shows the life expectancy in 13 Asian countries.
   (a) In Saudi Arabia, the life expectancy is 76.9 years. In South Korea, the life expectancy is 83.5 years. Copy the diagram and add these two values.
   (b) For these 15 countries, what is:
      (i) the median life expectancy,
      (ii) the modal life expectancy,
      (iii) the range?
   (c) What fraction of these countries have a life expectancy of greater than 80?

```
76 | 4
77 |
78 | 2 7 7 7 8
79 | 3 9
80 |
81 |
82 | 5
83 | 4
84 | 5
85 | 4 5
```
Key: 76 | 4 means 76.4 years

19.3 Pie Charts

A pie chart is a circular representation of data. It is a useful way to display data when you want to show how something is divided up.

The pie chart on the right shows how many students in a year group got each grade in an exam.

Drawing pie charts

The next example shows how to calculate the angle for each sector of the circle.

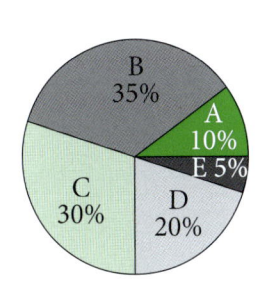

175

GCSE MATHEMATICS M3 AND M7

Example 3

Keith is an office worker. The table on the right shows how he spends a typical day.

Draw a pie chart to show this data.

Activity	Time	Angle
Sleeping	8 hours	
Travel	2 hours	
Working	8 hours	
Eating	2 hours	
Sport	1 hour	
Other	3 hours	

There are 360° in a full circle and 24 hours in a day.

Sleeping takes up $\frac{8}{24}$ of Keith's day.

To find the angle for the Sleeping sector in the pie chart, find $\frac{8}{24}$ of 360°

Use the third column in the table for the angle calculations.

Activity	Time	Angle
Sleeping	8 hours	$\frac{8}{24} \times 360° = 120°$
Travel	2 hours	$\frac{2}{24} \times 360° = 30°$
Working	8 hours	$\frac{8}{24} \times 360° = 120°$
Eating	2 hours	$\frac{2}{24} \times 360° = 30°$
Sport	1 hour	$\frac{1}{24} \times 360° = 15°$
Other	3 hours	$\frac{3}{24} \times 360° = 45°$

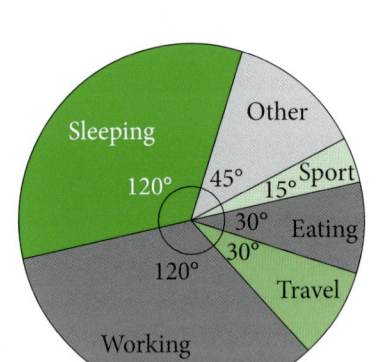

Then draw the pie chart. Measure the angles carefully with a protractor. The chart is shown above.

Exercise 19C

1. The favourite sports of the girls in year 11 are shown in the table on the right. Draw a pie chart to show this information. Use the third column in the table to calculate the size of the angle for each sector.

Sport	Number of girls	Angle (°)
Hockey	16	
Netball	10	
Swimming	12	
Football	8	
Other	2	

2. Draw pie charts to represent the following data.
   (a) Favourite subjects for the pupils in class 10F.
       Maths 7   English 8   Science 4   French 2   PE 10   Other 5
   (b) The colours of the cars on Paul's road.
       White 24   Black 6   Silver 16   Blue 8   Red 10   Other 16

3. The ingredients on a small tub of ice cream are shown in the table on the right. The tub contains 180 g of ice cream. A pie chart is then drawn to show this information.

   Note: You do not have to draw the pie chart in this question.

Ingredient	Weight (grams)	Angle
Milk		100°
Cream	40	
Sugar		70°
Egg	30	
Fruit		
Total	180	

   Use the information in the table to work out:
   (a) the weight of the milk,   (b) the weight of the sugar,
   (c) the weight of the fruit,   (d) the angle used for the fruit.

4. Georgia has a weekend job and earns £240 a month. She spends this money as follows.

   Savings £150   Clothes £30   Books £24   Cinema £20   Other £16

   Draw a pie chart to show how Georgia spends her money.

CHAPTER 19: STATISTICAL DIAGRAMS

5. The bar chart on the right shows the favourite flavour of fizzy drink of 60 students. Show this information on a pie chart.

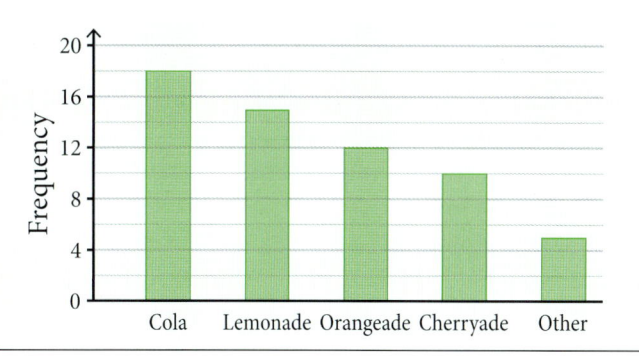

Interpreting pie charts

If you are given a pie chart, you can work out the number of items in each category. You need to know the total number of items.

Example 4

Between 2003 and 2022 only 4 players won the men's tennis championships at Wimbledon. In 2020 the championships were cancelled. The pie chart shows the percentage of wins for each of these 4 players during this 20 year period. 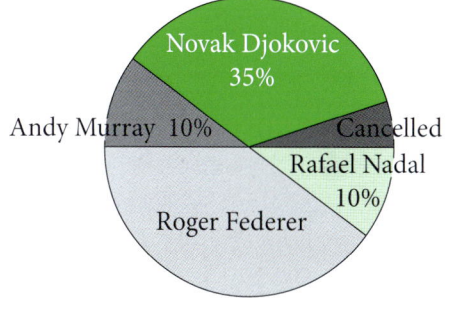
(a) The tournament was cancelled only once. What percentage was this?
(b) How many times did Novak Djokovic win the tournament during this time?
(c) Work out the percentage of wins for Roger Federer.
(d) Máire says 'If Roger Federer had won just one more time his sector would take up half of the pie chart.' Is she right? Explain your answer.

(a) One tournament out of 20 = $\frac{1}{20}$ = 5%
(b) To calculate Djokovic's wins: 35% of 20 = $\frac{35}{100} \times 20 = 7$
(c) To calculate the percentage of wins for Federer, add all the other percentages: 10% + 35% + 5% + 10% = 60%
Then Federer's percentage = 100% − 60% = 40%
(d) One year is $\frac{1}{20}$ or 5%. If Federer had one more win, his sector would increase in size from 40% to 45%. So, Máire is not correct. Half of the pie chart would be 50%.

Exercise 19D

1. The pie chart shows the types of shops in a busy shopping area of Belfast.
   (a) Which is the most common type of shop in this area?
   (b) Jenny says 'There are twice as many clothing shops as health and beauty shops.' Is she correct? Explain your answer.
   (c) If there are 60 shops altogether, estimate the number of clothing shops.

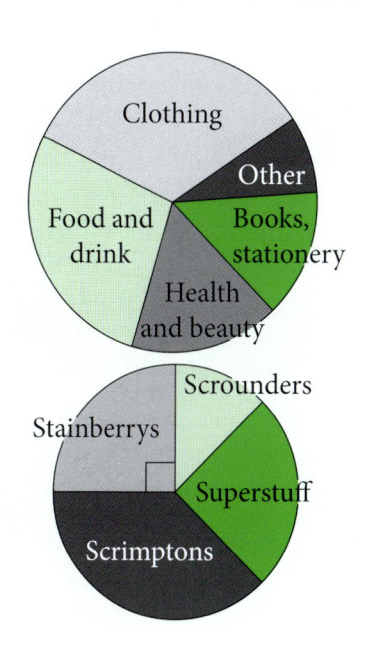

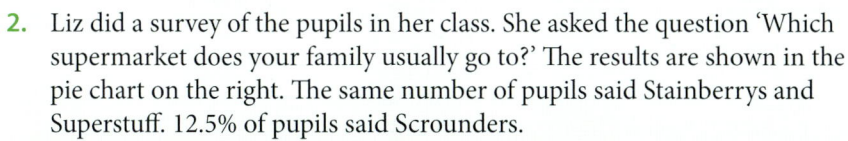

2. Liz did a survey of the pupils in her class. She asked the question 'Which supermarket does your family usually go to?' The results are shown in the pie chart on the right. The same number of pupils said Stainberrys and Superstuff. 12.5% of pupils said Scrounders.
   (a) What percentage said Scrimptons?
   (b) There are 32 pupils in the class. Find the number of pupils who said each of the four supermarkets.

3. Torin asked all the pupils in his class what they liked to drink with their school dinner or packed lunch. The results are shown in the pie chart on the right, but Torin has forgotten to put the numbers on the chart. Torin remembers that:
   - he asked 20 pupils altogether,
   - the same number of pupils said fruit juice and fizzy drink,
   - 15% said milk.

   Find the number of pupils who said:
   (a) fruit juice (b) fizzy drink (c) milk (d) water.

4. The pie chart shows how the UK generated its electricity in 2022. The angles for the sectors 'Other' and 'Coal and oil' are missing.
   (a) Given that the 'Other' sector is twice as large as the 'Coal and oil' sector, calculate both these missing angles.
   (b) What **percentage** of the UK's electricity came from gas in 2022?
   (c) Wind energy is usually defined as a type of renewable energy. The 'Other renewables' sector includes solar, tidal and hydro energy. Joel says that over 30% of the UK's electricity came from renewables in 2022. Is he right? Explain your answer.

5. The pie charts on the right show the constituent food types of cheese and eggs.
   (a) Calculate the angle for the 'Other' sector in the Eggs pie chart.
   (b) A café wants to give nutritional information on its menu. Copy and complete the information sheet (below the pie charts) for the café, giving your answers as percentages to the nearest whole number.
   (c) Which has more fat: 100 g of cheese or 200 g of eggs? Explain your answer.

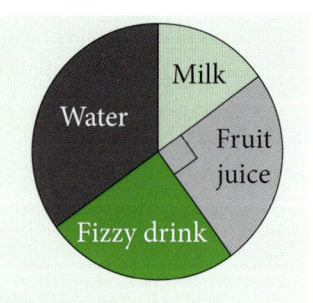

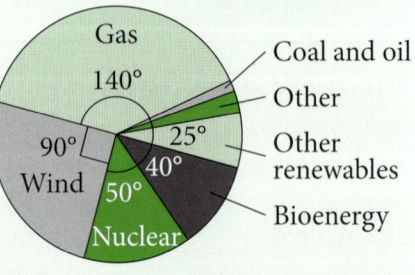

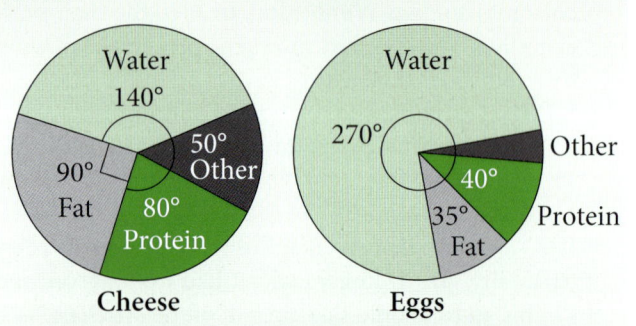

Food type	Cheese	Eggs
Water		
Fat	25%	
Protein		
Other		

19.4 Two-Way Tables

A two-way table is a table that shows information categorised in two different ways. The table's rows and columns are used for the different categories. A frequency tree is a diagram showing how a collection of people or items can be split up into two or more categories. You may be asked to fill in missing values in both two-way tables and frequency trees.

Example 5

In November each year, some people wear a poppy to remember those who have fought and died in wars. One November, Ryan carries out a survey into how many people on TV are wearing a poppy. He included 100 people on various TV programmes. Ryan's results are shown in the two-way table. Copy and complete the table, filling in the values for A, B, C, D and E.

	Wearing a poppy	Not wearing a poppy	Total
Men	A	17	B
Women	40	C	D
Total	78	E	100

Ryan surveyed 100 people. In total 78 were wearing a poppy, so 22 were not, so E = 22
In total 78 people wore a poppy. 40 were women, so 38 were men. A = 38
In total 22 were not wearing a poppy. 17 were men, so 5 were women. C = 5

Looking at the total column:
B = 38 + 17 = 55
D = 40 + 5 = 45

The completed table is shown on the right.

	Wearing a poppy	Not wearing a poppy	Total
Men	38	17	55
Women	40	5	45
Total	78	22	100

Exercise 19E

1. Carla rolls a dice and tosses a coin 30 times. The two-way table shows her results. Copy and complete this table.

	Heads	Tails	Total
Not a 6	11		
6	4	3	
Total			

2. Twenty-eight workers work in a small supermarket. The workers are asked when they could work over the Christmas period. The two-way table shows the results.
   (a) Copy and complete the table.
   (b) How many female workers can work at weekends?

	Weekdays only	Weekends only	Both	Total
Male	4			11
Female		6		
Total	6		14	28

3. An airline operates 100 flights from Belfast City airport. Look at this frequency tree. It shows the number of flights that leave and arrive early, on time and late.
   (a) Copy and complete the frequency tree, filling in the two missing numbers.
   (b) How many flights in total arrived on time?
   (c) What **percentage** of flights arrived late?
   (d) Copy and complete the two-way table below to show the same information.

	Arrive early	Arrive on time	Arrive late	Total
Leave on time				
Leave late				
Total				100

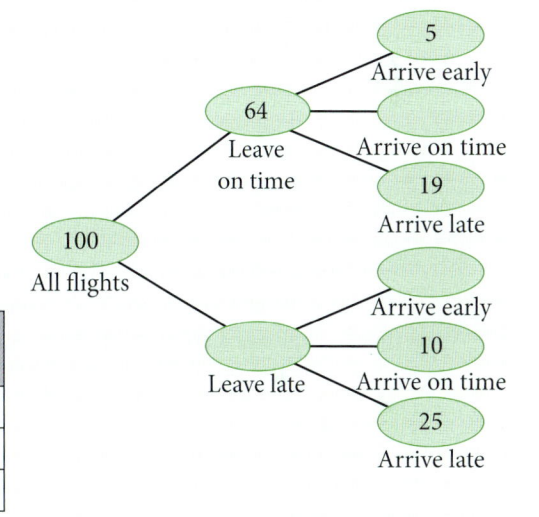

19.5 Venn Diagrams

A Venn diagram is a way in which we can present data. We use a rectangle to show all the data. Inside the rectangle are circles to show the data in particular categories. We show data that belongs in two or more different categories in the intersection (overlap) of the circles.

Two-circle Venn diagrams

In some cases, there is no intersection between two circles.

Example 6

In Class 11P there are 28 pupils. On Tuesday, 15 of them came to school by car. Ten of them walked.
(a) How many pupils neither walked nor came by car?
(b) Draw a Venn diagram to show this data.

(a) There are 28 pupils in the class.
Adding the numbers who walked or came by car gives: 15 + 10 = 25
To find the number who neither walked nor came by car: 28 − 25 = 3
(b) The Venn diagram contains two non-overlapping circles, because none of the pupils walked **and** came by car. The three pupils who neither walked nor came by car are inside the rectangle, but outside the two circles.

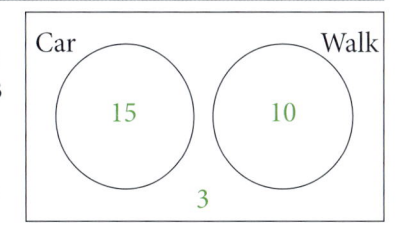

GCSE MATHEMATICS M3 AND M7

In other cases, two circles intersect.

Example 7

In Class 11P there are 28 pupils. Of these 28 pupils, 9 wear glasses and 15 have school dinners. Four pupils wear glasses **and** have school dinners. Draw a Venn diagram to show this information.

Four pupils wear glasses and have school dinners.
This number goes in the intersection.
There are 9 pupils altogether in the Glasses circle, so 5 are in the 'glasses only' section.
There are 15 pupils altogether in the School dinner circle, so 11 are in the 'school dinner only' section.
Adding up the numbers inside the circles: 5 + 4 + 11 = 20
There are 28 pupils, so 8 goes outside the circles.

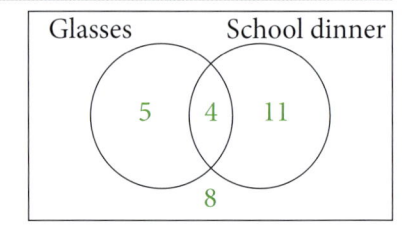

Exercise 19F

1. 60 people are in the audience at a concert. 20 are girls and 15 are boys. The rest are adults.
   (a) How many adults are in the audience?
   (b) Draw a Venn diagram showing this information.

2. At Kingsmead School, every member of Year 9 was asked whether they were a member of Scouts or Guides and a youth club. Look at the Venn diagram on the right showing the results.
   (a) How many pupils are in Year 9?
   (b) How many pupils are members of both Scouts or Guides and youth club?
   (c) How many pupils are members of only one of these organisations?
   (d) How many pupils are members of neither?

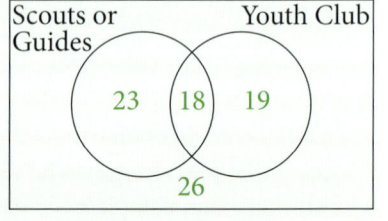

3. Maisie is a dog. When she sleeps, she sometimes dreams about walks and sometimes about bones. Sometimes she dreams about both. In one week, Maisie dreams about walks 25 times and about bones 18 times. 6 of her dreams featured both walks and bones. Only 7 of her dreams did not involve walks or bones.
   (a) Draw a Venn diagram to show this information.
   (b) How many dreams did Maisie have this week?

4. In an aquarium, there are 60 puffer fish and 25 catfish. There are 150 fish altogether.
   (a) How many fish are neither puffer fish nor catfish?
   (b) Draw a Venn diagram to show this information.
   (c) What percentage of the fish are puffer fish?

5. 100 people were asked whether they had been to the Giants Causeway, the Titanic Exhibition in Belfast, both or neither. 62 people said they had been to the Giants Causeway, while 38 said they had been to the Titanic Exhibition. 25 had been to both.
   (a) How many people had been to neither the Giants Causeway nor the Titanic Exhibition?
   (b) Draw a Venn diagram to show this information.
   (c) How many people had **not** been to the Giants Causeway?
   (d) What percentage of people had only been to one of these attractions?

6. On a Cub camp, the children are given the option to take part in archery or caving. Children could only choose one of these activities, or they could choose not to do either. There are 50 children on the Cub camp. 20 children choose archery, while 27 choose caving.
   (a) How many children choose to do neither activity?
   (b) Draw a Venn diagram to demonstrate this information.

   The following year, the rules are changed so that children can take part in both activities. This time, out of 50 children, 25 choose archery and 30 choose caving. Four children choose to do neither activity.
   (c) How many children choose to do both activities?
   (d) Draw a Venn diagram to show this information.

Three-circle Venn diagrams

When there are three overlapping circles, begin by filling in the central section of the diagram, where all three circles intersect.

CHAPTER 19: STATISTICAL DIAGRAMS

Example 8

In a hotel breakfast room one morning, the number of guests choosing to have cereal, a cooked breakfast and coffee was recorded. One hundred guests came for breakfast.
30 guests chose to have cereal, cooked breakfast and coffee.
18 chose to have cereal and coffee but no cooked breakfast.
5 had a cooked breakfast and coffee but no cereal.
In total 40 guests chose both cereal and cooked breakfast, including those having coffee.
In total 70 had cereal; 53 had a cooked breakfast and 66 had coffee.

(a) Copy and complete the Venn diagram on the right to show this information.

> **Note:** In the diagram, the regions are labelled A to H for clarity, but they may not be labelled in exercise or exam questions.

(b) How many guests in this survey **did not have** cereal, a cooked breakfast or coffee?
(c) How many guests did not have coffee?
(d) What **percentage** of guests chose not to have cereal or cooked breakfast?

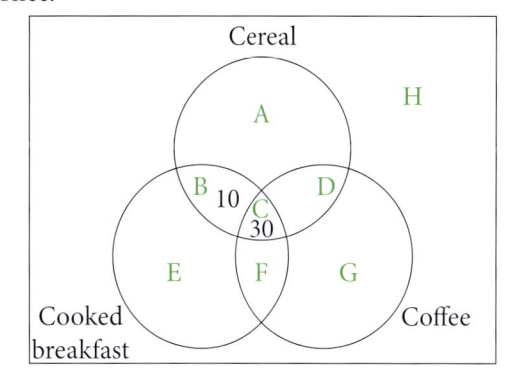

(a) 30 guests chose all three of cereal, a cooked breakfast and coffee, so this number goes in the middle of the diagram, region C.
18 chose to have cereal and coffee but no cooked breakfast, so this number goes in region D.
5 had a cooked breakfast and coffee but no cereal, so this number is in region F.
40 guests in **total** have both cereal and cooked breakfast. This is the total for regions B and C. Since region C has 30, region B must have 10
In total 70 had cereal. 12 must appear in region A, so that the sum of the numbers in the cereal circle is 70
Likewise, 8 must appear in region E to make the cooked breakfast circle add up to 53
And 13 must appear in region G so that the coffee circle adds up to 66

(b) Adding the numbers in regions A to G gives 96
Therefore, there must be 4 guests in region H to give a total of 100

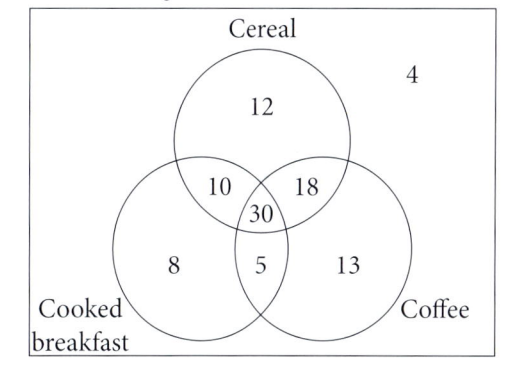

(c) 66 guests out of 100 chose coffee, so 100 − 66 = 34 did not.

(d) The guests having neither cereal nor a cooked breakfast are in regions G and H.
The total is 13 + 4 = 17 $\frac{17}{100} = 17\%$

Exercise 19G

1. In a sixth form college there are 100 students.
   83 of the students study chemistry, biology or physics.
   15 of these students study all three subjects.
   18 study both chemistry and biology but not physics.
   12 study both chemistry and physics but not biology.
   30 in total study both physics and biology.
   In total: 51 study chemistry; 52 study biology; and 55 study physics.
   (a) How many students do not study any of these three subjects?
   (b) Draw a Venn diagram to show this information.

2. In a doctor's waiting room, there are 16 people.
   One person is wearing a hat, glasses and a coat.

Two people are wearing a hat and glasses.
Three people are wearing a hat and a coat.
Four people are wearing glasses and a coat.
Six people are wearing a hat.
Eight people are wearing glasses.
Eight people are wearing a coat.
(a) Draw a Venn diagram to show this information.
(b) How many people are not wearing a hat, glasses or a coat?

3. There are 25 pupils in Lucy's class. Lucy conducts a survey on three traditional Christmas foods: turkey, Brussels sprouts and Christmas pudding. Of the 25 pupils:
Only 2 said they would eat all 3 types of food.
6 said they would eat Brussels sprouts and Christmas pudding.
4 said they would eat Brussels sprouts and turkey.
4 said they would eat turkey and Christmas pudding.
10 said they would eat turkey.
11 said they would eat Brussels sprouts.
13 said they would eat Christmas pudding.
(a) Draw a Venn diagram showing this information.
(b) How many pupils said they would eat none of these types of food?

4. Forty children are surveyed about whether they have seen the films Shrek, Paddington and Frozen. Copy the Venn diagram on the right.
(a) 7 said they had seen Shrek only, while 5 said they had seen Shrek and Paddington, but not Frozen. Complete the 'Shrek' circle on the diagram.
(b) In total, 16 said they had seen Paddington. Complete the 'Paddington' circle.
(c) In total, 18 said they had seen Frozen. Complete the 'Frozen' circle.
(d) How many of the 40 children had not seen any of these three films? Enter this number on your diagram.

5. The Venn diagram shown on the right has three circles representing even numbers, prime numbers, and multiples of 3, for all the numbers between 1 and 20

A zero has been placed at the centre of diagram, because there are no numbers that are even, prime and multiples of 3. The number '3' has been placed at the intersection of the even and multiples of 3 circles because there are three numbers that are in both these categories: 6, 12 and 18.
(a) How many numbers are both even and prime?
(b) Copy and complete the Venn diagram.
(c) What number should go outside all of the circles?

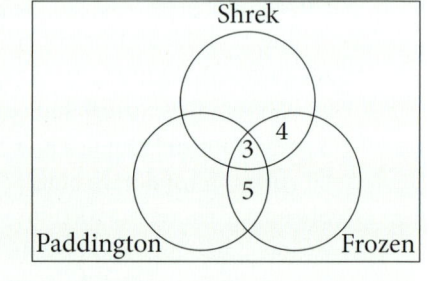

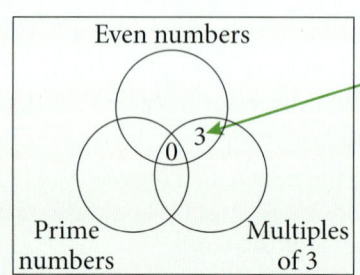

19.6 Summary

In this chapter you have learnt:
- About various types of graph: stem and leaf diagrams; pie charts; two-way tables and Venn diagrams.
- That a stem and leaf diagram has a single column for the 'stem', which is usually a single digit, and several columns for the 'leaves'. Each leaf represents a single value.
- That a pie chart is a circular representation of data. It is a useful way to display data when you want to show how something is divided up.
- That a two-way table is a table that shows information categorised in two different ways. The table's rows and columns are used for the different categories.
- That a Venn diagram is a way to present data in different categories. A rectangle contains all the data. Inside the rectangle are two or three circles to show the data in the different categories. If data belongs in two or more different categories it lies in the intersection (overlap) of the circles. In some cases, the circles do not intersect.

Chapter 20
Scatter Graphs

20.1 Introduction

In the news you may hear statements such as 'the high temperatures are causing a large number of wildfires in the countryside'. Such a statement may be supported by research aiming to show a connection between the temperature and the number of fires.

A **scatter graph** (or scatter diagram) is a way to compare two variables, for example the number of fires and the temperature. The scatter graph should allow us to see whether there is a relationship between them.

Key words
- **Correlation**: The relationship between two variables. In this chapter we discuss positive correlation, negative correlation and no correlation.
- **Line of best fit**: A straight line that passes through the middle of the points plotted on a scatter graph.
- **Outlier**: A point that doesn't follow the pattern followed by most of the other points on the scatter graph.
- **Causality**: Whether a change in one variable causes a change in the other.

Before you start you should:
- Know how to plot points on a coordinate grid.
- Understand the **gradient** of a straight line.

What you will learn
In this chapter you will learn:
- How to plot a scatter graph.
- About correlation (positive, negative and no correlation).
- How to add a line of best fit to a scatter graph.
- How to estimate the value of a variable using the line of best fit.
- About outliers.
- About causality.

Exercise 20A (Revision)

1. (a) Plot these points on a copy of the coordinate grid given on the right.
   A(0, 2), B(1, 0), C(2, 1), D(3, 0), E(4, 2)
   (b) Join each point to the next using straight lines. What letter do you get?

2. Look at the line plotted on the grid on the right.
   (a) Does the line have a positive gradient or a negative gradient? Give your reason.
   (b) Find the gradient of the line.

3. A straight line has the equation $y = 5x - 3$
   Write down the line's gradient.

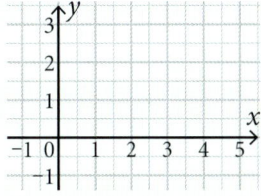

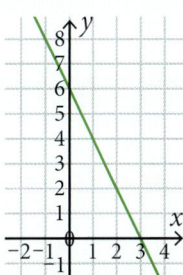

20.2 Correlation

The relationship between two sets of data is called the **correlation**. There are three types of correlation:

Positive correlation	Negative correlation	No correlation or little correlation
When one variable increases, the other variable increases.	When one variable increases, the other variable decreases.	There is no relationship between the variables.

In the previous section we discussed temperature and wildfires. The scatter graph on the right shows the daytime maximum temperature plotted against the number of wildfires reported in the Mourne Mountains over ten summer days.

There appears to be a **positive correlation** between the temperature and the number of wildfires in the Mournes.

You may be asked to plot or complete a scatter graph, as in the next example.

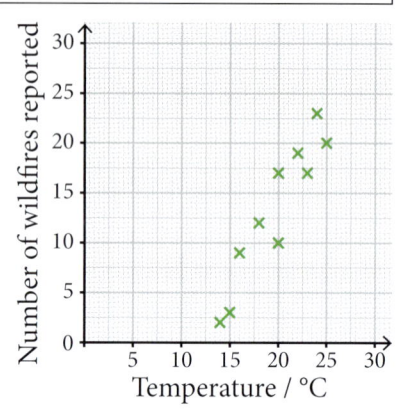

Example 1

In the table that follows:
- x is the finishing position of the fastest 10 runners in the Wallace Park Parkrun on Saturday, 1 June.
- y is the finishing position of the same runners in the same run on Saturday, 8 June.

x	1	2	3	4	5	6	7	8	9	10
y	2	1	3	5	4	6	7	10	8	9

(a) Plot a scatter graph for these data.

> **Note:** The variable in the first row of the table should be plotted in the x-direction.

(b) Suggest the type of correlation, if any, between these two variables.
(c) Suggest a reason for this type of correlation.

(a) A scatter graph is shown for these data on the right.
(b) The scatter graph shows a positive correlation.
(c) Each runner is likely to complete the race in a similar time from one week to the next; so a runner's two positions are likely to be similar.

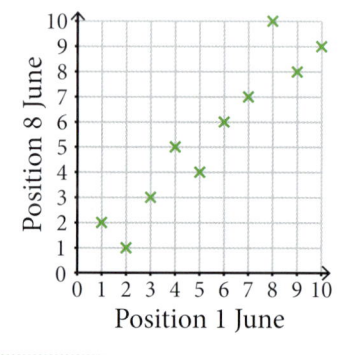

Example 2

The graph on the right shows the total sales of jam plotted against the sales of butter in Belfast during the years of World War II, 1939 to 1945. State the type of correlation between the sales of jam and butter and interpret your result.

There appears to be no correlation or little correlation between butter and jam sales during these years. Higher sales of butter are not associated with higher sales of jam during these years.

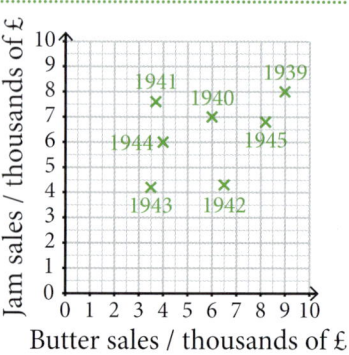

CHAPTER 20: SCATTER GRAPHS

Exercise 20B

1. In each part, state whether you would expect a positive correlation, a negative correlation or no correlation between the variables.
   (a) The weight and age of 10 toddlers.
   (b) The speed of 30 cars on the A1 and the time it takes them to get from Banbridge to Dromore.
   (c) The number of doors in a house and the number of windows.
   (d) The number of siblings and the distance travelled to school, for each person in a class.
   (e) The temperature and the number of cold drinks sold by a café.
   (f) The size of Sally's posters and the number she can fit onto her wall.
   (g) The height and weight of 20 police officers.

2. Look at the three graphs on the right. State which graph shows:
   (a) a positive correlation
   (b) a negative correlation
   (c) no correlation.

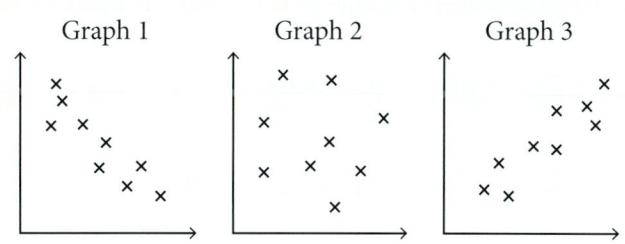

3. Look at the scatter graph below showing the age and value of 9 cars. Do you think the age of a car affects its value? If so, how?

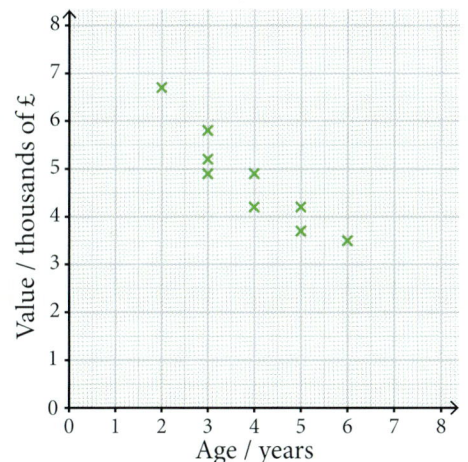

4. The graph below shows the number of hours of recorded sunshine in two towns, Alfreton and Belper, for the first 9 days of June 2021.

   State the type of correlation between the hours of sunshine in the two towns and interpret your result.

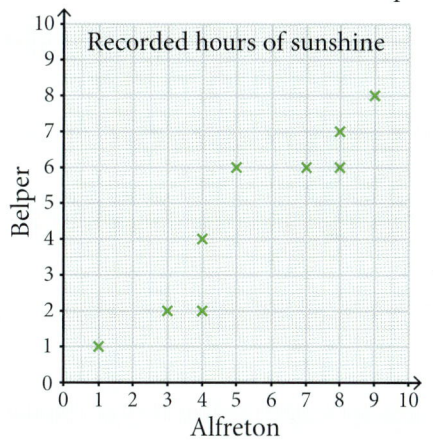

5. Marie and John both fly from Belfast to London regularly for meetings. The graph below shows the number of flights they each took for the years 2010 to 2019.
   (a) State the type of correlation between the number of flights Marie takes and the number of flights John takes.
   (b) Do you think Marie and John work together? Explain your answer.

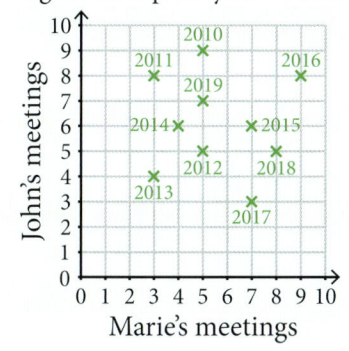

6. A shop sells second hand mobile phones. The manager thinks there is a relationship between the battery life and the age of the phone. He collects the appropriate data for 12 phones and plots a graph.
   (a) Considering only the phones that are less than 18 months old, what type of correlation exists between the age and the battery life?
   (b) Considering only the phones that are over 18 months old, what type of correlation exists?

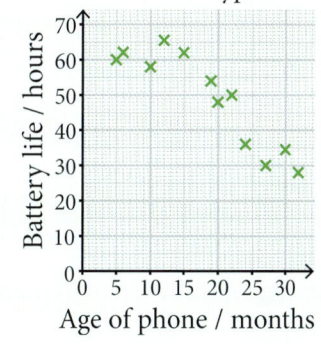

7. Ten pupils took a maths test and a science test. Their scores (out of 20) are shown in the table below.

Maths score	18	6	14	12	10	8	16	14	16	20
Science score	18	4	12	6	10	6	14	10	16	16

(a) Plot a scatter graph for these data.
(b) State what type of correlation exists, if any.

8. Ten families are asked to record, over many years, their spending **per child** on school uniform. The number of children in these families ranges from 1 to 5. The information is shown in the table below.

Number of children in family	2	4	3	1	5	3	4	1	2	5
Spending per child on school uniform (£)	1200	850	1050	1400	750	970	920	1650	1350	775

(a) Plot a scatter graph for these data.
(b) What type of correlation exists?
(c) Suggest a possible reason this type of correlation exists.

9. A café in Portstewart records the number of cold drinks and hot drinks it sells each day, along with the air temperature, for 10 days in August. The data is shown in the table.

Temperature (°C)	18	15	14	16	20	23	25	24	22	20
Number of cold drinks sold	16	15	14	20	34	32	30	35	31	25
Number of hot drinks sold	18	20	29	21	9	6	2	3	3	7

(a) Plot a scatter graph of temperature against the number of **cold** drinks sold.
(b) What type of correlation exists, if any, in this case?
(c) Plot a scatter graph of temperature against the number of **hot** drinks sold.
(d) What type of correlation exists, if any, in this case?
(e) Explain why such a correlation exists for (i) the cold drinks and (ii) the hot drinks.

20.3 The Line of Best Fit

If there is a correlation between two variables, a **line of best fit** can be drawn through the points on the scatter graph. There should be roughly the same number of points on either side of the line.

When drawing a line of best fit, you may find it useful to draw a 'bubble' around the points before drawing the line. The bubble tells you roughly in which direction the line should go.

Example 3

Look again at the graph of maximum daytime temperature plotted against the number of wildfires in the Mourne Mountains (start of section 20.2). Draw a line of best fit through the points on the graph.

The graph is reproduced on the right.
Firstly, draw a bubble around the points marked.
Then draw the line of best fit through the bubble.
The line should run along the length of the bubble.

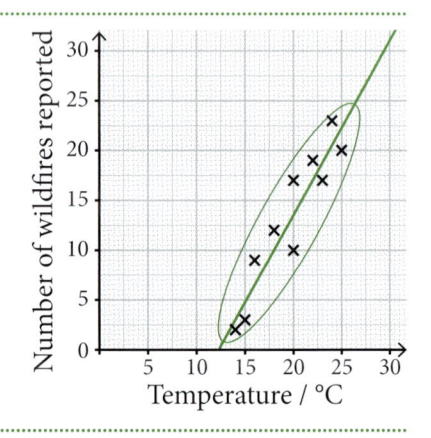

When drawing a line of best fit, remember:
- You should aim to have roughly the same number of points on each side of the line.
- You should ignore any points which obviously do not fit the pattern. Such points are called **outliers**.
- A line of best fit will not necessarily go through the origin.

It is also worth noting that the gradient of the line of best fit shows the type of correlation.
- A positive gradient shows there is a positive correlation.
- A negative gradient shows there is a negative correlation.

Estimation using interpolation and extrapolation

We can use the line of best fit to predict the value of one variable if we know the value of the other.

CHAPTER 20: SCATTER GRAPHS

In the next example, we use the line of best fit to estimate the number of wildfires on a day with a certain temperature. Note, however, that this will only be an estimate.

If the temperature is within the range of the points plotted on the graph, the estimate should be fairly accurate. This is called **interpolation**.

If the temperature lies outside of this range, the estimate may not be accurate. This is called **extrapolation**.

Example 4

Look again at the graph of the number of wildfires in the Mourne Mountains.
(a) Use the line of best fit to estimate the number of wildfires that would be expected for a temperature of 22°C.
(b) Do you think the answer for part (a) is a reliable estimate?
(c) Use the line of best fit to estimate the temperature if 28 wildfires occur in one day. Give an answer to the nearest degree Celsius.
(d) Do you think the answer for part (c) is a reliable estimate?

(a) The graph shows that for a temperature of 22°C, roughly 17 wildfires would be expected.
(b) This is a reliable estimate. The temperature of 22°C falls within the range of values used for the graph. This is called interpolation.
(c) The graph shows that roughly 28 wildfires would be expected for a temperature of 28°C.
(d) The answer to part (c) is an unreliable estimate, since the temperature of 28°C lies outside the range of values used for the graph. (The highest temperature used for the graph is 25°C) This is called extrapolation.

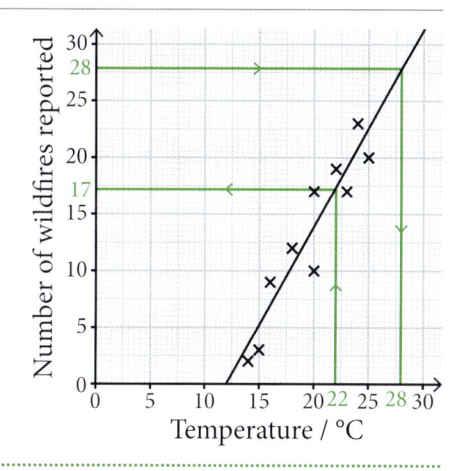

Example 5

Look at the scatter graph on the right.
(a) Identify any points that are outliers.
(b) Draw the line of best fit.
(c) State the type of correlation.

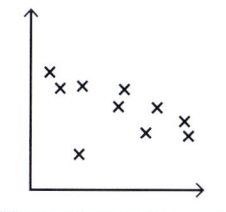

(a) Most of the points approximately follow the same straight line. The one outlier is marked on the diagram on the right.
(b) The line of best fit has been drawn through all the remaining points, ignoring the outlier.
(c) There is a negative correlation between the two variables plotted.

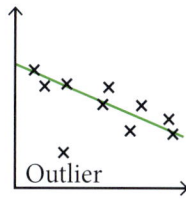

Exercise 20C

1. In each part, copy the scatter diagram and, if possible, add a line of best fit. Also state what type of correlation the graph shows.
   (a) (b) (c) (d)

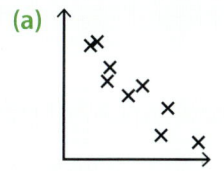

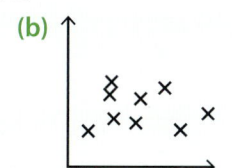

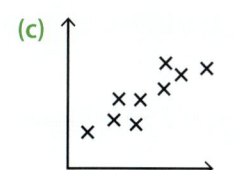

2. The marks for 8 pupils in their History and Geography exams are shown in the table below.

History (out of 40)	19	29	36	12	27	21	34	40
Geography (out of 30)	17	22	29	11	24	18	28	10

(a) Draw a scatter graph to show these scores.
(b) Identify one outlier.
(c) Add a line of best fit to your graph.
(d) Describe the correlation between the History and Geography scores.
(e) Use the line of best fit to estimate:
   (i) the Geography score for a pupil who scored 30 in the History exam,
   (ii) the History score for a pupil who scored 6 in the Geography exam.
(f) Which of the two estimates in part (e) do you think is more reliable? Explain briefly why.

3. Wesley makes a batch of ten scones. Sam eats them, putting jam on each one. The weights of the scones and the amounts of jam Sam uses are shown in the table.

Weight of scone (g)	40	45	38	40	42	50	37	55	52
Weight of jam (g)	10	12	8	9	11	13	8	16	2

(a) Draw a scatter graph to show this information.
(b) Identify one outlier on the graph and suggest a reason for it.
(c) Add a line of best fit to your graph.
(d) Describe the correlation between the weight of a scone and the amount of jam used.
(e) Use the line of best fit to estimate:
   (i) the amount of jam Sam would use for a scone weighing 35 grams,
   (ii) the weight of a scone for which Sam uses 15 grams of jam.
(f) Which of the two estimates in part (e) do you think is more reliable? Explain briefly why.

4. The table on the right shows the price of 17 electric cars and the range for each car. The range is the distance in miles you can drive on one charge of the battery.
(a) Draw a scatter graph to show this information.
(b) Add a line of best fit to the scatter graph.
(c) Describe the correlation between the price and range of the electric cars used in this survey.
(d) Use the line of best fit to estimate:
   (i) the range for a car costing £35 000,
   (ii) the cost of a car that has a range of 300 miles.
(e) Which of the two estimates in part (d) do you think is more reliable? Explain briefly why.

Price (thousand £)	Range (miles)
42	280
43	270
46	255
43	250
39	240
37	235
39	225
36	215
39	210
36.5	200
40	190
37	185
39	175
31	145
33	125
32.5	110
28	85

20.4 Causality

Suppose there is a correlation between two variables. In some cases, it is clear that an increase in one variable **causes** the increase in the other. However, this is not always the case.

Example 6

Look again at the scatter graph of the temperature plotted against the number of wildfires in the Mourne Mountains (start of section 20.2). There is a positive correlation between these two variables. Do you think an increase in one of these variables **causes** an increase in the other? If so, which way round?

Yes, higher temperatures can cause a higher number of wildfires.

In other cases, there is no causality. A positive correlation between the two variables does not mean that an increase in one causes an increase in the other.

Example 7

It has been discovered that there is a positive correlation between ice cream sales and the number of people drowning in the sea. As ice cream sales increase, the number of people drowning also increases. Is there causality: does higher ice cream consumption **cause** higher levels of drowning?

The suggestion that higher rates of ice cream consumption causes the higher number of people drowning is clearly false.

There is a third factor that causes both the increase in the number of ice creams sold and the number of people drowning: temperature. As temperature rises, the number of ice cream sales increases and the number of people swimming in the sea increases, leading to more accidents.

Exercise 20D

1. In each of the following cases there is a positive correlation between the two variables mentioned. State whether an increase in one variable **causes** the increase in the other. If so, which way round is the cause working?
   (a) The size of a house and its value.
   (b) The number of fish and the number of dolphins in the Yellow River in China over the last 20 years.
   (c) The length of Mr Walker's daily walk and the time he takes for it.
   (d) A child's height and their age in years.

2. Lucy carries out a survey in her school into the amount of time spent on social media and the amount of time spent on homework. She notices a negative correlation between these two variables. Do you think there is causality here?

3. Since the 1950s, both the level of CO_2 in the atmosphere and obesity levels have increased sharply. Hence, atmospheric CO_2 causes obesity. Discuss this conclusion.

4. Twenty people were surveyed at random in a doctor's surgery. They were asked two questions:
   - How many times have you had a cold or flu in the last year?
   - How many times have you taken medication for cold or flu symptoms in the last year?

   The chart on the right is a scatter graph displaying these data, with the number of colds plotted in the x-direction and the number of cold remedies in the y-direction.

   (a) What type of correlation exists between these two variables?
   (b) Is any causation taking place here? If so, which way does it work?

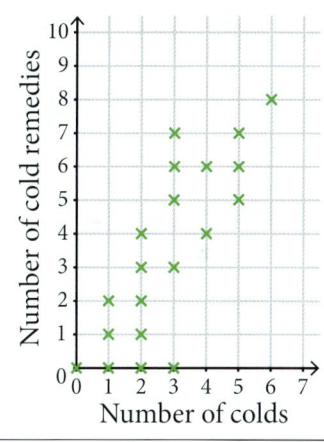

20.5 Summary

In this chapter you have learnt:
- How to plot a scatter graph, or add points to one. If you are given a table of values, the variable in the top row of the table should be plotted in the x-direction and the variable in the second row in the y-direction.
- That if there is a correlation between the two variables, a line of best fit can be drawn through the middle of the points on the scatter graph. Drawing a bubble around the points gives you an idea of the direction for the line. You should aim to have the same number of points on either side of the line.
- That the gradient of the line of best fit tells you what type of correlation exists: a positive gradient means there is a positive correlation; a negative gradient means a negative correlation.
- That the line of best fit can be used to estimate the value of one variable given the value of the other. This estimate will be most reliable if the value used lies within the range of values used on the graph.
- That an outlier is a point that does not follow the general shape or pattern of the other points. Outliers should be ignored when drawing a line of best fit.
- That a correlation may exist between two variables, but this does not necessarily mean that a change in one variable **causes** the change in the other.

Chapter 21
Cumulative Frequency

21.1 Introduction

In Chapter 18, on Statistical Averages and Spread, you learnt about frequency tables. That chapter also briefly discussed cumulative frequencies. In a frequency table (grouped or ungrouped), the cumulative frequency is the sum of the frequencies up to and including a group.

Key words
- **Mean**: The sum of all the values divided by the number of values.
- **Median**: The middle value when the values are listed in order from smallest to largest.
- **Mode**: The value that occurs most frequently.
- **Range**: The difference between the highest value and the lowest value.
- **Lower quartile**: The value that lies 25% of the way through the ordered list.
- **Upper quartile**: The value that lies 75% of the way through the ordered list.
- **Interquartile range**: The difference between the upper quartile and the lower quartile.
- **Cumulative frequency**: In a frequency table, the cumulative frequency is the sum of the frequencies up to and including a group.
- **Box plot**: A diagram that shows the median, the quartiles and the extreme values of a distribution.

Before you start you should know how to:
- Calculate the median, quartiles and interquartile range from an ungrouped set of data.
- Calculate percentages.

In this chapter you will learn how to:
- How to construct cumulative frequency curves.
- How to use a cumulative frequency curve to estimate the median, the quartiles and the interquartile range.
- How to draw a box plot.
- How to interpret a box plot.

Note: You will sometimes see:
- The lower quartile referred to as Q_1
- The median referred to as Q_2
- The upper quartile referred to as Q_3

Exercise 21A (Revision)

1. A vet weighs 11 dogs. The weights in kilograms are shown below.
   Find **(a)** the median **(b)** the lower and upper quartiles and **(c)** the interquartile range.

   19, 29, 11, 20, 13, 9, 17, 15, 16, 20, 18

2. There are 50 people on the number 38 bus from Lisburn to Belfast.
   Twenty-nine of them are school pupils.
   (a) What percentage of the people on the bus are school pupils?
   (b) What percentage are not school pupils?

21.2 Cumulative Frequency Curves

Calculating cumulative frequencies

To calculate the cumulative frequency for a row in a frequency table, find the sum of all the frequencies up to and including that row, as shown in the following example.

CHAPTER 21: CUMULATIVE FREQUENCY

Example 1

A number of pupils take a science test. All the pupils score between 17 and 20.

Copy the frequency table on the right and complete the cumulative frequency column.

Test score	Frequency	Cumulative frequency
17	3	3
18	8	
19	9	
20	2	

The first cumulative frequency is the same as the first frequency.
In the second row of the table, we add the second frequency. The cumulative frequency is 3 + 8 = 11
In the third row of the table, we add the third frequency, giving 11 + 9 = 20
In the fourth row of the table, we add the fourth frequency: 20 + 2 = 22

Test score	Frequency	Cumulative frequency
17	3	3
18	8	11
19	9	20
20	2	22

Note: The final cumulative frequency is the total number of items, in this case the number of pupils who took the test.

Drawing cumulative frequency curves

In the following example, the cumulative frequencies are calculated, then a cumulative frequency graph is drawn.

Example 2

The table on the right shows the number of defects in the items produced by a machine one day.
(a) Calculate the cumulative frequency for each row of the table.
(b) Draw a cumulative frequency curve for this data set.

Number of defects	Frequency
0	16
1	12
2	9
3	7
4	6
5	5
6	4
7	4
8	3
9	3
10	2
11	2
12	1

(a) To calculate the cumulative frequencies:
- The first cumulative frequency is equal to the first frequency, 16
- For the second cumulative frequency, add the second frequency: 16 + 12 = 28
- For the third cumulative frequency, add the third frequency: 16 + 12 = 28, etc.

The calculated cumulative frequencies are shown in the table on the right.

Number of defects	Frequency	Cumulative frequency
0	16	16
1	12	28
2	9	37
3	7	44
4	6	50
5	5	55
6	4	59
7	4	63
8	3	66
9	3	69
10	2	71
11	2	73
12	1	74

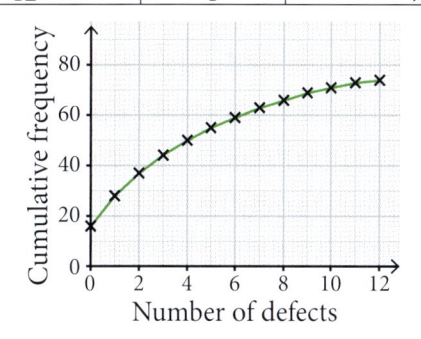

(b) Plot the number of defects on the x-axis against the cumulative frequency on the y-axis.

GCSE MATHEMATICS M3 AND M7

Interpreting cumulative frequency graphs

You may be asked to use a cumulative frequency curve to complete a cumulative frequency table.

Example 3

The cumulative frequency graph on the right shows the weights of babies born during six days in the Royal Victoria Hospital, Belfast.

(a) How many babies were born during the six days?
(b) Using the graph, copy and complete the cumulative frequency table below.

Weight, w (kg)	Frequency (Number of babies)	Cumulative frequency
$1.5 \leq w < 2$	1	1
$2 \leq w < 2.5$		
$2.5 \leq w < 3$		
$3 \leq w < 3.5$		
$3.5 \leq w < 4$		
$4 \leq w < 4.5$		

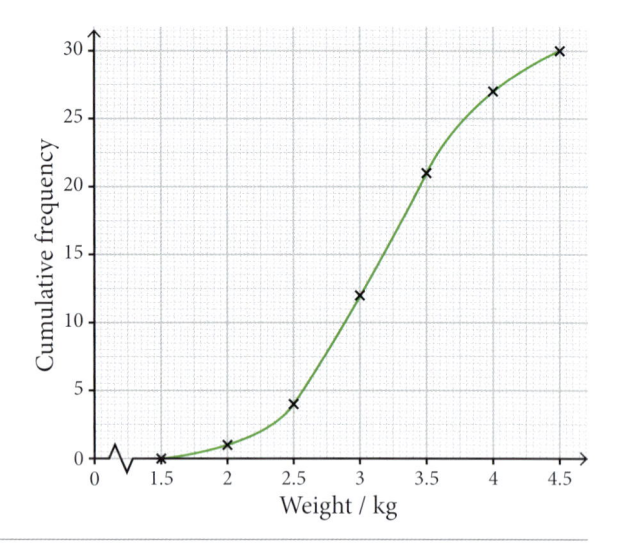

(a) This week 30 babies were born, since 30 is the highest cumulative frequency.

(b) To complete the table, start by filling in the cumulative frequency column from the graph.

The second step is to fill in the frequency column.

The frequency in the $2 \leq w < 2.5$ row is the difference between the cumulative frequencies of 4 and 1

The frequency in the $2.5 \leq w < 3$ row is the difference between the cumulative frequencies of 12 and 4, and so on.

The completed table is shown on the right.

Weight, w (kg)	Frequency (Number of babies)	Cumulative frequency
$1.5 \leq w < 2$	1	1
$2 \leq w < 2.5$	3	4
$2.5 \leq w < 3$	8	12
$3 \leq w < 3.5$	9	21
$3.5 \leq w < 4$	6	27
$4 \leq w < 4.5$	3	30

You may be asked to find the quartiles from a cumulative frequency curve. Quartiles divides the data set into four equal groups, with:
- 25% of the data below the lower quartile.
- 25% of the data between the lower quartile and the median.
- 25% of the data between the median and the upper quartile.
- 25% of the data above the upper quartile.

The following example shows how to identify the quartiles from a cumulative frequency curve.

Example 4

The cumulative frequency curve on the right shows the times, in seconds, taken to run 100 metres by the 60 members of an athletics club in June 2024.

Using the graph:
(a) Estimate the median time.
(b) Estimate the interquartile range for the times.
(c) Estimate the number of runners who took less than 12.5 seconds.

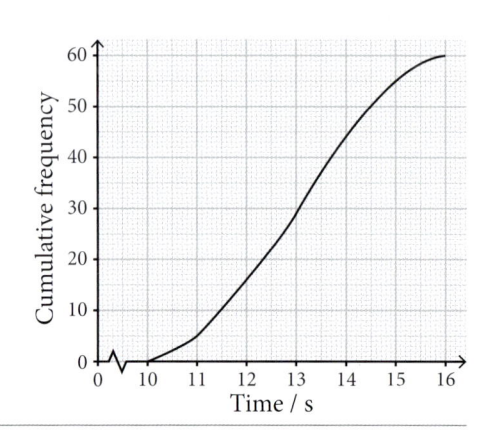

The maximum cumulative frequency is 60

(a) To find the median:

$\frac{1}{2} \times 60 = 30$

Using the dashed green construction lines on the graph, the median is 13.1 seconds.

(b) To find the lower quartile:

$\frac{1}{4} \times 60 = 15$

To find the upper quartile:

$\frac{3}{4} \times 60 = 45$

Using the grey construction lines on the graph, the lower quartile is 11.9 s and the upper quartile is 14.1 s.

The interquartile range
= 14.1 − 11.9 = 2.2 seconds

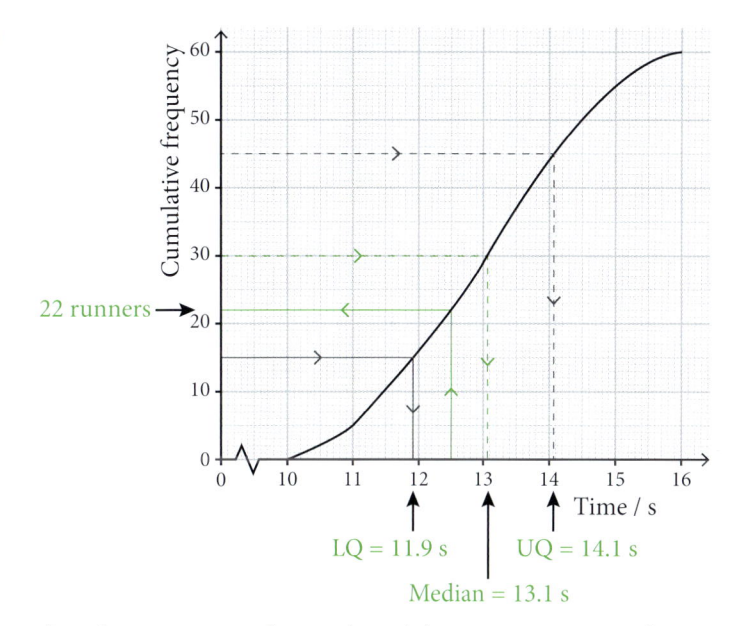

(c) To estimate the number of runners taking less than 12.5 seconds, use the solid green construction lines on the diagram, starting at 12.5 on the time axis.

This tells us that 22 runners took less than 12.5 seconds.

> **Note:** If the question had asked for the number of runners taking **more** than 12.5 s, you would first find the number taking less than 12.5 seconds and subtract from the total number of runners: 60 − 22 = 38
>
> You may instead be asked to express one of these numbers as a percentage or fraction of the total number of runners:
>
> For example, the fraction of runners taking more than 12.5 s is $\frac{38}{60}$ or $\frac{19}{30}$
>
> As a percentage this is 63.3%.

Note: Make your construction lines very clear on your diagram to show how the values have been found in each part of the question.

Exercise 21B

1. The table below shows the heights of the 80 students on a university course.

Height (cm)	140 ≤ h < 150	150 ≤ h < 160	160 ≤ h < 170	170 ≤ h < 180	180 ≤ h < 190	190 ≤ h < 200
Number of students	5	12	20	26	14	3
Cumulative frequency	5					

(a) Copy the table and complete the cumulative frequency row.
(b) Draw a cumulative frequency graph for this data.
(c) Use your cumulative frequency graph to estimate (i) the median height of the students and
 (ii) the interquartile range, giving your answers to the nearest integer.
(d) Estimate the number of students who are shorter than 1.55 m.

2. The table shows the times taken t by 120 runners on a park run.

Height (cm)	15 ≤ t < 20	20 ≤ t < 25	25 ≤ t < 30	30 ≤ t < 35	35 ≤ t < 40	40 ≤ t < 45	45 ≤ t < 50	50 ≤ t < 55	55 ≤ t < 60
Number of runners	5	10	17	20	23	20	15	6	4
Cumulative frequency									

(a) Copy the table and complete the cumulative frequency row.
(b) Draw a cumulative frequency graph for this data.
(c) Use your cumulative frequency graph to estimate (i) the median time and (ii) the interquartile range, giving your answers to the nearest integer.
(d) Estimate the number of runners who took more than 41 minutes.

3. Jordan surveyed the pupils in his year group, asking how far they travelled to school. The table on the right shows the results of the survey.
   (a) Copy the table and complete the cumulative frequency column of the table.
   (b) Draw a cumulative frequency graph for this data.
   (c) Use your cumulative frequency graph to estimate
       (i) the median distance travelled and (ii) the interquartile range, giving your answers to one decimal place.
   (d) Pupils who travel more than 18 km to school are entitled to a free bus pass. Use your cumulative frequency curve to estimate the number of pupils who are eligible for the bus pass.

Distance travelled	Number of pupils	Cumulative frequency
$0 \leq d < 2$	3	3
$2 \leq d < 5$	5	
$5 \leq d < 10$	20	
$10 \leq d < 15$	17	
$15 \leq d < 20$	4	
$20 \leq d < 25$	1	

4. Forty families were asked how much they spent on their last summer holiday. The results are shown in the table on the right.
   (a) Copy the table and complete the four empty spaces.
   (b) Draw a cumulative frequency graph to show this information.
   (c) Use your cumulative frequency graph to estimate
       (i) the median cost and (ii) the interquartile range, giving your answers to the nearest £10
   (d) Estimate the **fraction** of families that spent more than £1700 on their holiday.

Cost (£)	Number of families	Cumulative frequency
$0 \leq c < 250$	1	1
$250 \leq c < 500$	4	
$500 \leq c < 1000$	9	14
$1000 \leq c < 1500$		30
$1500 \leq c < 2000$	7	
$2000 \leq c < 3000$	2	39
$3000 \leq c < 4000$		40

5. An estate agent carries out a survey into house prices in Belfast. She uses the values of 100 houses and the results are shown in the table on the right.
   (a) In two rows of the table, the number of houses is missing. Find these missing numbers.
   (b) Draw a cumulative frequency graph for the data in the table.
   (c) Use your cumulative frequency graph to estimate
       (i) the median house price and (ii) the interquartile range, giving your answers to the nearest £10 000
   (d) Estimate the **percentage** of houses in Belfast that are worth more than £250 000

House value (thousands of £)	Number of houses	Cumulative frequency
$50 \leq v < 100$	2	2
$100 \leq v < 150$	7	9
$150 \leq v < 200$		19
$200 \leq v < 300$	21	40
$300 \leq v < 400$		67
$400 \leq v < 500$	14	81
$500 \leq v < 750$	12	93
$750 \leq v < 1000$	7	100

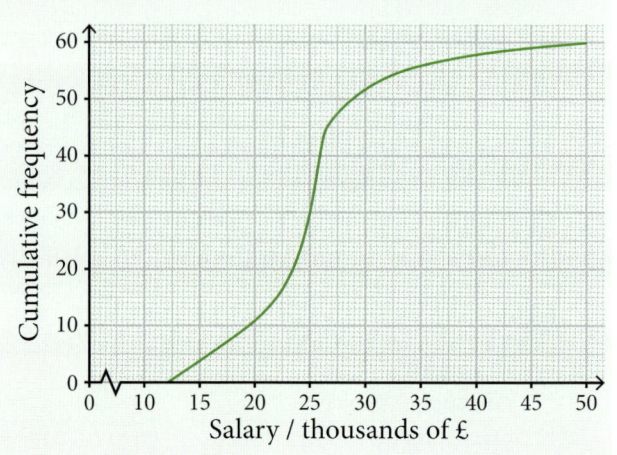

6. Seán owns a graphic design company, employing 60 people. A cumulative frequency graph for the salaries of his workers is shown on the right.
   (a) Use the graph to estimate:
       (i) The median salary.
       (ii) The interquartile range of the salaries.
   (b) Use the graph to estimate what **fraction** of the workers earns less than £35 000, giving your answer in its simplest form.

7. Mr Kinsella's class of 28 pupils took their end of year science exam. The cumulative frequency graph on the right shows the percentage scores.
   (a) Use the graph to estimate:
      (i) the median percentage score,
      (ii) the interquartile range of the percentage scores.
   (b) Use the graph to estimate the number of pupils who scored over 85% in the exam.

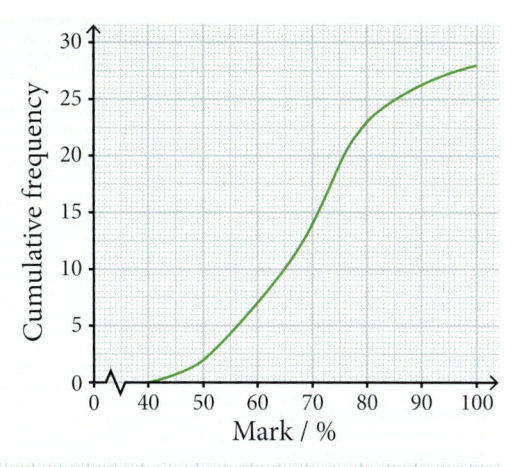

8. Kathy invites some friends to play ten-pin bowling for her birthday. For the party of 24 people, the cumulative frequency graph on the right shows their scores.
   (a) Use the graph to estimate:
      (i) the median score,
      (ii) the interquartile range of the scores.
   (b) Use the graph to estimate the number of people who got a score of 90 or more.

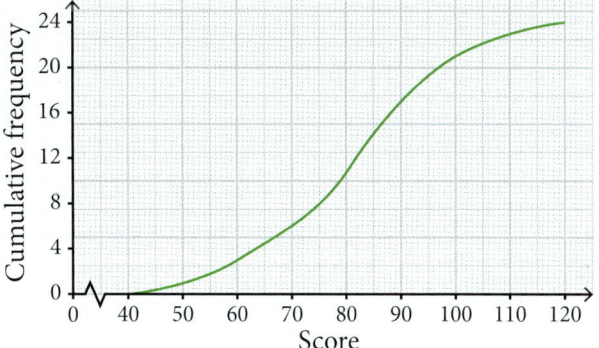

21.3 Box Plots

Box plots are a simple way to show the distribution of a dataset. The left and right-hand sides of the central box are the lower and upper quartiles. The left and right-hand ends of the plot are the lowest and highest values in the dataset. The following example illustrates a box plot.

Example 5

Draw a box plot using the following information:
- the lowest value is 50
- the lower quartile is 64
- the median is 68
- the upper quartile is 75
- the highest value is 89

Using the data given, the box plot is shown on the right.

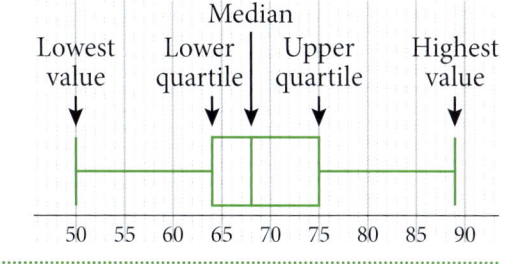

The following example demonstrates how to find the range and interquartile range from a box plot.

Example 6

For the following box plot, find the range, the interquartile range and the median value.

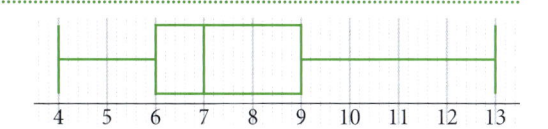

For the box plot shown:
- the lowest value is 4
- the lower quartile is 6
- the median is 7
- the upper quartile is 9
- the highest value is 13

195

GCSE MATHEMATICS M3 AND M7

Range = highest value – lowest value
      = 13 – 4
      = 9

Interquartile range = upper quartile – lower quartile
      = 9 – 6
      = 3

Exercise 21C

1. For each of (a) – (c), draw a box plot using the information in the table.

   (a)
Lowest value	15
Lower quartile	20
Median	27
Upper quartile	30
Highest value	36

   (b)
Lowest value	80
Lower quartile	100
Median	150
Upper quartile	220
Highest value	280

   (c)
Lowest value	7.2
Lower quartile	8.5
Median	9.1
Upper quartile	11.0
Highest value	12.9

2. Find (i) the median, (ii) the range, and (iii) the interquartile range for the following box plots.

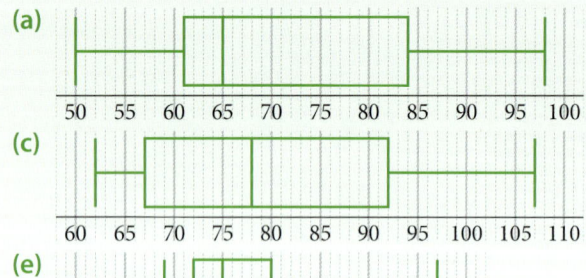

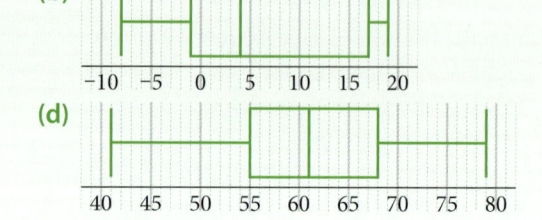

   (a)
   (b)
   (c)
   (d)
   (e)

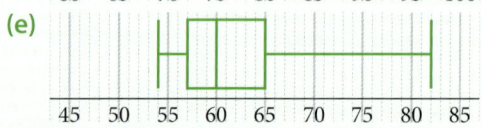

3. Draw a box plot for each of these sets of data.
   (a) 9, 11, 14, 15, 15, 16, 16, 17, 18, 20, 22, 23, 25, 31, 36
   (b) 30, 60, 80, 80, 90, 110, 130
   (c) 4.2, 5.6, 6.1, 6.2, 6.7, 7, 7.2, 7.2, 7.2, 7.3, 7.3, 7.4, 7.5, 7.6, 7.8, 8.1, 8.2, 8.5, 9.2

4. The box plot on the right shows the number of parking spaces used in a multi-storey car park each day during the month of June. The box plot is incomplete. 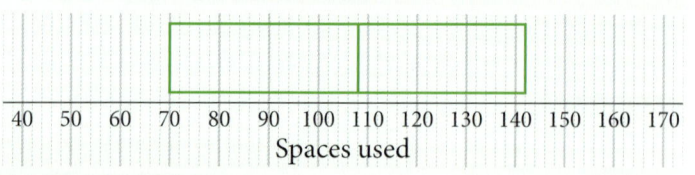

   Copy and complete the diagram using the following information:
   - The lowest number of spaces used is exactly half of the median number.
   - The range is 32 more than the interquartile range.

21.4 Cumulative Frequency Curves and Box Plots

Examination questions often involve both a cumulative frequency curve and a box plot. In the following example, a cumulative frequency graph is used to generate a box plot.

Example 7

Mr Magee's French class sits a test. There are 32 pupils in the class and the test has a maximum score of 20. The cumulative frequency graph on the right shows the distribution of scores.
(a) Use the cumulative frequency graph to find the lower quartile, the median and the upper quartile of the scores.
(b) Draw a box plot for the scores in the space below the cumulative frequency graph. You may assume that the lowest score in the test was 8 and the highest score was 20.

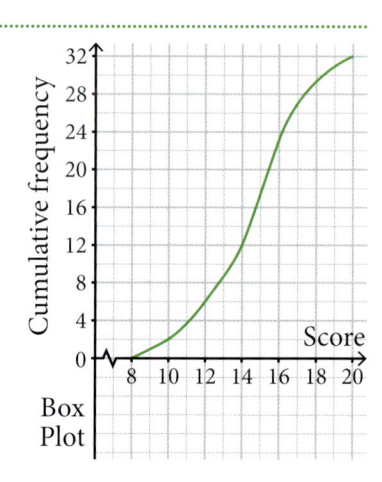

196

The construction lines added to the diagram on the right show the calculation of the lower quartile, median and upper quartile.
(a) The lower quartile is 12.7, the median is 14.6 and the upper quartile is 16.2
(b) The box plot has been drawn in the space provided using the calculated quartiles and the median. The lowest value of 8 and the highest value of 20, which were provided in the question, have also been used.

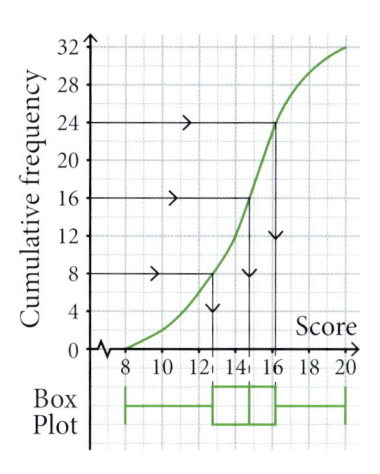

Note: Use your ruler to draw the constructions lines for the quartiles and the median. The vertical construction lines can then be extended downwards into the box plot area of the graph, as shown, to provide the three lines required for the box.

Comparing distributions

You may be asked to compare two datasets using a cumulative frequency table, a cumulative frequency graph, a box plot, or some combination of these. You will usually perform two comparisons:

- Compare the two median values.
- Compare the spread within the two datasets by comparing the ranges or the interquartile ranges.

Example 8

In Example 4, the time for runners to complete a 100 m race in June 2024 were recorded. The median and the interquartile range were calculated, as shown on the right.

June 2024
Median time: 13.1 s
Interquartile range: 2.2 s

In October 2024, the club records the times of the 60 runners again.
(a) Look at the box plot on the right, which shows the distribution of the 60 runners' times in October. Write down the lower quartile, the median and the upper quartile times. 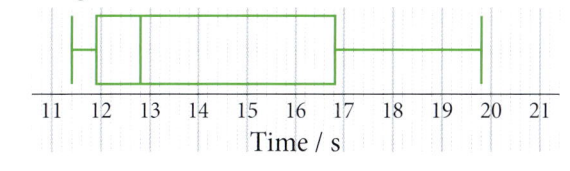
(b) Calculate the interquartile range for these times.
(c) Compare the distributions of the runners' times in June and October.

(a) The median is recorded as 12.8 seconds; the lower quartile is 11.9 s and the upper quartile 16.8 s.
(b) The interquartile range is 16.8 − 11.9 = 4.9 seconds
(c) The median time has decreased. This tells us that the average performance of the runners has improved. The interquartile range has increased. This tells us that there is now a larger spread in the times. This would suggest that there may be some faster times and some slower times.

Exercise 21D

1. Brigín is studying the lengths of the leaves on a particular tree. She takes a random sample of 48 leaves. The cumulative frequency curve on the right shows the results of her study.
   (a) Copy the graph carefully. Use the curve to find:
      (i) The lower and upper quartiles for these lengths.
      (ii) The median length.
   (b) In the area below the cumulative frequency curve add a box plot for the distribution. You may assume that the smallest leaf length was 4 cm and the greatest length was 16 cm.

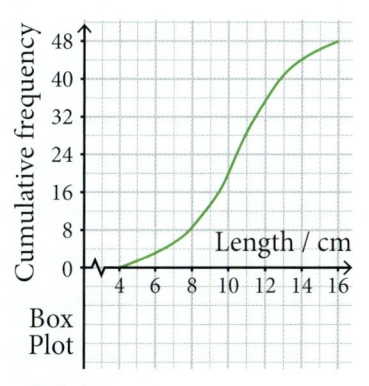

2. The table on the next page shows the ages of children and young adults at a holiday camp.
   (a) Copy and complete the table.
   (b) How many people were at the holiday camp altogether?

(c) Draw a cumulative frequency graph for this data set.
(d) Underneath your cumulative frequency graph, using the same *x*-axis, draw a box plot. Show clearly the construction lines for the quartiles and median. You may assume that the lowest age is 0 and the highest age is 19
(e) What is the median age?
(f) What is the interquartile range of the ages?

Age (years)	Frequency	Cumulative frequency
0 – 3	7	7
4 – 7	6	
8 – 11	11	24
12 – 15	14	
16 – 19	2	

3. Kirsty is carrying out a research project into the size of stag beetles. She studies a sample of 50 male and 50 female beetles. Look at the box plots on the right, which show the results of Kirsty's research. Make two comparisons of the sizes of the female and male stag beetles.

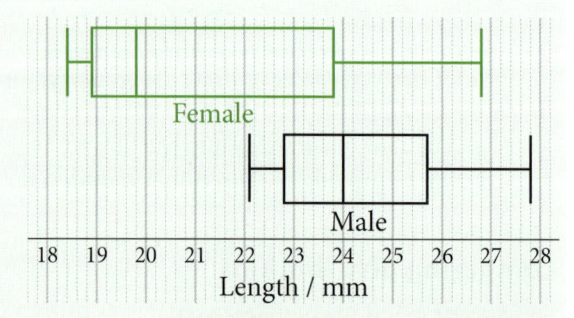

4. The coach of a rugby team analysed the performance of two of his players, Pete and Tadg. The table on the right relates to the number of tackles performed by each player over one season. For example, Pete's median number of tackles during the season was 9 tackles.
   (a) Copy and complete the table.
   (b) Draw box plots for both players.
   (c) Compare the performance of the two rugby players during the season. State which player should be included in the team if the coach can only choose one of these two players. State your reasons clearly.

	Pete	Tadg
Low value	3	1
Lower quartile	5	4
Median	9	7
Upper quartile	11	
High value		12
Interquartile range		7
Range	10	

5. The box plots show the average August rainfall totals, in millimetres, on two islands, Aran Island and the Isle of Man. The rainfall totals were measured every year from 1981 to 2020.
   (a) What is the median August rainfall total on Aran Island?
   (b) What is the highest August rainfall total on the Isle of Man? 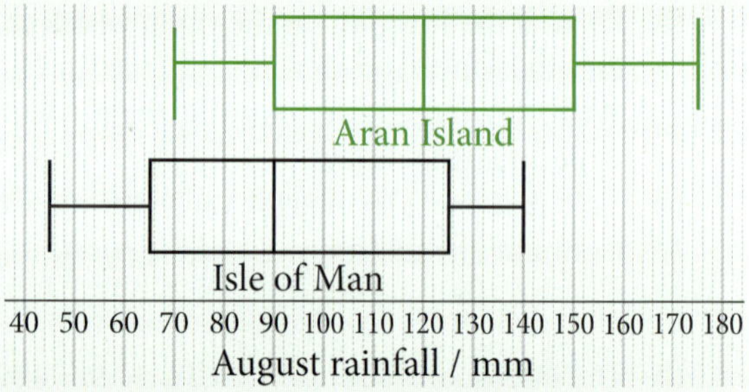
   (c) Rainfall totals were measured for 40 years.
      (i) For what percentage of these 40 years was the rainfall total more than 90 mm on Aran Island?
      (ii) For what percentage of these 40 years was the rainfall total more than 90 mm on the Isle of Man?
   (d) Write down two comparisons of the August rainfall data on these two islands.

6. Look at the two cumulative frequency curves shown in the diagram on the right. They show the heights of police officers at Lisburn police station in 1974 and 2024. There were 52 police officers at the station in both years.
   (a) Make a comparison of the median height in each distribution, giving your answer in the context of the question.
   (b) Make a comparison of the interquartile range for each distribution, giving your answer in the context of the question.

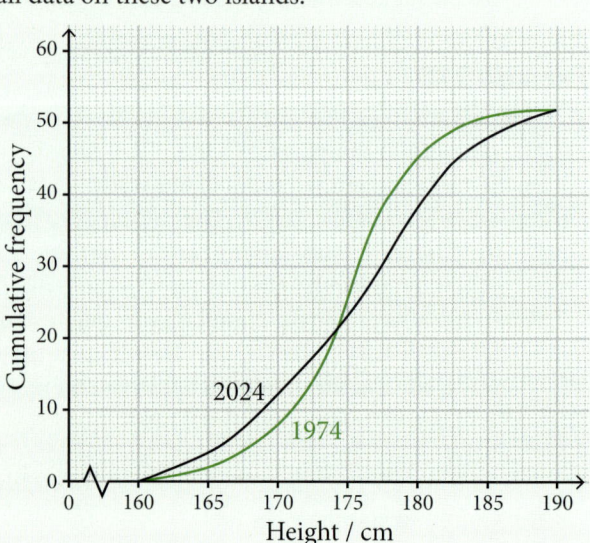

7. The manager of a multi-screen cinema carries out a survey into the ages of people watching two films, 'The Mystery of the Slow Hamster Wheel' and 'The Dangers of Kissing your Dragon'.
   (a) The results for 'The Mystery of the Slow Hamster Wheel' are shown in the table on the right. Copy the table and complete the empty spaces.
   (b) Which is the modal age group for the people watching 'The Mystery of the Slow Hamster Wheel'?
   (c) Why do you think the number of people in the $30 \leq a < 40$ group is higher than the number in the $15 \leq a < 20$ group?

The Mystery of the Slow Hamster Wheel

Age, a (years)	Number of people	Cumulative frequency
$6 \leq a < 8$	15	15
$8 \leq a < 10$	20	
$10 \leq a < 15$	8	43
$15 \leq a < 20$		46
$20 \leq a < 30$	5	51
$30 \leq a < 40$	9	
$40 \leq a < 50$	8	68

   (d) The cumulative frequency graph on the right shows the ages of people watching 'The Dangers of Kissing your Dragon'.

   Copy the cumulative frequency graph carefully. Plot the cumulative frequency curve for 'The Mystery of the Slow Hamster Wheel' on the same graph.

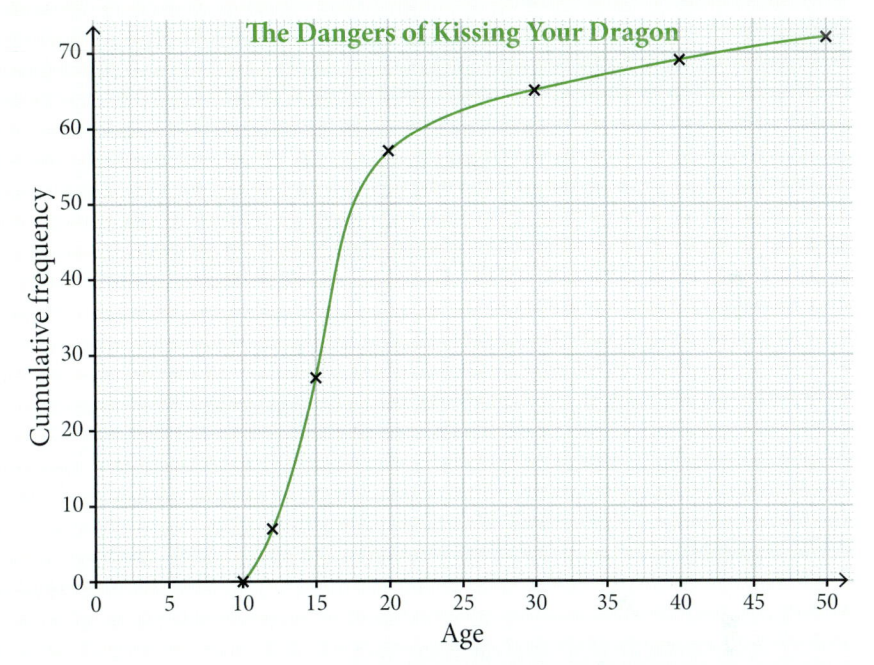

   (e) Copy the cumulative frequency table on the right for the ages of people watching 'The Dangers of Kissing your Dragon'. Complete the table using the graph.
   (f) Using the graph, calculate the interquartile range for the ages of the people watching both films.
   (g) Explain any difference in the two interquartile ranges you have calculated in part (f).

The Dangers of Kissing your Dragon

Age, a (years)	Number of people	Cumulative frequency
$10 \leq a < 12$	7	7
$12 \leq a < 15$		27
$15 \leq a < 20$	30	
$20 \leq a < 30$	8	65
$30 \leq a < 40$		
$40 \leq a < 50$	3	

21.5 Summary

In this chapter you have learned that:
- In a frequency table, you can calculate the cumulative frequency for a row by adding all the frequencies up to and including that row.
- A cumulative frequency curve can be used to find the median value, the lower and upper quartiles and the interquartile range.
- Box plots are a simple way to display the quartiles, the median, the lowest and highest values in a distribution.
- You may be asked to draw and/or interpret both cumulative frequency curves and box plots.
- You may be asked to compare two data sets by interpreting cumulative frequency curves and/or box plots.

Progress Review
Chapters 17–21

This Progress Review covers:

Chapter 17: Collecting Data

Chapter 18: Statistical Averages and Spread

Chapter 19: Statistical Diagrams

Chapter 20: Scatter Graphs

Chapter 21: Cumulative Frequency

1. Say whether each of these questions is suitable for a questionnaire and why.
   - (a) What time does the movie start?
   - (b) What do you look for when choosing a vet?
   - (c) Would you rather have a car or a bike?
   - (d) What led to the disagreement?

2. State one thing that is wrong with the response options for this question:
   What is your weight? Tick one box for your weight w in kilograms.
   $w < 50$ ☐ $50 \leq w < 60$ ☐ $60 \leq w < 70$ ☐ $w > 70$ ☐

3. State one thing that is wrong with the response options for this question.
   How many cars are owned by your family?
   0 ☐   1 ☐   3 ☐   More than 3 ☐

4. The following question appeared on a questionnaire about a dry cleaning service. State one thing that is wrong with the response options.
   How satisfied are you with the service provided?
   Completely dissatisfied ☐   Neither satisfied nor dissatisfied ☐   Somewhat satisfied ☐   Completely satisfied ☐

5. These questions appeared on a questionnaire about attitudes to music. For each question, state something that is wrong with either the question or the response options.
   - (a) **Don't you think classical music is rubbish?**
     Yes ☐   No ☐   Don't know ☐
   - (b) **Who is your favourite female artist?**
     Taylor Swift ☐   Dua Lipa ☐   Sabrina Carpenter ☐
     Chappell Roan ☐   Charli XCX ☐   Billie Eilish ☐
   - (c) **What do you like about her?**
   - (d) **How much time do you spend listening to music each day?**
     0–1 hours ☐   1–2 hours ☐   2–3 hours ☐   More than 3 hours ☐
   - (e) Read the statement below and tick one box to show how much you agree.
     I get a lot of satisfaction listening to music.
     Strongly disagree ☐   Neither agree nor disagree ☐   Agree ☐   Strongly agree ☐

6. The following two questions appeared on a questionnaire about pupils' attitudes to school sports. For each question, state one thing that is wrong with the response options.
   - (a) **What is your favourite sport to play in school?**
     Football ☐   Hockey ☐   Rugby ☐   Gaelic ☐
   - (b) **How long do you spend doing sport in school? Tick one box for the time t.**
     $t < 1$ hour ☐ 1 hour $\leq t < 2$ hours ☐ $t > 2$ hours ☐

7. Ed carries out a survey on driving.
   - (a) A question from his questionnaire is shown below.
     Q1. Do you own a car?
     Yes, petrol ☐   Yes, diesel ☐   Do not drive ☐

PROGRESS REVIEW: CHAPTERS 17–21

State two things that are wrong with the response options.

(b) For this question, Ed uses this section of his recording sheet. State one thing that is wrong with the table.

Type of car	Tally	Total
Petrol		
Diesel		

8. Jake has 9 sunflowers. Their heights in centimetres are: 55, 67, 109, 68, 101, 81, 80, 83, 35
   Find the mean height of Jake's sunflowers.

9. Twenty-five dogs visit a vet's surgery during one week. Each dog is weighed upon arrival. The weights are shown in the table on the right.
   (a) Find an estimate for the mean weight of these 25 dogs.
   (b) What is the modal class?
   (c) Find the median class.

Weight w (kg)	Number of dogs
$5 \leq w < 15$	3
$15 \leq w < 25$	10
$25 \leq w < 35$	8
$35 \leq w < 45$	4

10. The ages of 25 students on a college course are shown in the table on the right.
    (a) Calculate the mean age of the students, giving your answer to 1 decimal place.
    (b) What is the median age?
    (c) What is the modal age?

Age	Number of students
18	7
19	8
20	6
21	4

11. The table on the right shows the daily number of ships using a harbour during the month of August.
    (a) Find the mean number of ships using the harbour per day in August. Give your answer to 1 decimal place.
    (b) What is the modal number of ships visiting the harbour per day?
    (c) What is the median number?

Number of ships	Number of days
4	3
5	5
6	8
7	2
8	11
9	2

12. In the Hillsborough Castle Running Festival, runners are given the choice of entering the 5 km run, the 10 km run, or the half marathon, which is 20.8 km. The number of runners entering at each distance is shown in the table on the right.
    Find the mean distance run by the runners, giving your answer to 1 decimal place.

Distance	5 km	10 km	Half marathon
Number of runners	300	140	75

13. Patricia takes her holiday in County Kerry for 14 days. Each day she records the number of hours of sunshine. The frequency table on the right shows her data.
    (a) Find an estimate for the mean number of hours of sunshine during Patricia's two weeks away.
    (b) Does the mean fall within the median class or the modal class? Explain your answer.

Number of hours of sunshine, s	Number of days
$0 \leq s < 2$	2
$2 \leq s < 4$	4
$4 \leq s < 6$	3
$6 \leq s < 8$	2
$8 \leq s < 10$	2
$10 \leq s < 12$	1

14. An airline records the delays to its flights leaving Belfast International Airport during one week. The data are shown in the frequency table on the right. Find an estimate for the mean delay. Give your answer in minutes to 1 decimal place.

Number of minutes delay, t	Number of times
$0 \leq t < 15$	15
$15 \leq t < 30$	9
$30 \leq t < 45$	4
$45 \leq t < 60$	2
$60 \leq t < 75$	1

15. There are 18 pupils in a class: 10 boys and 8 girls. Each pupil grows a tomato plant. For the 10 boys, the heights of the plants in centimetres are:

    14 10 8 12 11 9 6 15 12 9

    (a) Find the mean height for the boys' plants.
    (b) For the 8 girls, the mean height is 12 cm. Find the mean height for all 18 plants, giving your answer to 1 decimal place.

201

16. The scores in a maths test for 16 boys are shown in the frequency table on the right.
    (a) Find the mean score for the boys.

    There are 10 girls in the same maths test. The total score for the 10 girls is 85
    (b) Find the mean score for the girls.
    (c) Find the total score for all the pupils.
    (d) Find the mean score for all the pupils, giving your answer to 1 decimal place.

Score	Number of boys
5	2
6	5
7	4
8	2
9	2
10	1

17. Sometimes Tomás has Shredded Wheat for breakfast. He records the number of Shredded Wheat he eats each morning for 3 weeks. The results are in the table on the right.
    (a) On how many days does Tomás have Shredded Wheat?
    (b) Calculate the mean number of Shredded Wheat Tomás has over all 21 days. Give your answer to 2 decimal places.
    (c) Calculate the mean number of Shredded Wheat for those days on which he does have Shredded Wheat. Give your answer to 2 decimal places.

Number of Shredded Wheat	0	1	2	3
Frequency	14	2	4	1

18. James works in a bike repair shop. For one week, he records the time it takes him, in minutes, for each job. The results are shown in the grouped frequency table on the right.
    (a) Find an estimate for the mean time taken for one of these jobs.
    (b) Explain why your answer to part (a) is an estimate rather than an exact value.

Time, t (minutes)	Number of jobs
$0 \leq t < 15$	6
$15 \leq t < 30$	18
$30 \leq t < 45$	7
$45 \leq t < 60$	4
$60 \leq t < 75$	1

19. The lengths of a number of leaves are measured and rounded to the nearest centimetre. The results are recorded in the table on the right.
    (a) In which group is the median?
    (b) In which interval is the lower quartile Q_1?
    (c) In which interval is the upper quartile Q_3?

Length (cm)	Frequency
1 – 3	12
4 – 7	24
8 – 9	5
10 – 12	4

20. A survey was carried out looking into the number of pets for the pupils in class 8C1. It found that the median number of pets was 3 and the interquartile range was also 3
    For a different class, 8C2, the median number of pets was 4 and the interquartile range was 9
    Make two comparisons of the number of pets owned by the pupils in the two classes.

21. Plot these values in a stem and leaf diagram. On the leaf side of the diagram, remember to keep the columns lined up and remember to use a key.

    146 151 179 140 149
    164 163 184 141 153
    152 150 135 169 170

22. The table on the right shows the number of people competing in each type of sport in a girls' athletics team.
    (a) Copy and complete the table, using the third column in the table to calculate the size of each sector of a pie chart.
    (b) Draw the pie chart to show this information.

Sport	Number of girls	Angle (°)
Running	24	120°
Jumping	21	
Throwing	18	
Mixed (e.g. heptathlon)	9	

23. In a company there are thirty-two office workers. The two-way table categorises the workers by age and gender.
    (a) Copy and complete the table.
    (b) The company boss wants to bring the number of workers up to 40. He also wants the number of under-30s to be equal to the number of over-30s. How many under-30s should he recruit?

	Under 30	Over 30	Total
Male	4		10
Female		13	
Total			

24. There are 30 pupils in a lesson. Twenty of them are talking and five are asleep. One pupil is both asleep and talking.
    (a) How many pupils are neither talking nor asleep?
    (b) Show this information on a Venn diagram.

25. The Venn diagram on the right shows the number of pupils visiting the school canteen one lunch time. Copy the diagram. 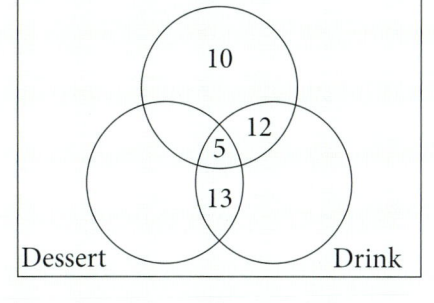
    (a) In total, 35 pupils took a main meal. How many took a main meal and a dessert, but no drink? Enter this number in the correct space on the diagram.
    (b) In total, 32 pupils took a dessert. How many took a dessert only? Enter this number in the correct space on the diagram.
    (c) The number of pupils taking a dessert only was equal to the number taking a drink only. Enter the number taking a drink only in the correct space on the diagram.
    (d) Seventy-six pupils visited the canteen altogether. How many bought nothing? Enter this number into the correct space on the diagram.

26. In each part (a)–(c) below, state whether you would expect a positive correlation, a negative correlation or no correlation between the variables.
    (a) The amount of rainfall in an area and the height of a river running through that area.
    (b) The number of people hired to clean up a festival site and the number of hours taken for the job.
    (c) The strength of the wind across Northern Ireland and the scores in GCSE Maths for pupils in Northern Ireland.

27. In Belfast, thirty families were surveyed. All of these families had:
    • a garden; and
    • children who liked playing football in the garden.
    Each family was asked to record:
    • the height of the fence, wall or hedge surrounding their garden; and
    • the percentage of times the children's football was kicked over the fence during the summer months.
    The scatter graph on the right shows the results of this survey. 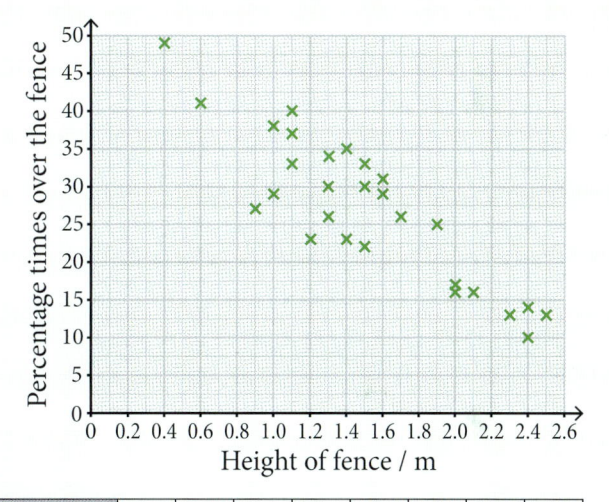
    (a) What type of correlation exists, if any?
    (b) Explain why such a correlation may exist.

28. The marks for 8 pupils in their Religious Studies and Home Economics exams are shown in the table on the right.

Religious Studies (out of 40)	10	19	10	16	13	15	16	13
Home Economics (out of 25)	11	22	26	35	32	33	27	11

    (a) Draw a scatter graph to show these scores.
    (b) Describe the correlation, if any, between the Religious Studies and Home Economics scores.
    (c) Suggest briefly why your answer in part (b) arises.

29. Nine patients visiting a health centre are chosen at random. The age and systolic blood pressure for each patient was recorded. The results are shown in the table below.

Age	11	52	21	24	18	47	25	32	40
Blood pressure (mm Hg)	91	151	110	110	105	138	124	126	190

    (a) Draw a scatter graph showing systolic blood pressure plotted on the vertical axis against age plotted on the horizontal axis.
    (b) On your scatter graph, identify one outlier. Suggest a reason for this measurement.
    (c) Add a line of best fit to your graph.
    (d) Describe the correlation between age and systolic blood pressure for these patients.
    (e) Use the line of best fit to estimate:
        (i) the systolic blood pressure for a patient with an age of 35,
        (ii) the age of a patient whose systolic blood pressure is 120,
        (iii) the age of a patient whose systolic blood pressure is 90.
    (f) Do you think your answer to part (e)(iii) is reliable? Explain briefly why.

30. A researcher discovers that there is a positive correlation between the number of ice lollies sold per day and the number of deaths by drowning at sea. Do you think there is any causality behind this positive correlation? If yes, which way round is it? If no, how else could you explain the positive correlation?

31. The cumulative frequency graph on the right shows the heights of a random sample of 200 trees in a forest park.
    (a) Use the graph to estimate, to the nearest metre:
        (i) the median height of the trees in the forest park,
        (ii) the interquartile range of the heights.
    (b) Using the graph, estimate what percentage of trees in the forest park grow to a height of 30 m or more.

32. Over 32 days, Thomas records the amount of time he has to wait for his bus to school. His results are shown in the table on the right.
    (a) Copy the table and complete the four empty spaces.
    (b) Draw a cumulative frequency for Thomas' data.
    (c) Using your graph, estimate:
        (i) the median time Thomas had to wait for his bus,
        (ii) the interquartile range for these times,
        (iii) the number of days Thomas waited **longer than** 8 minutes.

Wait, t (mins)	Number of days	Cumulative frequency
$0 \leq t < 2$	2	2
$2 \leq t < 5$	4	6
$5 \leq t < 10$		13
$10 \leq t < 15$	10	
$15 \leq t < 20$		29
$20 \leq t < 30$	3	

33. Compare the distributions for each of these pairs of box plots.
    (a) Class 12A and Class 12B's French test scores out of 50 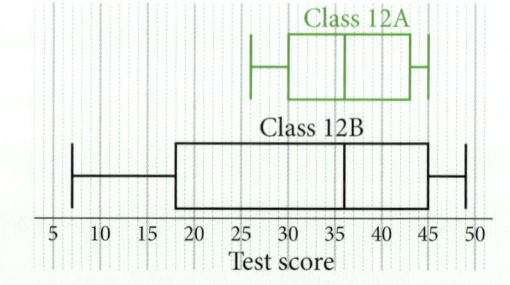
    (b) The lengths of earthworms in Northern Ireland and the Republic of Ireland. A sample of 100 earthworms was studied in each location. 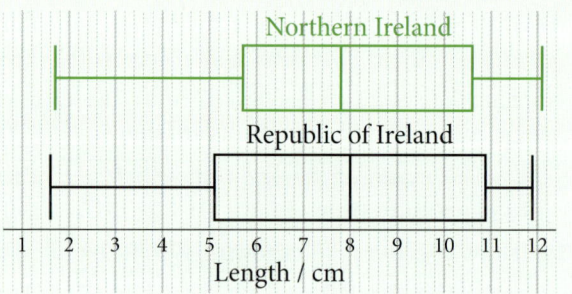
    (c) The times in seconds for a group of children and adults to complete a puzzle.

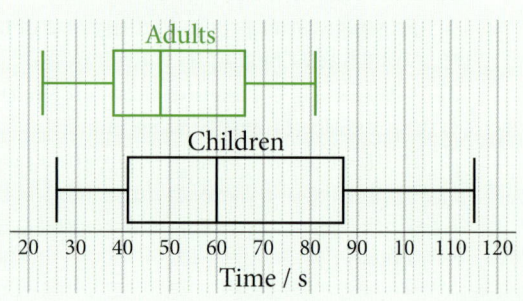

Chapter 22
Number Systems

22.1 Introduction

In this chapter we explore how different number systems work. We focus on the two main systems in our modern world: the decimal number system and the binary number system. The chapter covers:

- Understanding the two number systems decimal and binary.
- Converting between decimal and binary.

Key words

- **Decimal**: The name for the number system used in everyday life. It is a **base 10** number system, which means it uses the ten digits from 0 to 9
- **Binary**: The name for the number system used widely in computing and technology. It is a **base 2** number system, which means it uses only two digits, 0 and 1

Before you start you should:

- Remember the way in which numbers are arranged in place value columns: hundreds, tens and units, etc.
- Know the powers of 2:

2^1	2^2	2^3	2^4	2^5	2^6
2	4	8	16	32	64

- Know the powers of 10:

10^1	10^2	10^3	10^4	10^5	10^6
10	100	1000	10 000	100 000	1 000 000

In this chapter you will learn:

- About the decimal number system.
- About the binary number system.
- How to convert between the decimal and binary representations of a number.

Exercise 22A (Revision)

1. Look at the number 1403.97
   What is the name of the column that the four appears in?
2. Add the following two numbers without a calculator: 452.6 + 78.91
   Write the two numbers in columns, keeping the decimal points lined up.
3. Evaluate these powers of 2 and 10:
   (a) 2^3 (b) 10^2 (c) 2^5 (d) 10^4

22.2 The Decimal Number System

The decimal number system is based on the number ten, probably because human beings normally have ten fingers, so a number system based on the number ten came naturally.

> **Note:** Sometimes you may see the word **denary** instead of **decimal** number system.

The position of each digit in a decimal number tells you its value. For example, the 2 in 128 is in the tens column and means $2 \times 10 = 20$

In the earliest place value systems, it was common to leave a space to denote, for example, zero tens. So, 108 would be written as '1 8'. The digit 0 was a later invention and revolutionised mathematics.

GCSE MATHEMATICS M3 AND M7

The place values are all powers of ten:

1 million = 1 000 000 = 10^6
1 ten thousand = 10 000 = 10^4
1 hundred = 100 = 10^2
1 unit = 10^0

1 hundred thousand = 100 000 = 10^5
1 thousand = 1000 = 10^3
1 ten = 10 = 10^1

So, the column headings could be written as:

10^6 10^5 10^4 10^3 10^2 10^1 10^0

Example 1

(a) What value does the number 7 have in the number 2709?
(b) Write the number that means three tens, two hundreds and five units.
(c) Write twenty-two tens plus eight units as a single number.

(a) 700
(b) 235 (Be careful with the order of the digits.)
(c) 228 (Twenty-two tens equals two hundreds plus two tens.)

Example 2

Write out the value of the 3 in each of these numbers.
(a) 1432   (b) 109 301   (c) 8953   (d) 703 002   (e) 3 057 082

(a) 30   (b) 300   (c) 3   (d) 3000   (e) 3 000 000

The idea is extended to parts of a whole number with column headings $\frac{1}{10}, \frac{1}{100}, \frac{1}{1000}$, and so on. The column headings in the decimal number system become:

1000   100   10   1 $\frac{1}{10}$ $\frac{1}{100}$ $\frac{1}{1000}$

More columns can be added to the left and the right as necessary.

Example 3

Write out the value of the 4 in each of these numbers.
(a) 57.4   (b) 0.014   (c) 1967.342   (d) 7.4098   (e) 1.54

(a) 4 tenths or 0.4 or $\frac{4}{10}$
(b) 4 thousandths or 0.004 or $\frac{4}{1000}$
(c) 4 hundredths or 0.04 or $\frac{4}{100}$
(d) 4 tenths or 0.4 or $\frac{4}{10}$
(e) 4 hundredths or 0.04 or $\frac{4}{100}$

Exercise 22B

1. Write out the value of the 7 in each of the numbers below:
   (a) 207   (b) 702   (c) 678   (d) 71   (e) 27

2. Write each number out in digits:
   (a) Five tens plus 8 units
   (b) Eight units plus 6 tens
   (c) Three hundreds, nine tens and four units
   (d) Six hundreds plus seven units
   (e) Five hundreds plus two tens

3. In the decimal number system, what is the value of a 4 in the column to the left of the hundreds column?

4. In the number 9380571.642 what is the value of the:
   (a) 1   (b) 2   (c) 3   (d) 4   (e) 5
   (f) 6   (g) 7   (h) 8   (i) 9   (j) 0

5. Write out the value of the seven in each of the numbers below:
   (a) 2079   (b) 37.013   (c) 31.073   (d) 7 015 924   (e) 93.857

6. Write out the value of the four in each of these numbers:
   (a) 94   (b) 235.064   (c) 777.041   (d) 8.4   (e) 549 182

7. Write out the value of the five in each of these numbers.
   (a) 9856   (b) 98.56   (c) 10.95   (d) 5.0319   (e) 6 578 039.1

22.3 The Binary Number System

The modern world depends on a different number system, the **binary** system, which is widely used in computing. The base for the binary system is **two**, so it uses only two digits – 0 and 1

The binary system is suitable for computing because of the physical properties of some materials. A magnetic material can be magnetised in two different polarities: north or south. An electric switch can be either on or off. To use the properties of these materials to represent numbers, it is necessary to use a number system that only involves the two digits 0 and 1

The column headings are the powers of two (rather than the powers of ten used in the decimal system). So, the columns have the values:

$2^8 = 256$ $2^7 = 128$ $2^6 = 64$ $2^5 = 32$ $2^4 = 16$ $2^3 = 8$ $2^2 = 4$ $2^1 = 2$ 1

The first 16 binary numbers look like this:

Decimal Number	Equivalent Binary Number
1	1
2	10
3	11
4	100
5	101
6	110
7	111
8	1000

Decimal Number	Equivalent Binary Number
9	1001
10	1010
11	1011
12	1100
13	1101
14	1110
15	1111
16	10000

Converting from binary to decimal

To convert a binary number into decimal, we add the place values for every column that has a one in it.

Example 4

Convert the binary number 1110 into decimal.

Remember the place values for each column.

$2^3 = 8$ $2^2 = 4$ $2^1 = 2$ $2^0 = 1$
1   1   1   0

So, the decimal value of the binary number $1110 = 1 \times 8 + 1 \times 4 + 1 \times 2 + 0 \times 1$
$= 8 + 4 + 2$
$= 14$

Example 5

Find the decimal value of each binary number.
(a) 1010   (b) 101   (c) 110   (d) 1011   (e) 1111

(a) $1 \times 8 + 0 \times 4 + 1 \times 2 + 0 \times 1 = 8 + 2 = 10$ (b) $4 + 1 = 5$
(c) $4 + 2 = 6$ (d) $8 + 2 + 1 = 11$ (e) $8 + 4 + 2 + 1 = 15$

The following example demonstrates this method for a larger binary number.

Example 6

Find the decimal value of the binary number 10010101

There are 1s in the columns with the place values 128, 16, 4 and 1
So we calculate $1 \times 128 + 1 \times 16 + 1 \times 4 + 1 \times 1 = 128 + 16 + 4 + 1 = 149$

Example 7

A binary number has 4 digits. Find the smallest and largest possible decimal values for this number.

The smallest 4-digit binary number is 1000   This has a decimal value of 8
The largest 4-digit binary number is 1111   This has a decimal value of 15

Exercise 22C

1. Write the following base two numbers in decimal:
   (a) 111   (b) 10   (c) 1110   (d) 100110
2. Write the following base two numbers in decimal:
   (a) 1010   (b) 11101   (c) 1010110   (d) 1111100
3. Write the following base two numbers in decimal:
   (a) 110   (b) 101010   (c) 11101010   (d) 10111011
4. Convert these binary numbers into decimal:
   (a) 11   (b) 1111   (c) 111111   (d) 11111111
5. Copy and complete the following sentences.
   You may assume that all binary numbers start with the digit 1
   (a) A binary number has 3 digits. Its decimal value lies between 4 and _____
   (b) A binary number has 5 digits. Its decimal value lies between _____ and _____

22.4 Converting a Decimal Number into Binary

The following examples demonstrate the method used to convert a decimal number to its binary form.

Example 8

Write the decimal number 25 in binary.

- First, we need to write out the place values of the binary system until we come to a column heading larger than 25:

  32   16   8   4   2   1

  We won't need to use the 32s column because 25 is smaller than 32

- As 25 is bigger than 16, include a 1 in the 16 column:

  16   8   4   2   1
  1

  Subtracting 16 leaves 25 − 16 = 9

  As 9 is bigger than 8, include a 1 in the 8 column:

  16   8   4   2   1
  1   1

  Subtracting 8 leaves 9 − 8 = 1

- We are left with 1, so we finish by placing zeros in the 4 and 2 columns and a 1 in the units column:

  16   8   4   2   1
  1   1   0   0   1

  So, the decimal number 25 is 11001 in binary.

Example 9

Write the decimal number 54 in binary.

- Write out the binary place headings up to the first power of 2 above 54:

| 64 | 32 | 16 | 8 | 4 | 2 | 1 |

- Then, as 54 is larger than 32 but smaller than 64, place a one in the 32s column:

32	16	8	4	2	1
1					

This leaves: 54 − 32 = 22

- This is larger than 16, so we place a 1 in the 16s column:

32	16	8	4	2	1
1	1				

This leaves: 22 − 16 = 6

- This is smaller than 8, but larger than 4, so place a 0 in the 8s column and a 1 in the 4s column:

32	16	8	4	2	1
1	1	0	1		

This leaves: 6 − 4 = 2

Finish by placing a 1 in the 2s column and a zero in the units column:

32	16	8	4	2	1
1	1	0	1	1	0

So, the decimal number 54 is 110110 in binary.

Example 10

Write the following decimal numbers in binary.

(a) 4 (b) 12 (c) 15 (d) 7 (e) 6

(a) 100 (b) 1100 (c) 1111 (d) 111 (e) 110

Exercise 22D

1. Convert the following decimal numbers into binary:
   (a) 6 (b) 12 (c) 18 (d) 23 (e) 27
2. Convert these decimal numbers into binary:
   (a) 120 (b) 111 (c) 99 (d) 221 (e) 267
3. Convert these decimal numbers into binary:
   (a) 7 (b) 31 (c) 127 (d) 255

22.5 Summary

In this chapter you have learnt:
- That the decimal number system uses digits 0, 1, 2, 3, 4, 5, 6, 7, 8 and 9
- That the place values of the decimal number system are the powers of 10
- That the binary number system uses digits 0 and 1
- That the place values of the binary number system are the powers of 2
- How to convert a binary number into decimal.
- How to convert a decimal number into binary.

Chapter 23
Surds

23.1 Introduction

What is the solution to the equation $x^2 = 5$, given that x is a positive number?

Taking the square root of both sides of the equation gives $x = \sqrt{5}$

Until now you have then used your calculator and rounded the answer, e.g. $x = 2.24$ to 2 d.p.

However, because of the rounding, this is not an exact answer.

$\sqrt{5}$ is an example of a **surd**.

By leaving our answer in surd form, $x = \sqrt{5}$, we can give an exact answer to the question, and we do not lose any accuracy through rounding.

This chapter demonstrates how to simplify and work with surds.

Key words
- **Surd**: A surd is a number that is written using a root sign $\sqrt{}$ in its simplest form.

Before you start you should know:
- The square numbers.
- How to expand brackets.
- How to solve equations.
- How to use Pythagoras' Theorem to find the hypotenuse of a right-angled triangle or to find one of the shorter sides.

In this chapter you will learn how to:
- Simplify surds.
- Work with surds, e.g. adding and multiplying.

Exercise 23A (Revision)

1. Which of these numbers are square numbers?
   1, 2, 4, 8, 16, 32, 64

2. Expand these brackets:
   (a) $6x(2x^2 - 3)$ (b) $(2d + 3c)(d - 4c)$

3. Solve this equation for x, given that x is a positive number: $x^2 + 15 = 40$

4. Find the lengths marked x and y in the diagram on the right.

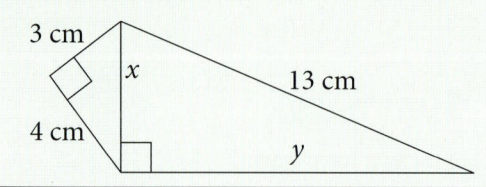

23.2 Simplifying Surds

When giving an answer in surd form, it is usual to use the surd's simplest form. To start this process, write $\sqrt{x}$ as $\sqrt{a \times b}$, where a is the largest square number factor of x.

Example 1

Write these surds in their simplest form.
(a) $\sqrt{80}$ (b) $3\sqrt{20}$ (c) $\sqrt{10}$

(a) Write 80 in the form $a \times b$. 16 is the largest square number factor of 80 $\quad = \sqrt{16 \times 5}$
You can split a surd up like this for the product of two numbers: $\quad = \sqrt{16} \times \sqrt{5}$
Replace $\sqrt{16}$ with 4 $\quad = 4 \times \sqrt{5}$
Write the surd in its simplest form: $\quad = 4\sqrt{5}$

(b) Replace 28 with 4×7 because 4 is the largest square number factor of 28 $\quad = 3\sqrt{4 \times 7}$
$\sqrt{4 \times 7}$ can be written as $\sqrt{4} \times \sqrt{7}$ $\quad = 3 \times \sqrt{4} \times \sqrt{7}$
Replace $\sqrt{4}$ with 2 $\quad = 3 \times 2 \times \sqrt{7}$
Write the surd in its simplest form: $\quad = 6\sqrt{7}$

(c) There are no square number factors of 10 (apart from 1), so this surd is already in its simplest form.

Note: Your calculator will simplify any expression involving surds. You should only use your calculator to check your answers in this chapter.

Questions involving surds will nearly always appear on the non-calculator exam paper.

Exercise 23B

1. Simplify the following surds. In some cases, the surd is already in its simplest form.
   (a) $\sqrt{19}$ (b) $\sqrt{40}$ (c) $\sqrt{12}$ (d) $\sqrt{27}$ (e) $\sqrt{70}$
   (f) $\sqrt{84}$ (g) $\sqrt{94}$ (h) $\sqrt{75}$ (i) $\sqrt{60}$ (j) $\sqrt{120}$

2. Simplify the following surds. In some cases, the surd is already in its simplest form.
   (a) $2\sqrt{48}$ (b) $5\sqrt{60}$ (c) $10\sqrt{18}$ (d) $3\sqrt{54}$ (e) $2\sqrt{99}$ (f) $4\sqrt{30}$

23.3 The Rules of Surds

When adding, subtracting, multiplying or dividing surds, you should be aware of some rules.

Rule 1 $\sqrt{a} \times \sqrt{b} = \sqrt{ab}$

Example 2

Write $\sqrt{6} \times \sqrt{3}$ in its simplest form.

Using Rule 1:	$= \sqrt{18}$
The largest square factor of 18 is 9	$= \sqrt{9} \times \sqrt{2}$
Replace $\sqrt{9}$ with 3; this is now in its simplest form.	$= 3\sqrt{2}$

Rule 2 $\sqrt{a} \times \sqrt{a} = a$

Example 3

Simplify $\sqrt{5} \times \sqrt{5}$

Using Rule 2: $\sqrt{5} \times \sqrt{5} = 5$

Rule 3 $\dfrac{\sqrt{a}}{\sqrt{b}} = \sqrt{\dfrac{a}{b}}$

Example 4

Write $\dfrac{\sqrt{40}}{\sqrt{2}}$ in its simplest form.

Using Rule 3: $= \sqrt{\frac{40}{2}}$
$= \sqrt{20}$

The largest square factor of 20 is 4: $= \sqrt{4} \times \sqrt{5}$

Since $\sqrt{4} = 2$: $= 2\sqrt{5}$

Rule 4 $a\sqrt{c} \pm b\sqrt{c} = (a \pm b)\sqrt{c}$

Example 5

Simplify **(a)** $\sqrt{12} + \sqrt{27}$ **(b)** $\sqrt{50} - \sqrt{32}$

(a) Simplify $\sqrt{12}$ and $\sqrt{27}$ separately:
$\sqrt{12} = \sqrt{4}\sqrt{3} = 2\sqrt{3}$
$\sqrt{27} = \sqrt{9}\sqrt{3} = 3\sqrt{3}$

So: $\sqrt{12} + \sqrt{27} = 2\sqrt{3} + 3\sqrt{3}$

Using Rule 4: $= 5\sqrt{3}$

(b) Simplify $\sqrt{50}$ and $\sqrt{32}$ separately:
$\sqrt{50} = \sqrt{25}\sqrt{2} = 5\sqrt{2}$
$\sqrt{32} = \sqrt{16}\sqrt{2} = 4\sqrt{2}$

So: $\sqrt{50} - \sqrt{32} = 5\sqrt{2} - 4\sqrt{2}$

Using Rule 4: $= 1\sqrt{2}$
$= \sqrt{2}$

> **Note:** Rule 4 is similar to the rule used when adding or subtracting like algebraic terms.
> For example: $2x + 3x = 5x$ and: $5y - 4y = y$
> It is not possible to simplify $5\sqrt{3} + 4\sqrt{2}$, just as you cannot simplify $5x + 4y$

The following example demonstrates more difficult simplification.

Example 6

Simplify **(a)** $3\sqrt{5} \times 6\sqrt{75}$ **(b)** $\dfrac{\sqrt{20} + \sqrt{8}}{2}$ **(c)** $\dfrac{\sqrt{10}}{\sqrt{2}} \times \sqrt{5}$

(a) Firstly, simplify $\sqrt{75}$:
$\sqrt{75} = \sqrt{25}\sqrt{3} = 5\sqrt{3}$

So: $6\sqrt{75} = 6 \times 5\sqrt{3} = 30\sqrt{3}$

So: $3\sqrt{5} \times 6\sqrt{75}$
$= 3\sqrt{5} \times 30\sqrt{3}$

Since $\sqrt{5} \times \sqrt{3} = \sqrt{15}$ using Rule 1: $= 90\sqrt{15}$

(b) $\dfrac{\sqrt{20} + \sqrt{8}}{2}$

First, simplify the surds. The largest square number factor in both cases is 4
$= \dfrac{\sqrt{4}\sqrt{5} + \sqrt{4}\sqrt{2}}{2}$

Because $\sqrt{4} = 2$
$= \dfrac{2\sqrt{5} + 2\sqrt{2}}{2}$

Divide numerator and denominator by 2 $= \sqrt{5} + \sqrt{2}$

(c) $\dfrac{\sqrt{10}}{\sqrt{2}} \times \sqrt{5}$

Using Rule 3: $= \sqrt{\dfrac{10}{2}} \times \sqrt{5}$
$= \sqrt{5} \times \sqrt{5}$

Using Rule 2: $= 5$

Exercise 23C

1. Simplify the following.
   (a) $\sqrt{3} \times \sqrt{3}$
   (b) $\sqrt{14} \times \sqrt{14}$
   (c) $\sqrt{5} \times \sqrt{8}$
   (d) $\sqrt{8} \times \sqrt{2}$
   (e) $\dfrac{\sqrt{24}}{\sqrt{3}}$
   (f) $\dfrac{\sqrt{28}}{\sqrt{7}}$

2. Simplify these expressions, if possible.
   (a) $2\sqrt{3} + 5\sqrt{3}$
   (b) $\sqrt{2} + 3\sqrt{2}$
   (c) $2\sqrt{3} + 3\sqrt{2}$
   (d) $5\sqrt{5} - 2\sqrt{5}$
   (e) $9\sqrt{7} - \sqrt{7}$
   (f) $5\sqrt{10} - 10\sqrt{5}$

3. Simplify the following.
   (a) $\dfrac{\sqrt{20}}{2}$
   (b) $\dfrac{\sqrt{15}}{\sqrt{5}} \times \sqrt{3}$
   (c) $2\sqrt{5} \times 5\sqrt{2}$
   (d) $2\sqrt{12} \div 4\sqrt{3}$

4. Simplify:
   (a) $4\sqrt{5} - 3\sqrt{5}$
   (b) $3\sqrt{3} + \sqrt{12}$
   (c) $\sqrt{8} + 4\sqrt{2}$
   (d) $4\sqrt{12} - 2\sqrt{3}$
   (e) $\sqrt{60} - 2\sqrt{15}$
   (f) $4\sqrt{50} - 3\sqrt{8}$

5. For the following rectangles, calculate the length of the diagonal.
   (a) Length = $\sqrt{29}$ cm, width = $\sqrt{7}$ cm
   (b) Length = $\sqrt{3}$ cm, width = $\sqrt{8}$ cm

6. A cuboid has a length of 5 cm, a width of 4 cm and a height of $2\sqrt{2}$ cm
   Find the volume of the cuboid as a surd in its simplest form.

7. A rectangular bed sheet has an area of $(2\sqrt{2} + 2\sqrt{3})$ m²
   If its width is 2 m, find its length.

23.4 Expanding Brackets With Surds

You can use the usual rules when expanding brackets with surds.

Example 7

Expand these brackets. **Hint**: In part (c) use FOIL (first, outer, inner, last).
(a) $5(3 - 2\sqrt{2})$
(b) $2\sqrt{2}(4\sqrt{2} + \sqrt{3})$
(c) $(2 - \sqrt{5})^2$

(a) Multiply each term inside the brackets by 5 $5(3 - 2\sqrt{2}) = 15 - 10\sqrt{2}$

(b) We must multiply each term inside the brackets by $2\sqrt{2}$
So multiplying the first term in the brackets: $2\sqrt{2} \times 4\sqrt{2}$
Because $2 \times 4 = 8$ and $\sqrt{2} \times \sqrt{2} = 2$ $= 8 \times 2$
    $= 16$

And then multiplying the second term in the brackets: $2\sqrt{2} \times \sqrt{3}$
Because $\sqrt{2} \times \sqrt{3} = \sqrt{6}$ $= 2\sqrt{6}$
Therefore we can write: $2\sqrt{2}(4\sqrt{2} + \sqrt{3}) = 16 + 2\sqrt{6}$

(c) $(2 - \sqrt{5})^2 = (2 - \sqrt{5})(2 - \sqrt{5})$
Using FOIL: $= 4 - 2\sqrt{5} - 2\sqrt{5} + 5$
Grouping the integer terms and grouping the terms involving $\sqrt{5}$ $= 9 - 4\sqrt{5}$

Exercise 23D

1. In each case, expand the brackets and simplify.
   (a) $2(1 + \sqrt{2})$
   (b) $8(2\sqrt{3} - 3)$
   (c) $6(\sqrt{2} - 3\sqrt{3})$
   (d) $-\sqrt{2}(\sqrt{2} - 4)$
   (e) $2\sqrt{3}(3 + 5\sqrt{3})$
   (f) $\sqrt{5}(2 - \sqrt{3})$
   (g) $\sqrt{3}(\sqrt{6} + 2)$
   (h) $2\sqrt{5}(3\sqrt{8} - 10)$
   (i) $-\sqrt{2}(\sqrt{3} - \sqrt{5})$
   (j) $\sqrt{10}(2\sqrt{2} - 5\sqrt{5})$

2. In each case, expand the brackets and simplify.
   (a) $(1 + \sqrt{2})^2$
   (b) $(4 + \sqrt{5})(4 - \sqrt{5})$
   (c) $(2 - \sqrt{3})(3 + \sqrt{3})$
   (d) $(3 + \sqrt{5})(3 - \sqrt{5})$
   (e) $(1 - 2\sqrt{2})(2 + \sqrt{2})$
   (f) $(3 + 3\sqrt{7})(3 - 2\sqrt{7})$

(g) $(1 - \sqrt{3})(2 + \sqrt{5})$ (h) $(2\sqrt{2} + 3\sqrt{3})(1 - 2\sqrt{2})$ (i) $(2 - 3\sqrt{2})^2$
(j) $(\sqrt{3} - \sqrt{2})(2\sqrt{3} + 3\sqrt{2})$

3. $(3 + 2\sqrt{5})^2$ can be simplified to $a + b\sqrt{5}$
   Find the values of a and b

4. Show that $(3\sqrt{2} + 2\sqrt{3})^2 = 30 + 12\sqrt{6}$ You must show every step of your working.

5. A TV screen has a width of $(20\sqrt{2} - 8\sqrt{11})$ m and a height of 125 cm
   Find the area of the screen, giving your answer in square metres as a simplified surd.

6. A children's soft play area has a length of $4\sqrt{3}$ m and a width of $(2 + \sqrt{2})$ m
   Find the area of the soft play area, giving your answer in square metres as a simplified surd.

7. A rectangular mirror has a width of $(1 + \sqrt{5})$ m and a height of $(2 + \sqrt{5})$ m
   Find the mirror's area, giving your answer as a simplified surd.

23.5 Exact Answers

In some problems, you may be asked to give an exact answer, or give an answer in terms of $\sqrt{2}$ or $\sqrt{3}$, etc. The question will state if it requires an answer in this form.

Example 8

Look at the right-angled triangle shown in the diagram. Its two shorter sides are $2\sqrt{5}$ cm and 3 cm in length. Find:
(a) The length of the hypotenuse.
(b) The area of the triangle.
Give both answers as simplified surds.

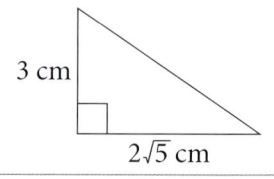

(a) Let $a = 3$, $b = 2\sqrt{5}$ and the hypotenuse be c cm
Then: $a^2 + b^2 = c^2$
$c^2 = 3^2 + (2\sqrt{5})^2$
$c^2 = 9 + (4 \times 5)$
$c^2 = 29$
$c = \sqrt{29}$ cm

(b) The area $A = \frac{1}{2} \times$ base $\times$ perpendicular height
$= \frac{1}{2} \times 2\sqrt{5} \times 3$
$= 3\sqrt{5}$ cm^2

Exercise 23E

1. A square has a side length of 1 cm. Find the length of the diagonal, giving your answer as a simplified surd.

2. Solve these equations to find positive values of x. In each case, give an exact answer in its simplest form.
   (a) $x^2 = 6$ (b) $x^2 = 20$ (c) $x^2 + 1 = 31$ (d) $x^2 - 4 = 36$ (e) $40 - x^2 = -35$

3. The longest side of a right-angled triangle is 9 cm and one of the other sides is 6 cm. Find the length of the third side, giving your answer in simplified surd form.

4. A rectangular snooker table ABCD measures 1 m by 2 m, as shown on the right. The white ball is hit from corner A of the table. It hits the centre of the opposite end, rebounds and reaches the pocket at corner D. How far has the snooker ball travelled? Give your answer as a simplified surd.

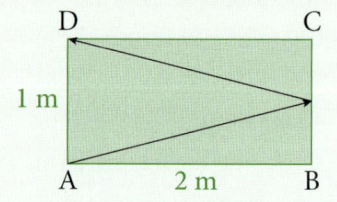

23.6 Summary

In this chapter you have learnt how to:
- Simplify a surd using the largest square factor of a number.
- Use the rules of surds for addition, subtraction, multiplication and division.
- Expand brackets involving surd terms.
- Give an exact answer to a question, involving surds, instead of rounding to a certain number of decimal places.

Chapter 24
Indices

24.1 Introduction

In this chapter we review the basic knowledge of the laws of **indices**, or powers. Remember an index is the quick way to write down the same value multiplied by itself a number of times, e.g. $3 \times 3 = 3^2$
We will then extend this concept by applying the laws of indices to negative indices.

Key words
- **Index**: A power, for example the 2 in 3^2, or the 3 in x^3
- **Squaring**: Squaring means multiplying a number by itself. For example, $3^2 = 3 \times 3 = 9$
- **Cubing**: Multiplying a number by itself and then by itself again. For example, $2^3 = 2 \times 2 \times 2 = 8$

Before you start you should know:
- How to use indices in basic arithmetic and algebra, e.g. that $5^3 = 5 \times 5 \times 5$ and $x^2 = x \times x$
- How to write squares as square roots and cubes as cube roots, e.g. if $2^3 = 8$ then $\sqrt[3]{8} = 2$

In this chapter you will learn how to:
- Use index notation.
- Use the index laws for zero, positive and negative powers.
- Use index laws in algebra for integer powers.

Exercise 24A (Revision)

1. Write out each of these products as a single square or cube, for example $5 \times 5 = 5^2$
   (a) 4×4 (b) $3 \times 3 \times 3$ (c) 11×11 (d) $13 \times 13 \times 13$
2. Write out each of these products as a single square or cube e.g. $y \times y = y^2$
   (a) $z \times z \times z$ (b) $p \times p$ (c) $5n \times 5n \times 5n$

24.2 What is an Index?

An index or power of a value equals how many copies of that value are being multiplied together.
We have already seen that : $\quad 2^2 = 2 \times 2 \quad$ where we say 2^2 as '2 squared'
Also: $\quad 9^3 = 9 \times 9 \times 9 \quad$ where we say 9^3 as '9 cubed'
In exactly the same way, we call 4 copies of 6 multiplied together '6 to the power 4' and write: $6^4 = 6 \times 6 \times 6 \times 6$

Example 1

What value is (a) the number 7 cubed (b) the number 3 to the power 4 (c) the number 2 to the power 6?

(a) $7 \times 7 \times 7 = 343$ (b) $3 \times 3 \times 3 \times 3 = 81$ (c) $2 \times 2 \times 2 \times 2 \times 2 \times 2 = 64$

Exercise 24B

1. Write each of these products as a number to a power:
   (a) $2 \times 2 \times 2 \times 2 \times 2$ (b) $8 \times 8 \times 8 \times 8 \times 8 \times 8$ (c) $9 \times 9 \times 9 \times 9$
   (d) $12 \times 12 \times 12 \times 12 \times 12 \times 12 \times 12$ (e) $6 \times 6 \times 6 \times 6$
2. Write out each of these powers using multiplication signs:
   (a) 8^3 (b) 4^5 (c) 11^6 (d) 3^7 (e) 2^5
3. Evaluate:
   (a) $6^3 - 4^3$ (b) $7^3 - 3^5$ (c) $3^2 + 4^3 + 5^4$ (d) $3^5 - 6^3$

215

24.3 The Use of Indices in Algebra

Indices are used in algebra in the same way as with numbers. Just like we can have $2^2 = 2 \times 2$:
We can have: $\quad\quad\quad\quad\quad\quad x^2 = x \times x \quad\quad$ where we say x^2 as 'x squared'
And we can have: $\quad\quad\quad\quad y^3 = y \times y \times y \quad$ where we say y^3 as 'y cubed'

We call 5 copies of y multiplied together 'y to the power 5' and write: $y^5 = y \times y \times y \times y \times y$
The 5 is what we call the power or index.

Example 2

Write the following using index notation:
(a) $t \times t \times t \times t$ (b) $y \times y \times y \times y \times y \times y \times y \times y$ (c) $p \times p \times p \times p \times p$
(d) $q \times q \times q \times q$ (e) $d \times d \times d \times d \times d \times d \times d$

(a) t^4 (b) y^8 (c) p^5 (d) q^4 (e) d^7

These terms can be combined just as with ordinary variables.

Example 3

Express the following products as a single term:
(a) $5t \times 2y$ (b) $y \times y \times y \times p \times p \times p \times p$ (c) $5y \times 7p \times p$
(d) $2p \times 7q \times 6p$ (e) $m \times m \times m \times m \times 7t$

(a) $5t \times 2y = 5 \times 2 \times t \times y = 10ty$ (b) $y^3 \times p^4 = y^3 p^4$
(c) $5 \times 7 \times y \times p^2 = 35yp^2$ (d) $2p \times 7q \times 6p = 2 \times 7 \times 6 \times p \times p \times q = 84p^2 q$
(e) $m \times m \times m \times m \times 7t = 7t \times m^4 = 7tm^4$

Example 4

Express the following as a single term:
(a) $m \times m \times m \div (p \times p)$ (b) $8y \div (2x^2)$
(c) $2p \times 3p \div (4q)$ (d) $t \times t \times t \times t \div (d \times d \times d \times d \times d)$

(a) $m^3 \div p^2 = \dfrac{m^3}{p^2}$ (b) $8y \div (2x^2) = \dfrac{8y}{2x^2} = \dfrac{4y}{x^2}$ (Remember to cancel wherever you can.)
(c) $2 \times 3 \times p \times p \div (4q) = 6p^2 \div (4q) = \dfrac{6p^2}{4q} = \dfrac{3p^2}{2q}$ (d) $t^4 \div d^5 = \dfrac{t^4}{d^5}$

Exercise 24C

1. Write each of the following using index notation:
   (a) $r \times r$ (b) $s \times s \times s$ (c) $p \times p \times p \times p$
   (d) $y \times y \times y \times y \times y$ (e) $t \times t \times t \times t \times t \times t \times t$

2. Write each of the following using index notation:
   (a) $q \times q \times q \times q \times q \times q \times q \times q \times q$ (b) $x \times x \times x \times x \times x \times x$ (c) $p \times p \times p \times p \times p \times p \times p$
   (d) $t \times t \times t \times t \times t \times t$ (e) $k \times k \times k \times k \times k \times k \times k \times k$

3. Express the following products as a single term:
   (a) $2x \times 3y$ (b) $5p \times 3p$ (c) $3w \times 7w \times w$ (d) $4m \times 3m \times 2n$ (e) $r \times 4t \times 2r \times t$

4. Perform the following divisions:
   (a) $x \times x \div y$ (b) $5t \div q$ (c) $6m \div 3n$ (d) $5x \times 12x \div 4y$ (e) $6q \times 8q \div (4x \times 3x)$

5. Simplify each of these expressions into a single term:
   (a) $6p \times 5q \times 3p \times 2q$ (b) $8x \times y \div (x \times 4y)$ (c) $7t \times 4r \times t \div (2t \times 14r)$
   (d) $3y \times 2w \times w \times 2y \div 3w^2$ (e) $2m \times 8n + 4n^2 \times m \div n$

24.4 The Laws of Indices

There are three laws that are always true when combining indices.

Each 'law' is just a way to remember a common-sense connection. Many people find the easiest way to remember the laws is to learn off a simple example of each.

First law of indices

The first power law concerns the multiplication of two expressions in index form.

Imagine working out:
$$2^2 \times 2^3$$
$$= 2 \times 2 \ \times \ 2 \times 2 \times 2$$
$$= 2 \times 2 \times 2 \times 2 \times 2$$
$$= 2^5$$

Now imagine working out: $a^3 \times a^4$
$$= a \times a \times a \ \times \ a \times a \times a \times a$$
$$= a \times a \times a \times a \times a \times a \times a$$
$$= a^7$$

Now think about the pattern in the powers (indices):

$2^2 \times 2^3 = 2^5$ Note the powers: $2 + 3 = 5$

and $a^3 \times a^4 = a^7$ Note the powers: $3 + 4 = 7$

The pattern is that the final power equals the other two powers added together.
We can summarise this using the formula:

$$a^p \times a^q = a^{p+q}$$

However, most people find the example $2^2 \times 2^3 = 2^5$ easier to remember.

> **Note:** To use this law, both the base numbers must be the same. In other words, something like $3^4 \times 2^5$ can't be simplified because the first part is a product of 3s and the second part is a product of 2s.

Example 5

Simplify:

(a) $x^2 \times x^4$ (b) $t^5 \times t^3$ (c) $y^7 \times y^5$ (d) $w^4 \times w^7$ (e) $2p^{21} \times 3p^{19}$

(a) Add the powers. $2 + 4 = 6$ So: $x^2 \times x^4 = x^6$ (b) Add the powers. $5 + 3 = 8$ So: $t^5 \times t^3 = t^8$
(c) $y^7 \times y^5 = y^{12}$ (d) $w^4 \times w^7 = w^{11}$ (e) $2p^{21} \times 3p^{19} = 6p^{40}$

Second law of indices

The next power law concerns the division of two expressions in index form.

Imagine working out:
$$2^3 \div 2^2$$
$$= 2 \times 2 \times 2 \div (2 \times 2)$$

We are dividing the product of three 2s by the product of two 2s:
$$2^3 \div 2^2$$
$$= \frac{2 \times 2 \times 2}{2 \times 2}$$
$$= 2$$

Now imagine working out:
$$p^5 \div p^3$$
$$= p \times p \times p \times p \times p \div (p \times p \times p)$$
$$= \frac{p \times p \times p \times p \times p}{p \times p \times p}$$
$$= p^2$$

Now think about the pattern in the powers (indices):

$2^3 \div 2^2 = 2$ Note the powers: $3 - 2 = 1$

and $p^5 \div p^3 = p^2$ Note the powers: $5 - 3 = 2$

The final power equals the difference between the other two powers.
We can summarise this using the formula:

$$a^p \div a^q = a^{p-q}$$

Again most people find the example $2^3 \div 2^2 = 2$ easier to remember.

> **Note:** To use this law, both the base numbers must be the same. In other words, something like $3^5 \div 2^4$ can't be simplified because the first part is a product of 3s and the second part is a product of 2s.

Example 6

Simplify:

(a) $x^6 \div x^4$ (b) $t^5 \div t^3$ (c) $y^9 \div y^4$ (d) $4w^8 \div w^3$ (e) $4p^{21} \div 2p^{11}$

(a) Subtract the powers. $6 - 4 = 2$ So: $x^6 \div x^4 = x^2$
(b) Subtract the powers. $5 - 3 = 2$ So: $t^5 \div t^3 = t^2$
(c) $y^9 \div y^4 = y^5$ (d) $4w^8 \div w^3 = 4w^5$ (e) $4p^{21} \div 2p^{11} = \frac{4}{2}p^{21} \div p^{11} = 2p^{10}$

Third law of indices

The third law of indices involves raising a power to a power.

Imagine working out: $(2^2)^3$

This means cubing two squared, i.e. cubing (2×2). So:
$$(2^2)^3 = (2 \times 2) \times (2 \times 2) \times (2 \times 2)$$
$$= 2^6$$

This shows that $(2^2)^3$ is six 2s multiplied together.

Now imagine working out: $(x^3)^4$

This involves multiplying four copies of x cubed. So:
$$(x^3)^4 = (x \times x \times x) \times (x \times x \times x) \times (x \times x \times x) \times (x \times x \times x)$$
$$= x^{12}$$

This shows that $(x^3)^4$ is 12 x's multiplied together.

The final power is the product of the first two powers.

We can summarise this using the formula:

$$(a^p)^q = a^{p \times q}$$

Most people find the example $(2^2)^3 = 2^{2 \times 3}$ easier to remember.

Example 7

Simplify:
(a) $(x^4)^2$ (b) $(t^2)^5$ (c) $(y^5)^3$ (d) $(w^7)^4$ (e) $(2p^4)^3$

(a) $(x^4)^2 = x^{4 \times 2} = x^8$ (b) $(t^2)^5 = t^{2 \times 5} = t^{10}$ (c) $(y^5)^3 = y^{5 \times 3} = y^{15}$ (d) $(w^7)^4 = w^{7 \times 4} = w^{28}$
(e) $(2p^4)^3 = 2p^4 \times 2p^4 \times 2p^4 = 2^3 \times (p^4)^3 = 2^3 p^{4 \times 3} = 8p^{12}$

> **Note:** There is a difference between $(2x^3)^4$ and $2(x^3)^4$
> $$(2x^3)^4 = (2x^3) \times (2x^3) \times (2x^3) \times (2x^3) = 16x^{12}$$
> However, $2(x^3)^4 = 2 \times x^{12} = 2x^{12}$

Exercise 24D

1. Simplify:
   (a) $p^3 \times p^4$ (b) $t^5 \times t^6$ (c) $w^3 \times w^8$ (d) $x^7 \times x^9$ (e) $y^9 \times y^4$

2. Simplify:
   (a) $m^7 \div m^3$ (b) $r^{12} \div r^3$ (c) $p^{19} \div p^{11}$ (d) $s^4 \div s^4$ (e) $y^9 \div y^8$

3. Simplify:
   (a) $(z^3)^6$ (b) $(x^7)^2$ (c) $(p^9)^5$ (d) $(w^3)^7$ (e) $(y^4)^7$

4. Simplify:
   (a) $2k^6 \times 3p^4$ (b) $5t^3 \times 3p^3$ (c) $4w^8 \times 12x^5$ (d) $3x^3 \times 3y^9$ (e) $y^4 \times 5y^4$

5. Simplify:
   (a) $w^3 \div (3w)^2$ (b) $2y^5 \div (3y)^2$ (c) $t^7 \div (2t)^3$ (d) $12y^3 \div (2y)^2$ (e) $6x^4 \div x^4$

24.5 The Laws of Indices With Negative Powers

The powers of 2 can be written in a table, as shown on the right. We can use this table to see how negative powers work.

Index	1	2	3	4	5
Power of 2	2^1	2^2	2^3	2^4	2^5
Value	2	4	8	16	32

If you start at the right-hand side of the table and move left, from one column to the next one, we can make three observations:
- the indices reduce by 1 every step,
- the powers of 2 reduce by one every step, and
- the values are divided by 2 every step across.

If we continue moving to the left in the table and follow the same patterns we can add the values of 2^0 and negative powers of 2, which are as follows:

Index	−4	−3	−2	−1	0	1	2	3	4	5
Power of 2	2^{-4}	2^{-3}	2^{-2}	2^{-1}	2^0	2^1	2^2	2^3	2^4	2^5
Value	$\frac{1}{16}$	$\frac{1}{8}$	$\frac{1}{4}$	$\frac{1}{2}$	1	2	4	8	16	32

We can immediately see that $2^0 = 1$

We can also see that $2^{-1} = \frac{1}{2}$, $2^{-2} = \frac{1}{2^2}$, $2^{-3} = \frac{1}{2^3}$, $2^{-4} = \frac{1}{2^4}$ and so on.

A general way to sum up these powers of two is by the following two statements:
- $2^0 = 1$ and
- $2^{-n} = \frac{1}{2^n}$ for any power, n

The table above used base 2, but the conclusions would be true no matter what base number was used. So we can write two extra index laws that apply to any base, a:
- $a^0 = 1$ and
- $a^{-n} = \frac{1}{a^n}$ for any power, n

Example 8

Write out the value of the expressions:
(a) 3^{-5} (b) 6^{-3} (c) 8^{-2} (d) 5^0

(a) $3^{-5} = \frac{1}{3^5}$ (b) $6^{-3} = \frac{1}{6^3}$ (c) $8^{-2} = \frac{1}{8^2}$ (d) $5^0 = 1$

The same rules apply when using algebra.

Example 9

Write out the value of the expressions:
(a) x^{-1} (b) y^{-4} (c) p^{-2} (d) x^0

(a) $x^{-1} = \frac{1}{x}$ (b) $y^{-4} = \frac{1}{y^4}$ (c) $p^{-2} = \frac{1}{p^2}$ (d) $x^0 = 1$

> **Note:** Notice we don't include an index of 1 in part (a). We write $\frac{1}{x^1}$ as $\frac{1}{x}$

Negative powers also obey the three rules discussed in section 24.4:
- $a^p \times a^q = a^{p+q}$
- $a^p \div a^q = a^{p-q}$
- $(a^p)^q = a^{p \times q}$

Example 10

Simplify:
(a) $3^2 \times 3^{-4}$ (b) $x^{-3} \div x^{-3}$ (c) $5^{-3} \div 5^{-7}$ (d) $y^{-8} \div y^{-2}$ (e) $(4^{-2})^{-3}$ (f) $(x^5)^{-2}$

Using the three rules of indices above:
(a) $3^2 \times 3^{-4} = 3^{2+(-4)} = 3^{2-4} = 3^{-2}$
(b) $x^{-3} \div x^{-3} = x^{(-3)-(-3)} = x^{-3+3} = x^0 = 1$
(c) $5^{-3} \div 5^{-7} = 5^{(-3)-(-7)} = 5^{-3+7} = 5^4$
(d) $y^{-8} \div y^{-2} = y^{(-8)-(-2)} = y^{-8+2} = y^{-6}$
(e) $(4^{-2})^{-3} = 4^{(-2)\times(-3)} = 4^6$
(f) $(x^5)^{-2} = x^{5\times(-2)} = x^{-10}$

Don't forget that powers in the denominator are equal to negative powers.

Example 11

Calculate:

(a) $\dfrac{1}{5^4} \times 5^3$ (b) $x^3 \div \dfrac{1}{x^5}$

(a) $5^{-4} \times 5^3 = 5^{(-4)+3} = 5^{-1}$ (b) $x^3 \div \dfrac{1}{x^5} = x^3 \div x^{-5} = x^{3-(-5)} = x^{3+5} = x^8$

Example 12

If x has a value between 0 and 1, place the following powers of x in ascending order of size:

(a) x^2 (b) x^{-2} (c) x^{-1} (d) x^0

If you are unsure, you may find it helpful to let x take a particular value and use it to calculate the value of each power of x. We are told that x lies between 0 and 1 so we could let $x = \dfrac{1}{2}$:

$x^2 = \left(\dfrac{1}{2}\right)^2 = \dfrac{1}{4}$ $x^{-1} = \left(\dfrac{1}{2}\right)^{-1} = (2^{-1})^{-1} = 2^1 = 2$ $x^{-2} = \left(\dfrac{1}{2}\right)^{-2} = (2^{-1})^{-2} = 2^2 = 4$

Don't forget that $x^0 = 1$

So we can see that the ascending order is $x^2 < x^0 < x^{-1} < x^{-2}$

> **Note:** You do not have to pick a value for x and calculate the values like we did in this example. However, doing so may make it easier for you to see what is going on and therefore easier to answer the question.

Exercise 24E

1. Evaluate:
   (a) 4^{-2} (b) 3^{-3} (c) 2^{-4} (d) 7^{-1} (e) 9^0 (f) 5^{-2} (g) 3^{-4} (h) 1^0

2. Write the following as single value:
   (a) $3^2 \times 3^{-1}$ (b) $2^{-5} \times 2^8$ (c) $4^4 \times 4^{-5}$ (d) $3^{-1} \times 3^{-2}$ (e) $5^6 \times 5^{-6}$

3. Write the following as single values:
   (a) $4^2 \div 4^{-1}$ (b) $3^{-1} \div 3^{-1}$ (c) $5^3 \div 5^{-1}$ (d) $7^{-4} \div 7^{-4}$ (e) $6^{-1} \div 6^{-2}$

4. Write each part as a single base to a power:
   (a) $x^3 \times x^{-7}$ (b) $y^1 \div y^{-1}$ (c) $p^{-5} \times p^{-3}$ (d) $x^{-2} \div x^3$ (e) $q \times q^{-4}$

5. Simplify:
   (a) $(x^2)^{-1}$ (b) $(x^3)^{-5}$ (c) $(x^{-2})^{-3}$ (d) $(x^0)^{-2}$ (e) $(x^6)^0$

6. If y has a value between 0 and 1, place the following powers of y in ascending order:
   (a) y^2 (b) y^{-3} (c) y^0 (d) y^{-2} (e) $(y^{-1})^{-1}$

24.6 Summary

In this chapter you have learnt about:

- Five power laws that help to simplify expressions involving powers or indices:

  - $a^p \times a^q = a^{p+q}$ for example $2^2 \times 2^3 = 2^5$
  - $a^p \div a^q = a^{p-q}$ for example $2^3 \div 2^2 = 2$
  - $(a^p)^q = a^{p \times q}$ for example $(2^2)^3 = 2^{2 \times 3} = 2^6$
  - $a^0 = 1$ for example $2^0 = 1$
  - $a^{-n} = \dfrac{1}{a^n}$ for example $2^{-3} = \dfrac{1}{2^3} = \dfrac{1}{8}$

Chapter 25
Standard Form

25.1 Introduction

The distance from Earth to the sun is roughly 149 200 000 000 metres.

The mass of a hydrogen atom is approximately 0.000 000 000 000 000 000 000 000 001 67 kg.

These numbers are quite awkward to write down accurately.

For this reason, we use a different notation called **standard form**.

A standard form number is written in the form:

$a \times 10^n$ where $1 \leq a < 10$ and n is any integer. a is known as the **coefficient**.

For example, 3.2×10^7 is a number written in standard form.

By contrast, 3 200 000 is a number written in decimal form.

Any number can be written in standard form, but it is most useful for very large or very small numbers.

Key words
- **Standard form**: A useful way to write a number if it is very small or very big.
- **Coefficient**: The part of a standard form number that comes before the multiplication sign.

Before you start you should:
- Know how to work with whole numbers and decimals.
- Know how to round numbers to a suitable level of accuracy.
- Be familiar with index notation and know how to use the rules of indices.

In this chapter you will learn how to:
- Write decimal numbers in standard form.
- Convert a standard form number to decimal form.
- Perform calculations with standard form numbers, both with and without a calculator.

> **Note:** When we refer to 'decimal numbers' and 'decimal form', this does not necessarily mean that the number has decimal places. It could be an integer. For example, the numbers 10, 672, 13, 3.14, and 0.00012 are all in decimal form. The number 5×10^6 is in standard form.

Exercise 25A (Revision)

1. Calculate the following.
   - (a) 98×10
   - (b) $6.5 \div 100$
   - (c) $610 + 56$
   - (d) $78 - 39$

2. Round these numbers to the given level of accuracy.
   - (a) 147 (nearest 10)
   - (b) 3521 (2 significant figures)
   - (c) 6.375 (1 decimal place)
   - (d) 4.468 (nearest whole number)
   - (e) 0.087135 (3 significant figures)

3. (a) Write the following as 2^n where n is an integer: $\dfrac{2^3 \times 2^{-2}}{2^7}$

   (b) Simplify the following: $\dfrac{4x^2y \times 3y^2x}{6x^2y^2}$

25.2 Converting to Decimal Form and Standard Form

You may be asked to convert a decimal number to standard form.

GCSE MATHEMATICS M3 AND M7

To do so, re-write the number, moving the decimal point so that it comes after the first non-zero digit. The power of 10 depends on the number of decimal places you have moved the decimal point.

When writing in standard form, note that for numbers greater than or equal to 10, such as 512 000, the power is positive. For numbers smaller than 1, such as 0.0021, the power is negative.

Example 1

Write the following in standard form.
(a) 237.5   (b) 512 000   (c) 0.0021   (d) 0.0006

(a) Reposition the decimal point so that it comes directly after the first non-zero digit. This tells us that the **coefficient** is 2.375

The decimal point moves 2 spaces, so:
$237.5 = 2.375 \times 10^2$

$2\,3\,7.5$

In this part of the question, the power on the ten is positive, since the number 237.5 is greater than or equal to 10 and the decimal point has been moved to the left.

(b) The decimal point moves 5 spaces, so:
$512\,000 = 5.12 \times 10^5$

$5\,1\,2\,0\,0\,0.0$

Again, the power on the ten is positive, since the number 512 000 is greater than or equal to 10 and the decimal point has been moved to the left.

(c) The decimal point moves 3 spaces to the right, so:
$0.0021 = 2.1 \times 10^{-3}$

$0.0\,0\,2\,1$

In this case the power on the ten is negative because 0.0021 is less than 1 and the decimal point has been moved to the right.

(d) To place the decimal point after the first non-zero digit, the 6, it has to move 4 spaces to the right:
$0.0006 = 6 \times 10^{-4}$

$0.0\,0\,0\,6$

If you are asked to convert a standard form number to decimal form, the process is reversed, as shown in the next example.

Example 2

Write the following standard form numbers in decimal form.
(a) 8×10^4 (b) 3.7×10^{-3}

(a) Rewrite the coefficient to have some zeroes after the decimal point 8.0000×10^4
Then move the decimal point 4 spaces to the right:
$8 \times 10^4 = 80\,000$

$8.0\,0\,0\,0\,0$

> **Note:** A standard form number with a power of 4 is always in the tens of thousands. You may find it useful to remember this.

(b) Rewrite the coefficient to have some leading zeroes: $3.7 \times 10^{-3} = 0003.7 \times 10^{-3}$
Then move the decimal point 3 spaces to the left:
$3.7 \times 10^{-3} = 0.0037$

$0\,0\,0\,3.7$

In the following example, the numbers given are in neither decimal form nor standard form.

Example 3

Write these numbers in standard form.
(a) 0.047×10^3 (b) 1053×10^4 (c) 0.022×10^{-5} (d) 268×10^{-6}

(a) The coefficient is multiplied by 100

To keep the value unchanged, the power must be decreased by 2, so: $0.047 \times 10^3 = 4.7 \times 10^1$
(b) The coefficient is divided by 1000
To keep the value unchanged, the power must be increased by 3, so: $1053 \times 10^4 = 1.053 \times 10^7$
(c) The coefficient is multiplied by 100 and the power decreased by 2, so: $0.022 \times 10^{-5} = 2.2 \times 10^{-7}$
(d) The coefficient is divided by 100 and the power increased by 2, so: $268 \times 10^{-6} = 2.68 \times 10^{-4}$

Exercise 25B

1. Which of the following numbers is in standard form? If not, give a reason.
   (a) 142 000
   (b) 3×10^5
   (c) 10×10^9
   (d) 1.175×10^{-3}
   (e) 6×10
   (f) $2.3 \div 10^8$
   (g) 5.08×10^1
   (h) $4.89 \times 10^{6.8}$
   (i) 0.31×10^5
   (j) 6×10^{-99}
   (k) 10^{13}

2. Write the following numbers in standard form.
   (a) 14 000
   (b) 2576
   (c) 0.039 81
   (d) 6 000 124
   (e) 0.000 595
   (f) 0.000 000 765
   (g) 2 000 000 000
   (h) 65 536
   (i) 12.1
   (j) 0.8

3. Write the following standard form numbers in decimal form.
   (a) 2×10^5
   (b) 1.6×10^6
   (c) 3.2×10^{-2}
   (d) 8×10^{-4}
   (e) 9.8×10^{-1}
   (f) 1.61×10^2
   (g) 7.252×10^{-5}
   (h) 3.14×10^1
   (i) 8.56×10^{-6}
   (j) 9.99×10^9

4. Change these numbers to standard form.
   (a) 0.7×10^5
   (b) 128×10^4
   (c) 0.0031×10^5
   (d) 74000×10^{-6}

5. Some female spiders can lay up to 3000 eggs at a time. Write this number in standard form.

6. The circumference of the Earth is 40 075 km. Write this in (a) metres (b) standard form.

7. A honey bee hive contains 55 230 bees. Write this in standard form.

8. Some football clubs have a lot of money!
   (a) At the time of publication, the record fee for a male footballer was set by the transfer of Neymar Júnior to Paris Saint-Germain from Barcelona. Paris Saint-Germain paid €222 million for the player in August 2017. Write this in standard form.
   (b) The female record at the time of publication was set by the transfer of Racheal Kundananji from San Diego to Chelsea for €1 100 000 in February 2025. Write this in standard form.

9. An average male human body contains roughly 36 000 000 000 000 (36 trillion) cells. Write this in standard form.

10. The table on the right shows the populations of Northern Ireland, Republic of Ireland, England, Scotland and Wales. Write the population of each one in standard form.

Northern Ireland	1.9 million
Republic of Ireland	5.25 million
England	57.7 million
Scotland	5.49 million
Wales	3.1 million

25.3 Ordering Numbers in Standard Form

You may be asked to put standard form numbers in order, from smallest to largest, or largest to smallest.

Example 4

Put these standard form numbers in order from smallest to largest:

2.3×10^3 6.7×10^{-2} 4.5×10^7 3.2×10^3

Method 1

Convert each number to decimal form:
$2.3 \times 10^3 = 2300$ $6.7 \times 10^{-2} = 0.067$ $4.5 \times 10^7 = 45\,000\,000$ $3.2 \times 10^3 = 3200$
From this, we can see that the correct order is: 6.7×10^{-2}, 2.3×10^3, 3.2×10^3, 4.5×10^7

Method 2

Look at the powers of 10
The smallest power gives the smallest number, so the smallest number is 6.7×10^{-2}

The two numbers 2.3×10^3 and 3.2×10^3 both involve 10^3 (meaning they are both in the thousands), with 2.3×10^3 being the smaller.

The number 4.5×10^7 has the largest power of 10, so this is the largest number.

So the correct order is: 6.7×10^{-2}, 2.3×10^3, 3.2×10^3, 4.5×10^7

Exercise 25C

1. In each part (a) to (d), state which of the two standard form numbers is the smaller.
   (a) 5.4×10^6 and 9.8×10^4
   (b) 1.7×10^3 and 7.1×10^{-3}
   (c) 5.52×10^{-4} and 4.66×10^{-2}
   (d) 1.523×10^{-9} and 2.153×10^{-9}

2. The table on the right shows the radii of different stars. Put these stars in order of size from largest to smallest.

Star name	Radius (km)
Proxima Centauri	1.07×10^5
Sirius A	1.19×10^6
The Sun	6.957×10^5
WOH G64	1.07×10^9
Betelgeuse	5.32×10^8

3. The table below shows the tallest buildings in some countries around the world. Put them in order of height, from smallest to largest.

Country	Building	City	Height (metres)
Armenia	Elite Plaza Business Center	Yerevan	8.51×10^1
China	Shanghai Tower	Shanghai	6.32×10^2
Croatia	Dalmatia Tower	Split	1.1×10^2
England	The Shard	London	3.096×10^2
Japan	Azabudai Hills Mori JP Tower	Tokyo	3.25×10^2
Northern Ireland	Obel Tower	Belfast	8.6×10^1
United Arab Emirates	Burj Khalifa	Dubai	8.28×10^2
United States	One World Trade Center	New York City	5.41×10^2

4. Answer the following questions. One tonne is 1000 kg.
   (a) A cruise ship has a mass of 125 000 tonnes. Write the mass of the cruise ship in kilograms in standard form.
   (b) A ferry crossing the Irish Sea has a mass of 20 000 tonnes. Write this in kilograms in standard form.
   (c) A speedboat has a mass of 530 kg. Write this in standard form.
   (d) A cargo ship has a mass of 1.9×10^5 tonnes. Write this mass in kilograms in standard form.
   (e) A sailing yacht has a mass of 2300 kg. Write this in standard form.
   (f) Put the five ships/boats in parts (a) to (e) in order from smallest mass to largest mass.

25.4 Calculations With Standard Form Numbers

You may be asked to add, subtract, multiply and divide standard form numbers, either with or without a calculator.

Addition and subtraction without a calculator

To add or subtract numbers in standard form without a calculator, convert the numbers to decimal form before adding or subtracting. Remember to change your answer back to standard form if the question asks you to do so.

Example 5

On a building site there is 4.3×10^5 kg of rubble, to be used as foundations. A truck delivers an additional 7.2×10^4 kg of rubble. How much rubble is there on the site now? Give your answer in standard form.

Convert the numbers to decimal form: $4.3 \times 10^5 = 430\ 000$
$7.2 \times 10^4 = 72\ 000$

And then add:
```
   430000
+   72000
   ------
   502000
```

So there is now 502 000 kg of rubble on the site, which is 5.02×10^5 kg in standard form

CHAPTER 25: STANDARD FORM

Example 6

A flask contains 9.5×10^{-1} litres of coffee. 7.8×10^{-2} litres of coffee is poured out of the flask. How much is left in the flask? Give your answer in litres in standard form.

Convert the numbers to decimal form: $9.5 \times 10^{-1} = 0.95$ litres $\quad 7.8 \times 10^{-2} = 0.078$ litres
And then subtract:
$$\begin{array}{r} 0.950 \\ -\ 0.078 \\ \hline 0.872 \end{array}$$
The flask now contains 8.72×10^{-1} litres of coffee.

Multiplication and division without a calculator

To multiply or divide standard form numbers, you will need to remember the rules of indices.

Example 7

Find:
(a) $(2.5 \times 10^7) \times (3 \times 10^{-3})$
(b) $(-4.8 \times 10^{-4}) \div (2 \times 10^3)$
(c) $(6 \times 10^{-5}) \times (3 \times 10^2)$
(d) $(2.8 \times 10^4) \div (7 \times 10^2)$

(a) Multiply the two coefficients together: $\quad 2.5 \times 3 = 7.5$
And multiply the powers of 10: $\quad 10^7 \times 10^{-3} = 10^4$
So: $(2.5 \times 10^7) \times (3 \times 10^{-3}) = 7.5 \times 10^4$

Note: Remember that you add the indices when multiplying. The base stays as 10

(b) Divide the coefficients: $\quad -4.8 \div 2 = -2.4$
And divide the powers of 10. $\quad 10^{-4} \div 10^3 = 10^{-7}$
So $(-4.8 \times 10^{-4}) \div (2 \times 10^3) = -2.4 \times 10^{-7}$

Note: Remember that you subtract the indices when dividing. The base stays as 10

(c) Using the rules described in part (a) for multiplication: $\quad (6 \times 10^{-5}) \times (3 \times 10^2) = 18 \times 10^{-3}$
However, this is not in standard form. We require the coefficient to be less than 10: the number can be rewritten in standard form by dividing the coefficient by 10 and increasing the power by 1:
$18 \times 10^{-3} = 1.8 \times 10^{-2}$

The effect of increasing the power by 1 is to multiply by 10, so overall the value is unchanged.

(d) Using the rules described in part (b) for division: $\quad (2.8 \times 10^4) \div (7 \times 10^2) = 0.4 \times 10^2$
However, this is not in standard form. The coefficient should be greater than or equal to 1: we multiply the coefficient by 10 and decrease the power by 1, keeping the value unchanged: $0.4 \times 10^2 = 4 \times 10^1$

You should know how to multiply a standard form number by a number in decimal form.

Example 8

Calculate the following:
(a) $(6.52 \times 10^7) \times 10$
(b) $8 \times (2.5 \times 10^{-3})$

(a) The effect of multiplying a standard form number by 10 is to increase the power of 10 by 1
So: $\quad (6.52 \times 10^7) \times 10 = 6.52 \times 10^8$
(b) Multiply the coefficient by 8: $\quad 8 \times 2.5 = 20$
So: $\quad 8 \times (2.5 \times 10^{-3}) = 20 \times 10^{-3} = 2 \times 10^{-2}$

We have divided the coefficient by 10 to put this in standard form. At the same time, the power is increased the power by 1 to keep the value unchanged.

Exercise 25D

You must **not** use a calculator in this exercise.

1. Calculate these, giving your answers in standard form, to 4 significant figures where appropriate.
   (a) $6.9 \times 10^2 + 4.3 \times 10^1$
   (b) $6 \times 10^7 + 3 \times 10^4$
   (c) $4.1 \times 10^4 + 9.2 \times 10^5$
   (d) $7.415 \times 10^8 + 2.8 \times 10^3$
   (e) $1.9 \times 10^{-3} + 3.45 \times 10^3$
   (f) $1 \times 10^{-4} + 6.38 \times 10^{-2}$
   (g) $7.3 \times 10^6 - 3 \times 10^5$
   (h) $2.5 \times 10^{-2} - 5 \times 10^{-3}$
   (i) $6 \times 10^5 - 2.3 \times 10^3$
   (j) $7 \times 10^6 - 1.54 \times 10^4$

2. Calculate these, giving your answers in standard form.
   - (a) $(2 \times 10^8) \times (3 \times 10^2)$
   - (b) $(2.6 \times 10^1) \times (2 \times 10^{-6})$
   - (c) $(6 \times 10^{-3}) \times (1.5 \times 10^{-9})$
   - (d) $(2.5 \times 10^4) \times (3 \times 10^{-7})$
   - (e) $(1 \times 10^7) \times (9.1 \times 10^{-4})$
   - (f) $(6 \times 10^5) \div (3 \times 10^9)$
   - (g) $(4.9 \times 10^5) \div (7 \times 10^2)$
   - (h) $(9.9 \times 10^{-2}) \div (3 \times 10^1)$
   - (i) $(8.4 \times 10^{-3}) \div (4 \times 10^{-4})$
   - (j) $(6.12 \times 10^{-4}) \div (6 \times 10^{-4})$

3. Calculate these, giving your answers in standard form.
   - (a) $10 \times (3.2 \times 10^6)$
   - (b) $(4.8 \times 10^4) \div 10$
   - (c) $6 \times (1.5 \times 10^3)$
   - (d) $(2.2 \times 10^{-2}) \times 2$
   - (e) $(3.6 \times 10^6) \div 3$
   - (f) $(2 \times 10^{-3}) \times 7$
   - (g) $(1.6 \times 10^8) \div 2$
   - (h) $(2.5 \times 10^{-2}) \div 5$

4. A road is 4×10^3 m long. Seamus drives the length of this road 5 times. How far does he drive? Give your answer in metres in standard form.

5. A plane can carry 1.59×10^5 litres of fuel. After its first flight of the day, the plane has 9.9×10^4 litres of fuel left. The plane is then completely refuelled before its second flight. How much fuel is used to refuel the plane? Give your answer in litres in standard form.

6. When two electrical resistors are placed in series, as shown in the diagram, the combined resistance can be calculated by adding each resistance together.
   Calculate the combined resistance, giving your answer in ohms in standard form.

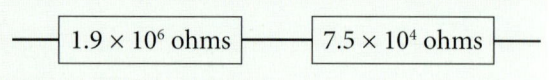

   1.9×10^6 ohms —— 7.5×10^4 ohms

7. A pumpkin contains 4.3×10^2 seeds. One tenth of these seeds are removed and cooked. Giving your answers in standard form, calculate how many seeds are (a) cooked (b) not cooked.

8. Each day, 1.2×10^4 tomatoes arrive at a factory in Newtownards that makes tomato soup and ketchup.
   - (a) 9.8×10^3 tomatoes are used for soup and the rest for ketchup. How many tomatoes are used for ketchup each day? Give your answer in standard form.
   - (b) The factory operates seven days a week. How many tomatoes arrive at the factory in one week? Give your answer in standard form.
   - (c) How many tomatoes arrive at the factory in two weeks? Give your answer in standard form.

9. A landfill site in Belfast already contains 3.7×10^6 tonnes of waste. During October, another 1.2×10^4 tonnes of waste is brought to the site. What mass of waste is at the site at the end of October? Give your answer in tonnes in standard form.

10. Given that $p = 4 \times 10^6$ and $q = 8 \times 10^9$ find, in standard form, the value of (a) $\frac{p}{q}$ (b) pq

Calculations with a calculator

To enter a standard form number on a calculator, use the button that looks like one of these $\boxed{\times 10^x}$ $\boxed{\times 10^{\square}}$ $\boxed{\text{EXP}}$

Example 9

Enter 2×10^5 on the calculator.

Press these buttons: $\boxed{2}$ $\boxed{\times 10^x}$ $\boxed{5}$

Then press: $\boxed{=}$ or $\boxed{\text{EXE}}$

In this case, the calculator converts your number to decimal form, 200 000
Note that you do not need to enter a multiplication sign.

Calculator tip

On the Casio fx–83GT CW, you can adjust the settings so that your answers appear in standard form by default. Take these steps:
- Press Settings
- Select Calc Settings
- Scroll down and select Number Format
- Scroll down to Scientific Form and select it with OK
- Scroll down and choose the number of significant figures, e.g. 4

CHAPTER 25: STANDARD FORM

Clear the screen. Try a calculation, for example: 3.1 × 1000 and press EXE. The calculator should display 3.100×10^3
To return to normal mode:
- Press Settings
- Select Calc Settings
- Scroll down and select Number Format
- Scroll down to Norm and select it with OK
- Scroll down to Norm2 and select it with OK

Clear the screen and try the same calculation: 3.1 × 1000 and press EXE. The calculator should display 3100

Example 10

(a) An unmanned spacecraft travels 4.2 billion km. Write this distance in kilometres in standard form.
(b) For each kilometre travelled, the spacecraft uses up 1.27×10^{-10} kg of its nuclear power source. What mass of the power source has been used altogether? Give your answer in kilograms in standard form.

(a) 4.2 billion = 4 200 000 000 = 4.2×10^9 km
(b) The total mass of the power source used is: $(4.2 \times 10^9) \times (1.27 \times 10^{-10})$
On your calculator, enter: (4 . 2 ×10ˣ 9) × ((1 . 2 7 ×10ˣ − 1 0)
This will give a result of 0.5334 kg = 5.334×10^{-1} kg

Example 11

The planet Mars has a diameter of 6.974×10^3 km while Jupiter has a diameter of 1.398×10^5 km. How many times wider is Jupiter than Mars? Round your answer to the nearest whole number.

This question can be answered using division. $(1.398 \times 10^5) \div (6.974 \times 10^3) = 20.04...$
So Jupiter is about 20 times wider than Mars.

> **Note:** On the calculator, it is safest to use brackets around standard form numbers. Leaving out the brackets does not usually make a difference, but it may do when dividing, depending on your calculator. For example, use: $(4 \times 10^3) \div (2 \times 10^5)$

Exercise 25E

1. Carry out the following calculations on your calculator. Give your answers in standard form, rounded to 4 significant figures where necessary.
   (a) $8.93 \times 10^3 + 8.81 \times 10^5$
   (b) $2.7 \times 10^4 + 5.243 \times 10^6$
   (c) $4.056 \times 10^0 + 4.5984 \times 10^1$
   (d) $7.014 \times 10^{-5} + 7.139 \times 10^{-7}$
   (e) $2.747 \times 10^2 + 5.71272 \times 10^3$
   (f) $4.5 \times 10^6 - 7 \times 10^5$
   (g) $8 \times 10^6 - 2.14 \times 10^4$
   (h) $1.15 \times 10^8 - 3.246 \times 10^7$
   (i) $1.9 \times 10^{-2} - 6.172 \times 10^{-3}$
   (j) $8.898 \times 10^5 - 7.4 \times 10^{-2}$

2. Carry out the following calculations on your calculator. Give your answers in standard form, rounded to 4 significant figures where necessary.
   (a) $(6 \times 10^{-7}) \times (8.5 \times 10^3)$
   (b) $(4.3 \times 10^5) \times (9.7 \times 10^{-5})$
   (c) $(1.8 \times 10^6) \times (6 \times 10^{-8})$
   (d) $(7.8 \times 10^{-6}) \times (7.5 \times 10^1)$
   (e) $(9.46 \times 10^{-8}) \times (2 \times 10^5)$
   (f) $(1.6439 \times 10^1) \div (9.296 \times 10^4)$
   (g) $(2.068 \times 10^1) \div (1.3 \times 10^{-5})$
   (h) $(2 \times 10^{-1}) \div (2.8 \times 10^5)$
   (i) $(8.774 \times 10^{-4}) \div (6.267 \times 10^2)$
   (j) $(5.94 \times 10^9) \div (3 \times 10^{-3})$

3. A construction company uses an average of 1.1×10^4 bricks for each house. A housing development in Cookstown is to have 44 houses. Find the total number of bricks to be used in the development, giving your answer in standard form.

4. A Norwegian cross-country skier has a total distance of 65.71 km to travel.
   (a) Write the total journey distance in metres in standard form.
   (b) He travels 3.72×10^4 metres in one day. The next day he skis another 9.81×10^3 metres. Find the total distance travelled by the skier during these two days.
   (c) The skier completes his journey on the third day. How far does he ski this day? Give your answer in metres in standard form.

5. An adult human with a full head of hair usually has between 1×10^5 and 2×10^5 hairs on their head, depending on many factors.
   - People with blonde hair have an average of 1.35×10^5 hairs.
   - People with black or brown hair have an average of 1.05×10^5 hairs.
   - People with red hair have an average of 9×10^4 hairs.
   (a) On average, how many more hairs does a blonde person have when compared with a person with brown hair? Give your answer in standard form.
   (b) On average, how many times more hair does a blonde person have when compared with a person with red hair? Give your answer in standard form.

6. A kitchen is 5 m long, 4 m wide and 2 m high.
   (a) Calculate the volume of the kitchen in cubic metres (m³).
   (b) Convert your answer to cubic centimetres (cm³) and give your answer in standard form. There are 1 million cubic centimetres in a cubic metre.
   (c) How many baked beans would it take to fill the kitchen? You may assume that one baked bean has a volume of 0.2 cm³. Give your answer in standard form.

7. A particular virus is 20 nanometres in diameter, or 2.0×10^{-8} m. A bacterial cell measures 5 microns, or 5×10^{-6} m in length.
   (a) Which one is bigger: the virus or the bacterial cell?
   (b) How many times bigger is it?

8. (a) The distance from Earth to the Sun is roughly 149 200 000 kilometres. Write this in standard form.
   (b) The distance from Earth to the Moon is roughly 384 400 kilometres. Write this in standard form.
   (c) How many times further away from the Earth is the Sun than the Moon? Round your answer to 1 significant figure.

9. Stan is a writer, but he refuses to use a computer, so he gets through a lot of pencils.
   (a) On average Stan writes 1500 words per day with his pencil. Assuming each word is equivalent to a line of length 0.04 m, how many metres of writing does Stan do each day? Give your answer in metres in standard form.
   (b) Stan's pencil lasts 125 days. Calculate the approximate total length of the pencil line that Stan has put onto paper during this time.
   (c) The company that makes Stan's pencils claims that a single pencil could draw a line of length 56 km. Write this in metres in standard form.
   (d) Comment on the company's claim.

10. Liam is building a planetary path for people to walk along. The planetary path will have a model of the Sun and models of the planets. Liam uses two different scales. He uses:
    - 1 cm to 1000 km for the diameter of each planet
    - 1 m to 1 000 000 km for the distance from the Sun to each planet
    (a) Liam makes a model of the planet Mercury. The model has a diameter of 4.9 cm. Work out the real diameter of the planet Mercury. Give your answer in standard form.
    (b) Liam works out the distance from the model of the Sun to the model of the planet Uranus. The real distance from the Sun to Uranus is 2.926×10^9 km. Work out the distance from the model of the Sun to the model of the planet Uranus. Give your answer in kilometres, correct to 1 decimal place.

25.5 Summary

In this chapter, you have learnt how to:
- Write decimal numbers in standard form.
- Convert between standard form and decimal form.
- Perform calculations with standard form numbers, both with and without a calculator.

Chapter 26
Simultaneous Equations

26.1 Introduction

In Chapter 8 you learned how to solve linear equations in one variable, for example $2x - 1 = 7$

Simultaneous equations are a set of two or more equations involving more than one variable, for example:
$\quad 2a - 3b = 47 \quad$ and $\quad 4a + b = 45$

In this chapter you will learn how to solve two linear simultaneous equations involving two variables, to find the values of each of the variables. Simultaneous equations can be useful in solving many real-life problems.

Key words
- **Linear equation**: A linear equation has terms in x (and/or y, etc), and possibly some number terms, but does not have any terms with higher powers, such as x^2
- **Simultaneous equations**: A system of equations involving more than one variable.
- **Elimination**: A method involving adding or subtracting two equations to eliminate one of the variables.
- **Substitution**: A method that involves rearranging one equation and substituting into the other.

Before you start
- You should know your times tables up to 12
- You should know how to solve a linear equation in one variable, for example $2x - 1 = 7$

In this chapter you will learn how to:
- Solve two simultaneous equations in two variables using the methods of elimination and substitution.
- Form a pair of simultaneous equations from a worded problem.

Exercise 26A (Revision)

1. Give the answer to each of the following multiplications without using a calculator.
   (a) 2×6 (b) 6×7 (c) 7×8 (d) 8×4 (e) 4×7
   (f) 7×9 (g) 9×3 (h) 3×6 (i) 6×8 (j) 9×5

2. Solve the following.
   (a) $2a - 1 = 7$ (b) $2 + 3b = 20$ (c) $4c - 5 = -21$ (d) $5d + 6 = -14$
   (e) $10 - 2e = 8$ (f) $8 - 3f = -4$ (g) $6 - 4g = 10$ (h) $-7 - 8h = -47$

26.2 Solving Simultaneous Equations Using Elimination

The method of elimination involves adding or subtracting the two equations to eliminate one of the variables. In the first example, we add the two equations to eliminate x.

Example 1

Solve the following pair of simultaneous equations.
$2x + 3y = -2 \quad\quad (1)$
$-2x - y = 6 \quad\quad (2)$

Adding equations (1) and (2), the x terms cancel out and we have:
$\quad 2y = 4 \quad$ giving $\quad y = 2$
Substitute the value of y into either of the original equations. We choose to use Equation (1):
$2x + 3(2) = -2$
$\quad 2x + 6 = -2$
$\quad\quad 2x = -8 \quad$ giving $\quad x = -4$

You may need to change one of the equations to eliminate one of the variables. You can do this by multiplying both sides by a value, as shown in the next example.

Example 2

Solve the following pair of simultaneous equations.
$2a - 3b = 47$ (1)
$4a + b = 45$ (2)

In this example, we choose to eliminate b
Keep Equation (1) unchanged, but multiply Equation (2) by 3
This gives positive or negative $3b$ in both equations:
$2a - 3b = 47$ (3)
$12a + 3b = 135$ (4)
Add the equations, since the b terms have different signs:
$14a = 182$
$a = 13$
Substitute into Equation (2). This may be easier than using Equation (1), since the b term is positive:
$4(13) + b = 45$
$52 + b = 45$
$b = 45 - 52$
$b = -7$

Note: An alternative method would be to multiply Equation (1) by 2
Then both equations would involve $4a$; the next step would be subtraction to eliminate a

You may need to multiply both equations by different numbers to eliminate one of the variables.

Example 3

Solve the following pair of simultaneous equations.
$5x + 3y = 11$ (1)
$3x + 2y = 6$ (2)

To eliminate one of the variables, we need either the x terms or the y terms to be the same in each equation. In this example we choose to make the y terms the same by multiplying Equation (1) by 2 and Equation (2) by 3:
(1) × 2 gives: $10x + 6y = 22$ (3)
(2) × 3 gives: $9x + 6y = 18$ (4)
Now subtract (4) from (3). On the left-hand side, we are left with x since the y terms cancel out.
On the right-hand side we get 4:
$x = 4$
Substitute $x = 4$ into (2):
$3(4) + 2y = 6$
$12 + 2y = 6$
$2y = -6$
$y = -3$

In Example 2 above, Equation (3) involves $-3b$ and Equation (4) involves $+3b$
The two equations were added to eliminate b

In Example 3, both Equations (3) and (4) involve $+6y$
In this case the equations were subtracted to eliminate y

Some people like to remember: Same sign subtract (SSS)
Different sign add (DSA)

'Same sign subtract' also applies in the case of two negative terms.
In the following example the two original equations must be rearranged so that the variables x and y appear on the left, with the number terms on the right.

CHAPTER 26: SIMULTANEOUS EQUATIONS

Example 4

Solve the following pair of simultaneous equations.
$4x = 3y$
$4y = 25 - 3x$

Firstly, rearrange both equations so that the x and y terms appear on the left-hand side. The right-hand sides involve only numbers.
$4x - 3y = 0$ (1)
$3x + 4y = 25$ (2)

In this example, we choose to eliminate x
Multiply Equation (1) by 3 and Equation (2) by 4:
$12x - 9y = 0$ (3)
$12x + 16y = 100$ (4)

This time we choose to subtract (3) from (4) (rather than the other way round) in order to give a positive y term. Be careful with the subtraction because it involves a negative number:
$16y - (-9y) = 100$
$ 25y = 100$
$ y = 4$

Substitute $y = 4$ into (1):
$4x - 3(4) = 0$
$4x - 12 = 0$
$4x = 12$
$x = 3$

Exercise 26B

1. Solve the following pairs of simultaneous equations using the method of elimination.
   (a) $-2x + 5y = -14$
        $-3x - 5y = 4$
   (b) $3a + 4b = -19$
        $3a + 2b = -11$
   (c) $3c + 5d = -20$
        $3c + 7d = -22$
   (d) $3e + 7f = 51$
        $-2e - 7f = -48$
   (e) $-7g - 4h = 27$
        $7g + 5h = -32$
   (f) $-2j - 7k = 45$
        $4j - 7k = 57$
   (g) $4m + 3n = 9$
        $4m - 7n = 59$
   (h) $7p - 5q = 26$
        $7p - 4q = 25$

2. Solve the following pairs of simultaneous equations by elimination.
   (a) $-6x - y = -11$
        $x + 3y = -1$
   (b) $-2a + 7b = -40$
        $-4a - 5b = 34$
   (c) $-2c - d = -4$
        $-5c + 4d = -23$
   (d) $e - 7f = 27$
        $-5e + 2f = -3$
   (e) $g - 3h = 12$
        $3g - 4h = 11$
   (f) $-3j - 4k = -1$
        $-j + 6k = -37$
   (g) $-m + 3n = 1$
        $-3m + n = 3$
   (h) $-2p + q = 11$
        $-p - 2q = 8$

3. Solve the following pairs of simultaneous equations.
   (a) $-5x + 4y = 0$
        $2x - 7y = 27$
   (b) $6a + 2b = -8$
        $-4a + 7b = -3$
   (c) $4c - 3d = -29$
        $-3c - 2d = 9$
   (d) $2e + 3f = 1$
        $3e - 5f = 30$
   (e) $3g - 5h = 6$
        $-7g - 4h = -61$
   (f) $-5j + 4k = 4$
        $-3j + 5k = 5$
   (g) $2m + 5n = -3$
        $7m + 6n = 24$
   (h) $6p + 5q = 25$
        $5p + 6q = 19$

4. Solve the following pairs of simultaneous equations. Give answers as fractions where appropriate.
   (a) $-3x + 4y = -15$
        $2x + 2y = 3$
   (b) $15x - 20y = -2$
        $-10x - 2y = 9$
   (c) $15x + 18y = 38$
        $3x + y = 5$
   (d) $4x + 10y = 7$
        $-4x + 28y = -7$
   (e) $5x - 5y = -4$
        $x + 6y = 9$
   (f) $10x + 10y = -11$
        $20x + 5y = -28$
   (g) $x + 6y = -3$
        $-3x + 12y = -11$
   (h) $6x - 7y = 4\frac{1}{2}$
        $x + y = 4$

5. Solve the following pairs of simultaneous equations. As a first step, rearrange the equations so that the x and y terms appear on the left-hand side, with the number terms on the right.

(a) $-7x - 3y + 22 = 0$
$x = 8 + 2y$

(b) $4a = 7b - 30$
$-a = 5b - 33$

(c) $c = d$
$2d = 7c - 25$

(d) $5e = f - 23$
$e + f + 13 = 0$

(e) $3 + 4g + 7h = 0$
$4h - g = -5$

(f) $6k = 7j - 31$
$k = 3j - 7$

(g) $5n = 39 + 7m$
$7n = 7 + 3m$

(h) $4p = 8 - 3q$
$7q = 10 - 5p$

26.3 Solving Simultaneous Equations Using Substitution

The method of substitution is an alternative to elimination. Any pair of simultaneous equations can be solved in this way, as shown in the example below. In an examination, you can choose whether to use elimination or substitution, unless the question requires you to use a particular method.

Example 5

Solve the following pair of simultaneous equations using substitution.
$-7x + 2y = 61$ (1)
$-7x - 6y = 13$ (2)

Take Equation (1) and rearrange to make y the subject:
$$-7x + 2y = 61$$
$$2y = 61 + 7x$$
$$y = \frac{61 + 7x}{2}$$

First, substitute this expression for y into Equation (2):
$$-7x - 6y = 13$$
$$-7x - 6\left(\frac{61 + 7x}{2}\right) = 13$$
$$-7x - \left(\frac{366 + 42x}{2}\right) = 13$$
$$-7x - (183 + 21x) = 13$$
$$-7x - 183 - 21x = 13$$
$$-28x = 196$$
$$x = -7$$

Then, substitute $x = -7$ into either of the original equations. We choose to use Equation (1), which should be slightly simpler.
$$-7x + 2y = 61$$
$$-7(-7) + 2y = 61$$
$$49 + 2y = 61$$
$$2y = 12$$
$$y = 6$$

Note: You could instead make x the subject in one equation and substitute this into the other equation.

Exercise 26C

1. Solve the following pairs of simultaneous equations using the method of substitution.

(a) $x - y = -6$
$x - 4y = -27$

(b) $-5a - 7b = 0$
$-a - b = 2$

(c) $-5c - 2d = -1$
$5c - 7d = 19$

(d) $-4e - 5f = 7$
$-3e + 5f = -56$

(e) $-7g - h = -14$
$-5g - 4h = -10$

(f) $-4j + 5k = -23$
$3j - 2k = 12$

(g) $7m + 5n = 3$
$-3m - 4n = 8$

(h) $4p - 3q = -33$
$6p - 5q = -51$

26.4 Problem Solving

Simultaneous equations can be used in a wide range of real-life problems.

Example 6

The sum of 2 numbers is 1633 and their difference is 35
Find the two numbers.

CHAPTER 26: SIMULTANEOUS EQUATIONS

Let the numbers be x and y. Then: $y + x = 1633$ (1)
$y - x = 35$

Subtract: $2x = 1598$
$x = 799$

Substitute into (1) to find y: $y + 799 = 1633$
$y = 1633 - 799$
$y = 834$

Example 7

Tom is 3 years older than Ciaran. Adding twice Tom's age and three times Ciaran's age gives 51
Find the age of each boy.

Let Tom's age be t years and Ciaran's age be c years.
Tom is 3 years older than Ciaran, so: $t - c = 3$ (1)

Adding twice Tom's age and three times Ciaran's age gives 51
As an equation: $2t + 3c = 51$ (2)

Multiplying Equation (1) by 3 gives: $3t - 3c = 9$ (3)

Add Equations (2) and (3) to eliminate c: $5t = 60$
$t = 12$

Substituting into Equation (1): $12 - c = 3$
$c = 9$

Example 8

An apple and three bananas cost £1.70
Two apples and four bananas cost £2.50
Find the cost of an apple and the cost of a banana.

We choose to work in pence to avoid decimal values, but working in pounds is also possible. Let a be the cost of an apple and b be the cost of a banana, in pence.
Then: $a + 3b = 170$ (1)
$2a + 4b = 250$ (2)

Multiply Equation (1) by 2: $2a + 6b = 340$ (3)
Leave Equation (2) unchanged: $2a + 4b = 250$ (2)

Subtracting Equation (2) from Equation (3): $2b = 90$
$b = 45$

Substitute this into Equation (1): $a + 3(45) = 170$
$a + 135 = 170$
$a = 170 - 135 = 35$

So an apple costs 35p and a banana costs 45p

Note: In questions involving money, you can often choose whether to work in pounds or pence. Working in pounds will often involve decimals, whereas working in pence may involve some large numbers. Select whichever method that makes the numbers in the question easier to work with.

Note: In many worded problems you know there is a mistake in your working if either of your answers is negative. For example, negative prices are not possible. Neither are negative measurements, ages, and some other quantities.

Note: There are other checks you can do to help decide whether an answer looks sensible. For example, a cost of £300 for a packet of crisps is unrealistic. A price that has several decimal places is also probably wrong.

Exercise 26D

1. Two numbers x and y have a sum of 69 and a difference of 23. Given that $x > y$, find x and y.
2. Philip says 'My mum is three times as old as me. But in 15 years' time she'll be twice as old as me!'
   Find the ages of Philip and his mum.
3. Look at the right-angled triangle on the right.
   (a) What is the sum of angles Q and R? Write this information as an equation.
   (b) The difference between angles Q and R is 22°
       Write down a second equation involving Q and R.
   (c) Solve this pair of equations to find the sizes of angles Q and R.
4. In the shop *Top Tat*, you can buy 4 glowsticks and one stress ball for £3.40
   You could instead buy 2 glowsticks and 3 stress balls for £3.70
   Find the cost of a glowstick and a stress ball.
5. In the school tuck shop Aya buys 2 bags of crisps and 3 drinks for £3.25
   Brett buys 3 bags of crisps and 2 drinks for £3
   Find (a) the price of a bag of crisps and (b) the price of a drink.
6. Twice a number added to four times another number gives 28
   The difference between the two numbers is 5
   Find the two numbers.
7. The Robinson family visits the *Best Café* and orders 2 cups of tea and a bun. Their order comes to £6.90
   The Johnston family orders 5 cups of tea and 3 buns. Their order comes to £18.50
   Find (a) the cost of a cup of tea; and (b) the cost of a bun.
8. The perimeter of a rectangular car park is 220 m.
   The difference between the length and width is 30 m.
   (a) Form and solve two simultaneous equations to find the length and width of the car park.
   (b) Find the area of the car park.
9. Reuben empties his money box, which contains only 10p coins and 50p coins.
   There are 20 coins in total, with a total value of £6.40
   How many 10p coins and how many 50p coins are there?
10. Look at the rectangle on the right, which is made up of 5 identical tiles. The width of the entire rectangle is 10 cm.
    (a) Using simultaneous equations, find the width and height of the tiles.
    (b) Find the area of the entire rectangle.
11. The straight line $y = mx + c$ passes through the points (4, 5) and (1, −4). Using simultaneous equations, find the equation of the line.
12. Look at the trapezium below. The three shorter sides are all equal in length.

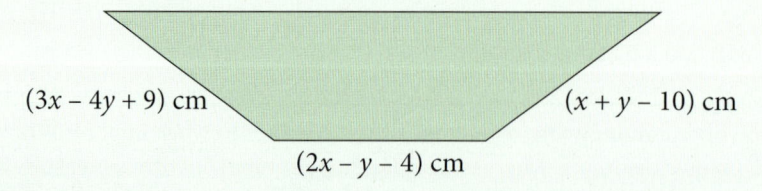

    $(3x - 4y + 9)$ cm $(x + y - 10)$ cm
    $(2x - y - 4)$ cm

    (a) By forming and solving two simultaneous equations, find the values of x and y.
    (b) The longest side has a length of $(x + y - 2)$ cm. Find this side length using your values of x and y.

26.5 Summary

In this chapter you have learned how to:
- Solve two simultaneous equations in two variables using the methods of elimination and substitution.
- Form and solve a pair of simultaneous equations from a worded problem.

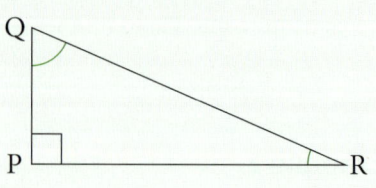

Chapter 27
Inequalities

27.1 Introduction

You will remember that combining = (equals) with < (less than) creates the sign ≤ meaning 'less than or equal to'. Likewise, combining = (equals) with > (greater than) makes the sign ≥ meaning 'greater than or equal to'. All of these signs, with the exception of = (equals), are known as 'inequalities'.

Key words
- **Inequality**: A mathematical statement involving an inequality sign, for example $2x - 1 \leq 5$
- **Strict inequality**: A mathematical statement involving either of the strict inequality signs < or >, e.g. $2x - 1 < 5$
- **Combined inequality**: Two inequalities written in a single statement, for example $2 < x < 7$
- **Solution set**: The set of values of x that satisfy an inequality.
- **Integer**: The set of integers includes the positive and negative whole numbers and zero.
- **Real number**: The set of real numbers includes the set of integers and all values between them.

Before you start you should know how to:
- Use the signs < = > ≤ and ≥ to compare numbers.
- Use a number line to represent the solutions to inequalities: 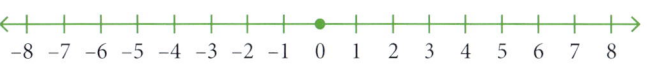
- Solve simple equations.

In this chapter you will learn how to:
- Solve a linear inequality in one variable.
- Represent the solution set on the number line.
- Draw and interpret an inequality in two variables graphically.
- Solve linear inequalities in two variables representing the solution set on a graph.

Exercise 27A (Revision)

1. Copy each pair of numbers inserting the correct sign <, = or > between each pair.
   (a) 6, 17   (b) 8, 12   (c) 6, 3   (d) 20, 20   (e) 7, −5
   (f) −3, −8   (g) −2, 0   (h) 12, −9   (i) −3, 3   (j) −5, −5

2. Draw a number line and plot on it the following numbers:
   (a) 6   (b) −4   (c) 2   (d) −1   (e) 5   (f) −1.5   (g) −4.5   (h) 0

3. Solve for x:
   (a) $3x - 7 = x + 11$ (b) $6x - 5 = 2x + 13$ (c) $7x - 2 = 10 - 3x$ (d) $3x - 14 = 37$
   (e) $6 - 5x = 2x - 8$ (f) $4x + 11 = 5x + 19$ (g) $6x + 13 = 21$ (h) $9x - 35 = 24 + 12x$

27.2 Variables and Inequalities for Real Numbers

The number line shown on the right has negative numbers on the left increasing to positive numbers on the right.

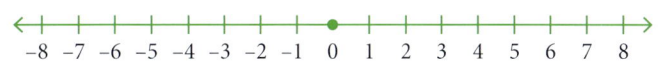

$-3 < 5$
-3 is less than 5

$5 > -3$
5 is greater than -3

When we compare two numbers by plotting their positions on the number line, the number on the left is smaller than the number on the right. An example is shown on the right.

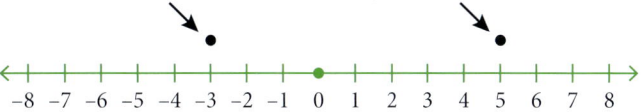

235

GCSE MATHEMATICS M3 AND M7

While an equation always involves the equals sign =, an inequality always involves one of the inequality symbols <, >, ≤ or ≥.

Their meanings are summarised in the table on the right.

<	is less than
≤	is less than or equal to
≥	is greater than or equal to
>	is greater than

For example, $x < 5$ means 'x is less than 5', and $x \geq 3$ means 'x is greater than or equal to 3'.

Let us consider: what values of x make $x \leq 5$ true? Look at the number line:

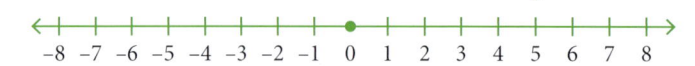

Clearly $4 < 5, 3 < 5, \ldots, 0 < 5, -1 < 5, -2 < 5, \ldots$ Additionally, $5 = 5$, so $5 \leq 5$ is also true. We can see that any number to the left of 5 on the number line is less than 5, and 5 is equal to 5 of course.

We call the values of x that make the inequality true the **solution set** of the inequality.

The solution set for the inequality $x \leq 5$ includes every real number on the line to the left of 5
This means, for example, that $x = 2.5$ and $x = -\frac{22}{7}$ are also in the solution set.

The solution set of an inequality is almost always a range of values like this. We represent the solution set by a range of numbers on a number line. The solution to the inequality $x \leq 5$ is represented by the dot and arrow above the number line:

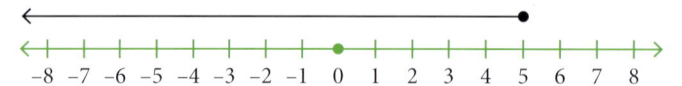

Example 1

Represent the solution to the inequality $x \leq 6$ on a number line, given that x is a real number.

As x is a real number it will be represented by a solid line over the number axis.
As the solution set is values 'less than' a number, the solid line will go to the left.
As the inequality includes an equality part, 6 will also be a solution, so will have a solid dot above it.
So the number line solution is:

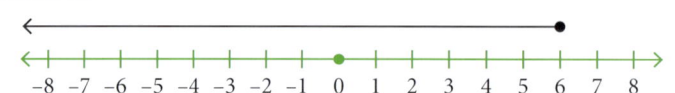

When we have a **strict** inequality (that is, an inequality that has no 'equals sign' part, for example $x < 3$) we have to be careful with the value 3 lying at the boundary.

It is possible to list an endless sequence of numbers, with each one larger and closer to 3 than the last one but never equal to 3:

2, 2.5, 2.9, 2.99, 2.999, 2.9999, 2.99999,…. going on forever.

To show this idea on a diagram, we draw an **open** circle over 3, indicating that the number 3 itself is excluded, but we join the circle above 3 to a solid line on the left. So $x < 3$ is represented by the number line shown on the right.

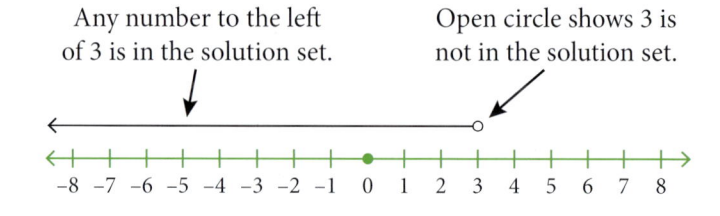

Any number to the left of 3 is in the solution set.

Open circle shows 3 is not in the solution set.

Example 2

Represent the solution to the inequality $x > -2$ on a number line, given that x is a real number.

As x is a real number, it will be represented by a solid line over the number axis.
As the solution set is values 'greater than' a number, the solid line will go to the right.
As the inequality is strict (i.e. it does not have an equality part) -2 will have an open circle above it.
The number line solution is:

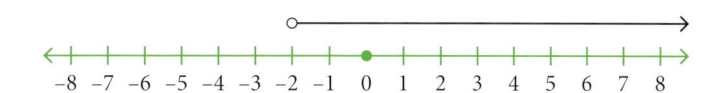

CHAPTER 27: INEQUALITIES

Exercise 27B

1. Write down the inequalities corresponding to these solution sets:
   (a) 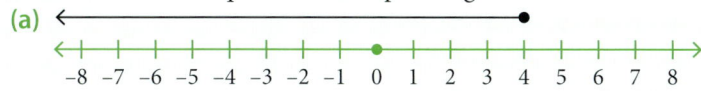
   (b) 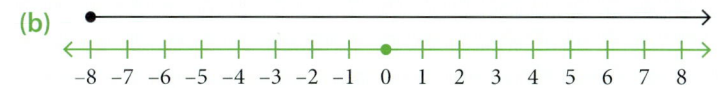
   (c) 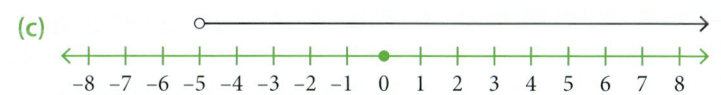
   (d) 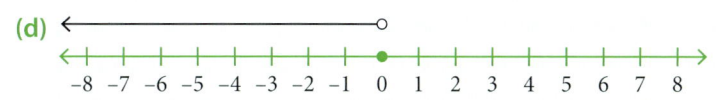
   (e) 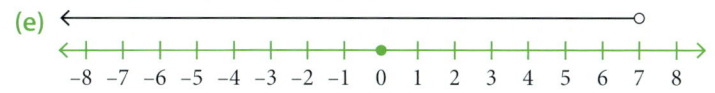
   (f) 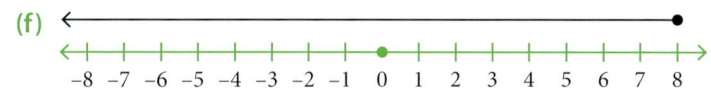
   (g) 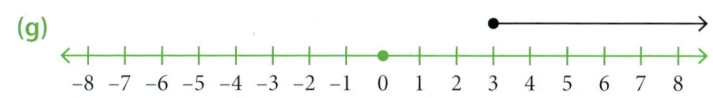
   (h) 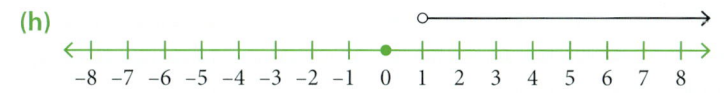
   (i) 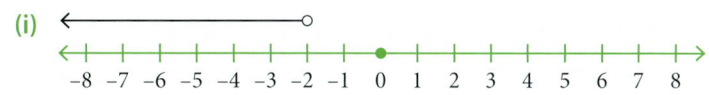
   (j)

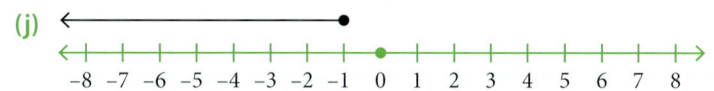

2. Draw the solution sets of the following inequalities on a number line.
   (a) $x \leq -2$
   (b) $x > 3$
   (c) $x \geq -4$
   (d) $x < 5$
   (e) $x \geq 1$
   (f) $x > 6$
   (g) $x \geq -7$
   (h) $x < 6$
   (i) $x < 0$
   (j) $x \geq 0$

27.3 Variables and Inequalities for Integers

We also need to consider the type of number the variable represents. When x is a real number, values like $x = 2.5$ or $x = -\frac{22}{7}$ can be a part of the solution set. On the other hand, if x in an integer, we illustrate this by using dots above the integers in the number line:

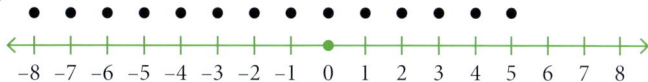

Example 3

Represent the solution to the inequality $x > -1$ on a number line, given that x is an integer.

As x is an integer, it will be represented by a row of dots.
As the solution set is values 'greater than' a number, the row of dots will go to the right.

As the inequality doesn't include −1, then 0 will be the first number with a dot above it.

The number line solution is: (number line from −8 to 8 with dots above 0, 1, 2, 3, 4, 5, 6, 7, 8)

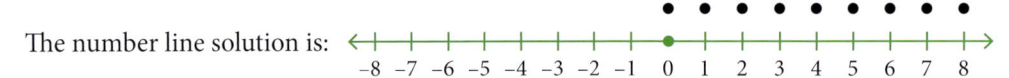

Note that the number line above is also the solution set to the inequality $x \geq 0$

Example 4

Represent the solution to the inequality $x < 5$ on a number line, given that x is an integer.

As x is an integer, it will be represented by a row of dots.
As the solution set is values 'less than' a number, the row of dots will go to the left.
As the inequality doesn't include 5, then 4 will be the first number with a dot above it.

The number line solution is: (number line from −8 to 8 with dots above −7, −6, −5, −4, −3, −2, −1, 0, 1, 2, 3, 4)

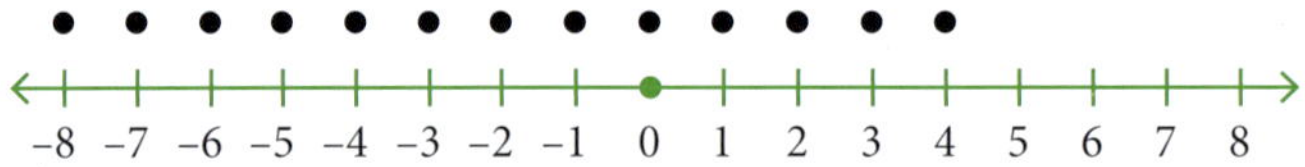

Note that the number line above is also the solution set to the inequality $x \leq 4$

Exercise 27C

1. Write down the inequalities corresponding to these solution sets, using x as an integer variable:

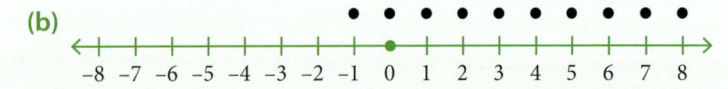

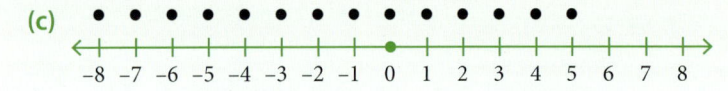

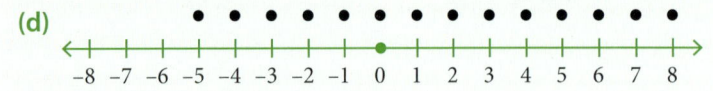

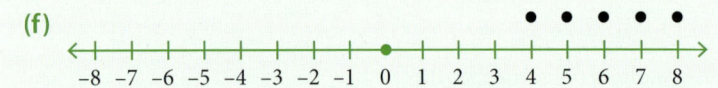

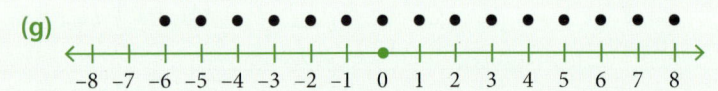

   (a) (dots above −7 to 2)

   (b) (dots above 0 to 7)

   (c) (dots above −8 to 3)

   (d) (dots above −5 to 7)

   (e) (dots above −8 to −4)

   (f) (dots above 3 to 7)

   (g) (dots above −6 to 7)

   (h) (dots above −3 to 7)

   (i) (dots above −7 to 6)

   (j) (dots above −6 to 7)

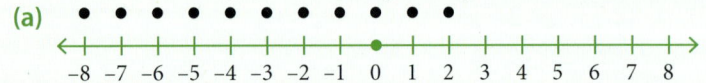

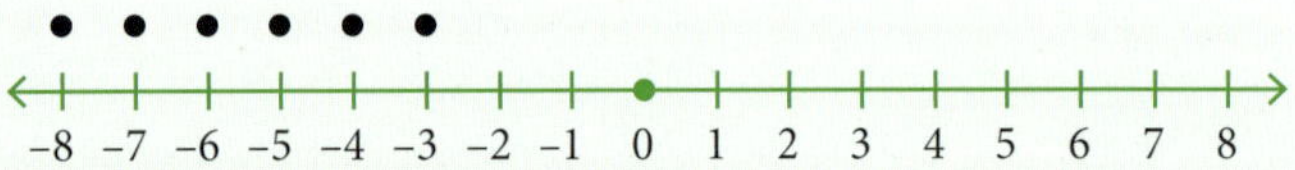

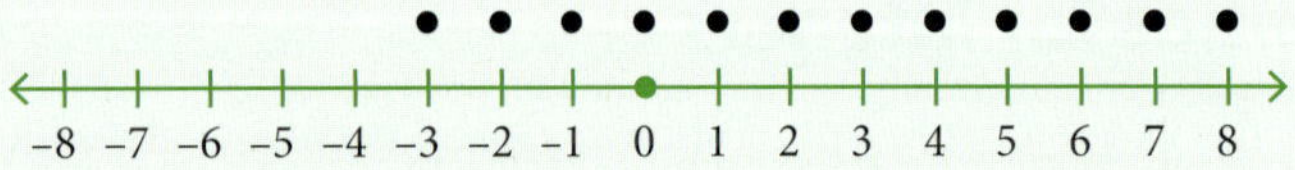

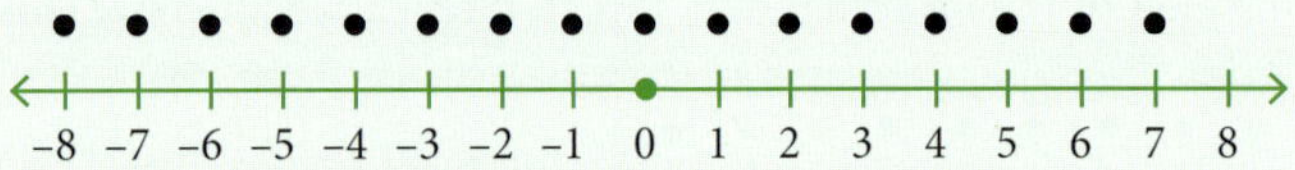

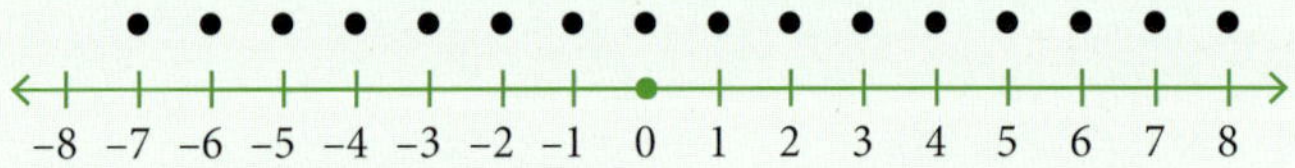

2. Draw the solution sets of the following inequalities on a number line, where x is an integer.

   (a) $x \leq -2$ (b) $x \geq 3$ (c) $x < 7$ (d) $x \leq 5$ (e) $x > 2$
   (f) $x \leq 1$ (g) $x > -6$ (h) $x \geq -2$ (i) $x \leq 0$ (j) $x > 5$

CHAPTER 27: INEQUALITIES

27.4 Combining Inequalities

Sometimes an inequality describes a limited piece of the number line. This usually takes the form:

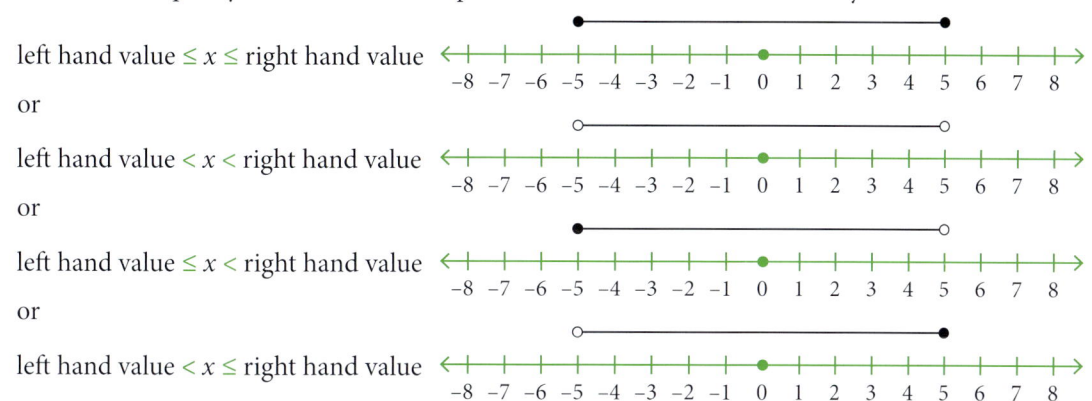

left hand value $\leq x \leq$ right hand value

or

left hand value $< x <$ right hand value

or

left hand value $\leq x <$ right hand value

or

left hand value $< x \leq$ right hand value

Example 5

If x is a real number, represent on the number line the combined inequalities $-2 < x \leq 5$

As x is a real number there will be a solid line.
The left hand end of the line will be an open circle over -2, as -2 is not in the solution set.
The right hand end of the line will be a solid dot over 5, as 5 is in the solution set.

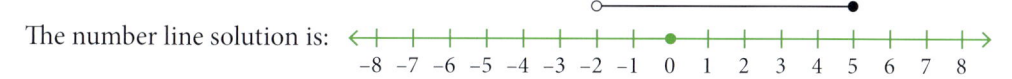

The number line solution is:

When x is representing integers, the solution set is just a line of dots for each permitted value of x.

Example 6

Represent on a number line the solution set of the combined inequalities $-1 < x < 3$, where x is an integer.

This means that x can take any value bigger than -1 and smaller than 3
So, because it is an integer, x can be 0, 1 or 2 only.

This is represented by the number line:

Other types of combined inequalities include two separate ranges, at both ends of the number line.

Example 7

If x is a real number, represent on the number line the inequalities $x \leq -4$ and $x > 3$

As x is a real number, there will be two solid lines.

The number line solution is:

Exercise 27D

1. Describe each of these solution sets by an appropriate inequality:
   (a)

   (b)

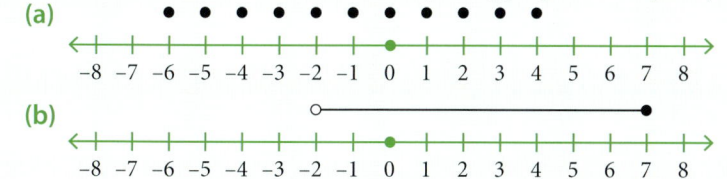

(c) number line showing filled dots at −5, −4, −3, −2

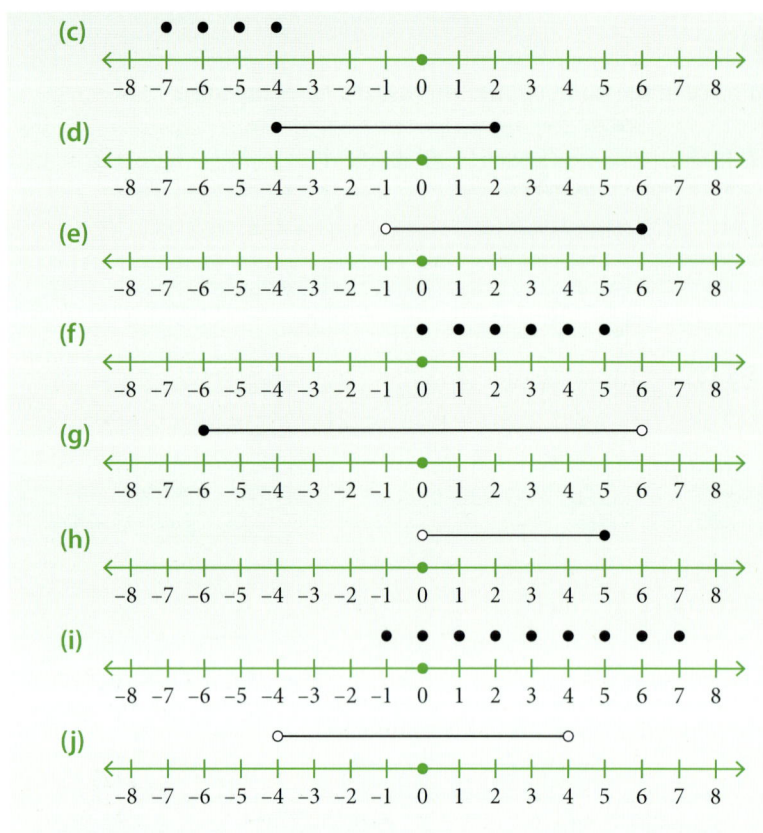

(d) number line with closed segment from −3 to 2

(e) number line with open circle at −1 and closed circle at 5 (segment between)

(f) number line showing filled dots at 0, 1, 2, 3, 4, 5

(g) number line with closed dot at −6 and open circle at 5 (segment between)

(h) number line with open circles at −1 and 4 (segment between)

(i) number line showing filled dots at −1, 0, 1, 2, 3, 4, 5

(j) number line with open circles at −4 and −3... (segment between)

2. Draw the solution set for each of these inequalities on a number line, treating x as a real number.
   (a) $4 < x < 6$ (b) $-6 \leq x < 1$ (c) $-3 \leq x \leq 7$ (d) $-5 \leq x < 4$ (e) $2 < x < 3$
   (f) $0 \leq x \leq 8$ (g) $-7 \leq x < -1$ (h) $-1 < x \leq 4$ (i) $-3 < x < 0$ (j) $-4 \leq x \leq -3$

27.5 Rearranging Inequalities

Sometimes we need to begin solving an inequality by rearranging algebra to simplify it. This is done in exactly the same way that we solve equations. We make identical changes to each side of the inequality making it simpler every step.

In the following examples we solve each inequality beside the solution of a similar equation to show you how this works.

Example 8

Solve for the real number x the:

Equation:	Inequality:
$4x - 5 = 23$	$4x - 5 < 23$

Add 5 to each side:
$4x - 5 + 5 = 23 + 5$ $4x - 5 + 5 < 23 + 5$

This simplifies to:
$4x = 28$ $4x < 28$

Dividing both sides by 4 we obtain:
$x = 7$ $x < 7$

The solution set of the inequality can then be shown on a number line:

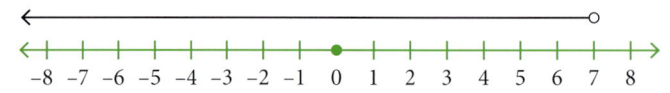

number line from −8 to 8 with open circle at 7 and arrow extending left

CHAPTER 27: INEQUALITIES

Example 9

Solve for the real number x the:

Equation:	Inequality:
$7x + 2 = 4(x + 5)$	$7x + 2 \leq 4(x + 5)$

First step is to multiply out the brackets:

$7x + 2 = 4x + 20$	$7x + 2 \leq 4x + 20$

Then subtract 2 from each side:

$7x + 2 - 2 = 4x + 20 - 2$	$7x + 2 - 2 \leq 4x + 20 - 2$

And simplify:

$7x = 4x + 18$	$7x \leq 4x + 18$

Now subtract $4x$ from each side:

$7x - 4x = 4x + 18 - 4x$	$7x - 4x \leq 4x + 18 - 4x$

And simplify:

$3x = 18$	$3x \leq 18$

Finally divide by 3:

$x = 6$	$x \leq 6$

The solution set of the inequality can then be shown on a number line:

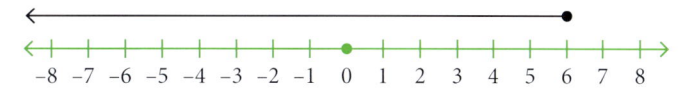

$$\xleftarrow{\qquad\qquad\qquad\qquad\qquad\bullet}$$
$$-8 \; -7 \; -6 \; -5 \; -4 \; -3 \; -2 \; -1 \; 0 \; 1 \; 2 \; 3 \; 4 \; 5 \; 6 \; 7 \; 8$$

There is one thing you need to watch out for when solving inequalities. When an inequality is multiplied or divided by a negative number, the direction of the inequality changes. That is, '<' becomes '>' and vice versa.

It is safer to avoid multiplying by a negative sign at all by moving negative x terms to the other side of the inequality. The next example shows a question answered in two ways: first by changing signs and then by avoiding changing signs.

Example 10

Solve $2x + 5 < x + 3$

Method changing signs	Method avoiding changing signs
$2x + 5 - 2x < x + 3 - 2x$	$2x + 5 - x < x + 3 - x$
$5 < 3 - x$	$x + 5 < 3$
$5 - 3 < 3 - x - 3$	$x + 5 - 5 < 3 - 5$
$2 < -x$	$x < -2$
$-2 > x$ by multiplying both sides by -1	

Note: $-2 > x$ is the same as $x < -2$

Exercise 27E

1. Find the solution of the following inequalities in the form (for example) $x < 3$ or $x \geq -2$
   - (a) $3x < 24$
   - (b) $4x - 5 \geq 11$
   - (c) $6x + 12 < 15$
   - (d) $4x \leq 2x + 22$
   - (e) $7x - 13 < 22$
   - (f) $5x - 13 \leq 3x + 1$
   - (g) $6x + 5 < 3x + 26$
   - (h) $7x - 8 \geq 5x + 4$
   - (i) $10x - 4 < 7x - 10$
   - (j) $23 - 4x > 14 - x$

2. Display the solution sets to these inequalities on a number line:
   - (a) $3x - 7 < 11$
   - (b) $8x + 14 \geq 38$
   - (c) $15x + 2 < 6x - 25$
   - (d) $12x - 5 \geq 3x + 31$
   - (e) $4 - 3x < 24 - 7x$
   - (f) $20 - 3x \leq x + 6$
   - (g) $5(3 + x) \geq 2x$
   - (h) $12 + 2(3x - 2) < 4(x + 4)$

27.6 Inequalities in Two Variables

Just as equations may be in one variable (for example $3x = 12$) or in two variables (for example $2x - y = 4$), so inequalities may also be in two variables. In this section we study how to draw two-variable inequalities on a graph.

To plot the solution set of an inequality we must indicate the boundary line and the region (which lies on one side of the line) where the inequality is true. We do this with both of the following:

- Mark the region where the inequality is true with a capital R.
- Shade the region where the inequality is **not** true.

Example 11

Plot the solution set of the inequality $y \geq x + 1$

First plot the line $y = x + 1$
Draw a solid line because the inequality includes an equality, so points on the line itself are included.

Second, we must find on which side of the line the inequality is true. To do this, we substitute one point (that does not lie on the line) into the inequality to see whether the inequality is true or false on that side of the line.

The easiest point to substitute is the origin (0, 0). Substituting $x = 0$ and $y = 0$ into the inequality gives $0 \geq 0 + 1$

As this is **not** true, the origin must lie outside the solution set of the inequality.

Therefore we shade below the line to exclude that region and mark above the line with a capital R to show that region is the solution set for the inequality.

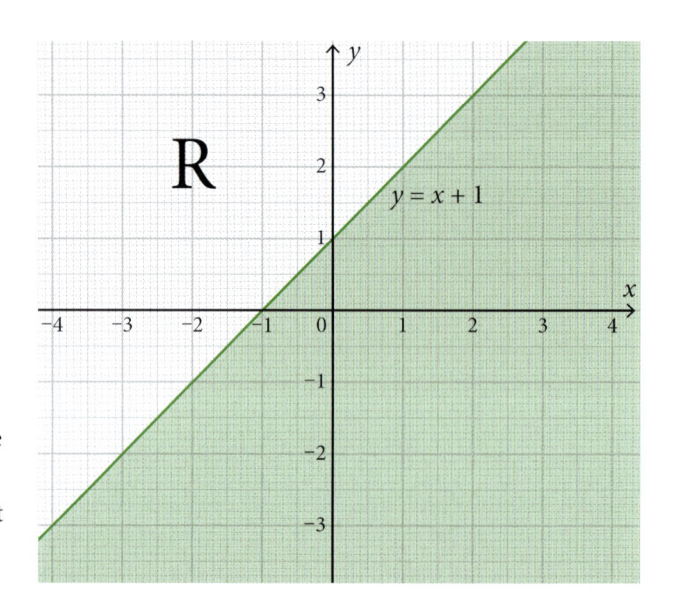

> **Note:** In these examples the entire excluded area is shaded. However, in your own work, you only need to shade an area close to the line. This indicates that the entire region below the line is not part of the solution set. In addition, you do not need to use colour; shading is sufficient.

Example 12

Draw the solution set of the inequality $4x + 5y \leq 20$

First, we plot the line $4x + 5y = 20$
When a line has this form, an easy way to plot it is to find the points where it cuts both axes.

- Setting $x = 0$ we find $5y = 20$ so $y = 4$ giving the point (0, 4)
- Setting $y = 0$ we find $4x = 20$ so $x = 5$ giving the point (5, 0)

This technique is often called the 'cover up' method, as we 'cover up' y finding $4x = 20$ etc. Then we 'cover up' x finding $5y = 20$, etc.

Second, we must find on which side of the line the inequality is true. We must substitute one point, not on the line, into the inequality to see whether the inequality is true or false on that side of the line.

The easiest point to substitute is the origin (0, 0). Substituting $x = 0$ and $y = 0$ into the inequality gives $0 + 0 \leq 20$

As this is true, the origin must lie inside the solution set of the inequality.

So we shade above the line to exclude that region and mark the area below the line with a capital R to show that region is the solution set for the inequality.

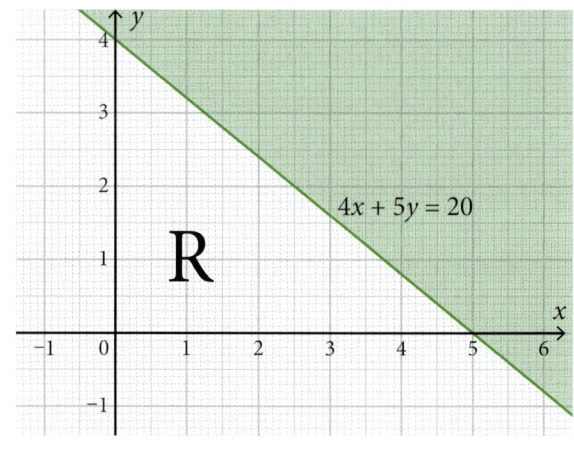

CHAPTER 27: INEQUALITIES

The inequalities in Examples 11 and 12 were both of the form ≤ or ≥ as they included an = sign. This represented the fact that any point **on** the line made the inequality true, and so was in the solution set.

When we draw a strict inequality, involving < or >, to show that points on the line are **excluded** from the solution set, we instead draw a dashed line.

Example 13

Draw the solution set of the inequality $3x + 2y > 6$

Using the cover up method, the line crosses the axes at (2, 0) and (0, 3).

Draw $3x + 2y = 6$ with a dashed line.

Substituting (0, 0) into the inequality we get $0 + 0 > 6$ which is false, so the solution set lies above this line.

Shade the excluded region and add an R.

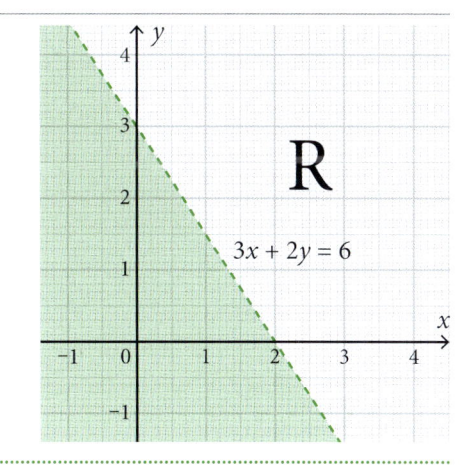

Often, we are required to draw the solution set to more than one inequality. We achieve this by drawing and shading each inequality separately on the same graph. The solution set will then be represented by the unshaded region and marked with an R.

Example 14

Represent the solution set of the following inequalities on a coordinate graph:
$x \geq 0, y \geq 0, y < 2x + 1, 3x + 4y \leq 12$

We must shade below the x-axis and to the left of the y-axis as x and y must be positive or zero.

To draw $y = 2x + 1$, we join (0, 1) with any other point on the line, eg. (1, 3)
To decide on which side of $y = 2x + 1$ the inequality is true, just substitute the coordinates of a point and see if the inequality is true or not at that point.

Pick (0, 0) and substitute $x = 0$ and $y = 0$ into $y < 2x + 1$, giving $0 < 0 + 1$

This is clearly true so (0, 0) lies in the solution set of the inequality. Thus, we shade above the line.

To draw $3x + 4y = 12$ we join (4, 0) and (0, 3) to create the line.

Again substituting (0, 0) into $3x + 4y \leq 12$ we obtain $0 + 0 \leq 12$

This is clearly true, so (0, 0) is in the solution set of that inequality. We shade above it.

The remaining unshaded area is the solution set of all four inequalities.

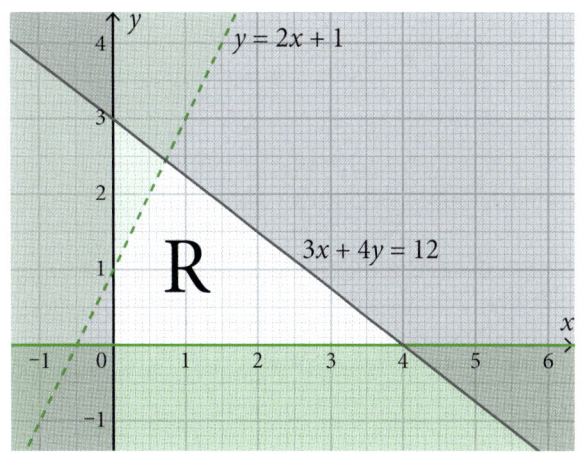

Example 15

(a) Represent the solution set of the following inequalities on a coordinate graph:
$x \geq 0, y \geq 0, x + 4y < 8, 4x + y \geq 8$

(b) Represent the solution set of the following inequalities on a coordinate graph:
$x \geq 0, y \geq 0, x + 4y > 8, 4x + y \leq 8$

(a) We must shade below the x-axis and to the left of the y-axis. Draw $x + 4y = 8$ dashed through (8, 0) and (0, 2). Shade above the line. Draw $4x + y = 8$ solid through (2, 0) and (0, 8). Shade below the line. The unshaded area is the solution set.

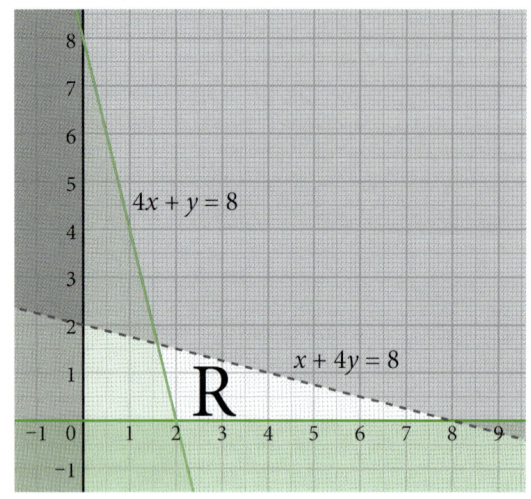

(b) We must shade below the x-axis and to the left of the y-axis. Draw $x + 4y = 8$ dashed through (8, 0) and (0, 2). Shade below the line. Draw $4x + y = 8$ solid through (2, 0) and (0, 8). Shade above the line. The unshaded area is the solution set.

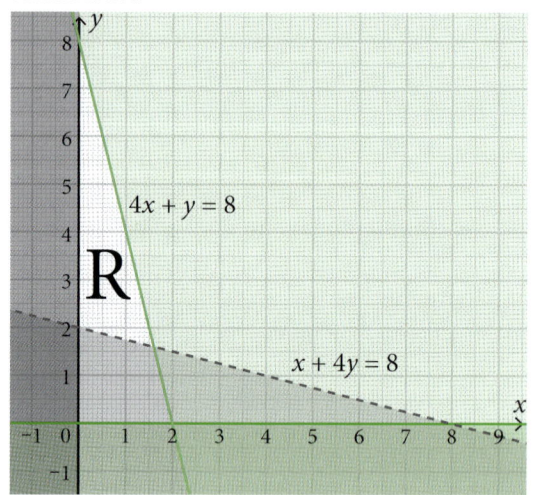

Exercise 27F

1. Represent the solution set for these inequalities on separate graphs:
   (a) $y \geq x + 2$
   (b) $3x + 5y \leq 15$
   (c) $4x + 3y \leq 24$
   (d) $x - y < 3$
   (e) $2x - 5y > 10$
   (f) $x > 4$
   (g) $y - 6x \leq 6$
   (h) $y < 3x$
   (i) $2 > x + y$
   (j) $7x + 8y \geq 28$

2. Show the solution sets of these collections of inequalities on separate graphs:
   (a) $x \geq 0, y \geq 0, x + y \leq 8, 2x + y \geq 8$
   (b) $x \geq 0, x - y < 1, x + y \leq 7$
   (c) $y \geq 0, x - y > -1, x + y \leq 5$
   (d) $x \geq 0, y \geq 0, x + 2y \leq 8, 2x + y \leq 8, x + y \geq 2$
   (e) $x \geq -1, y \geq -1, x \leq 6, y \leq 6$
   (f) $x \geq 0, y \geq 0, 3x + 4y \leq 12, 5x + 3y \leq 15$
   (g) $x \geq 0, y \geq 0, x + y \geq 1, 2x + y \leq 6, x + 2y \leq 6$
   (h) $x \geq 0, y \geq 0, y \geq x - 2, x + 6 > 3y, x + y \leq 3$

3. Describe these solution sets by a collection of inequalities:

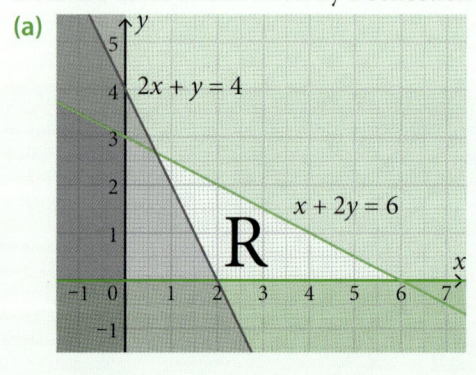

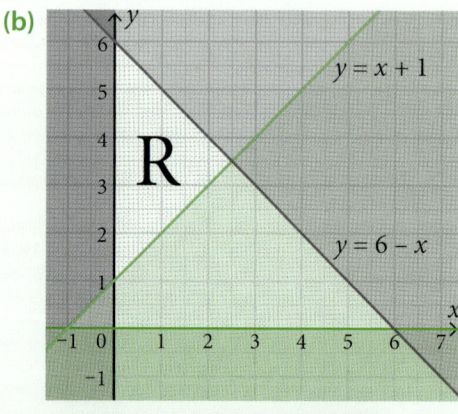

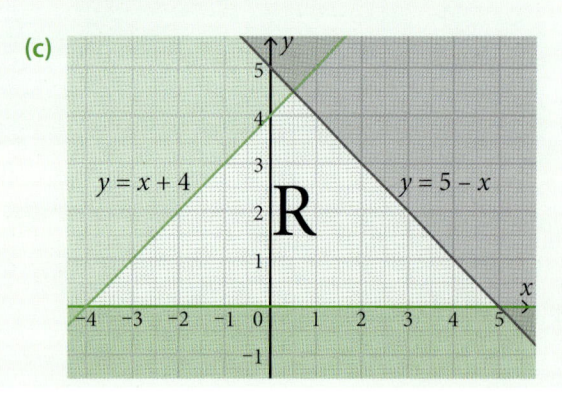

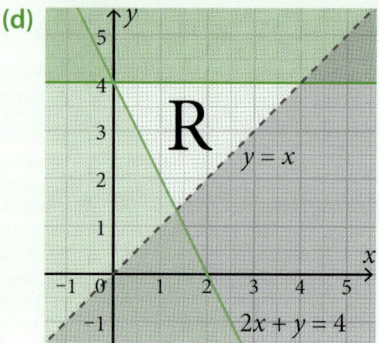

CHAPTER 27: INEQUALITIES

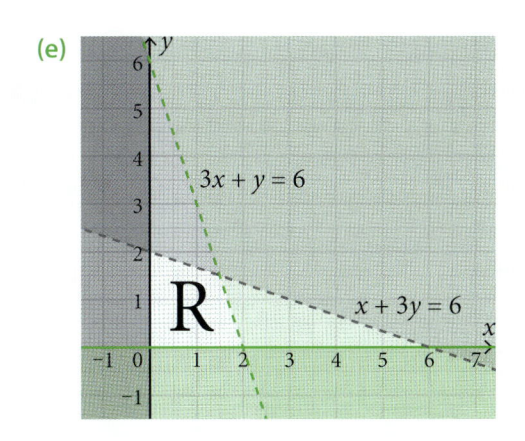

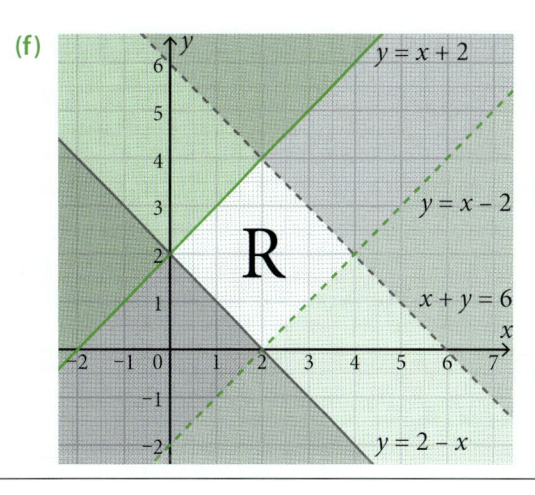

(e) [graph showing $3x + y = 6$ and $x + 3y = 6$ with region R]

(f) [graph showing $y = x + 2$, $y = x - 2$, $x + y = 6$, $y = 2 - x$ with region R]

27.7 Maximising or Minimising an Expression Over a Solution Set

Often we are asked to maximise (or minimise) the value of an expression in the solution set of a collection of inequalities. This means that we must find which point (x, y) in the solution set has the greatest (or least) value when the expression is evaluated at that point. This expression is sometimes called an **objective function**.

Example 16

Find the maximum value of the sum $x + y$ on the points of the solution set of the inequalities:
$x \geq 0, y \geq 0, 2x + y \leq 6, x + 2y \leq 6$

First, we draw the solution set of the inequalities as shown on the right.

The maximum and minimum values of an expression in x and y usually occur at the corners of the solution set. So we evaluate $x + y$ at the corners of the solution set:

(0, 0) $x + y = 0$
(0, 3) $x + y = 3$
(3, 0) $x + y = 3$
(2, 2) $x + y = 4$

So the maximum value of $x + y$ is 4 at the point (2, 2) in the solution set.

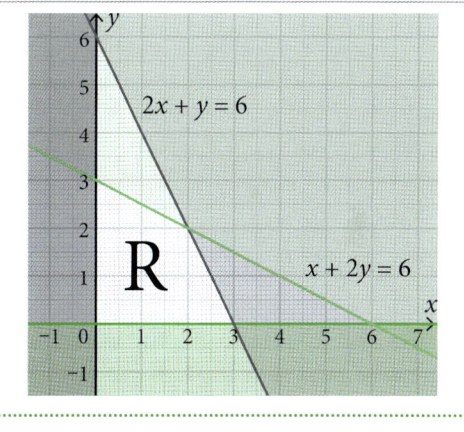

The next example demonstrates a different approach to finding a maximum or minimum value of an expression in a solution set. Some students may find this approach helpful.

Example 17

Find the maximum value of the sum $x + 3y$ on the points of the solution set of the inequalities:
$x \geq 0, y \geq 0, x + y \leq 8, x + 2y \leq 10, x + 4y \leq 16$

We will first draw the solution set. Then we will pick some lines of the form $x + 3y =$ [a constant] and add them to the solution set.

So first, we draw the solution set of the inequalities, adding the lines $x + 3y = 6$, $x + 3y = 9$ and $x + 3y = 12$ as shown on the right.

The maximum and minimum values of an expression in x and y usually occur at the corners of the solution set.

We can identify which corner will produce the largest value of $x + 3y$ by drawing some lines of equal value of $x + 3y$

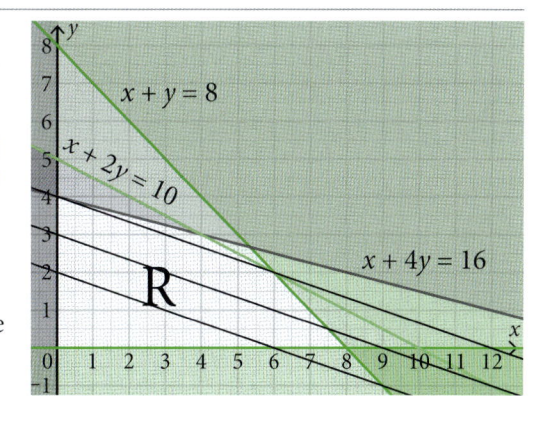

245

Here we have drawn $x + 3y = 6$, $x + 3y = 9$ and $x + 3y = 12$ in the middle of the solution set. These are shown by the three parallel black lines.

You can imagine that the last line that is (a) parallel to and above $x + 3y = 12$ and (b) that still touches the solution set will go through the point (4, 3). Thus the largest value of $x + 3y$ inside the solution set will occur at the point (4, 3).

This maximum value is $x + 3y = 4 + 3 \times 3 = 13$

The lines of equal value, as in this example, are called 'search lines'. We imagine moving a line parallel to these lines until it is about to leave the solution set. That indicates where the maximum (or minimum) value of the search expression occurs within the solution set.

Exercise 27G

1. Draw the solution set of the inequalities: $x \geq 0, y \geq 0, x + 3y \leq 12, 3x + y \leq 12$
   Then find the maximum of $x + y$ in the solution set.

2. Draw the solution set of the inequalities: $x \geq 0, y \geq 0, x + 2y \leq 12, 2x + y \leq 12$
   Then find the maximum of $x + 3y$ in the solution set.

3. Draw the solution set of the inequalities: $x \geq 0, y \geq 0, x + 4y \geq 12, x + y \leq 8$
   Then find the minimum of $x + y$ in the solution set.

4. Draw the solution set of the inequalities: $x \geq 1, y \geq 2, 2x + y \geq 10$
   Then find the minimum of $x + 2y$ in the solution set.

5. (a) Draw the solution set of the inequalities: $x \geq 0, y \geq 0, 2x + y \geq 8, x + 3y \leq 12$
   (b) By solving the simultaneous equations, find the point of intersection of $2x + y = 8$ and $x + 3y = 12$
   (c) Then find the maximum of $x + 4y$ in the solution set.

6. (a) Draw the solution set of the inequalities: $x \geq 0, y \geq 0, 3x + 2y \geq 12, x + 4y \geq 10$
   (b) By solving the simultaneous equations, find the point of intersection of $3x + 2y = 12$ and $x + 4y = 10$
   (c) Then find the minimum of $x + y$ in the solution set.

7. (a) Draw the solution set of the inequalities: $x \geq 0, y \geq 0, x + 3y \leq 21, 4x + y \leq 24$
   (b) Then find the maximum of $x + 4y$ in the solution set.

8. (a) Draw the solution set of the inequalities: $x \geq 0, y \geq 0, x + 4y \leq 15, 5x + y \leq 18$
   (b) Then find the maximum of $4x + 5y$ in the solution set.

9. (a) Draw the solution set of the inequalities: $x \geq 0, y \geq 0, y \leq 5 - 2.5x, 3y \leq 8 - 2x$
   (b) Then find the maximum of $x + y$ in the solution set, given that both x and y must be integers.

27.8 Summary

In this chapter you have learnt that:
- Inequalities may be rearranged algebraically, in the same manner as equations.
- The final line should give the range of possible values for x, for example $x <$ something or $x \geq$ something.
- There is usually more than one number that makes the inequality true. All these possible solutions are together called the **solution set** of the inequality.
- Solution sets in one variable can be represented on the number line.
- The solution set of a two-variable inequality can be represented as all the points on one side of the line. For example the inequality $5x + 2y \leq 10$ can be represented as all the points on one side of the line $5x + 2y = 10$
- The solution set of a collection of inequalities in x and y can be represented as an area of the coordinate plane.

Progress Review
Chapters 22 – 27

This Progress Review covers:

Chapter 22: Number Systems
Chapter 23: Surds
Chapter 24: Indices
Chapter 25: Standard Form
Chapter 26: Simultaneous Equations
Chapter 27: Inequalities

1. Find the decimal value for each of the following binary numbers.
   (a) 10101111
   (b) 10011001
   (c) 101101
   (d) 110011
   (e) 10101010

2. Convert the following decimal numbers into binary.
   (a) 11
   (b) 36
   (c) 45
   (d) 59
   (e) 61

3. Simplify the following surds. In some cases, the surd may already be in its simplest form.
   (a) $\sqrt{40}$
   (b) $\sqrt{45}$
   (c) $\sqrt{30}$
   (d) $\sqrt{98}$
   (e) $\sqrt{108}$

4. Simplify the following surds. In some cases, the surd may already be in its simplest form.
   (a) $3\sqrt{44}$
   (b) $3\sqrt{50}$
   (c) $5\sqrt{8}$
   (d) $2\sqrt{48}$
   (e) $3\sqrt{180}$

5. Simplify the following.
   (a) $\sqrt{3} \times \sqrt{3}$
   (b) $\dfrac{\sqrt{36}}{\sqrt{12}}$
   (c) $\sqrt{27} \times \sqrt{3}$
   (d) $\sqrt{12} \times \sqrt{5}$

6. Simplify $3\sqrt{18} \times \sqrt{6}$

7. Simplify the following.
   (a) $\sqrt{75} - \sqrt{48}$
   (b) $\sqrt{12} - \sqrt{3}$
   (c) $\sqrt{8} + 3\sqrt{2}$
   (d) $\sqrt{60} + \sqrt{15}$
   (e) $4\sqrt{50} - 3\sqrt{8}$

8. In each part, expand the brackets and simplify as far as possible.
   (a) $3(4 + \sqrt{5})$
   (b) $-2(2\sqrt{7} - 3\sqrt{11})$
   (c) $\sqrt{2}(3 - 3\sqrt{2})$
   (d) $3\sqrt{5}(6 - \sqrt{5})$

9. Expand the brackets and simplify.
   (a) $(2 + \sqrt{2})(1 - \sqrt{2})$
   (b) $(2\sqrt{3} - 1)(3 - \sqrt{3})$
   (c) $(\sqrt{3} + \sqrt{5})(\sqrt{5} - \sqrt{3})$
   (d) $(2\sqrt{7} + 7\sqrt{2})(1 - 7\sqrt{2})$

10. The length and width of a rectangle are both surds. Find possible values for the length and width so that:
    (a) The area is also a surd.
    (b) The area is an integer.

11. For the following rectangles, calculate the length of the diagonal.
    (a) Length = $\sqrt{13}$ cm, width = $\sqrt{3}$ cm
    (b) Length = $\sqrt{4}$ cm, width = $\sqrt{9}$ cm

12. Find the perpendicular height of a triangle that has an area of $(\sqrt{32} + \sqrt{48})$ cm² and a base length of 4 cm. Give your answer in simplified surd form.

13. Evaluate without the use of a calculator:
    (a) 6^2
    (b) 2^4
    (c) 8^{-1}
    (d) 3^{-3}
    (e) 5^{-2}

14. Write as a power of x:
    (a) $x \times x \times x \times x \times x \times x$
    (b) $\dfrac{1}{x \times x \times x \times x}$
    (c) $\dfrac{x \times x \times x \times x}{x \times x \times x}$
    (d) $\dfrac{x \times x \times x \times x}{x \times x \times x \times x \times x \times x}$

15. Express as a single power of x:
    (a) $x^4 \times x^7$
    (b) $x^3 \times x^{-5}$
    (c) $x^0 \times x^6$
    (d) $x^9 \times x^3 \times x$

16. Express as a single power of x:
    (a) $x^5 \div x^2$
    (b) $x^3 \div x^5$
    (c) $x^6 \div x^6$
    (d) $x^7 \div x^{11}$

17. Calculate, leaving the answer as a mixed number where necessary:
    (a) $(2\frac{1}{4})^{-1}$
    (b) $(1\frac{1}{2})^{-2}$
    (c) $(1\frac{1}{3})^3$
    (d) $(7\frac{3}{4})^0$

18. Simplify:
    (a) $(x^5)^{-3}$
    (b) $(x^{-2})^{-2}$
    (c) $(x^{-5})^3 \times (x^3)^4$
    (d) $(x^9)^2 \div (x^4)^{-5}$

19. Find the value of the following, without a calculator:
    (a) $(3^2)^{-1}$
    (b) $(5^{-1})^{-2}$
    (c) $(4^3)^{-2}$
    (d) $(5^3)^{-6} \div (5^5)^{-4}$

20. If p has a value greater than 1, place the following powers of p in ascending order:
    (a) p^3
    (b) p^{-1}
    (c) $(p^{-1})^2$
    (d) $p^{-2} \times p^3$
    (e) $(p^{-2})^{-3}$

21. Match each expression (a) – (d) with an answer from (p) – (s):
    Expressions:
    (a) $(x^{-3})^{-6}$
    (b) $(x^{-6})^3 \times (x^{-3})^{-5}$
    (c) $(x^4)^{-3} \times x^{-6}$
    (d) $(x^{-5})^6 \div (x^{-3})^{12}$
    Answers:
    (p) x^{-3}
    (q) x^{18}
    (r) x^6
    (s) x^{-18}

22. Place the following expressions (a) – (e) in descending order:
    (a) $(4^3)^{-6}$
    (b) $(8^{-5})^{-6}$
    (c) $(x^{-8})^3 \times (x^{-4})^{-6}$
    (d) $(2^3)^2$
    (e) $(-16)^3$

23. Which of the following numbers is in standard form? For those that are not, give a reason.
    (a) 51 500
    (b) 2×10^3
    (c) 10×10^4
    (d) 2.1×10^{-8}
    (e) 5×10
    (f) $4.2 \div 10^{-3}$
    (g) 5.08×10^{-1}
    (h) 0.53×10^4

24. Write the number 6.178×10^{-4} in decimal form:

25. Write these numbers in standard form:
    (a) 269×10^3
    (b) 0.049×10^3

26. The current human population of the Earth is roughly 8.2 billion, or 8 200 000 000
    Write this in standard form.

27. Write these numbers in order of size from smallest to largest:
    0.058×10^2 5800×10^{-4} 58 0.58×10^{-1}

28. Given that $a = 5 \times 10^5$ and $b = 2 \times 10^{-4}$ find, in standard form:
    (a) $\frac{a}{b}$
    (b) $\frac{b}{a}$

29. During the 2024 presidential election campaign in the USA, the two main political parties are thought to have spent about $15 901 000 000 in total. Write this in standard form.

30. Calculate the following without a calculator. Give your answers in standard form.
    (a) $7.6 \times 10^6 + 3.8 \times 10^4$
    (b) $8.9 \times 10^{-1} + 2.5 \times 10^2$
    (c) $9.3 \times 10^4 - 2.7 \times 10^3$
    (d) $4.7 \times 10^{-1} - 3 \times 10^{-2}$
    (e) $(4 \times 10^2) \times (1.2 \times 10^5)$
    (f) $(3.5 \times 10^{-2}) \times (2 \times 10^{-4})$
    (g) $(6.3 \times 10^7) \div (3 \times 10^3)$
    (h) $(8.8 \times 10^{-5}) \div (2 \times 10^{-2})$
    (i) $(6 \times 10^5) \times (2 \times 10^3)$
    (j) $(1.2 \times 10^7) \div (6 \times 10^{-3})$

31. Carry out the following calculations on your calculator, giving your answers in standard form.
    (a) $2.8 \times 10^3 + 6.79 \times 10^4$
    (b) $3.6 \times 10^9 - 2.5 \times 10^7$
    (c) $(8.57 \times 10^{-6}) \times (3 \times 10^4)$
    (d) $(2.6 \times 10^7) \div (1.3 \times 10^{-5})$

32. There are approximately 1.67×10^{21} water molecules in a droplet of water. On your calculator, calculate the number of water molecules in 120 droplets, giving your answer in standard form.

33. Solve the following pairs of simultaneous equations.
    (a) $-7x + 6y = 3$
    $-7x + 2y = -13$
    (b) $-2x + 5y = 4$
    $7x + 5y = -59$
    (c) $7x + 4y = 12$
    $-7x + 4y = -44$
    (d) $5x - 7y = 23$
    $5x - 6y = 19$
    (e) $-4x + 5y = 18$
    $-2x + 5y = 4$
    (f) $-4x - 7y = -15$
    $6x - 7y = 5$
    (g) $-5x - 6y = -12$
    $-7x - 6y = 0$
    (h) $2x + 3y = 4$
    $2x + y = -4$
    (i) $-x - 2y = 8$
    $x - 2y = 8$
    (j) $3x + 7y = -25$
    $3x - 2y = 11$

34. Solve the following pairs of simultaneous equations.
    (a) $-5x + 7y = 24$
    $3x - y = 8$
    (b) $7x + 5y = -49$
    $2x - y = 3$
    (c) $7x - 3y = 0$
    $x - y = 4$
    (d) $-2x + y = -7$
    $-6x + 5y = -35$
    (e) $-3x + 7y = -22$
    $2x - y = 11$
    (f) $x + y = 3$
    $-7x + 4y = -43$
    (g) $-x - 2y = 0$
    $5x - 7y = 0$

35. Solve the following pairs of simultaneous equations.
    (a) $4x + 7y = 63$
    $-6x + 5y = -17$
    (b) $-5x - 6y = -4$
    $-6x - 7y = -4$
    (c) $-6x - 5y = 1$
    $-7x + 6y = 13$

248

PROGRESS REVIEW: CHAPTERS 22–27

36. Solve the following pairs of simultaneous equations. Give answers as fractions where appropriate.
    (a) $-14x - 12y = 19$
        $6x + 8y = -11$
    (b) $15x - 5y = -2$
        $35x + 10y = 4$
    (c) $10x - 10y = 3$
        $5x - 4y = \dfrac{7}{10}$

37. Solve the following pairs of simultaneous equations. As a first step, rearrange the equations so that the x and y terms appear on the left-hand side, with the number terms on the right.
    $2x + 3y = -35$
    $y = x$

38. Two numbers, p and q, have a sum of 95 and a difference of 13
    Given that $p > q$, find p and q

39. The perimeter of a rectangle is 52 cm. The difference between the length and width is 6 cm.
    (a) Form and solve two simultaneous equations to find the length and width of the rectangle.
    (b) Find the area of the rectangle.

40. Jo runs a café. In October she buys 4 boxes of tea and 5 boxes of coffee for £50
    In November she buys 2 boxes of tea and another 5 boxes of coffee, coming to £40
    (a) Using t for the price of a box of tea and c for the price of a box of coffee, form two simultaneous equations.
    (b) Solve the simultaneous equations to find the cost of a box of tea and the cost of a box of coffee.

41. Abacus Industries is a small engineering company hiring 5 men and 5 women. By chance, the men are all the same age, and the women are all the same age, although the men are older than the women.
    On Monday, only 3 men and 2 women are in the office. The total age of these workers is 140
    On Tuesday, all the men were in the office and only 1 woman was absent.
    The total age of the workers was 250
    (a) Using m for the age of a man and w for the age of a woman, form two simultaneous equations.
    (b) Solve the equations to find the age of the men and the age of the women at Abacus Industries.

42. What is the inequality with the solution set below?

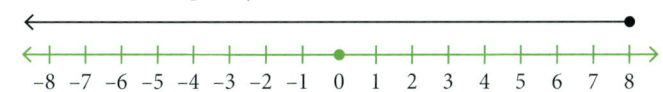

43. Draw the solution sets of the following inequalities on the number line, where x is a real number:
    (a) $x \geq -2$
    (b) $x < 7$
    (c) $x > -2$
    (d) $x \leq 0$

44. What is the inequality with the solution set below?

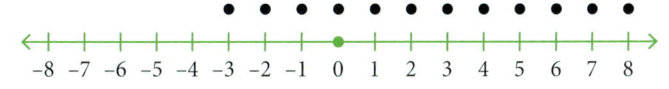

45. Draw the solution sets of the following inequalities on the number line, where x is an integer:
    (a) $x < 0$
    (b) $x \geq 1$
    (c) $x > -6$
    (d) $x \leq 5$

46. Draw the solution sets of the following inequalities on the number line:
    (a) $-3 < x < 2$, x is a real number
    (b) $8 > x \geq 1$, x is an integer
    (c) $1 \leq x \leq 5$, x is an integer
    (d) $-2 < x \leq 6$, x is a real number

47. Solve the following inequalities, where x is a real number:
    (a) $5x - 7 < 8x + 5$
    (b) $4(5 - x) > 8 + 2x$
    (c) $9 + 7(2x - 3) \geq 3(4x + 5)$
    (d) $8 + 3x < 5x - 2$

48. Solve the following inequalities, then write out the first three integers in each solution set.
    (Use ascending order if the inequality is of the form '$x > ...$' and descending order if the inequality is of the form '$x < ...$')
    (a) $8x - 3 < 14$
    (b) $3x - 4 \geq 13 - 7x$
    (c) $37 - 4x \geq 2x - 9$
    (d) $5 - (7 - x) \leq 2x - 3(4 - 3x)$

49. Show the solution sets of these inequalities. Mark each solution set as R.
    (a) $x \geq 1$ $y \geq 2$ $5x + 4y \leq 20$
    (b) $x \geq 0$ $y > x$ $x + y > 1$ $4x + 8y \leq 24$
    (c) $x < 4$ $y \geq 0$ $x + y \geq 2$ $x + 2y \leq 6$

50. Describe the inequalities that have the following solution sets:

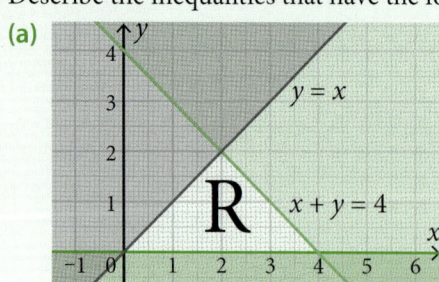

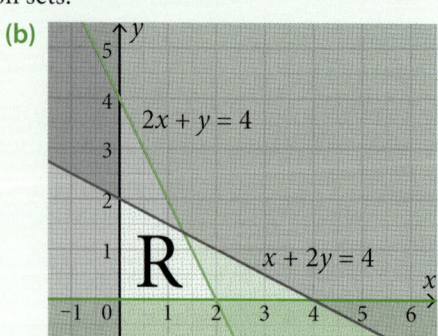

(a)

(b)

51. (a) Plot, on a pair of coordinate axes, the solution set of the inequalities:

$x \geq 2$ $y \geq 0$ $x + 2y \leq 12$ $3x + 4y \leq 30$

(b) Find the maximum value of the expression $2x + 3y$ on the points of this solution set.

Chapter 28
Formulae

28.1 Introduction

In a formula, letters are used to represent quantities or variables. It is a concise way of writing down a mathematical statement.

For example, the formula: $\text{speed} = \dfrac{\text{distance}}{\text{time}}$

represents the relationship between the speed of an object, the distance it travels, and the time taken.

It may also be written as: $s = \dfrac{d}{t}$

The **subject** of a formula is the variable which is written in terms of the other variables. In the formula above, for example, speed is the subject.

A formula can be used to find the value of one of the variables by substituting in values for the other variables. This allows you to calculate the value of the subject.

> **Note:** It is not necessary at GCSE level to learn any of the formulae mentioned in this chapter.

Key words
- **Formula**: A relationship between two or more variables, for example $F = ma$
- **Substitute**: Substituting a value for a variable in a formula involves replacing that variable with its numerical value.
- **Subject**: The subject of a formula is the variable which is written in terms of the other variables. For example, in the formula $F = ma$, F is the subject.

Before you start you should know how to:
- Expand brackets.
- Factorise algebraic expressions using a common factor.
- Solve linear equations.
- Substitute values into expressions.
- Substitute values into formulae, which may be expressed in words or algebraically.

In this chapter you will learn how to:
- Rearrange formulae to change the subject.

Exercise 28A (Revision)

1. Expand the brackets.
   (a) $6(9x - 3)$ (b) $a(3b - 2a)$

2. Factorise the following.
   (a) $2c - 8$ (b) $5x^2 - 10x$

3. Solve the following linear equations.
   (a) $6x + 3 = 21$ (b) $2 - 7c = -12$

4. Find the value of each expression when $p = 2$ and $q = -3$
   (a) $3p + 4q$ (b) $q^2 + 2pq$

5. Substitute into each formula to find the value of the subject.
   (a) $F = ma$ Find the value of F when $m = 5$ and $a = -6$
   (b) $s = \dfrac{d}{t}$ Find the value of s when $d = 5$ and $t = 15$
   (c) $E = mc^2$ Find the value of E when $m = 2$ and $c = 3$

(d) $A = \pi r^2$ Find the value of A when $r = 6.2$, giving your answer to 1 decimal place.

6. The formula for the total surface area of a cylinder is $A = \pi r^2 + 2\pi rh$
   Find the total surface area for a cylinder if $r = 6.2$ cm and $h = 10$ cm. Use the value of π from your calculator. Give your answer in square centimetres, rounded to one decimal place.

28.2 Changing the Subject of Formulae

The subject of a formula is the variable that is written in terms of the other variables. For example, in $F = ma$ the subject is F. The subject is usually written on the left-hand side of the equals sign.

You will be asked to **change the subject** of a formula. For example, using $F = ma$, you could be asked to make m the subject. This is useful if you want to find the value of m.

The steps involved in rearranging a formula are similar to those involved in solving an equation. You must always perform the same operation on each side of the formula, for example subtracting 2 from both sides.

Another way of thinking about this is that you use inverse operations when rearranging. For example, if the left-hand side involves +2, this term can be moved to the other side of the formula, where it becomes –2 on the right-hand side. Subtraction is the inverse of addition; division is the inverse of multiplication.

Eventually, we aim to have the required subject on one side of the equals sign, and all the other variables collected on the other side.

> **Note:** Changing the subject of a formula is a very important skill in mathematics. You will find it helpful in Chapter 33 on Proportion and Variation, and elsewhere. If you go on to study A-Level mathematics, it is a vital skill.

Simple formulae

This section covers the process of rearranging simple formulae. The rearranging involves addition, subtraction, multiplication and division. In Example 1, below, we do the subtraction before the division.

Example 1

Make x the subject of the formula: $Ax + B = C$

First subtract B from both sides of the formula:	$Ax = C - B$
Divide both sides by A:	$x = \dfrac{C - B}{A}$

In Example 2 we must do the addition before the division.

Example 2

Make x the subject of the formula: $E = Fx - G$

First add G to both sides of the formula:	$E + G = Fx$
Divide both sides by F:	$\dfrac{E + G}{F} = x$
Finally, write x on the left-hand side:	$x = \dfrac{E + G}{F}$

Example 3

Make q the subject of the formula: $pr - 3q = 7s$

Note that the term involving q is a negative term. In this case, add it to both sides:	$pr = 7s + 3q$
Subtract 7s from both sides:	$pr - 7s = 3q$
Divide by 3:	$\dfrac{pr - 7s}{3} = q$
Write the formula with q on the left-hand side:	$q = \dfrac{pr - 7s}{3}$

CHAPTER 28: FORMULAE

Example 4

Make b the subject of the formula: $V = lbh$

This could be rewritten:	$V = lhb$
Divide both sides by lh:	$\dfrac{V}{lh} = b$
Write b on the left-hand side:	$b = \dfrac{V}{lh}$

Exercise 28B

1. Make x the subject of each of these formulae.
   - (a) $a = bx$
   - (b) $x - c = d$
   - (c) $ex - f = g$
   - (d) $hx + j = k - m$
   - (e) $n = p - x$
   - (f) $3x - 2 = q$
   - (g) $r = 4x - s$
   - (h) $5u + 6v = w - zx$
   - (i) $Ab - x = c$
   - (j) $C - Dx = F$

2. Make y the subject of each of these formulae.
   - (a) $y + a = b$
   - (b) $cy = d$
   - (c) $e + fy = g$
   - (d) $2 - y = h$
   - (e) $jy + k = m$
   - (f) $n - py = q$
   - (g) $ry + s = t$
   - (h) $3 - uy - v = w$
   - (i) $x = y - z$
   - (j) $Ay = B$

3. Make h the subject of each of these formulae:
   - (a) $V = lhb$
   - (b) $A = 2\pi rh$
   - (c) $V = \pi hr^2$

4. Make a the subject of the formula: $A = 2r^2 + 2ar$

5. Make M the subject of the formula: $mN = Mn$

Formulae involving brackets

When a formula involves brackets, follow one of these rules:
- If the variable you want as the subject is inside the brackets, you should expand the brackets first. See Examples 5 and 6.
- Otherwise, the brackets do not need to be expanded. See Example 7.

Example 5

Make a the subject of the formula: $b(a - c) = d$

Expand the brackets:	$ba - bc = d$
Add bc to each side:	$ba = d + bc$
Divide both sides by b:	$a = \dfrac{d + bc}{b}$

Example 6

Make a the subject of the formula: $e = f(2 - a) + 3$

Expand the brackets:	$e = 2f - fa + 3$
Add fa to both sides:	$fa + e = 2f + 3$
Subtract e from each side:	$fa = 2f + 3 - e$
Divide both sides by f:	$a = \dfrac{2f + 3 - e}{f}$

Example 7

Make a the subject of the formula: $g = a(h + 4) - j$

Add j to both sides:	$g + j = a(h + 4)$
Divide both sides by $(h + 4)$:	$\dfrac{g + j}{h + 4} = a$
Write the formula with a on the left:	$a = \dfrac{g + j}{h + 4}$

253

GCSE MATHEMATICS M3 AND M7

Exercise 28C

1. Make a the subject of each formula.
   (a) $e(a + f) = g$
   (b) $b(c - a) = d$
   (c) $h(a + j) = k$
   (d) $2(a - 3) = m$
   (e) $n = p(a + q)$
   (f) $4(1 - a) = r$
   (g) $s = t(a - u)$
   (h) $v(w + a) = xy$
   (i) $z = A(5 + ab)$
   (j) $B = 10(C - a)$

2. Make a the subject of each of these formulae.
   (a) $7(F - a) = G - 2H$
   (b) $J = a(K + 8 - b)$
   (c) $L = a(M - 9) - 10$
   (d) $N + P = 11 + a(Q + R)$

Formulae involving fractions

A formula may involve fractions. If the variable you want as the subject is in a fraction, you should eliminate the fraction. This involves multiplying both sides by the denominator.

Example 8

Make x the subject of the formula: $A = \dfrac{x + y}{2}$

Multiply both sides by 2
This eliminates the fraction on the right-hand side: $2A = x + y$
Subtract y from both sides: $2A - y = x$
Write x on the left-hand side: $x = 2A - y$

In the following example, b must be added to both sides before the multiplication by 3

Example 9

Make x the subject of the formula: $\dfrac{x}{3} - b = c$

Add b to both sides: $\dfrac{x}{3} = c + b$
Multiply both sides by 3: $x = 3(c + b)$

In the following example, the variable we wish to make the subject starts in the denominator of the fraction.

Example 10

Make x the subject of the formula: $d = \dfrac{4e}{x - f}$

Eliminate the fraction by multiplying both sides by the denominator: $d(x - f) = 4e$
Since x is now inside the brackets, we must expand the brackets: $dx - df = 4e$
Add df to both sides: $dx = 4e + df$
Divide both sides by d: $x = \dfrac{4e + df}{d}$

If there is a single fraction on each side of the equation, the simplest method to eliminate the fractions is cross-multiplying, as shown in the next example.

Example 11

Make d the subject of the formula: $\dfrac{2a}{b} = \dfrac{5}{d}$

Cross-multiply to eliminate the fractions: $2ad = 5b$
Divide both sides by $2a$: $d = \dfrac{5b}{2a}$

If an expression involves a multiplying fraction, begin by rewriting the entire expression as a fraction, as shown in the next example.

CHAPTER 28: FORMULAE

Example 12

Make v the subject of the formula: $s = \frac{t}{2}(u + v)$

Rewrite the right-hand side as a single fraction: $\quad s = \frac{t(u+v)}{2}$

Multiply both sides by 2 and expand the brackets: $\quad 2s = tu + tv$

Subtract tu from both sides: $\quad 2s - tu = tv$

Divide both sides by t: $\quad \frac{2s - tu}{t} = v$

Write v on the left-hand side: $\quad v = \frac{2s - tu}{t}$

Exercise 28D

1. Make a the subject of these formulae:

   (a) $A = \frac{a}{5}$ (b) $C = \frac{e}{a}$ (c) $B = \frac{\pi a}{3}$ (d) $G = \frac{na - nb}{s}$ (e) $\frac{7}{p} = \frac{2r}{a}$

   (f) $W = \frac{2F}{B + a}$ (g) $L = f + \frac{ab}{10}$ (h) $H = \frac{4}{5}ma$ (i) $b = \frac{(15 + a)c}{4}$ (j) $j = \frac{a + b}{2b}$

2. (a) The formula for the area of a triangle is $A = \frac{1}{2}bh$, where b is the length of the base and h is the perpendicular height. Rearrange the formula to make b the subject.

   (b) The formula for the area of a trapezium is $A = \frac{1}{2}(a + b)h$, where a and b are the lengths of the two parallel sides and h is the perpendicular distance between them. Rearrange the formula to make a the subject.

   (c) The formula for the volume of a cone is $V = \frac{1}{3}\pi r^2 h$, where r is the base radius and h is its height. Rearrange the formula to make h the subject.

Formulae involving powers and roots

You may be asked to change the subject of a formula where a power or a root of the subject appears. Squaring and taking a square root are inverse operations. This means that a square root sign can be removed by squaring both sides, as in the next example.

Example 13

Make d the subject of the formula: $c = \sqrt{d - e}$

Square both sides: $\quad c^2 = d - e$
Add e to both sides: $\quad c^2 + e = d$
Write d on the left-hand side: $\quad d = c^2 + e$

In the following example, we take the square root of both sides. It is important to remember that taking a square root gives two answers – one positive and one negative.

Example 14

Make w the subject of the formula: $(Bw - C)^2 = D$

Take the square root of both sides: $\quad (Bw - C) = \pm\sqrt{D}$

Add C to both sides: $\quad Bw = \pm\sqrt{D} + C$

Divide both sides by B: $\quad w = \frac{\pm\sqrt{D} + C}{B}$

Note: On the left-hand side, this removes the squaring. On the right-hand side, don't forget to include the plus or minus sign.

When taking a cube root, you do not need to include a plus or minus sign, since there is only one answer.

Example 15

The formula for the volume of a sphere is $V = \frac{4}{3}\pi r^3$, where r is the radius. Rearrange this formula to make r the subject.

GCSE MATHEMATICS M3 AND M7

Rewrite the right-hand side as a single fraction: $V = \dfrac{4\pi r^3}{3}$

Multiply both sides by 3 to eliminate the fraction: $3V = 4\pi r^3$

Divide both sides by 4π: $\dfrac{3V}{4\pi} = r^3$

Take the cube root of both sides: $\sqrt[3]{\dfrac{3V}{4\pi}} = r$

Write r on the left-hand side: $r = \sqrt[3]{\dfrac{3V}{4\pi}}$

Example 16

The formula for the area of a circle is $A = \pi r^2$, where r is the radius. Rearrange this formula to make r the subject.

Divide both sides by π: $\dfrac{A}{\pi} = r^2$

Take the square root of both sides: $\sqrt{\dfrac{A}{\pi}} = r$

Write r on the left-hand side: $r = \sqrt{\dfrac{A}{\pi}}$

Note: In this example, we do not need to include a ± sign when taking the square root. This is because r is the radius of the circle and so cannot be a negative number.

Exercise 28E

1. Make a the subject of each formula.
   (a) $a^2 - 6 = b$
   (b) $a^2 + c = d$
   (c) $\dfrac{a^2}{3} = e$
   (d) $\sqrt{a} = f$
   (e) $g = \sqrt{h + a}$
   (f) $j = a\sqrt{k}$
   (g) $m = \sqrt{a - n}$
   (h) $\dfrac{pa^2}{q} = r$
   (i) $sa^3 = t$
   (j) $\pi\sqrt{\dfrac{a}{w}} = x$

2. Make b the subject of the following formulae.
   (a) $b^2 = C$
   (b) $b^3 = D$
   (c) $E^2 = \dfrac{1}{b^2}$
   (d) $\sqrt{b} + F = G$
   (e) $H\sqrt{b + I} = J$
   (f) $K = L - b^2$
   (g) $4\pi\sqrt{b + M} = N$
   (h) $\sqrt{P - b} = Q$
   (i) $b^3 + R = S$
   (j) $\dfrac{1}{2}\pi b^3 = T$

3. The formula for the time taken for a pendulum to swing is: $T = 2\pi\sqrt{\dfrac{L}{g}}$

   where L is the length of the pendulum and g is the acceleration due to gravity.
   Make L the subject of this formula.

4. In a right-angled triangle, the length of the hypotenuse can be found using the formula: $h = \sqrt{a^2 + b^2}$
   Rearrange the formula to make a the subject.

   Note: In this question you will take a square root. However, since a side length cannot be negative, you do not have to include the plus or minus sign. There is only one answer, which is positive.

Formulae where the subject appears more than once

You may be asked to change the subject of a formula in which the subject appears in more than one term. The method is:

1. Eliminate fractions and brackets from the formula.
2. Aim to group all the terms involving the subject on the left-hand side of the formula. Group all the other terms on the right-hand side.
3. Factorise the left-hand side by taking the subject out of brackets.
   Divide both sides by the expression in the brackets.

Example 17

Make a the subject of the formula: $ab = cd - ae$

There are no fractions or brackets to eliminate here.
Group the terms involving a on the left-hand side: $ab + ae = cd$

Next, factorise the left-hand side by taking a out of brackets: $\quad a(b + e) = cd$

Divide both sides by $(b + e)$: $\quad a = \dfrac{cd}{b + e}$

Example 18

Make d the subject of the formula: $Q = \dfrac{10(t - d)}{d}$

Multiply both sides by d to eliminate the fraction:	$Qd = 10(t - d)$
Expand the brackets:	$Qd = 10t - 10d$
Group all terms involving d on the left-hand side:	$Qd + 10d = 10t$
Factorise the left-hand side by taking d out of brackets:	$d(Q + 10) = 10t$
Divide both sides by the expression in the brackets:	$d = \dfrac{10t}{Q + 10}$

Example 19

Make x the subject of the formula: $x + 2 = \dfrac{x + w}{y}$

Multiply both sides by y to eliminate the fraction:	$y(x + 2) = x + w$
Expand brackets:	$xy + 2y = x + w$
Group terms involving x on the left-hand side and the other terms on the right:	$xy - x = w - 2y$
Factorise the left-hand side:	$x(y - 1) = w - 2y$
Divide both sides by the expression in the brackets:	$x = \dfrac{w - 2y}{y - 1}$

Exercise 28F

1. Make a the subject of each of these formulae.
   (a) $ba + ca = d$
   (b) $ea = f(a + g)$
   (c) $h - ja = a(k - m) + na$
   (d) $p = \dfrac{2 + a}{2 - a}$
   (e) $q + ra = \dfrac{s - a}{2}$
   (f) $a + ta = ua - 4$

2. Make b the subject of each of these formulae.
   (a) $vb = w - 5b$
   (b) $b(x + y) = z(6 - b)$
   (c) $A = \dfrac{b + C}{b - D}$
   (d) $7b - 8 = Eb - 9$
   (e) $F = \dfrac{50(G - b)}{b}$
   (f) $H + b = 2Ib - 3$

3. In a chemical reaction, the amount of substance P depends on the time t, and also on the amounts of substances Q and R. The formula is: $P = 3Qt + 4Rt$
   Make t the subject of this formula.

4. The electrical current in a circuit varies with time t seconds according to the formula: $I = \dfrac{t}{t + 3}$
   Make t the subject of the formula.

5. The manager of a sandwich café works out a formula for the amount of time taken in minutes to prepare one sandwich. It is: $T = \dfrac{5(f + 1)}{2(f + 3)}$ where f is the number of sandwich fillings.
   Rearrange the formula to make f the subject.
   Hint: Expand both sets of brackets before trying to eliminate the fraction.

28.3 Summary

In this chapter, you have learnt how to:
- Change the subject of formulae. These formulae include:
  - Simple formulae in which the rearranging involves addition, subtraction, multiplication and division.
  - Formulae involving brackets.
  - Formulae involving fractions.
  - Formulae involving powers and roots.
  - Formulae in which the subject appears more than once.

Chapter 29
Sequences

29.1 Introduction

Sequences are everywhere. The most famous sequence is:

   1, 2, 3, 4, 5, 6, 7, 8, 9, 10, ...

which is just the list of natural numbers. Every sequence is just a list of numbers.

Of course, the numbers don't have to form a pattern. For example 4, 0, 2, −5, 999, 66.3, ... is a sequence.

However, at GCSE we are interested in sequences that have a pattern. For example, in the sequence:

   2, 4, 6, 8, 10, ...

you can spot that each number is 2 more than the one before. It goes up in steps of 2

So the next three terms are 12, 14, 16 and the sequence looks like this:

   2, 4, 6, 8, 10, 12, 14, 16, ...

This is an example of an ascending sequence.

Other sequences go down in steps, for example:

   20, 17, 14, 11, ...

In this case, each number is 3 less than the one before. It goes down in steps of three. So the next three terms are 8, 5, 2 and the sequence looks like this:

   20, 17, 14, 11, 8, 5, 2, ...

We call these descending sequences.

Key words
- **Sequence**: A list of numbers or **terms**.
- **Term**: One item in a sequence.
- **Arithmetic sequence**: A sequence that has a constant difference between consecutive terms, e.g. 2, 4, 6, 8, ... or 10, 7, 4, 1, ...

Before you start you should know how to:
- Add terms to a simple sequence by spotting the pattern.

In this chapter you will learn to:
- Identify the next number(s) in a sequence.
- Identify some famous sequences.
- Find the next number in a sequence using differences.
- Write out terms in a sequence using the nth term formula.
- Find the formula for the nth term of a linear sequence.
- Work with non-linear sequences.

Exercise 29A (Revision)

1. Write down the next two numbers in the following sequences:
   (a) 3, 4, 5, 6, ...
   (b) 4, 6, 8, 10, ...
   (c) 1, 4, 7, 10, ...
   (d) 13, 15, 17, 19

2. Write down the next two numbers in these descending sequences:
   (a) 30, 29, 28, 27, ...
   (b) 35, 30, 25, 20, ...
   (c) 20, 18, 16, 14, ...
   (d) 49, 46, 43, 40, ...

3. Write down the next three numbers in these sequences:
   (a) 12, 16, 20, 24, ...
   (b) 60, 54, 48, 42, ...
   (c) 2, 9, 16, 23, ...
   (d) 81, 72, 63, 54, ...

CHAPTER 29: SEQUENCES

29.2 Identifying the Next Numbers in a Sequence

This is a skill you should already have. We will apply it in this section to some of the most famous sequences. You should be familiar with these sequences. The main idea is to look for a pattern in the numbers in the sequence. This pattern can be in the steps from each number in the sequence to the next.

Example 1

Write down the next 6 numbers in the sequence of **even numbers**: 2, 4, 6, 8, ...

We look for the step from each number to the next. We can see easily that it's always 2 units.
Then we add 2 to each previous number to get the next:
8 + 2 = 10   10 + 2 = 12   12 + 2 = 14
14 + 2 = 16   16 + 2 = 18   18 + 2 = 20

So, the sequence continues:
2, 4, 6, 8, 10, 12, 14, 16, 18, 20, ...

Some sequences are illustrated by counting squares in a sequence of diagrams that follow a pattern. One such pattern is the **triangular numbers**, where we create a sequence of triangles of squares. Each new diagram is created by adding an extra row with one more square onto the bottom of the previous one as shown in the diagram.

Example 2

Write down the first five numbers in the sequence of **triangular numbers** by counting the number of squares in each of the previous diagram.

Look at the sequence of triangles in the picture above.

The first diagram has   1 square
Second diagram has   1 + **2** = 3 squares
Third diagram has   3 + **3** = 6 squares
Fourth diagram has   6 + **4** = 10 squares
Fifth diagram has   10 + **5** = 15 squares

So, the triangular numbers are the sequence: 1, 3, 6, 10, 15, ...

Do we need to use diagrams to write down more of this sequence? The answer is no, because you can see that the number we add to make the next number in the sequence just increases by 1 each time.

So, the 6th number in the triangular sequence is   15 + **6** = 21
The 7th number is   21 + **7** = 28   and so on...
However, the pattern in a sequence can be calculated in a different way. Let us look at another famous sequence, the **square numbers**. This sequence has this name because they are formed from the number of small squares there are in squares shown in the diagram.

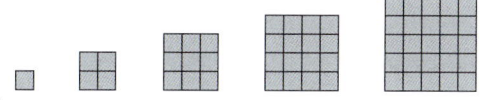

Example 3

Write down the first five numbers in the sequence of square numbers.

Look at the sequence of squares in the diagram above:

The first diagram has   1 square   Second diagram has   4 squares
Third diagram has   9 squares   Fourth diagram has   16 squares
Fifth diagram has   25 squares

Of course, we don't need the diagrams to work out the square numbers. We just multiply a value by itself to get its square:
1 × 1 = 1   2 × 2 = 4   3 × 3 = 9   4 × 4 = 16
5 × 5 = 25   and so on...

So, the square numbers are the sequence: 1, 4, 9, 16, 25, ...

In some sequences the next number in the sequence is calculated by combining the two previous terms in the sequence in some way. The **Fibonacci numbers** (also called the Fibonacci sequence) is famous because it describes many patterns in nature. It begins with two number 1s. Then each following value is calculated by adding together the two previous numbers in the sequence.

Example 4

Write down the first eight numbers in the Fibonacci sequence.

The 1st number is:	1
The 2nd number is:	1
The 3rd number is obtained by adding together the two previous numbers:	1 + 1 = 2
The 4th number is obtained by adding together the two previous numbers:	1 + 2 = 3
The 5th number is obtained by adding together the two previous numbers:	2 + 3 = 5
The sequence continues:	3 + 5 = 8
	5 + 8 = 13
	8 + 13 = 21 and so on…

So, the Fibonacci sequence is: 1, 1, 2, 3, 5, 8, 13, 21, …

Exercise 29B

1. Write out the first eight odd numbers. (Hint: it starts 1, 3, 5, 7, …)
2. Write down the first ten triangular numbers. (Hint: look back at Example 2)
3. Write down the first ten square numbers. (Hint: look back at Example 3)
4. The **cube numbers** are the number of small cubes in the diagrams on the right. Write down the first seven cube numbers.

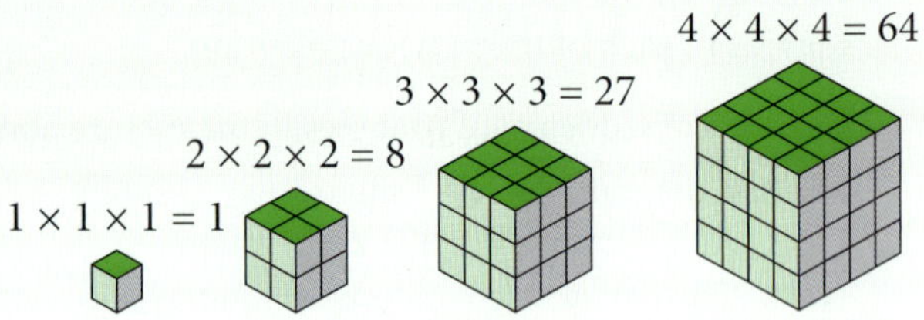

   $1 \times 1 \times 1 = 1$
   $2 \times 2 \times 2 = 8$
   $3 \times 3 \times 3 = 27$
   $4 \times 4 \times 4 = 64$

5. Write out the first ten Fibonacci numbers. (Hint: look back at Example 4)
6. The **Lucas numbers** start with 2, 1, 3, … and each successive term is calculated by adding the two previous terms (as in the Fibonacci sequence). Write out the first ten Lucas numbers.

29.3 Arithmetic Sequences

Each number in a sequence is called a **term** of the sequence. In this section we will consider sequences where the next term is calculated by adding or subtracting a fixed value from the previous term. Such sequences are called **arithmetic** or **linear sequences**. This name describes most of the sequences in this chapter. The next example illustrates how to recognise an arithmetic sequence.

Example 5

Which of the following sequences are arithmetic?
   (a) 1, 4, 9, 16, …
   (b) 2, 4, 6, 8, …
   (c) 90, 85, 80, 75, …
   (d) 1, 1, 2, 3, 5, 8, …
   (e) 13, 16, 19, 22, …

(a) The way to spot an arithmetic sequence is to look for the same gaps between the terms.

   Calculate the gaps between the terms of the first sequence:

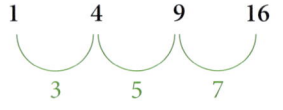

   1   4   9   16
     3   5   7

   The gaps between the terms in the sequence are 3, 5, 7 which are **not** the same.
   So, this is **not** an arithmetic sequence.

(b) Calculate the gaps between the terms:

   2   4   6   8
     2   2   2

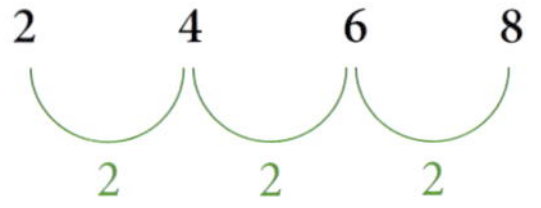

CHAPTER 29: SEQUENCES

The gaps between the terms in the sequence are all the same (they are all 2).
So, this **is** an arithmetic sequence.

(c) Calculate the gaps between the terms:

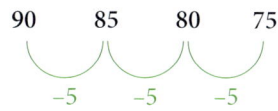

90   85   80   75
  -5   -5   -5

The gaps between the terms in the sequence are all the same (they are all −5).
So, this **is** an arithmetic sequence.

(d) Calculate the gaps between the terms:

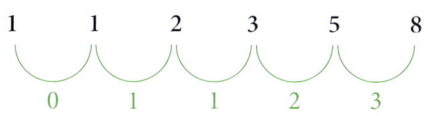

1   1   2   3   5   8
  0   1   1   2   3

The gaps between the terms in the sequence are 0, 1, 1, 2, 3 which are **not** the same.
So, this is **not** an arithmetic sequence.

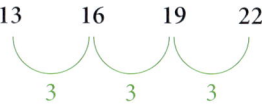

(e) Calculate the gaps between the terms:

13   16   19   22
  3   3   3

The gaps between the terms in the sequence are all the same (they are all 3).
So, this **is** an arithmetic sequence.

As the step between every pair of terms in an arithmetic sequence is the same, you just need to be given two terms to know the whole sequence. Equally it is enough to be told the first term and the size of the gap. The following examples illustrates how to write out arithmetic sequences when you are given these pieces of information.

Example 6

Write out the first 5 terms in the following arithmetic sequences:
(a) The first term is 12 and step between terms is 7
(b) The first two terms are 55 and 49

(a) The first term is 12   The second term is 12 + 7 = 19
    The third term is 19 + 7 = 26   In the same way the fourth and fifth terms are 33, 40
    So the first five terms are 12, 19, 26, 33, 40
(b) The step can be calculated by subtracting the first term from the second: 49 − 55 = −6
    So the third term is 49 − 6 = 43   In the same way, the fourth and fifth terms are 37 and 31
    So the first five terms are 55, 49, 43, 37, 31

Example 7

(a) Write out the third term in the sequence 1, 4, 7, 10, …
(b) Write out the 20th term in the arithmetic sequence that starts with 3 and has difference equal to 5
(c) Which term in the sequence 3, 6, 9, 12, … has the value 39?

(a) 3rd term = 7
(b) From the first term to the 20th term there will be 19 gaps. That makes an overall difference of 19 × 5 = 95
    So the 20th term = 3 + 95 = 98
(c) Note that each term is a multiple of three: 3 = 1 × 3, 6 = 2 × 3, 9 = 3 × 3, 12 = 4 × 3
    So the term that has the value 39 is the 13th term since 13 × 3 = 39

Exercise 29C

1. Write the next two terms in these arithmetic sequences:
   (a) 3, 7, 11, 15, …   (b) 6, 12, 18, 24, …   (c) 1, 5, 9, 13, …   (d) 76, 72, 68, 64, …
   (e) 4, 3, 2, 1, …   (f) 25, 31, 37, 43, …   (g) 101, 97, 93, 89, … (h) 7, 2, −3, −8, …
2. Write out the first four terms in the arithmetic sequence that:
   (a) Starts 5, 16, …   (b) Has first term 57 and a step of −2
   (c) Starts 85, 74, …   (d) Starts with 7 and has a step of −8

261

3. Which term of the sequence starting 5, 10, 15, … has the value 200?
4. The first term of an arithmetic sequence is 3 and the third term is 7
   Write down the fourth term.

29.4 Finding the nth Term of an Arithmetic Sequence

The nth **term** formula of an arithmetic sequence is given by:
   n^{th} term = [first term − step] + step × n

Example 8

Find the nth term formula for the following arithmetic sequences:
(a) 7, 11, 15, 19, …   (b) 2, 5, 8, 11, …   (c) 16, 23, 30, 37, …

(a) Look at the first four terms 7, 11, 15, 19. We can see that the step between every pair of values = 4
So n^{th} term   = [first term − step] + step × n
                   = [7 − 4] + 4 × n
                   = 3 + 4n

We can check that this is correct by using our formula to calculate the first term and making sure it matches the actual first term, and so on.

1st term	= 3 + 4n	= 3 + 4 × 1	= 7 which is correct
2nd term	= 3 + 4n	= 3 + 4 × 2	= 11 which is correct
3rd term	= 3 + 4n	= 3 + 4 × 3	= 15 and so on

(b) Looking at 2, 5, 8, 11 we can see that the step between every pair of values = 3
So n^{th} term   = [first term − step] + step × n
                   = [2 − 3] + 3 × n
                   = −1 + 3n

In this case the nth term formula starts with a negative number. This happens quite often. For neatness, when this happens, we normally swap the terms over and write it as:
                   = 3n − 1

Check our formula:

1st term	= 3n − 1	= 3 × 1 − 1	= 2
2nd term	= 3n − 1	= 3 × 2 − 1	= 5
3rd term	= 3n − 1	= 3 × 3 − 1	= 8

(c) Looking at 16, 23, 30, 37 we can see that the step between every pair of values = 7
So n^{th} term   = [first term − step] + step × n
                   = [16 − 7] + 7 × n
                   = 9 + 7n

Check our formula:

1st term	= 9 + 7n	= 9 + 7 × 1	= 16
2nd term	= 9 + 7n	= 9 + 7 × 2	= 23
3rd term	= 9 + 7n	= 9 + 7 × 3	= 30

The same method works when the arithmetic sequence is a list of decreasing values. Just remember that the step has to be a negative number.

Example 9

Find the nth term formula for the following arithmetic sequences:
(a) 50, 44, 38, 32, …   (b) −5, −13, −21, −29, …   (c) 16, 7, −2, −11, …

(a) Looking at 50, 44, 38, 32 we can see that the step between every pair of values = 44 − 50 = −6
So n^{th} term   = [first term − step] + step × n
                   = [50 − (−6)] + (−6) × n
                   = 56 − 6n

Check our formula:
1st term	= 56 − 6n	= 56 − 6 × 1	= 50
2nd term	= 56 − 6n	= 56 − 6 × 2	= 44
3rd term	= 56 − 6n	= 56 − 6 × 3	= 38

(b) Looking at −5, −13, −21, −29 we can see that the step between every pair of values = −13 − (−5)
= −13 + 5 = −8

So n^{th} term = [first term − step] + step × n
= [(−5) − (−8)] + (−8) × n
= 3 − 8n

Check our formula:
1st term	= 3 − 8n	= 3 − 8 × 1	= −5
2nd term	= 3 − 8n	= 3 − 8 × 2	= −13
3rd term	= 3 − 8n	= 3 − 8 × 3	= −21

(c) Looking at 16, 7, −2, −11 we can see that the step size between every pair of values = 7 − 16 = −9

So n^{th} term = [first term − step] + step × n
= [16 − (−9)] + (−9) × n
= 25 − 9n

Check our formula:
1st term	= 25 − 9n	= 25 − 9 × 1	= 16
2nd term	= 25 − 9n	= 25 − 9 × 2	= 7
3rd term	= 25 − 9n	= 25 − 9 × 3	= −2

Exercise 29D

1. Find the nth term for the arithmetic sequences:
   (a) 65, 69, 73, 77, ...   (b) 34, 45, 56, 67, ...   (c) 144, 132, 120, 108, ...
   (d) −25, −21, −17, −13, ...   (e) 10.3, 11.5, 12.7, 13.9, ...

2. Find the nth term for the arithmetic sequences:
   (a) 73, 66, 59, 52, ...   (b) 3, 1, −1, −3, ...   (c) 48, 40, 32, 24, ...
   (d) 8, 3, −2, −7, ...   (e) 5.4, 3.3, 1.2, −0.9, ...

3. Find the nth term and the 25th term for the arithmetic sequences:
   (a) 3, 8, 13, 18, ...   (b) 14, 20, 26, 32, ...   (c) 23, 33, 43, 53, ...
   (d) 98, 96, 94, 92, ...   (e) 25, 34, 43, 52, ...

29.5 Using the nth Term Formula for Sequences

Let's think about the sequence of even numbers again:

1st term	2nd term	3rd term	4th term	5th term	
2	4	6	8	10	...

If you were asked for the 10th term you could count forward five more terms to get 12, 14, 16, 18, 20

So, the 10th term is 20 But if you were asked for the 123rd term this would be a very time-consuming way of working out the answer.

A more powerful way to calculate terms in a sequence is by using the **n^{th} term** formula. For the sequence of even numbers, the nth term is given by the formula:

n^{th} term = 2n

We use this formula by substituting the place of a term, n, to work out its value. So for example, for the sequence of even numbers:

1st term	2nd term	3rd term	4th term	5th term	
2n = 2 × 1 = 2	2n = 2 × 2 = 4	2n = 2 × 3 = 6	2n = 2 × 4 = 8	2n = 2 × 5 = 10	...

GCSE MATHEMATICS M3 AND M7

Different sequences have different n^{th} term formulae. The nth term formulae for the famous sequences you met earlier in this chapter are shown in the table on the right.

Sequence	n^{th} term formula
Even numbers	n^{th} term = $2n$
Odd numbers	n^{th} term = $2n - 1$
Square numbers	n^{th} term = n^2
Cube numbers	n^{th} term = n^3
Triangular numbers	n^{th} term = $\frac{1}{2}n(n + 1)$

Note: The nth term formula for the Fibonacci numbers is too complex to discuss here.

The following examples show how to use an nth term formula.

Example 10

Write out the 100th term for the sequences whose n^{th} term is given by:
(a) $n(n + 1)$ (b) $(n + 1)(n - 1)$ (c) $n^2 - n$ (d) $3n - 2$

(a) We are looking for the 100th term, which means $n = 100$ so:
100th term = $n(n + 1) = 100 \times (100 + 1) = 100 \times 101 = 10\,100$
(b) 100th term = $(n + 1)(n - 1) = (100 + 1)(100 - 1) = 101 \times 99 = 9999$
(c) 100th term = $n^2 - n = 100^2 - 100 = (100 \times 100) - 100 = 9900$
(d) 100th term = $3n - 2 = 3 \times 100 - 2 = 300 - 2 = 298$

Example 11

Which term of the sequence with the nth term = $2n - 7$ equals the value 43?

We want to find the **place** in the sequence. That means finding the value of n.
So: $2n - 7 = 43$
Add 7 to each side: $2n - 7 + 7 = 43 + 7$
$2n = 50$
Which gives: $n = 25$ So the term that equals 43 is the 25th term.

Exercise 29E

1. Find **(i)** the 10th and **(ii)** 50th terms for the sequences whose n^{th} term formulas are given by:
   (a) n^{th} term = $5n + 7$ (b) n^{th} term = $100 - 2n$ (c) n^{th} term = $n^2 - 1$
   (d) n^{th} term = $n(n + 2)$ (e) n^{th} term = $53 - 3n$ (f) n^{th} term = $n^2 - 2$
   (g) n^{th} term = $27 - 4n$ (h) n^{th} term = $(6 + n)(3 + n)$

2. Find the 15th term for the sequences whose n^{th} term formulas are given by:
   (a) n^{th} term = $2n - 1$ (b) n^{th} term = n^2 (c) n^{th} term = $(n - 10)(n + 5)$
   (d) n^{th} term = $100 - 3n$ (e) n^{th} term = $5n + 7$ (f) n^{th} term = n^3
   (g) n^{th} term = $2n^2$ (h) n^{th} term = $62 + 4n$

3. Which term of the sequence with n^{th} term = $2n + 8$ equals the value 50?
4. Which term of the sequence with n^{th} term = $50 - 3n$ equals the value 29?
5. Which term of the sequence with n^{th} term = $n^2 - 1$ equals the value 143?

29.6 Non-Linear Sequences

So far we have considered linear sequences which have the same difference between each term. However, many sequences do not follow an arithmetic pattern like this.

A non-linear sequence does not have the same difference between each of its terms. We have already met some famous non-linear sequences (see Section 29.2) such as the square numbers, the triangular numbers and the Fibonacci numbers.

To extend any sequence by a few terms, you must look for a pattern in its differences. Once you have identified the pattern, you can then extend the sequence by following the same pattern.

The following examples demonstrate how this is done.

CHAPTER 29: SEQUENCES

Example 12

Find the next term in the following sequences:
(a) 3, 7, 13, 21, 31 (b) 2, 10, 22, 38, 58

(a) First work out the differences between terms. Then work out the differences *between* the differences. These are called the **second differences**. Note that the next term must have a second difference of 2 to follow the pattern.

So, the next difference must be:
10 + 2 = 12

And the next number in the sequence must be:
31 + 12 = 43

Number	3	7	13	21	31	?
Difference		4	6	8	10	
Second difference			2	2	2	

Number	3	7	13	21	31	43
Difference		4	6	8	10	12
Second difference			2	2	2	2

(b) First work out the differences between terms and the second differences.

Note that the next term must have a second difference of 4 to follow the pattern.

So, the next difference must be: 20 + 4 = 24
The next number in the sequence must be: 58 + 24 = 82

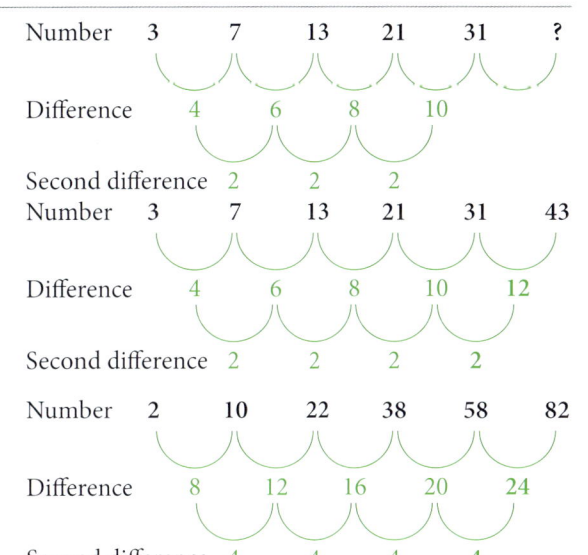

Number	2	10	22	38	58	82
Difference		8	12	16	20	24
Second difference			4	4	4	4

Sequences whose 2nd differences are identical are called **quadratic sequences**. Both sequences in Example 12 were quadratic. A quadratic sequence may be defined by a formula for its nth term. Note that the formula for the nth term of a quadratic sequence is a quadratic expression, as shown in the following examples.

Example 13

Write down the first 4 terms and the 10th term of the quadratic sequences with these n^{th} terms:
(a) $n^2 - 1$ (b) $(n + 2)(n + 3)$

In both (a) and (b) we substitute $n = 1$, then 2, then 3 etc. into the formula:

(a) For the first 4 terms we substitute into $n^2 - 1$:
$1^2 - 1 = 0$, $2^2 - 1 = 4 - 1 = 3$, $3^2 - 1 = 9 - 1 = 8$, $4^2 - 1 = 16 - 1 = 15$ …
and the 10th term is: $10^2 - 1 = 99$

(b) For the first 4 terms we substitute into $(n + 2)(n + 3)$:
$(1 + 2)(1 + 3) = 3 \times 4 = 12$, $(2 + 2)(2 + 3) = 4 \times 5 = 20$,
$(3 + 2)(3 + 3) = 5 \times 6 = 30$, $(4 + 2)(4 + 3) = 6 \times 7 = 42$ …
and the 10th term is: $12 \times 13 = 156$

Note: The formula for the nth term of this sequence multiplies out to give a quadratic expression:
$(n + 2)(n + 3) = n^2 + 5n + 6$

At GCSE you will not be asked to find n^{th} term for a general quadratic sequence, but you may be asked to find the nth term for a simple quadratic sequence, for example a sequence whose n^{th} term is given by $n^2 + 1$ The following example is a question of this kind.

Example 14

Find the nth term for the sequences:
(a) 2, 5, 10, 17, … (b) 0, 3, 8, 15, …

(a) First work out the differences between terms and the second differences.

As the second differences are the same value, 2, we expect a quadratic sequence, i.e. it will contain an n^2 term.

You should notice that each element of the sequence is one digit higher than the sequence of square numbers, given by n^2

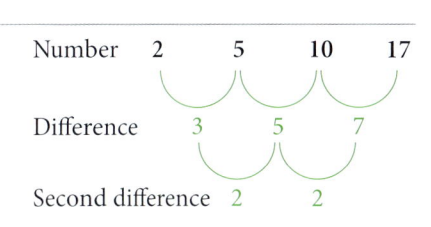

Number	2	5	10	17
Difference		3	5	7
Second difference			2	2

265

GCSE MATHEMATICS M3 AND M7

Knowing this, think how you can change the sequence of squares 1, 4, 9, 16 to get the given sequence 2, 5, 10, 17   The answer is to add 1 to each term: 2 = 1 + 1, 5 = 4 + 1, 10 = 9 + 1, 17 = 16 + 1
As each term is 1 more than n^2, the nth term formula is: $n^2 + 1$

(b) Work out the differences as before. As the second differences are the same value, 2, we again expect a quadratic sequence with an n^2 term. Think how you can change the sequence of squares 1, 4, 9, 16 to get the given sequence 0, 3, 8, 15

Number	0		3		8		15
Difference		3		5		7	
Second difference			2		2		

The answer is to subtract 1 from each term:
0 = 1 − 1, 3 = 4 − 1, 8 = 9 − 1, 15 = 16 − 1
As each term is 1 less than n^2, the nth term formula is: $n^2 − 1$

A different type of non-linear sequence is a list of fractions. Often, the numerators and the denominators are two separate sequences, as shown in the next example.

Example 15

Write out the first four terms and the 8th term of the sequences whose n^{th} terms are:

(a) $\dfrac{2n + 1}{n^2}$ (b) $\dfrac{n - 1}{n + 3}$

(a) Substituting $n = 1, 2, 3, 4$ and 8 into the formula we get:

$\dfrac{3}{1}, \dfrac{5}{4}, \dfrac{7}{9}, \dfrac{9}{16}$ which should be simplified to $3, \dfrac{5}{4}, \dfrac{7}{9}, \dfrac{9}{16}$ The 8th term = $\dfrac{17}{64}$

(b) Substituting $n = 1, 2, 3, 4$ and 8 into the formula we get:

$0, \dfrac{1}{5}, \dfrac{2}{6}, \dfrac{3}{7}$ which should be simplified to $0, \dfrac{1}{5}, \dfrac{1}{3}, \dfrac{3}{7}$ The 8th term = $\dfrac{7}{11}$

Spotting the nth term for fraction-type sequences involves looking for separate formulas for the numerator and denominator, as shown in the following example.

Example 16

Find the nth term formula for the following non-linear sequences:

(a) $\dfrac{1}{1}, \dfrac{3}{8}, \dfrac{5}{27}, \dfrac{7}{64}, \ldots$ (b) $\dfrac{4}{1}, \dfrac{9}{4}, \dfrac{16}{9}, \dfrac{25}{16}, \ldots$

(a) The numerators are the sequence 1, 3, 5, 7 which are the sequence of odd numbers.
This sequence has constant differences = 2 and first term = 1
Thus, this is a linear sequence, so we use the equation:
   n^{th} term = [first term − difference] + difference × n
So: n^{th} term = $[1 − 2] + 2n$
               = $−1 + 2n$

The denominators are the cube numbers, $1^3 = 1, 2^3 = 8, 3^3 = 27, 4^3 = 64$
The nth term formula for this sequence is therefore = n^3

So the overall n^{th} term for the sequence is $\dfrac{-1 + 2n}{n^3}$ or $\dfrac{2n - 1}{n^3}$

(b) The denominators are the square numbers, so the nth term of these will be n^2
We can also see that the numerators are the squares of the *next* values of n
The first term is $4 = 2^2 = (1 + 1)^2$
The second term is $9 = 3^2 = (2 + 1)^2$
The third term is $16 = 4^2 = (3 + 1)^2$
The fourth term is $25 = 5^2 = (4 + 1)^2$
Continuing the pattern, the nth term for this sequence is $(n + 1)^2$

So the overall n^{th} term for the sequence is $\dfrac{(n + 1)^2}{n^2}$

CHAPTER 29: SEQUENCES

Another type of sequence is the Fibonacci-type sequence. In this sequence, each term is the sum of the two previous terms. Recall the differences in the Fibonacci sequence, shown on the right.

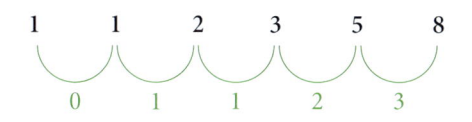

1 1 2 3 5 8
 0 1 1 2 3

This type of sequence can be identified because the sequence repeats itself in its differences, as you can see above.

Exercise 29F

1. Using differences, find the next 2 terms for the following sequences:
   (a) 4, 8, 14, 22, 32, …
   (b) 99, 96, 91, 84, 75, …
   (c) 6, 11, 18, 27, 38, …
   (d) 0, −2, −6, −12, −20, …
   (e) 1, 2, 3, 5, 8, …

2. Write down the first 4 terms and the 10th term of the quadratic sequences with n^{th} term:
   (a) $3n - n^2 - 1$
   (b) $2n^2 - 1$
   (c) $(n + 1)(n + 4)$
   (d) $n^2 - n - 2$
   (e) $(4n + 3)(n - 2)$

3. Find the nth term for the non − linear sequences:
   (a) 3, 6, 11, 18, 27, …
   (b) −1, 2, 7, 14, 23, …
   (c) −1, −4, −9, −16, −25, …
   (d) 4, 1, −4, −11, −20, …
   (e) 2, 9, 28, 65, 126, …

4. Write out the first four terms and the 8th term of the sequences with n^{th} term:
   (a) $\dfrac{3n - 1}{n}$
   (b) $\dfrac{5}{n + 2}$
   (c) $\dfrac{2 + 5n}{2n - 5}$
   (d) $\dfrac{2n^2}{n^3}$
   (e) $\dfrac{4n}{n^2 + 1}$

29.7 Summary

In this chapter you have learnt to:
- Be familiar with the sequences of squares, cubes, triangular numbers and the Fibonacci sequence.
- Identify the next number(s) in a sequence by using differences and second differences.
- Describe the rule for finding the next number in a sequence.
- Write out terms in a sequence using the nth term formula, for example the nth square number = $n \times n$
- Find the formula for the nth term of an arithmetic sequence using the rule:
  n^{th} term = [first term − difference] + difference × n
- Identify and use formulas for non-linear sequences.

Chapter 30
Non-Linear Graphs

30.1 Introduction

Not all graphs are straight lines. In this chapter, we will look at three types of non-linear graph: quadratic, cubic and reciprocal.

Key words
- **Quadratic graph**: A graph of a quadratic function. A quadratic function is of the form $y = ax^2 + bx + c$, where a, b and c are constants and $a \neq 0$
- **Cubic graph**: A graph of a cubic function. A cubic function is of the form $y = ax^3 + bx^2 + cx + d$, where a, b, c and d are constants and $a \neq 0$
- **Reciprocal graph**: A graph of a reciprocal function. A reciprocal function has the form $y = \frac{a}{x}$, where a is a constant.

Before you start you should:
- Know how to plot and read the coordinates of points in all four quadrants.
- Know how to substitute numbers into a formula.
- Understand reciprocals.

In this chapter you will learn how to:
- Recognise and plot the graphs of:
  - quadratic functions,
  - cubic functions,
  - reciprocal functions.

Exercise 30A (Revision)

1. (a) Draw a coordinate grid. Use values of x from −3 to 7 and values of y from −2 to 8
   Plot these points on the grid: A(1, 2), B(−2, 7), C(−3, −2) and D(6, −1).
   (b) Add these line segments to your diagram: AB, BC, CD, DA.
   (c) What shape have you drawn?

2. Find the value of T by substituting the values of P, Q and R given below into the formula $T = \sqrt{P} - QR^2$
   $P = 16, Q = -2, R = 3$

3. The **reciprocal** of a number x is $\frac{1}{x}$. Find the reciprocal of each of the numbers given in (a)–(e), giving each answer in its simplest form. **Hint**: to find the reciprocal of a fraction, the easiest method is to turn the fraction upside-down.
   (a) 6   (b) $\frac{1}{3}$ (c) −4   (d) $\frac{2}{5}$
   (e) 2.4   **Hint**: first turn this number into an improper fraction.

4. A number multiplied by its reciprocal always gives an answer of 1
   In each case (a)–(d), what number must you multiply by to get an answer of 1?
   Give each answer in its simplest form.
   (a) 5   (b) $\frac{1}{7}$ (c) −8   (d) 1.6

5. Explain briefly why the number zero has no reciprocal.

30.2 Quadratic Graphs

A quadratic function is of the form $y = ax^2 + bx + c$, where a, b and c are constants and $a \neq 0$

The graph of a quadratic function takes the shape of a **parabola**, as shown in the following three examples.

- If $a > 0$ then the curve is **u-shaped**, as shown in Example 1.
- If $a < 0$ then the curve is **n-shaped**, as shown in Example 2.

For a quadratic curve there is always one **turning point**, where y reaches either its maximum or minimum value.

Example 1

(a) Copy and complete the following table of values using the equation $y = x^2 - 2x + 1$

x	-2	-1	0	1	2	3	4
y							

(b) Plot the curve $y = x^2 - 2x + 1$, using values of x from -2 to 4
(c) Write down the coordinates of the turning point of this curve.

(a) To find the value of y when $x = -2$, use the equation of the curve: $y = x^2 - 2x + 1$
$y = (-2)^2 - 2(-2) + 1$
$= 4 + 4 + 1$
$= 9$

Repeat for each value of x. The completed table of values is as follows:

x	-2	-1	0	1	2	3	4
y	9	4	1	0	1	4	9

(b) Plot these points on the graph: (-2, 9), (-1, 4), etc. Join the points with a smooth curve. Make sure the curve extends to the edges of the grid. The completed graph is shown on the right.

> **Note:** For this quadratic function, $y = x^2 - 2x + 1$
> $a = 1$, $b = -2$ and $c = 1$
> Since $a > 0$, the curve takes a u-shape.

(c) The turning point is (1, 0), where y reaches its minimum value.

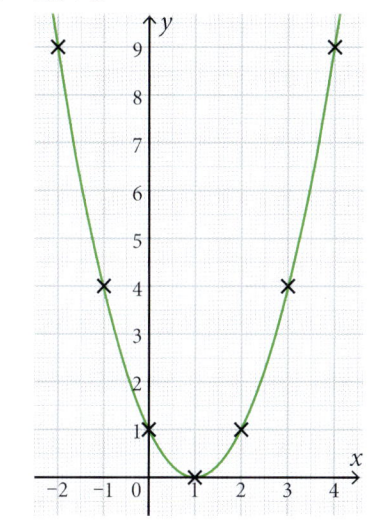

Example 2

(a) Plot the curve $y = -2x^2 + 4$, for $-2 \leq x \leq 2$
(b) Using your graph, answer the following questions:
  (i) For what value of x does y have a maximum value?
  (ii) For what values of x does $y = 0$?

(a)
> **Note:** For this quadratic function, $a = -2$, $b = 0$ and $c = 4$
> Since $a < 0$, we expect an n-shaped curve.

Draw a table of values, using x values from -2 to 2
Calculate the y values using the equation of the curve, $y = -2x^2 + 4$
For example, when $x = -2$:
$y = -2(-2)^2 + 4$
$= -8 + 4$
$= -4$

x	-2	-1	0	1	2
y	-4	2	4	2	-4

Plot the points and join them with a smooth curve, as shown.

(b) From the graph we can see that:
  (i) y reaches its maximum value of 4 when $x = 0$

> **Note:** This is equivalent to stating that the coordinates of the turning point are (0, 4)

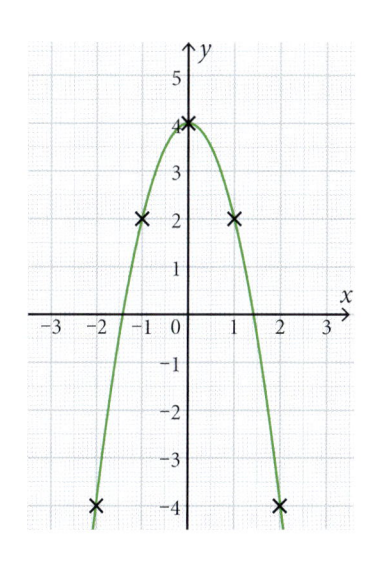

  (ii) $y = 0$ when x is roughly -1.4 or 1.4

Example 3
  (a) Plot the curve $y = x^2 - 3x - 2$ using integer values of x from -1 to 4
  (b) Using your curve:
     (i) Estimate the two x-intercepts of the curve; and
     (ii) Write down the y-intercept.
  (c) Using the symmetry of the curve, state the x-coordinate of the turning point.
  (d) Using your graph, estimate the y-coordinate of the turning point.

  (a) Using x values from -1 to 4, we obtain the following table of values:

x	-1	0	1	2	3	4
y	2	-2	-4	-4	-2	2

  Using these values, the graph is plotted as shown on the right.

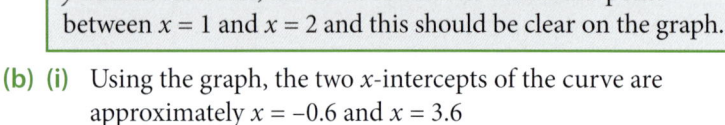

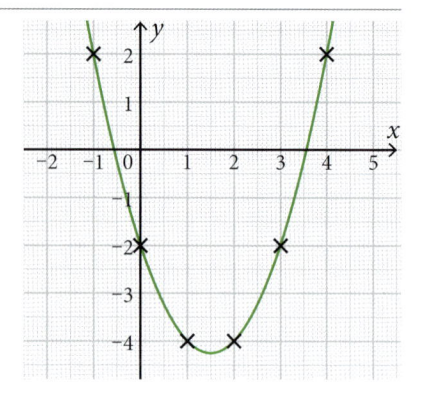

  Note: In the table, the points $(1, -4)$ and $(2, -4)$ have the lowest y-values. However, the curve reaches its minimum point between $x = 1$ and $x = 2$ and this should be clear on the graph.

  (b) (i) Using the graph, the two x-intercepts of the curve are approximately $x = -0.6$ and $x = 3.6$
     (ii) The y intercept is -2
  (c) From the graph, the turning point lies between $x = 1$ and $x = 2$
     Since all quadratic curves are symmetrical, we conclude that the x-coordinate of the turning point is $x = 1.5$
  (d) From the graph we can estimate the y-coordinate of the turning point as $y = -4.2$

Exercise 30B

1. A curve has the equation $y = x^2 - 3x$
   (a) Copy and complete the table of values on the right.

x	-1	0	1	2	3	4
y		0	-2		0	4

   (b) Draw a coordinate grid with x values from -1 to 4 and y values from -2 to 4
       Plot the points from the table.
   (c) Join the points with a smooth curve. **Hint**: There are two points with a y-coordinate of -2: the lowest point on the curve lies between these two points.

2. Draw each quadratic curve (a) – (d) using the x-values given.
   (a) $y = x^2 + 3x + 1$ for $-4 \leq x \leq 1$
   (b) $y = x^2 - 2x - 1$ for $-2 \leq x \leq 4$
   (c) $y = 2x^2 + 3x - 5$ for $-3 \leq x \leq 1$
   (d) $y = x^2 - 8x + 10$ for $1 \leq x \leq 7$

3. A curve has the equation $y = 2 - 3x - x^2$
   (a) Using a table of values with values of x from -4 to 1, draw the curve.
   (b) For what value of x does y have its maximum value?
   (c) Using your graph, estimate this maximum y-value.
   (d) For what approximate values of x does $y = 0$? **Hint**: There are two answers.

4. A curve has the equation $y = 8 + x - 2x^2$
   (a) Using a table of values with values of x from -3 to 3, draw the curve.
   (b) For what value of x does y have its maximum value?
   (c) For what value of x does $y = 0$?

5. (a) Plot the curve $y = 3x^2 - 5x + 3$ using values of x between -1 and 3
   (b) Using your graph, estimate the minimum y-value.

6. Plot the curve $y = x(x - 4)$ using $-1 \leq x \leq 5$

7. (a) Plot the curve $y = (x + 1)(x - 3)$ for $-2 \leq x \leq 4$
   (b) State where on the curve $y = 0$

30.3 Cubic Graphs

A cubic function is of the form $y = ax^3 + bx^2 + cx + d$, where a, b, c and d are constants and $a \neq 0$

The graph of a cubic function takes one of the following shapes:

CHAPTER 30: NON-LINEAR GRAPHS

If $a > 0$ Shape 1   Shape 2   Shape 3   If $a < 0$ Shape 1   Shape 2   Shape 3

Note: If $a > 0$ then the curve goes from the bottom left of the graph to the top right.
If $a < 0$ then the curve goes from the top left of the graph to the bottom right.

You can plot a cubic curve using a table of values, as with quadratic graphs.

Example 4

(a) Copy and complete the table below for the cubic function $y = x^3 - x - 1$

x	-2	-1	-0.5	0	0.5	1	2
y		-1		-1		-1	

(b) Using your table of values, draw the graph for $y = x^3 - x - 1$

(a) The equation of the curve is $y = x^3 - x - 1$
Use this to find the missing y-values:

When $x = -2$,
$y = (-2)^3 - (-2) - 1$
$= -8 + 2 - 1$
$= -7$

Follow a similar process for the other missing y-values:

x	-2	-1	-0.5	0	0.5	1	2
y	-7	-1	-0.625	-1	-1.375	-1	5

Note: It is usually sufficient to use integer values of x. In this example, however, $x = -0.5$ and $x = 0.5$ are also used. This allows us to gain a better picture of the curve in this region.

(b) The points are plotted, and the curve is shown on the right.

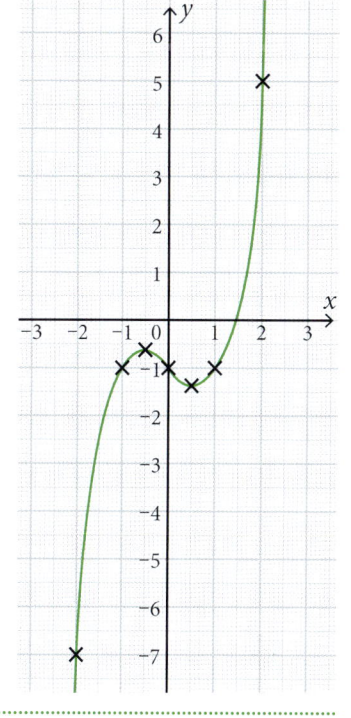

Example 5

(a) Using a table of values, with x-values from -3 to 3, draw the graph of:
$y = -x^3 + 4x - 4$

(b) Using your graph, estimate the solution to the equation
$-x^3 + 4x - 4 = 0$

(a) Use integer values of x from -3 to 3 in the table of values. Calculate the y-values using the equation of the curve.

For example, when $x = -3$,
$y = -(-3)^3 + 4(-3) - 4$
$= 27 - 12 - 4$
$= 11$

x	-3	-2	-1	0	1	2	3
y	11	-4	-7	-4	-1	-4	-19

The points are plotted, and the curve is shown on the right.

(b) We can find the solution to
$-x^3 + 4x - 4 = 0$
by finding where the curve passes through the x-axis.
This solution is roughly $x = -2.4$

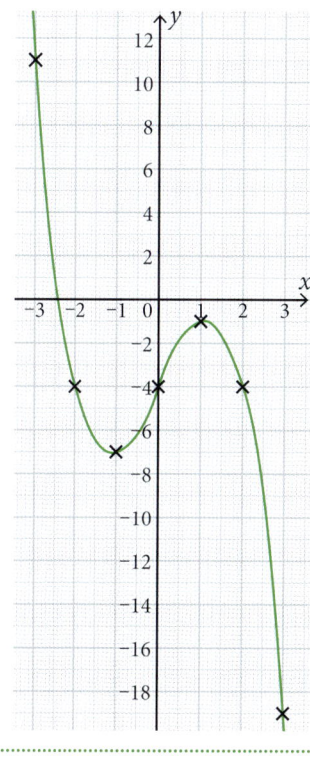

271

Exercise 30C

1. For each cubic functions (a) – (f):
   (i) Complete the table of values using the x-values given. (ii) Draw a graph of the function.

   (a) $y = x^3 + 2$ for $-2 \leq x \leq 2$

x	-2	-1	-0.5	0	0.5	1	2
y		1		2		3	10

   (b) $y = 4 - x^3$ for $-2 \leq x \leq 2$

x	-2	-1	0	1	2
y		5	4		-4

   (c) $y = 2x^3 - 2$ for $-2 \leq x \leq 2$

x	-2	-1	0	1	2
y	-18		-2		14

   (d) $y = x^3 - 2x$ for $-3 \leq x \leq 3$

x	-3	-2	-1	0	1	2	3
y		-4	1		-1		21

   (e) $y = x^3 + x - 2$ for $-2 \leq x \leq 2$

x	-2	-1	0	1	2
y		-4	0		

   (f) $y = -x^3 + 2x^2 - x + 2$ for $-2 \leq x \leq 3$

x	-2	-1	0	1	2	3
y	20		2		0	

2. A cubic curve has the equation $y = x^3 + 3x + 2$
   (a) Copy and complete the table on the right using the equation of the curve.

x	-1.5	-1	-0.5	0	0.5	1
y	-5.875		0.375		3.625	

   (b) Using the values in your table, plot the graph of the curve.
   (c) Estimate the solution to the equation $x^3 + 3x + 2 = 0$

3. (a) Draw the graph of the cubic function $y = \frac{1}{2}x^3 - 4x^2 + 2x + 12$ using all integer values of x from -2 to 7.
   (b) Using your graph, estimate for which value of x between -2 and 7:
       (i) y has its minimum value (ii) y has its maximum value.

30.4 Reciprocal Graphs

A reciprocal function has the form $y = \frac{a}{x}$, where a is a constant. For example, $y = \frac{1}{x}$ and $y = \frac{-4}{x}$ are both reciprocal functions.

Example 6

(a) Copy and complete the table of values below for the function $y = \frac{2}{x}$.
Round the y-values to 2 decimal places where appropriate.

x	-4	-3	-2	-1	-0.5	0.5	1	2	3	4
y		-0.67	-1		-4	4	2			0.5

(b) Using your table, plot the graph of $y = \frac{2}{x}$

(a) Using the equation of the function, $y = \frac{2}{x}$, we obtain the following table of values:

x	-4	-3	-2	-1	-0.5	0.5	1	2	3	4
y	-0.5	-0.67	-1	-2	-4	4	2	1	0.67	0.5

Note: We cannot use $x = 0$ in the table, since $y = \frac{2}{0}$ is undefined.

(b) Plotting the points in the table gives the curve shown on the right.

Note:
- Reciprocal curves have two branches, which appear in opposite quadrants of the graph.
- If the equation of the curve is $y = \frac{a}{x}$, where a is a positive number, the two branches of the curve lie in the bottom-left and top-right quadrants.
- If the curve was plotted for a wider range of x- and y-values, you would see that it gets closer and closer to the x- and y-axes, but it never touches them. The two axes are known as **asymptotes** to the curve.

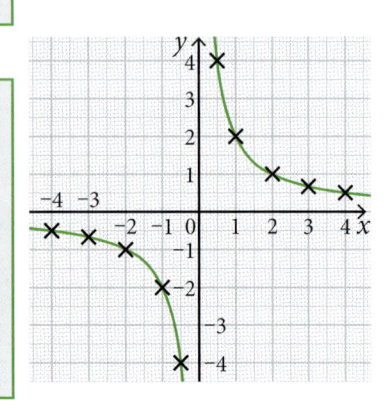

Example 7

(a) Copy and complete the table of values below for the function $y = \frac{-5}{x}$. Round the y-values to 2 decimal places where appropriate.

x	−4	−3	−2	−1	−0.5	0.5	1	2	3	4
y		1.67	2.5		10		−5	−2.5		−1.25

(b) Using your table, plot the graph of $y = \frac{-5}{x}$.

(a) Using the equation of the function, $y = \frac{-5}{x}$, we obtain the following table of values:

x	−4	−3	−2	−1	−0.5	0.5	1	2	3	4
y	1.25	1.67	2.5	5	10	−10	−5	−2.5	−1.67	−1.25

Note: As in the previous example, $x = 0$ cannot be used in the table, since $y = \frac{-5}{0}$ is undefined.

(b) Plotting the points in the table gives the curve shown on the right.

Note: If the equation of the curve is $y = \frac{a}{x}$ where a is a negative number, the two branches of the curve lie in the top-left and bottom-right quadrants.

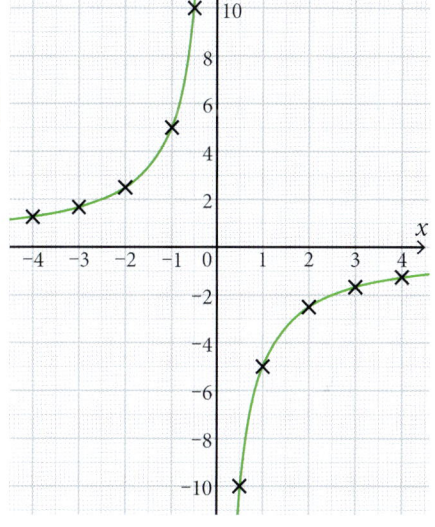

Exercise 30D

1. A reciprocal curve has the equation $y = \frac{1}{x}$

   (a) Copy and complete the following table of values using the equation of the curve. Give y-values to 2 decimal places where appropriate.

x	−3	−2	−1	−0.5	−0.25	0.25	0.5	1	2	3
y	−0.33	−0.5		−2				1	0.5	

   (b) Hence plot the graph of $y = \frac{1}{x}$

2. Draw the graph of $y = \frac{4}{x}$ using values of x from −4 to 4, excluding 0

3. Draw the graph of $y = \frac{-2}{x}$ using values of x from −4 to 4, excluding 0

4. (a) A reciprocal curve has the equation $y = \frac{-10}{x}$
   Copy and complete this table of values using the equation of the curve.

x	−10	−5	−4	−2	−1	1	2	4	5	10
y	1	2		5		−10		−2.5		−1

   (b) Use your table of values to draw the graph of this function.

   (c) Use your graph to determine whether there are any solutions to the equation $\frac{-10}{x} = 0$

30.5 Matching a Graph With its Equation

You may be given a sketch of a graph and asked to match it to an equation.

Example 8

Four curves are sketched on the next page. Match each curve with one of the following equations.

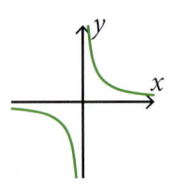

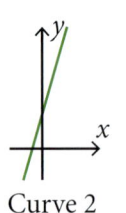

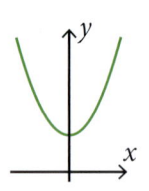

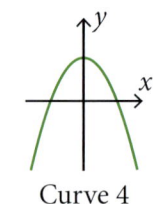

GCSE MATHEMATICS M3 AND M7

Curve 1	Curve 2	Curve 3	Curve 4	Equations
				A $y = 3x + 1$
				B $y = \dfrac{3}{x}$
				C $y = -x^2 + 1$
				D $y = x^2 + 1$

Curve 1 is a reciprocal curve. This must have Equation B, which is the only one with the form $y = \dfrac{a}{x}$

Curve 2 is a straight line. This must have Equation A, which is the only one with the form $y = mx + c$

Curves 3 and 4 both appear to be quadratic curves.

Curve 3 is u-shaped, so its equation must have a positive x^2 term. It must have Equation D.

Curve 4 is n-shaped, so its equation must have a negative x^2 term. It must have Equation C.

Exercise 30E

1. Match the curve sketches below with an appropriate equation.

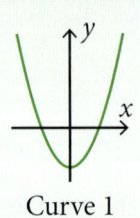

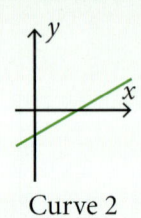

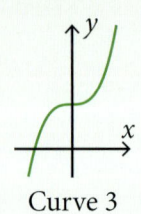

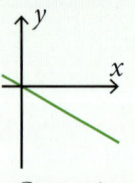

   Curve 1   Curve 2   Curve 3   Curve 4

   Equations
   A $y = -\dfrac{1}{2}x$
   B $y = x^3 + 1$
   C $y = x^2 - 1$
   D $y = \dfrac{1}{2}x - 1$

2. Match the curve sketches below with an appropriate equation.

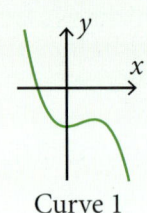

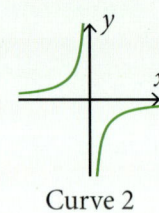

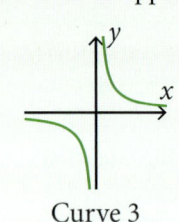

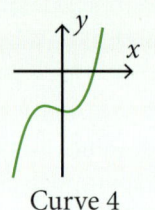

   Curve 1   Curve 2   Curve 3   Curve 4

   Equations
   A $y = \dfrac{-3}{x}$
   B $y = x^3 + x^2 - 1$
   C $y = -x^3 + x^2 - 1$
   D $y = \dfrac{2}{x}$

3. Match the curve sketches below with an appropriate equation.

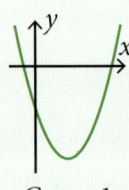

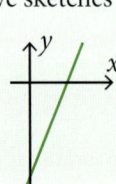

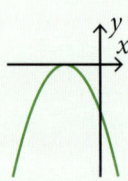

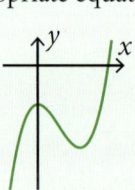

   Curve 1   Curve 2   Curve 3   Curve 4

   Equations
   A $y = -x^2 - 2x - 1$
   B $y = x^2 - 2x - 1$
   C $y = x^3 - 2x^2 - 1$
   D $y = 2x - 2$

30.6 Summary

In this chapter you have learnt how to:
- Recognise and draw a quadratic curve.
- Recognise and draw a cubic curve.
- Recognise and draw a reciprocal curve.
- Match a sketch of a graph with its equation.

Chapter 31
Real-Life Graphs

31.1 Introduction

Graphs are a common way to represent data in the real world. They can be used to show how things change over time, or to compare two different quantities. When we use an equation or a graph to represent a real-life situation, we say we are **modelling** that situation.

Key words
- **Real-life graph**: A graph that represents some real-life process.
- **Linear graph**: A straight-line graph.
- **Distance-time graph**: A graph on which a journey is represented. Time is plotted on the horizontal axis and distance travelled on the vertical axis. This is sometimes called a **travel graph**.
- **Gradient**: A measure of the steepness of a line.
- **Modelling**: Representing a real-life situation using an equation or a graph.

Before you start you should:
- Know how to plot a straight-line graph using a table of values.
- Know how to find the gradient of a straight-line graph and the y intercept, both from the equation of the line and from the graph.
- Understand that speed $= \dfrac{\text{distance}}{\text{time}}$. You should know how to use this formula to find any of the three quantities speed, distance or time, given values for the other two.

In this chapter you will learn how to:
- Use graphs to model real-world situations.
- Plot and take readings from distance-time graphs (travel graphs).
- Take readings from real-life graphs to solve problems.

Exercise 31A (Revision)

1. The equation of a straight line is given by $y = -2x + 1$
   (a) Write down the gradient and y-intercept of the line.
   (b) Copy and complete this table of values using the equation of the line.
   (c) Plot the line, using values of x from -2 to 3

x	-2	-1	0	1	2	3
y	5		1		-3	

31.2 Real-life Linear Graphs

In Chapter 11 you learnt how to plot a straight-line graph from its equation, for example $y = 2x + 1$
In this section, you will learn how to use linear graphs to model real-life situations and solve problems.
Here are some examples of how linear graphs can be used to model real-world situations:
- To convert from one currency to another.
- To show the relationship between the number of items bought and the total price.
- To calculate the distance travelled by a car using different amounts of petrol.

Example 1

A car can travel 40 miles on one gallon of petrol.
(a) Copy and complete the table below.

Gallons of petrol used	1	2	4	5	7.5	10
Miles travelled	40					400

(b) Draw coordinate axes, using values of 0 to 10 on the horizontal axis and from 0 to 400 in steps of 50 on the vertical axis. Draw a straight-line graph of the number of miles travelled on the vertical axis against the number of gallons of petrol used on the horizontal axis.
(c) Use your graph to estimate the number of miles travelled if 3 gallons of petrol are used.
(d) Use your graph to estimate the number of gallons of petrol used for a journey of 250 miles.
(e) A car is going on a long journey of 1000 miles. How could you use your graph to estimate the amount of petrol used?

(a)

Gallons of petrol used	1	2	4	5	7.5	10
Miles travelled	40	80	160	200	300	400

(b) The points are plotted on the graph on the right and the straight-line graph is drawn between the points.
(c) From the dashed construction lines on the graph, when 3 gallons of petrol are used, a journey of roughly 120 miles can be completed.
(d) From the grey construction lines on the graph, when 250 miles are travelled, the petrol consumption is about 6.2 gallons.
(e) There are many different methods that you could use to answer this question. For example, from the graph we can see that a journey of 100 miles uses about 2.5 gallons of petrol. Multiplying by 10, a journey of 1000 miles uses about 25 gallons of petrol.

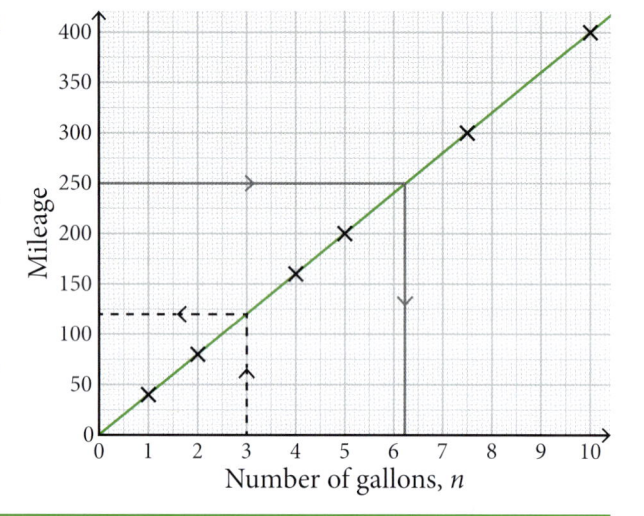

Note: This method of multiplying can only be used when the straight-line graph passes through the point (0, 0).

Exercise 31B

1. The conversion graph from pounds (£) to euro (€) is shown below.
   (a) How many euro is £10?
   (b) How many euro is £50?
   (c) How many pounds is €77?

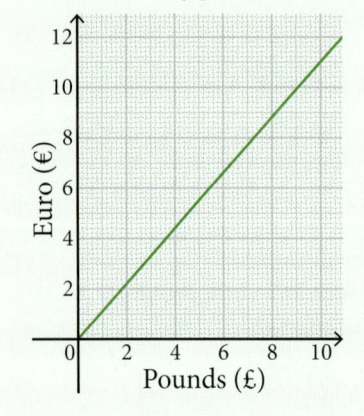

2. Siona can work out the cost of her monthly mobile phone bill using the graph shown below. It shows the monthly cost, £C plotted against the number of minutes, m of calls.
   (a) What is the cost of using 200 minutes on this phone?
   (b) If Siona pays £40, how many minutes has she used?

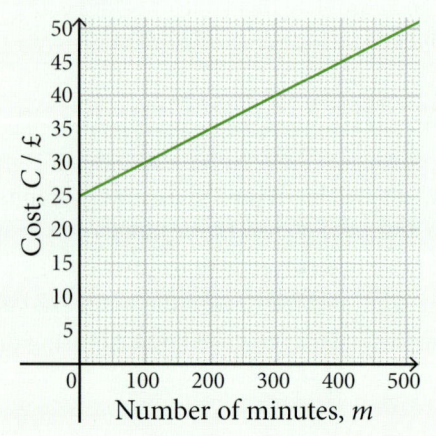

3. A lake is being drained. The graph on the right shows the depth, d metres, of water in the lake after t hours.
   (a) What is the initial depth of water?
   (b) What is the depth after 8 hours?
   (c) When will the depth be 25 metres?
   (d) After how many hours does the lake empty?

4. Paul plants a cherry tree. At the time of planting, the tree is 3 m high. The height, h (metres) of the tree increases by 0.5 m each year for the next 5 years.
   (a) Copy and complete the table of values below.

t (years)	0	1	2	3	4	5
h (m)	3					

   (b) Draw a straight line graph of the cherry tree's height over time. Use t values from 0 to 5 on the horizontal axis and h values from 0 to 6 on the vertical axis.
   (c) Use your graph to estimate the tree's height after 2½ years.
   (d) Do you think that a straight-line model for the tree's growth would continue over the entire lifetime of the tree?

5. The population of San Francisco between 1860 and 1930 can be modelled using a straight-line graph, with the population increasing by 85 000 every 10 years.
   (a) Copy and complete the table below.

Year	1860	1870	1880	1890	1900	1910	1920	1930
Population (thousands)	55	140						

   (b) Use the information from the table to draw a straight-line graph. Plot the year on the horizontal axis and the population on the vertical axis.
   (c) Use your graph to estimate:
      (i) The population of San Francisco in 1875 (ii) The year in which the population was 500 000

31.3 Distance-Time Graphs

A distance-time graph shows the distance of a person or an object from its starting point.

Reading distance-time graphs

You may be asked to read information from a distance-time graph.

Note: A **distance-time graph** is sometimes called a **travel graph**.

Example 2

Look at this graph showing the height of a hot air balloon.
(a) What is the maximum height the balloon reaches?
(b) What is shown by the lines BC and DE?
(c) For how long is the balloon at a height of 200 m?
(d) After it starts to descend, how long does the balloon take to land?
(e) At what two times is the balloon at a height of 300 m?

(a) The balloon reaches a maximum height of 600 m.
(b) The lines BC and DE represent times when the balloon remains at the same height.

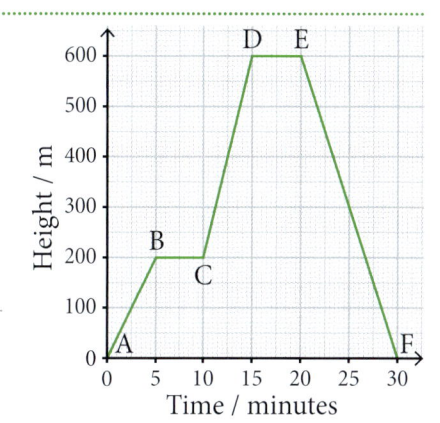

Note: Horizontal lines on the graph represent times when the balloon is not moving upwards or downwards.

(c) The balloon is at a height of 200 m for 5 minutes (between points B and C on the graph).
(d) The balloon takes 10 minutes to reach the ground.

Note: The line EF begins 20 minutes into the flight and finishes at 30 minutes.

(e) It is at a height of 300 m at roughly 11 minutes on the way up and at 25 minutes on the way down.

> **Note:** Draw a line horizontally from 300 m on the vertical axis and look for the two points where this line meets the graph.

Drawing a distance-time graph

You may be asked to construct a distance-time graph.

Example 3

Liz, Sarah and Thomas leave Lisburn and travel in the same direction.

- Liz sets off at 10:45 am and travels 40 km by electric car. She drives at a constant speed and it takes her 1 hour.
- Sarah leaves at 10:15 am. She walks 3 km at a steady pace in one hour. Then she catches a bus which travels another 15 km at a constant speed. The bus journey takes 1 hour.
- Thomas cycles, setting off at 10 am. He travels 20 km in 2 hours at a constant speed.

(a) Draw axes from 10 a.m. to 12:30 p.m. on the horizontal axis and from 0 km to 40 km on the vertical axis. Plot the travel graphs for all three people.
(b) At what time does Liz pass Sarah?
(c) At what time does Liz pass Thomas?
(d) Does Thomas pass Sarah? Explain your answer.
(e) At what speed is Thomas cycling?
(f) At what speed is Liz driving?

(a) The graph is shown on the right.
(b) Liz's green line and Sarah's black line cross at about 10:50 am. This is the time at which Liz passes Sarah.

> **Note:** When you read from a graph it is not always possible to get exact values. Here it is only possible to read an approximate time. There may be a range of answers that are acceptable.

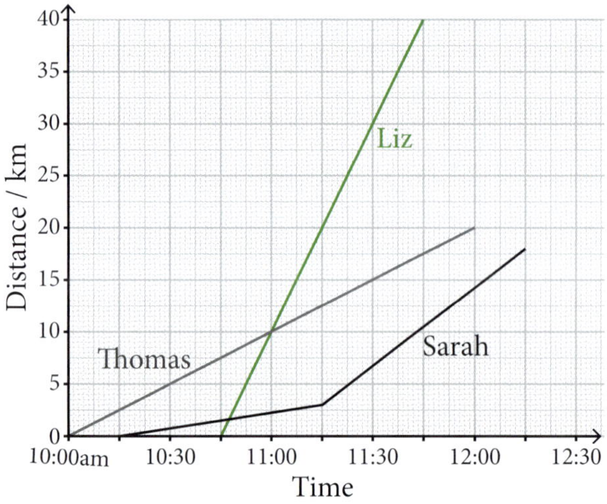

(c) Liz passes Thomas at about 11 a.m.
(d) Thomas' grey line and Sarah's black line do not meet, so Thomas does not pass Sarah. Thomas leaves Lisburn earlier than Sarah and stays in front of her the whole time.
(e) Thomas cycles 20 km in 2 hours. Speed = $\frac{\text{distance}}{\text{time}} = \frac{20}{2}$ = 10 km/h
(f) Liz drives 40 km in 1 hour. Speed = $\frac{\text{distance}}{\text{time}} = \frac{40}{1}$ = 40 km/h

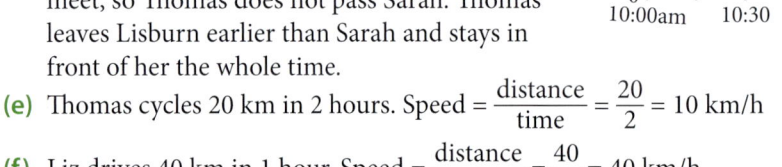

Exercise 31C

1. (a) For each of the travel graphs below, work out the speed in kilometres per hour.
   Hint: The speed is the gradient of a distance-time graph. It can be calculated using: Speed = $\frac{\text{distance}}{\text{time}}$

(i)   (ii) 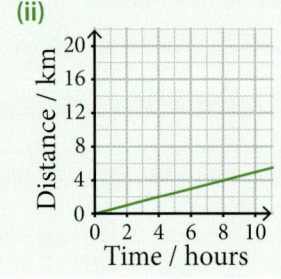   (iii) 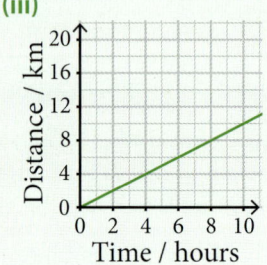   (iv)

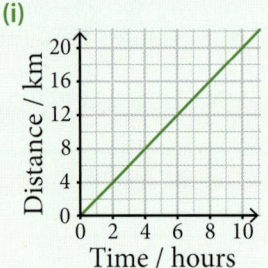

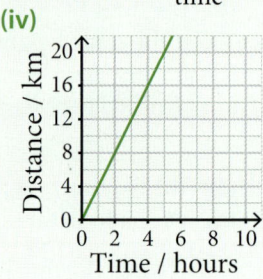

(b) In which of the travel graphs is the speed greatest?
(c) Copy and complete these sentences:
The steepest graph shows the object travelling at the _____ speed.
The graph with the _____ gradient shows the object travelling at the lowest speed.

2. Grace walks to her local shop to buy some milk and then returns home. A distance-time graph for her journey is shown on the right.
   (a) How far is the shop from Grace's home?
   (b) How long does Grace spend in the shop?
   (c) How long does it take Grace to walk home from the shop?

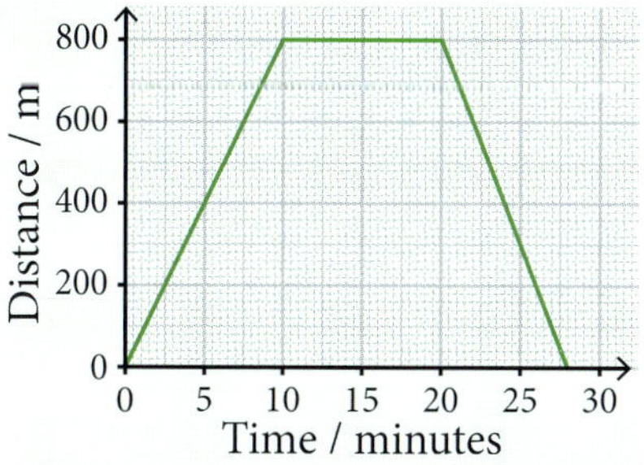

3. A chef tosses a pancake. The graph on the right shows the pancake's distance above the ground at different times, measured in seconds.
   (a) How high is the pancake after 0.1 seconds?
   (b) The chef tosses the pancake too high and it hits the ceiling. How many seconds does the pancake take to hit the ceiling after being tossed?
   (c) How high is the ceiling?
   (d) When the pancake hits the ceiling, it sticks there. How long is it stuck for?

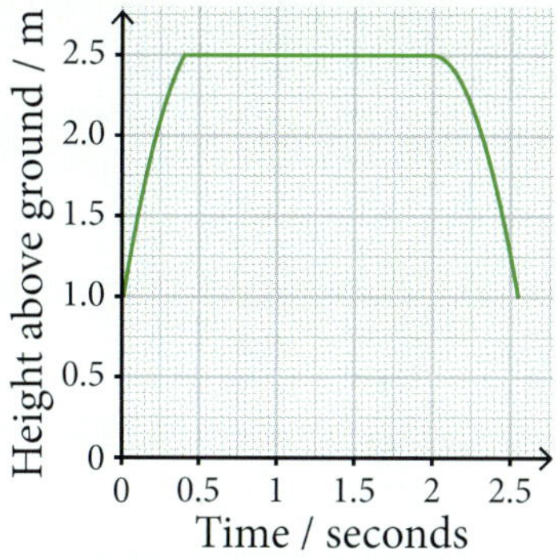

4. A bear is climbing a very tall tree to find some honey in a bees' nest at the top. The graph on the right shows the height of the bear above the ground throughout its climb.
   (a) How high is the tree?
   (b) How long does it take the bear to climb the tree?
   (c) How high is the bear after 6 minutes?
   (d) How long does the bear spend eating the honey at the top of the tree?
   (e) How long does it take the bear to climb down the tree?

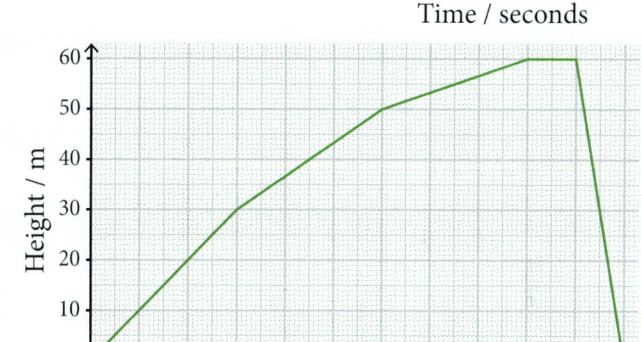

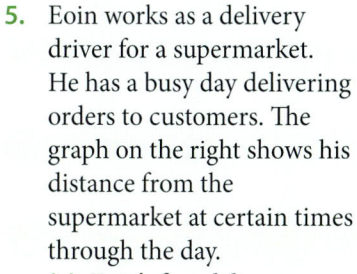

5. Eoin works as a delivery driver for a supermarket. He has a busy day delivering orders to customers. The graph on the right shows his distance from the supermarket at certain times through the day.
   (a) Eoin's first delivery was 12 km away from the supermarket. What time did he make this delivery?
   (b) How far was Eoin from the supermarket at 7:30 am?
   (c) Eoin made his third home delivery of the day at 7:45 am.
      (i) How far was this house from the supermarket?
      (ii) How long did he take to unload this delivery?

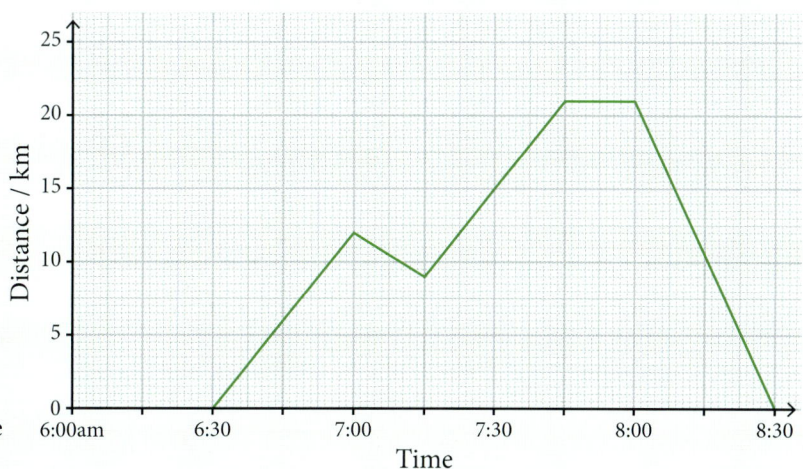

6. A cat jumps from the ground onto a bin. The graph on the right shows the cat's height above ground level plotted against the time after it jumped.
   (a) What is the cat's greatest height above the ground? Use the graph to make an approximate reading.
   (b) After roughly how many seconds does it reach this height?
   (c) What happens to the cat 1 second after it jumped?

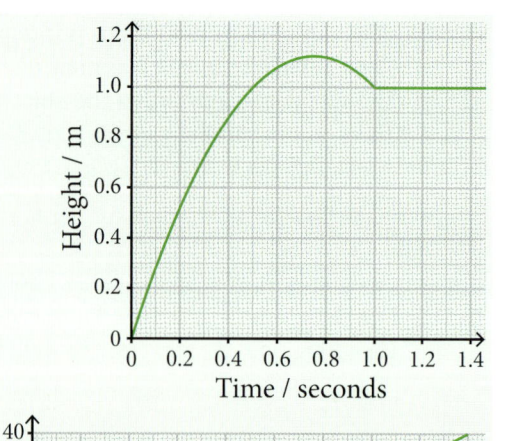

7. Una runs the Belfast Marathon. The graph on the right shows her distance from the start line at certain times throughout the race.
   (a) At what time does the marathon start?
   (b) How far has Una run by 11:00 am?
   (c) The marathon is 40 km long. At what time does Una finish the race?
   (d) Between which times does Una run fastest? Suggest why she might run faster between these times.
   (e) Una's friend Paula is running in the same race. Paula runs at a steady speed of 10 km/h. Copy the graph. Draw a line representing Paula's journey on the same diagram.
   (f) Between which times are Una and Paula running together?
   (g) When do they next meet?
   (h) How long does it take Paula to complete the race?
   (i) Find Una's average speed. Give your answer to 1 decimal place.

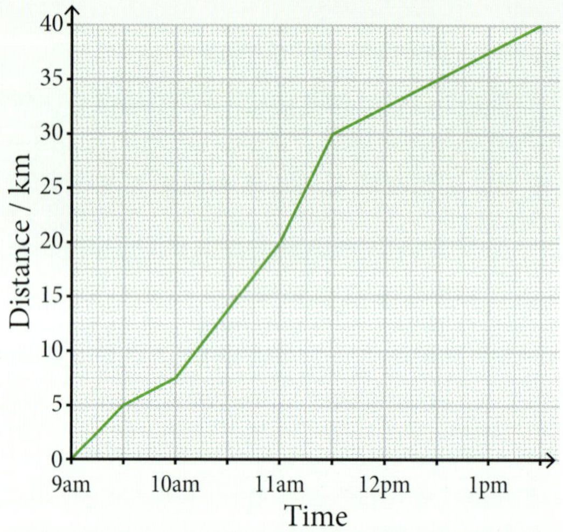

8. Cassie travels to work, heading south along the Ormeau Road in Belfast.
   - She leaves her house at 9:00 am.
   - She walks 0.4 km from her house to the bus stop in 5 minutes.
   - She then waits 3 minutes for the bus number eight.
   - The bus travels 2.4 km in 8 minutes to a bus stop near her workplace.
   - Cassie walks the last 0.3 km to her workplace in 5 minutes.
   (a) Copy and complete the distance-time graph on the right to represent Cassie's journey to work.
   (b) What is the average speed of bus number 8 in kilometres per hour?
   (c) At the time Cassie leaves her house a different bus, number nine, starts its journey **north** along the Ormeau Road, travelling at a constant speed of 12 km/h. How far does bus number nine travel in 5 minutes?
   (d) Bus number 9 starts 3 km away from Cassie's house. Copy and complete the table on the right for the distance of bus number nine from Cassie's house.

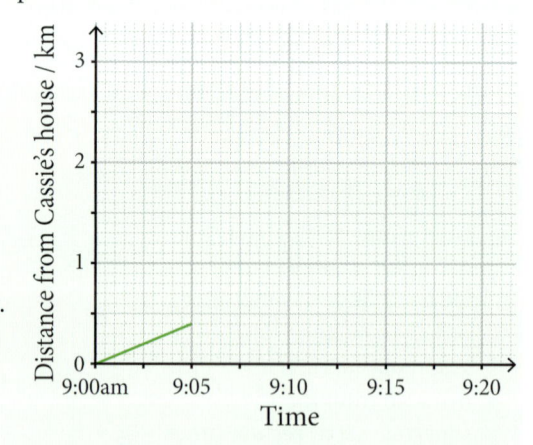

Time (mins)	0	5	10	15
Distance from Cassie's house (km)	3			

   (e) Add a straight line to your graph to show the journey of bus number nine.
   (f) Using your graph, find the time at which bus number eight and bus number nine pass each other.
   (g) How far is Cassie from her house at this time?

9. Simon walks to his local shop to buy some scones.
   - The shop is 1800 m from his home and it takes Simon 10 minutes to get there at a steady pace.
   - Simon then spends 12 minutes in the shop.

- He then sets off to return home and he covers the first 1000 m in 10 minutes.
- At that point, he sits down to eat one of the scones for 5 minutes.
- The final 800 m of his walk home takes him 5 minutes.

The first stage of Simon's journey is shown on the graph on the right. Copy the diagram and draw lines to show the rest of his journey.

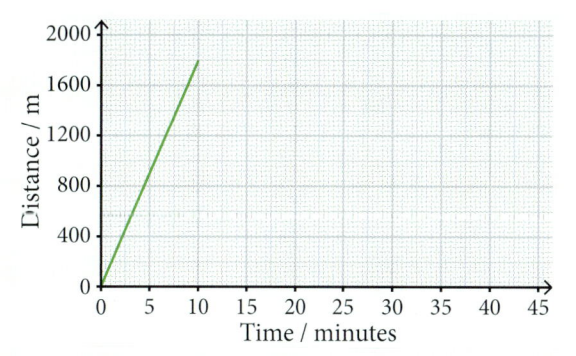

10. Reuben cycles to school every day. Some days he gets to school more quickly than others. The distance-time graph below shows his journeys each day this week.
    (a) On Friday Reuben took a different route, which had a slightly longer distance. Which line on the graph shows Friday's journey?
    (b) How far does Reuben **usually** cycle when he cycles to school?
    (c) The black (number 1) and grey (number 2) graphs have roughly the same shape. There is a steeper part of the graph, followed by a less steep part, followed by another steeper part. Suggest a reason for this shape.
    (d) Reuben arrived at school at the earliest time on Tuesday. Which line shows Tuesday's journey.
    (e) On Thursday Reuben stopped at the shop on the way to school. Which number line shows Thursday's journey? 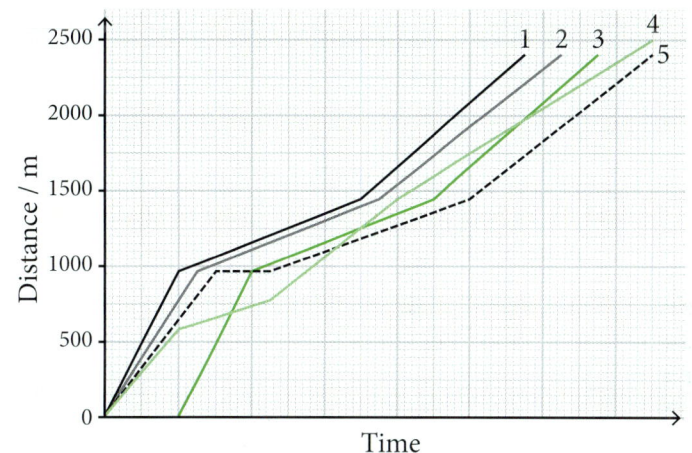
    (f) On Wednesday Reuben left his house a bit later than usual. Which line shows Wednesday's journey?

11. Katherine is 68 metres from her bus stop and sees the bus approaching. She runs towards the bus stop at a steady speed of 4 m/s for 15 seconds. She then takes 3 seconds to slow down and stop uniformly as she approaches the bus stop. Sketch a graph of Katherine's **distance** against time.

31.4 Other Real-life Graphs

Real-life graphs can be straight-line graphs, as discussed in Section 31.2. In Section 31.3, we also saw how the modelling of a real-world situation may give a curved graph, or a graph with curved sections.

Real-life graphs can also be any of the non-linear graphs discussed in Chapter 30.

Sometimes the equation of a real-life curve may not be known, but if you are given the equation, you may be asked to plot the curve. In all cases, you may be asked to take readings from a real-life graph.

Example 4

The height h metres of a carriage on a fairground ride is modelled using the quadratic function $h = -\frac{1}{100}t^2 + \frac{3}{5}t$ where t is the time in seconds from the start of the ride.

(a) Copy and complete the table of values below for the height of the carriage at the times shown.

t (s)	0	5	10	15	20	25	30	35	40	45	50	55	60
h (m)	0	2.75		6.75	8		9	8.75		6.75	5	2.75	

(b) Draw a graph of the height of the carriage from 0 to 60 seconds.
(c) Using your graph, find:
  (i) The maximum height of the carriage.
  (ii) The height of the carriage after 9 seconds.
  (iii) The two times when the carriage is 7 m above the ground.

(a) To calculate the height of the carriage when $t = 10$:
$h = -\frac{1}{100} \times 10^2 + \frac{3}{5} \times 10$
$h = 5$
Repeat for $t = 25, 40$ and 60

The completed table of values is shown below.

t (s)	0	5	10	15	20	25	30	35	40	45	50	55	60
h (m)	0	2.75	5	6.75	8	8.75	9	8.75	8	6.75	5	2.75	0

(b) The graph of the carriage's height against time is shown on the right.

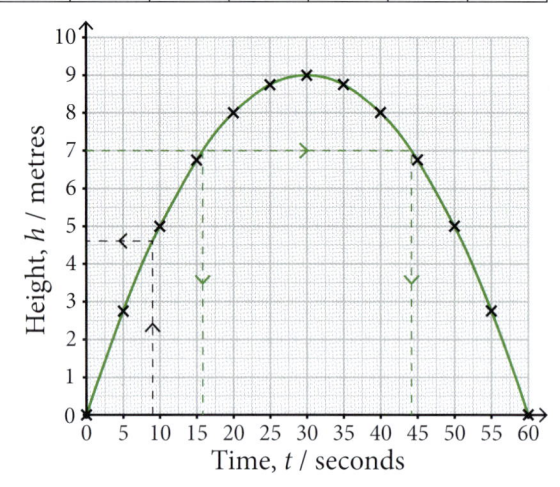

(c) From the graph we can see that:
   (i) The maximum height of the carriage is 9 m.
   (ii) The height of the carriage after 9 seconds is 4.6 m (shown by the black construction lines).
   (iii) The two times when the carriage is 7 m above the ground are 16 seconds and 44 seconds (shown by the green construction lines).

Exercise 31D

1. It rains every day for the first part of September. Carla has a water butt in her garden, which fills up with rainwater. The maximum height of water in the water butt is 80 cm. Carla measures the level of the water in the water butt at the end of each day. These measurements are shown in the graph on the right. 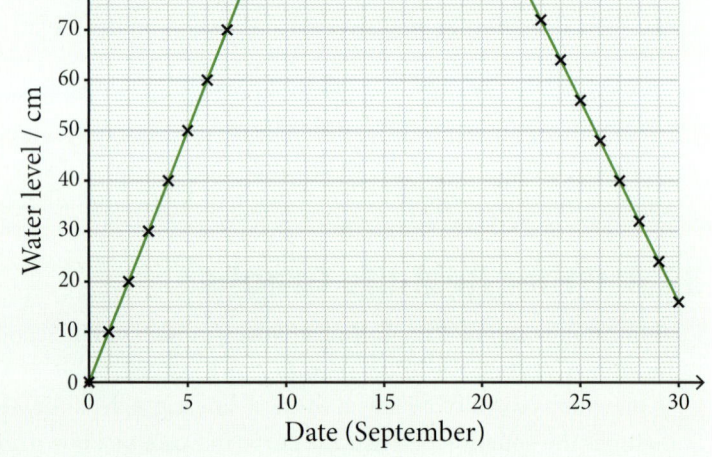
   (a) On which day did the water butt fill up?

   After a while, Carla begins to use the water for her garden and the water level starts to fall.
   (b) On which date did Carla begin using the water?
   (c) How much does the water level fall each day when Carla waters the garden?
   (d) Does Carla use up all the water by the end of September? If not, what is the water level at the end of the month?

2. An Olympic ski jump slope is in the shape of the graph shown on the right. 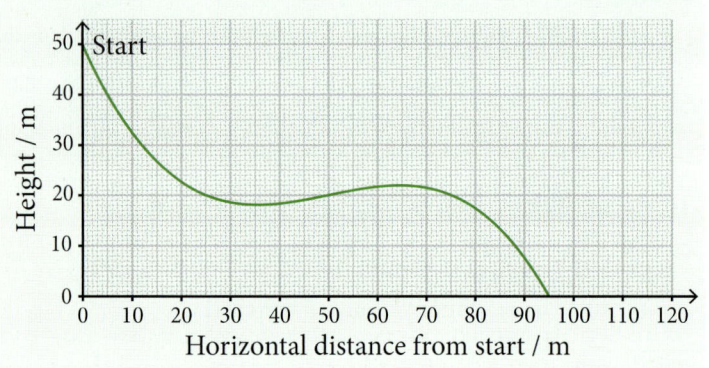
   (a) How high above ground are the skiers at the start of the slope?
   (b) At what horizontal distance is the slope 20 metres above ground level?
   Hint: There is more than one answer.

CHAPTER 31: REAL-LIFE GRAPHS

3. Mia cycles from her house to the nearest post box. Starting from rest she increases her speed steadily until she reaches a top speed of 10 m/s. This takes her 20 seconds. She then stays at this speed for 70 seconds, when she is close to the post box. She brakes and slows down steadily, coming to rest in 10 seconds.

   (a) Using the information above, copy and complete this table of values.

Time (seconds)	0	20	30	50	70	90	100
Speed (m/s)	0		10				

   (b) Draw axes with:
   - time on the x-axis running from 0 to 100 seconds; and
   - speed on the y-axis running from 0 to 10 m/s.

   Plot the points from your table on the graph. Join the points using straight lines. **Hint**: This graph should be made up of three straight lines sections.

   Use your graph to answer the following questions:
   (c) What was Mia's speed 60 seconds after leaving home?
   (d) At what two times after leaving home was Mia travelling at 5 m/s?

4. A farmer needs a rectangular field with an area of 360 m² for his sheep. There are many ways to do this. Some of them are shown in the table below.

   (a) Copy and complete the table.
   (b) Write down another width and length that would give an area of 360 m²

Width (m)	5	10	15		20		36	72
Length (m)	72		24	20		15		

   (c) Plot a graph of the height against the width, with the width on the x-axis.
   Use values from 0 to 80 metres on both axes. Use 1 cm for 10 metres in both directions.

   Use your graph to answer these questions:
   (d) How long is the field if the width is 30 metres?
   (e) Estimate the width of the field if its length is 25 metres.

   Note: The graph you have drawn is a reciprocal curve, discussed in Chapter 30, with the equation $y = \frac{360}{x}$

5. Sam throws a ball from an upstairs window of his house. The height of the ball can be calculated using this equation: $h = -\frac{1}{5}x^2 + 1.2x + 4$
   where h (metres) is the ball's height and x (metres) is its horizontal distance from the house.

   (a) Copy and complete this table of values for the ball's height.

Distance from house, x (m)	0	0.5	1	1.5	2	2.5	3	3.5	4	4.5	5	5.5	6	6.5	7	7.5	8
Height, h (m)	4		5	5.35		5.75		5.75	5.6	5.35	5		4	3.35		1.75	

   (b) Plot a graph of the ball's height against its distance from the house. Use x values from 0 to 9 metres on the horizontal axis and h values from 0 to 6 metres on the vertical axis.

   Using your graph, answer the following, giving your answers to 1 decimal place where appropriate.
   (c) How high above the ground is the window from which the ball was thrown?
   (d) What is the maximum height above ground the ball reaches?
   (e) Roughly how far from the house does the ball land?

31.5 Summary

In this chapter you have learnt that:
- Graphs can be used to model a variety of real-world relationships between two quantities.
- Real-life graphs can be linear (straight-line graphs) or some other shape.
- For any given situation, you may be asked to complete a table of values and then draw the graph.
- You can then use your graph to make an estimate of one quantity given the value of the other, or to find the maximum or minimum value of a quantity.
- Distance-time graphs are a common type of real-life graph. This type of graph shows a person or object's distance travelled over time.
- In some cases, the distances of two or more people or objects are shown on the same graph. In these cases, you can use your graph to find when the two people or objects are at the same distance.

Chapter 32
Graphical Solutions

32.1 Introduction

In Chapter 26, you learnt how to solve two linear simultaneous equations. In this chapter you will learn how to solve such equations graphically by finding the intersection point of two straight lines.

In Chapter 30, you learnt how to plot quadratic curves. In this chapter you will use these graphs to find approximate points of intersection of quadratic curves and straight lines of the form $y = mx + c$

Key words

Linear function: A function that involves a term in x, but no higher powers, for example $y = 3x + 1$

Linear equation: An equation that involves x and y (or other variables), but no terms in x^2 or y^2, or higher powers, for example $3x + 1 = 7$

Quadratic function: A function that involves an x^2 term, for example $y = x^2 + 2x + 1$

Quadratic equation: An equation that involves an x^2 term, for example $2x^2 + 5x - 1 = 0$

Before you start you should know how to:
- Solve two linear simultaneous equations.
- Plot a quadratic curve.

In this chapter you will learn how to:
- Solve two linear simultaneous equations graphically.
- Generate points and plot graphs of simple quadratic functions, and use these to find approximate solutions for points of intersection with lines of the form $y = mx + c$

Exercise 32A (Revision)

1. Solve this pair of simultaneous equations:
   $2x + 5y = -4$ $3x - 4y = 17$

2. A quadratic curve has the equation $y = 2x^2 + 5x - 1$
   (a) Copy and complete this table of values, using the equation of the curve.
   (b) Plot the graph of the curve.

x	−4	−3	−2	−1	0	1
y	11		−3		−1	

32.2 Solving Linear Simultaneous Equations Graphically

In Chapter 26, you learnt how to solve two linear simultaneous equations. In this section, you will learn how to solve such equations graphically by finding the intersection point of two straight lines.

Consider this pair of simultaneous equations:
$3x + y = 16$
$x - y = 4$

These are both **linear equations** as they involve x and y, but no higher powers, such as x^2

The graph of a linear equation is a straight line, sometimes called a linear graph. In Chapter 11, you learnt how to plot linear graphs.

To solve two linear simultaneous equations in x and y graphically, begin by plotting a graph for each equation. Then identify the intersection point. The x- and y-coordinates of the intersection point are the solutions to the simultaneous equations.

CHAPTER 32: GRAPHICAL SOLUTIONS

Example 1

(a) Copy and complete the table on the right for the equation $y = 2x + 2$

x	−3	−2	−1	0	1	2
$y = 2x + 2$	−4		0		4	

(b) Copy and complete the table on the right for the equation $y = 3x + 4$

x	−3	−2	−1	0	1	2
$y = 3x + 4$	−5		1		7	

(c) On the same graph, plot the straight lines $y = 2x + 2$ and $y = 3x + 4$
(d) Hence solve this pair of simultaneous equations:
$y = 2x + 2 \qquad y = 3x + 4$

(a) For the equation $y = 2x + 2$

When $x = -2$:   When $x = 0$:   When $x = 2$:
$y = 2(-2) + 2$ $y = 2(0) + 2$ $y = 2(2) + 2$
$y = -2$ $y = 2$ $y = 6$

x	−3	−2	−1	0	1	2
$y = 2x + 2$	−4	−2	0	2	4	6

(b) For the equation $y = 3x + 4$

When $x = -2$:   When $x = 0$:
$y = 3(-2) + 4$ $y = 3(0) + 4$
$y = -2$ $y = 4$

x	−3	−2	−1	0	1
$y = 3x + 4$	−5	−2	1	4	7

(c) The graph is plotted on the right.

(d) From the graph, we can see that the two lines intersect at the point (−2, −2). Therefore, the solution to this pair of simultaneous equations is:
$x = -2$
$y = -2$

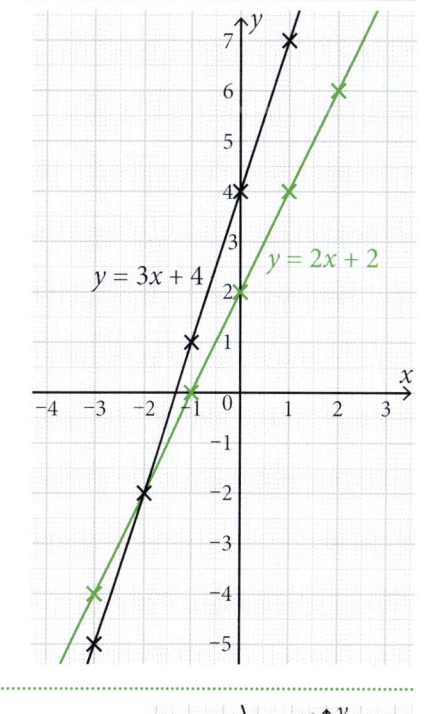

Example 2

Copy the graph on the right.

By adding a straight line to the graph, solve this pair of simultaneous equations:

$y = -3x + 2$
$y = x + 6$

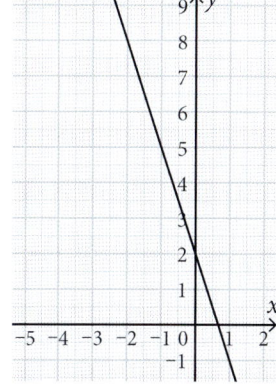

The line on the graph has a y-intercept of 2 and a gradient of −3, so its equation is:
$y = -3x + 2$

Since the first one of the simultaneous equations is already shown on the graph, we must draw a line representing the second equation and find the intersection point.

The second line $y = x + 6$ is shown in green. (This line could be plotted using a table of values, as in Example 1.)

The intersection point is (−1, 5), so the solutions to the simultaneous equations are:
$x = -1, y = 5$

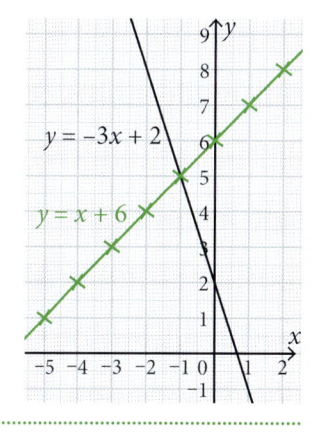

285

GCSE MATHEMATICS M3 AND M7

Exercise 32B

1. The graph on the right shows the lines given by various equations. Use the graph to write down the solution to each of these pairs of simultaneous equations.
   (a) $x = 1$ and $y = 1 - 2x$
   (b) $y = -\frac{1}{2}x + 1$ and $y = 1 - 2x$
   (c) $y = 3$ and $y = -\frac{1}{2}x + 1$
   (d) $y = 3$ and $y = 1 - 2x$

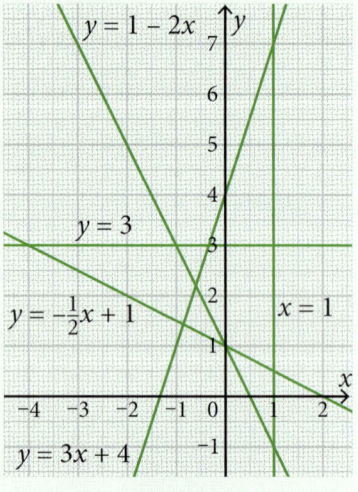

2. Look at the graph on the right.

   Using the graph, copy this sentence filling in the blank spaces:

   The solutions to the simultaneous equations

   $y = $ _____ and $y = $ _____ are:

   $x = $ _____ and $y = $ _____

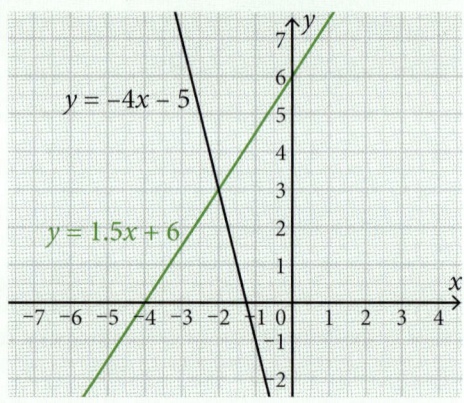

3. (a) Copy and complete the table on the right for the equation $y = 3 - 2x$

x	-2	-1	0	1	2	3	4
$y = 3 - 2x$	7		3		-1		-5

   (b) Copy and complete the table on the right for the equation $y = 2x - 1$

x	-2	-1	0	1	2	3	4
$y = 2x - 1$		-3		1		5	

   (c) On the same graph, plot the straight lines $y = 3 - 2x$ and $y = 2x - 1$
   (d) Hence solve this pair of simultaneous equations:
   $y = 3 - 2x \qquad y = 2x - 1$

4. (a) Copy and complete the table on the right for the equation $y = -x + 2$

x	-4	-3	-2	-1	0	1	2	3	4	5
$y = -x + 2$	6		4	3		1	0		-2	-3

   (b) Copy and complete the table on the right for the equation $y = 2x - 7$

x	1	2	3	4	5
$y = 2x - 7$	-5		-1		3

   (c) On the same graph, plot the straight lines $y = -x + 2$ and $y = 2x - 7$
   (d) Hence solve this pair of simultaneous equations:
   $y = -x + 2 \qquad y = 2x - 7$

5. Using a graphical method, solve this pair of simultaneous equations:
   $y = x + 2 \qquad y = 3x$

6. By drawing a table of values and a suitable graph, solve each pair of simultaneous equations for x and y.
   (a) $y = -x; y = x - 4$
   (b) $y = x - 2; y = -x - 4$
   (c) $y = 2x - 3; y = 9 - 2x$
   (d) $y = 2x + 5; y = -2x + 1$
   (e) $y = \frac{1}{2}x + 2; y = 2x - 1$
   (f) $y = 2x + 1; y = -\frac{1}{2}x + 6$
   (g) $y = -\frac{1}{2}x; y = \frac{1}{2}x + 2$
   (h) $y = 2x - 7; y = -\frac{1}{2}x + 3$
   (i) $y = \frac{1}{2}x + 5; y = -1.5x + 1$
   (j) $y = 1.5x - 4; y = -3.5x - 4$

7. Copy the graph on the right.

   By adding a second straight line to the graph, solve the following pair of simultaneous equations:

   $y = -x + 4$

   $y = -2x + 2$

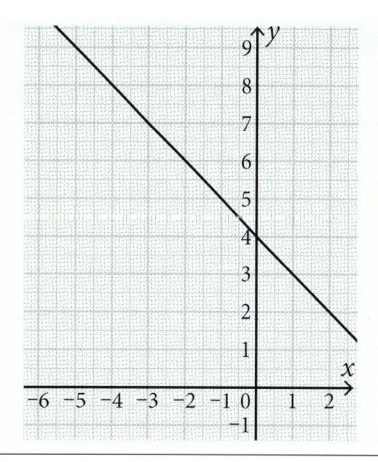

32.3 Intersection of Quadratic Curves and Straight Lines

In Chapter 30, you learnt how to plot the graph of a quadratic curve. In this section you will use such a graph to find the points of intersection of a quadratic curve and a straight line.

A straight line in this section may be:
- A horizontal line with the equation $y = c$, for example $y = 5$
- A vertical line with the equation $x = c$, for example $x = 5$
- A line of the form $y = mx + c$, for example $y = 2x + 5$

The coordinates of the intersection point may be integer values. If they are not integers, you will usually estimate both coordinates to one decimal place.

Example 3

A curve has the equation $y = x^2 + 2$

(a) Copy and complete the table of values on the right using the equation of the curve.

x	−3	−2	−1	0	1	2	3
y	11			3	2		6	11

(b) Plot the curve using the values in the table.
(c) Find approximate intersection points of the curve with the line $y = 4$
(d) Find the exact intersection points of the curve and the line $y = x + 4$

(a) When $x = -2$, $y = (-2)^2 + 2 = 6$
When $x = 1$, $y = 1^2 + 2 = 3$
The completed table of values is as follows:

x	−3	−2	−1	0	1	2	3
y	11	6	3	2	3	6	11

(b) The curve $y = x^2 + 2$ is plotted on the graph on the right.

The line $y = 4$ is also shown.
This is a horizontal line.

The line $y = x + 4$ is also shown.
This can be plotted using a table of values.

(c) The approximate coordinates of the points of intersection are (−1.4, 0) and (1.4, 0).

(d) The points of intersection are (−1, 3) and (2, 6).

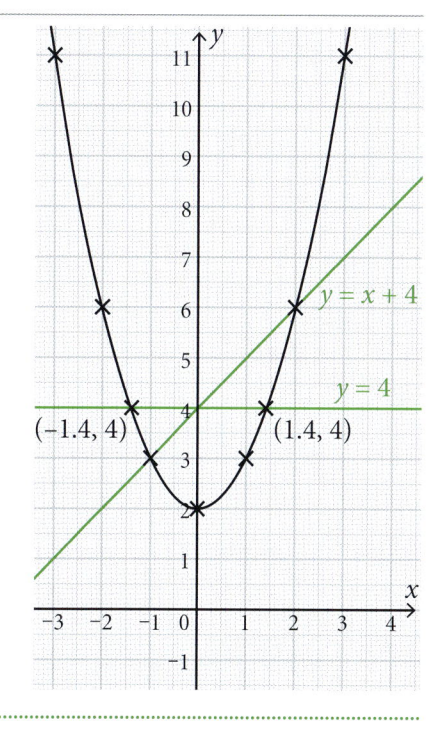

GCSE MATHEMATICS M3 AND M7

Exercise 32C

Throughout this exercise, if you are asked to estimate coordinates, give answers to one decimal place.

1. The curve shown on the right has the equation $y = x^2$
   (a) Copy the graph carefully and add the line $y = 5$
   (b) Find the coordinates of the two points of intersection of the curve and the straight line.

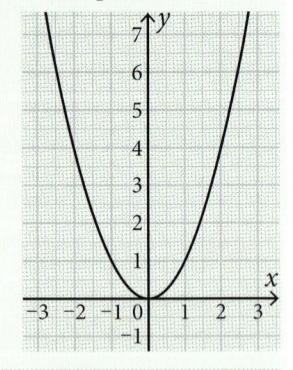

2. A curve has the equation $y = x^2 - 3x - 1$
   (a) Using the equation of the curve, find the missing y-value in the table below.

x	-2	-1	0	1	2	3	4	5
y	9	3	-1	-3	-3		3	9

   (b) Copy the coordinate grid on the right, using x-values from -2 to 5 and y-values from -3 to 9

   Plot the points in the table and join them with a smooth curve.

   (c) Use your graph to estimate the values of x for which $y = 2$

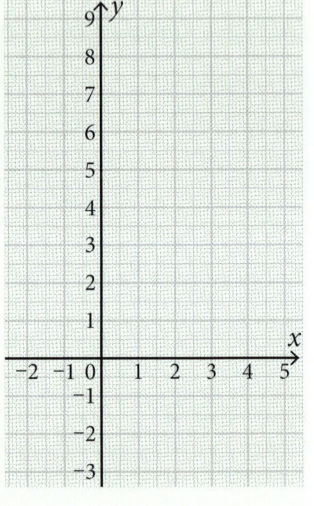

3. A curve has the equation $y = -x^2 + 2x$
   (a) Copy and complete the table of values, on the right using the equation of the curve.

x	-2	-1	0	1	2	3	4
y	-8		0	1	0		-8

   (b) Plot the curve using values of x from -2 to 4
   (c) Add the straight line $x = 1.5$ to your graph. Write down the coordinates of the single point of intersection of the straight line and the curve.

4. The curve shown on the right has the equation $y = x^2 + 4x - 7$
   By drawing appropriate straight lines on a copy of this graph, write down the coordinates of any points of intersection of this curve with:
   (a) The straight line $y = -2$
   (b) The straight line $y = -3$
   (c) The straight line $x = -3$

5. A curve has the equation $y = x^2 - 2$
   (a) Copy and complete the table of values below using the equation of the curve.

x	-3	-2	-1	0	1	2	3
y	7		-1	-2		2	7

   (b) Plot the curve using the values in the table.
   (c) Find approximate intersection points of the curve with the line $y = 3$
   (d) Find the exact intersection points of the curve and the line $y = 2x + 1$

6. A quadratic function is defined by the equation $y = x^2 + 4x + 1$
   (a) Using a table of values with values of x from -5 to 1, plot the curve.
   (b) By adding the appropriate straight lines to your diagram, find where the curve $y = x^2 + 4x + 1$ intersects with:
       (i) The straight line $y = -3$
       (ii) The straight line $y = x + 5$

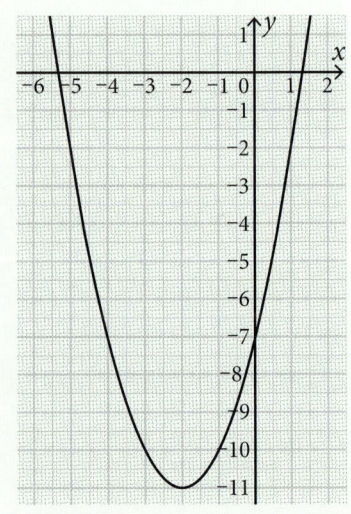

CHAPTER 32: GRAPHICAL SOLUTIONS

7. A quadratic function is defined by the equation $y = 2x^2 - 4x + 1$
   (a) Using a table of values with values of x from -1 to 3, plot the curve.
   (b) By adding appropriate straight lines to your diagram, find where the curve $y = 2x^2 - 4x + 1$ intersects with:
      (i) The straight line $y = 2$
      (ii) The straight line $y = -2x + 5$

8. (a) Copy and complete the table of values on the right for the quadratic curve $y = -x^2 + x + 2$

x	-2	-1	0	1	2	3
y	-4	0		2	0	

   (b) Using your table of values, plot the curve.
   (c) Add the straight lines $y = 1$ and $y = \frac{1}{2}x - 2$ to your graph.
   (d) Find all points of intersection of the curve with the straight line $y = 1$
   (e) Find all points of intersection of the curve with the straight line $y = \frac{1}{2}x - 2$

9. The curve shown in the diagram on the right has the equation $y = -\frac{1}{2}x^2 + 3x + 1$
   Copy the graph carefully.
   (a) Add the straight line $x = 5$ to your diagram.
   (b) Add the straight line $y = 2x - 1$ to your diagram.
   (c) Find all the points of intersection of the curve with the two straight lines.

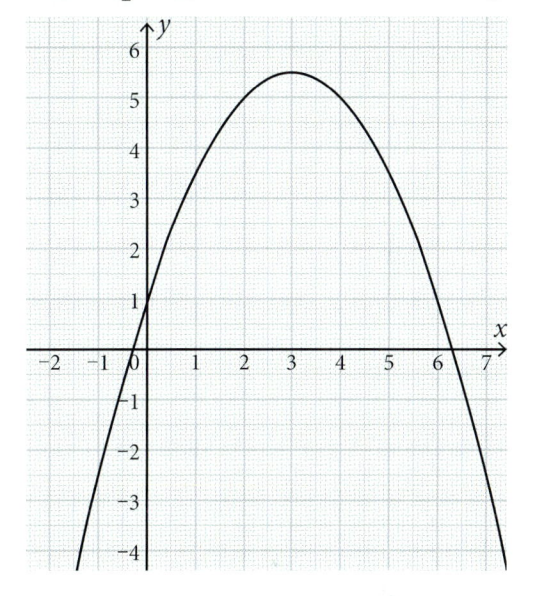

10. The curve C has the equation $y = -\frac{1}{2}x^2 + 2x + 5$
    (a) Copy and complete the table of values on the right using the equation of the curve.

x	-2	-1	0	1	2	3	4	5	6
y		2.5	5	6.5		6.5	5		-1

   (b) Plot the curve using the table of values.
   (c) The straight line L_1 has the equation $y = x + 1$
       Add this straight line to your graph.
   (d) The straight line L_2 has the equation $y = 2x - 1$
       Add this straight line to your graph.
   (e) Using your graph, find the two intersection points of the curve C and the line L_1
   (f) There are two intersection points between the curve C and the line L_2
       One of these points has a positive x-coordinate. Using your graph, find the approximate coordinates of this intersection point.
   (g) **Using your graph**, find the solution to the following pair of simultaneous equations:
       $x - y = -1$ $2x - y = 1$

32.4 Summary

In this chapter you learnt:
- How to solve simultaneous equations graphically by finding the intersection points of two straight lines.
- That the x- and y-coordinates of the intersection point are the solutions to the simultaneous equations.
- How to find approximate points of intersection of quadratic curves and straight lines. These lines may take the form $y = c$, $x = c$ or $y = mx + c$

Chapter 33
Proportion and Variation

33.1 Introduction

Sometimes one variable increases at the same rate as another. We say these two variables are **proportional** to each other. For example, if lemons are all priced at 30p, then the number of lemons you buy is proportional to the price. If the number of lemons doubles, so does the price; if the number of lemons is multiplied by 10, the price will also be 10 times larger, and so on.

In this chapter you will learn how to set up equations relating to variables that are in proportion. You will also solve problems involving proportion, including graphical and algebraic representations.

> **Note:** For the M7 module, you are only required to know about **direct proportion**, where an increase in one variable takes place with an increase in the second variable. You are not required to know about inverse proportion, where an increase in one variable takes place with a decrease in the second variable.

> **Note:** You will also come across questions involving the words 'variation' or 'varies as …' This terminology is an alternative to 'is proportional to …'

Key words
- **Proportion**: Two variables are said to be in proportion if they increase at the same rate, for example if one doubles the other one also doubles.
- **Variation**: An equivalent term to proportion. For example, we could say y is directly proportional to x or y varies directly as x.
- **Constant of proportionality**: The value that links the two variables. In the example involving the price of lemons, the constant of proportionality is 30, since the cost in pence is always 30 times the number of lemons.

Notation
- The symbol for proportionality is $\propto$

Before you start you should know how to:
- Solve problems involving ratio.
- Solve linear equations.
- Change the subject of a formula.

In this chapter you will learn how to:
- Find and use formulas to represent direct proportion.
- Find and use formulas to represent proportion between one variable and a power or a root of a second variable.
- Solve worded problems involving direct proportion.
- Recognise and use graphs that represent direct proportion between two variables.

Exercise 33A (Revision)

1. A farm has an area of 35 acres. It is to be divided into two smaller farms in the ratio 3:2 Find the area of each of these smaller farms.

2. Solve each of these linear equations.
   (a) $3x - 8 = 13$ (b) $2w + 14 = 8$ (c) $23 - 4y = 7$ (d) $6 + 4z = 28$

3. Make a the subject of each of these formulae.
   (a) $2a + 1 = b$ (b) $x^2 - a = y^2$ (c) $ba + 6 - c = da$

CHAPTER 33: PROPORTION AND VARIATION

33.2 Direct Proportion

Given that two variables are in direct proportion, you can find one of the variables given the value of the other. The steps are as follows:

1. Write down a mathematical statement using the proportional symbol $\propto$
2. Convert this statement to an equation using a constant of proportionality. For this constant we often use the letter k.
3. Use information given in the question to find the value of k.
4. Rewrite your equation, substituting in the value of k.
5. Use the equation to find one of the variables, given the other.

Example 1

In a greengrocer's shop, each apple costs k pence. Let n be the number of apples bought and let C be the total cost in pence. Given that 7 apples costs £2.45, find:
(a) The total cost of 11 apples.
(b) The number of apples purchased if the total cost is £1.40

Since each apple costs the same amount, the total cost of your apples is **proportional** to the number you buy. The information that 7 apples cost £2.45 allows you to find the **constant of proportionality**.

(a) Step 1: Write down a statement using the proportional symbol. In this case the cost C is proportional to the number of apples.

$$C \propto n$$

Step 2: Write this as an equation involving a constant of proportionality:

$$C = kn$$

Step 3: Using the information in the question, calculate k:
When $n = 7$, $C = 245$ so:

$$245 = k \times 7$$
$$k = \frac{245}{7} = 35$$

Step 4: Rewrite the equation using our value of k:

$$C = 35n$$

Step 5: Use the equation to find C when $n = 11$:

$$C = 35 \times 11$$
$$= 385 \text{ pence or £3.85}$$

(b) Use the equation $C = 35n$ again to find n when $C = 140$:

$$140 = 35n$$
$$n = \frac{140}{35}$$
$$n = 4$$

Four apples were purchased.

> **Note:** If variables y and x are in direct proportion, then $\frac{y}{x}$ is a constant value.
> In the example above, C and n are in direct proportion and $\frac{C}{n} = k$, which is 35 in this case.

Example 2

y varies directly as x. Given that $y = 6$ when $x = 4$, find:
(a) y when $x = 12$
(b) x when $y = 3$

The phrase 'y varies directly as x' means y is proportional to x. The other piece of information given allows you to find the constant of proportionality.

(a) Step 1: Write down a statement using the proportional symbol.

$$y \propto x$$

Step 2: Write this as an equation involving a constant of proportionality:

$$y = kx$$

GCSE MATHEMATICS M3 AND M7

Step 3: Using the information in the question, calculate k:
when $y = 6$, $x = 4$. So:

$6 = k \times 4$

$k = \dfrac{6}{4} = 1.5$

Step 4: Rewrite the equation using our value of k: $\quad y = 1.5x$

Step 5: Use the equation to find y when $x = 12$:

$y = 1.5 \times 12$
$y = 18$

(b) Use the equation $y = 1.5x$ again to find x when $y = 3$:

$3 = 1.5x$

$x = \dfrac{3}{1.5}$

$x = 2$

Example 3

Year 11 are going on a school trip. For a group of 18 students, the total cost would be £3150
What is the total cost of the trip if 23 students go? Assume that the cost per student is a fixed amount.

Since the cost per student is a fixed amount, the total cost is
proportional to the number of students going on the trip.
So, if n is the number of students and C is the total cost, then: $\quad C \propto n$

This can be rewritten as: $\quad C = kn$

When $n = 18$, $C = 3150$ so: $\quad 3150 = k \times 18$

$k = \dfrac{3150}{18} = 175$

Now that we have found the value of k, rewrite the formula: $\quad C = 175n$

When $n = 23$: $\quad C = 175 \times 23$

$C = £4025$

Exercise 33B

1. Each tomato in a supermarket has a fixed priced. Twenty two tomatoes cost £2.86
In the following, n is the number of tomatoes being purchased and C is the total cost in pence. Copy and complete the blank spaces in each part below to find the cost of 15 tomatoes.
   (a) Find the constant of proportionality:

   $C \propto n$
   $C = kn$
   $286 = k \times 22$
   $k = \dfrac{286}{22} = \underline{}$

   (b) Now that we have the value of k, the formula for the total cost can be written: $\quad C = \underline{} n$

   (c) Use the formula to find the cost C when $n = 15$: $\quad C = \underline{} \times \underline{}$

   $C = \underline{}$ pence
   or $C = £\underline{}$

2. It takes 6 hours for Ian to drive 252 miles. Assuming that Ian drives at a constant speed, copy and complete the following.
   (a) If Ian drives at a constant speed, then the distance d miles travelled is proportional to the time t taken.

   $d \propto t$
   $d = kt$
   $\underline{} = k \times \underline{}$
   $k = \underline{}$

   (b) Now that we have the value of k, the formula for the distance travelled can be written: $\quad d = \underline{} t$

   (c) Use the formula to find the time it would take for Ian to drive 189 miles:

   $189 = \underline{} \times t$
   $t = \underline{}$

CHAPTER 33: PROPORTION AND VARIATION

3. The variables y and x are in direct proportion.
   (a) Write down a formula for y in terms of x and a constant of proportionality k.
   (b) When $x = 10$, $y = 130$
      Use this information to find the value of k.
   (c) Using your answer to part (a) and your value of k, find y when $x = 2$
   (d) Find x when $y = 104$

4. p varies directly as q. When $p = 119$, $q = 17$
   (a) Find a formula for p in terms of q. (b) Find p when $q = 5$. (c) Find q when $p = 329$

5. Given that m and n are in direct proportion, copy and complete the table on the right.

m	1	2	3	4	5	10
n			16.5	22		

6. A cruise liner with a mass of 100 000 tons has an engine failure. It takes 5 tug boats to pull the cruise liner into harbour. Assuming that the number of tug boats needed is proportional to the mass of the ship, calculate the number of tug boats needed if a ship of mass 40 000 tons experiences engine failure.

7. Nine coaches can carry 1215 fans to a football match. Assuming all of the coaches are the same size:
   (a) Form a formula for the number of fans, F in terms of the number of coaches, C.
   (b) Use your formula to find the total number of fans that could travel to the match on 17 coaches.
   (c) How many coaches would be needed to transport 3000 fans? Hint: your answer should be a whole number, since a fraction of a coach is not possible.

8. In a widget factory, there are 8 robots, each making widgets. In one day, the 8 robots can make 512 widgets between them.
   (a) Assuming that all of the robots work at the same rate, find a formula for the number of widgets, N in terms of the number of working robots, R.
   The factory owner increases the number of robots to 10 to increase widget production.
   (b) How many widgets can the factory produce per day using all ten robots?
   (c) On Friday 13th December, 8 of the 10 robots break down. How many widgets can be produced per day using the remaining robots?
   (d) The factory owner wants to increase output again to 960 widgets per day. How many extra robots would he need in this second expansion of the factory?

9. Geostationary satellites orbit Earth at a fixed distance. The gravitational force, F newtons acting on a geostationary satellite varies directly as its mass, m kilograms.
   (a) Write down a formula linking F, m and a constant of proportionality k.
   (b) For a geostationary satellite of mass 1000 kg, the gravitational force is 981 newtons.
      Find the value of k.
   (c) Using your formula from part (a) and your value of k from part (b):
      (i) Calculate the gravitational force experienced by a geostationary satellite with a mass of 2500 kg.
         Give your answer in newtons to 3 significant figures.
      (ii) Calculate the mass of a geostationary satellite experiencing a gravitational force of 2000 newtons.
         Give your answer in kilograms to 3 significant figures.

33.3 Proportion Involving a Power or Root of a Variable

You will be asked questions in which one variable is proportional to a power of the second, for example $y \propto x^2$
Another possibility is that one variable is proportional to a root of the second, for example $y \propto \sqrt{x}$

Example 4

y is proportional to the cube of x.
(a) Write down an equation linking y, x and a constant of proportionality, k.
(b) When $y = 54$, $x = 3$ Find the value of k.
(c) Find y when $x = 2$
(d) Find x when $y = 2$

(a) y is proportional to the cube of x can be written: $y \propto x^3$

   or: $y = kx^3$

293

(b) When $y = 54$, $x = 3$, so:
$$54 = k \times 3^3$$
$$54 = k \times 27$$
$$k = \frac{54}{27} = 2$$

(c) $k = 2$, so the formula can be written as: $y = 2x^3$
When $x = 2$:
$$y = 2 \times 2^3$$
$$y = 16$$

(d) The formula linking y and x is $y = 2x^3$ When $y = 2$:
$$2 = 2x^3$$
$$x^3 = \frac{2}{2} = 1$$
$$x = \sqrt[3]{1} = 1$$

Note: If y is directly proportional to x^3, then $\frac{y}{x^3}$ is a constant value.
In the example above, $\frac{y}{x^3}$ is always equal to 2

Example 5

The power P (measured in watts) varies as the square of the current I (measured in amps).
When $I = 4$, $P = 2000$
Find P when $I = 15$

P varies as the square of I, so: $P \propto I^2$
or: $P = kI^2$
When $I = 4$, $P = 2000$ so:
$$2000 = k \times 4^2$$
$$2000 = 16k$$
$$k = \frac{2000}{16} = 125$$
So: $P = 125I^2$
When $I = 15$:
$$P = 125 \times 15^2$$
$$P = 28\,125 \text{ watts (or 28.125 kilowatts)}$$

Note: If P is directly proportional to I^2, then $\frac{P}{I^2}$ is a constant value.
In the example above, $\frac{P}{I^2}$ is always equal to 125

Example 6

The distance, d (miles) that you can see from a tower of height, h (feet) is proportional to the square root of the height. You can see 4.88 miles from a tower that is 16 feet high.
(a) Form an equation linking d and h.
(b) How far can you see from a tower 50 feet high?
(c) How high a tower would be needed to see at least 10 miles?

(a) Since d is proportional to the square root of the height, we can write: $d \propto \sqrt{h}$
or: $d = k\sqrt{h}$
When $d = 4.88$, $h = 16$, so:
$$4.88 = k\sqrt{16}$$
$$4.88 = k \times 4$$
$$k = \frac{4.88}{4} = 1.22$$
So, the formula linking distance and height is: $d = 1.22\sqrt{h}$

(b) When $h = 50$:
$$d = 1.22\sqrt{50}$$
$$d = 8.63 \text{ miles (3 s.f.)}$$

(c) When $d = 10$:

$$10 = 1.22\sqrt{h}$$
$$\sqrt{h} = \frac{10}{1.22}$$
$$h = \left(\frac{10}{1.22}\right)^2$$
$$h = 67.2 \text{ feet (3 s.f.)}$$

Exercise 33C

1. s is directly proportional to the square of v. When $v = 10$, $s = 500$
   (a) Express s in terms of v.
   (b) Find s when $v = 3$
   (c) Find v when $s = 4500$

2. T varies as the square of r. When $r = 0.15$, $T = 21.6$
   (a) Express T in terms of r.
   (b) Find the value of T when $r = 0.2$
   (c) Find a value of r for which $T = 15$

3. J is directly proportional to the cube of L. When $L = 5$, $J = 1000$
   (a) Find a formula linking J and L.
   (b) Find J when $L = 2$
   (c) Find L when $J = 27$

4. In a child's set of building blocks, the blocks are all cubes of different sizes. Each block is numbered. Three of the blocks are shown in the diagram. The volume, V of each cube is directly proportional to the cube of the number, n on it.
   (a) Write down a formula involving V, n and a constant of proportionality k.
   (b) Given that block number 4 has a volume of 96 cm³ find the value of k.
   (c) Find the volume of the block numbered 3
   (d) What number block has a volume of 324 cm³?

5. When a body is moving through the air, the air resistance, R newtons is proportional to the square of the velocity, v m/s. At a velocity of 20 m/s, the air resistance is 15 newtons.
   (a) Find a formula for R in terms of v.
   (b) Find the value of R when $v = 60$ m/s
   (c) Find the velocity, v of the object when it is experiencing an air resistance of 2.4 newtons.

6. The height, h metres of a weather balloon varies as the square root of its surface area, A square metres.
   (a) When the weather balloon's surface area is 36 m² its height is 18 m. Find a formula linking h and A.
   (b) What is the weather balloon's height when its surface area is 64 m²?
   (c) Find the balloon's surface area if it has a height of 300 m

7. The radius, r (cm) of an object varies as the cube root of its volume, V (cm³).
   (a) Write down a formula for r in terms of V and k, where k is a constant of proportionality.
   (b) Given that when $V = 1000$, $r = 40$ find the value of k.
   (c) Find r when $V = 614.125$ cm³

8. A marble is dropped into a large jar of honey. The speed of the marble (in centimetres per second) is proportional to the square root of its distance (in centimetres) below the surface of the honey. When the marble is 4 cm below the surface, the marble is moving at 2 cm/s.
   (a) Find a formula linking the marble's speed, s and its distance below the surface, d.
   (b) Find the speed of the marble when it is 6 cm below the surface.
       Give your answer in cm/s to 2 decimal places.
   (c) Find how far the marble has moved below the surface when it has a speed of 3 cm/s.
   (d) The depth of honey in the jar is 12 cm. Find the speed of the marble when it reaches the bottom of the jar. Give your answer in cm/s to 2 decimal places.

9. The number of dead birds that Fluffy the cat brings in from the garden each week depends on the temperature. Rob, Fluffy's owner, has worked out that the number of birds, N is proportional to the square of the temperature, T in °C. Given that Fluffy brings home 18 birds a week when the temperature is 6°C:
   (a) Find a formula for N in terms of the temperature T.
   (b) Copy and complete the table on the right.

T (°C)	0	2	4	6	8	10
N				18		

GCSE MATHEMATICS M3 AND M7

10. A ball is dropped from a balcony in an apartment block. The distance, d (metres) that the ball has fallen is directly proportional to the square of the time, t (seconds) since it was dropped. The balcony is 19.6 m above the ground and the ball takes 2 seconds to hit the ground.
    (a) Find a formula for d in terms of t.
    (b) What distance has the ball fallen after 1.5 seconds?
    (c) How long has the ball been falling when it has fallen 2.5 metres?

33.4 Harder Problems

Given some values for two variables, from an experiment for example, you may be asked to determine a formula linking the two variables.

Example 7

A formula links x and y. When $x = 5$, $y = 250$ and when $x = 2$, $y = 16$
(a) Which of these formulae is correct?
   $y = kx$ $\qquad$ $y = kx^2$ $\qquad$ $y = kx^3$
(b) Find the value of k.

(a) Firstly, check $y = kx$:
   If y is proportional to x, then $\frac{y}{x}$ is a constant value, k.
   When $x = 5$, $y = 250$ and $\frac{y}{x} = \frac{250}{5} = 50$
   When $x = 2$, $y = 16$ and $\frac{y}{x} = \frac{16}{2} = 8$
   Since $\frac{y}{x}$ is not a constant value, y is not proportional to x

   Next, check $y = kx^2$:
   If y is proportional to x^2, then $\frac{y}{x^2}$ is a constant value, k.
   When $x = 5$, $y = 250$ and $\frac{y}{x^2} = \frac{250}{5^2} = \frac{250}{25} = 10$
   When $x = 2$, $y = 16$ and $\frac{y}{x^2} = \frac{16}{2^2} = \frac{16}{4} = 4$
   Since $\frac{y}{x^2}$ is not a constant value, y is not proportional to x^2

   Then check $y = kx^3$
   If y is proportional to x^3, then $\frac{y}{x^3}$ is a constant value, k.
   When $x = 5$, $y = 250$ and $\frac{y}{x^3} = \frac{250}{5^3} = \frac{250}{125} = 2$
   When $x = 2$, $y = 16$ and $\frac{y}{x^3} = \frac{16}{2^3} = \frac{16}{8} = 2$
   Since $\frac{y}{x^3}$ is a constant value, y is proportional to x^3
   So $y = kx^3$ is the correct formula.

(b) Since $\frac{y}{x^3} = 2$, $y = 2x^3$
   So $k = 2$

If two variables that are in direct proportion are plotted on a graph, the graph will take the form of a straight line passing through the origin.

Example 8

y is directly proportional to x. When $x = 4$, $y = 24$
(a) Copy and complete the table of values shown on the right.
(b) Which of the following sketches could represent the graph of y against x?

x	0	1	2	3	4	5
y					24	

296

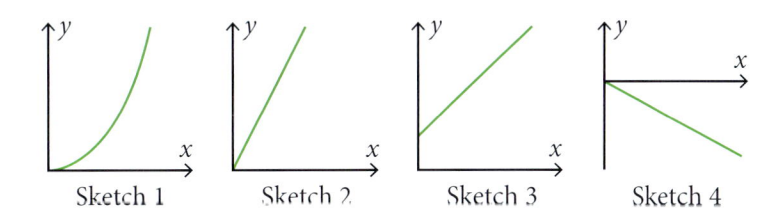

CHAPTER 33: PROPORTION AND VARIATION

Sketch 1   Sketch 2   Sketch 3   Sketch 4

(a) y is directly proportional to x. So: $y = kx$

When $x = 4$, $y = 24$, so: $24 = k \times 4$

$$k = \frac{24}{4} = 6$$

So the equation is: $y = 6x$

Using the formula, we can complete the table of values:

x	0	1	2	3	4	5
y	0	6	12	18	24	30

(b) If two variables that are in direct proportion are plotted on a graph, the graph will take the form of a straight line passing through the origin. This rules out Sketch 1 and Sketch 3.

From the table of values, we can see that positive x values always give positive y values. So, Sketch 4 is also ruled out.

The correct answer is Sketch 2.

If y is proportional to x^2, then the graph of y against x will take the form of a positive quadratic curve passing through the origin.

If y is proportional to x^3, then the graph of y against x will take the form of a positive cubic curve passing through the origin.

Example 9

S is directly proportional to T^2.
Which of the sketches below could represent the graph of S against T?

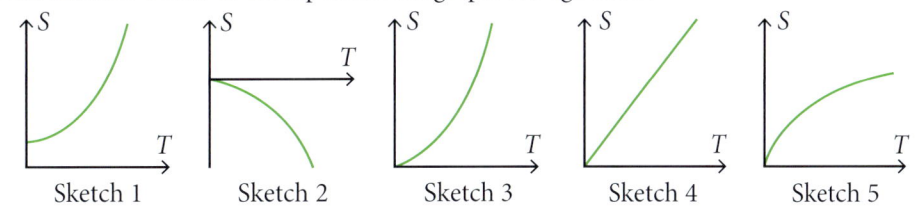

Sketch 1   Sketch 2   Sketch 3   Sketch 4   Sketch 5

If S is directly proportional to T^2, then the graph of S against T should be a positive quadratic curve passing through the origin.

Sketch 1 does not pass through the origin. Sketch 2 appears to be a negative quadratic curve. Neither Sketch 4 nor Sketch 5 are the shape of a quadratic curve.

The only possibility is Sketch 3.

Exercise 33D

1. In an experiment, the results in the table on the right were obtained. Which of the following laws connects the two variables P and R?

R	5	10	20	30	40
P	12.5	50	200	450	800

   $P \propto R$ $P \propto R^2$ $P \propto R^3$

2. I is directly proportional to V. When $V = 1.5$, $I = 18$
   (a) Copy and complete the table of values on the right.
   (b) Which of the sketches that follow could represent the graph of I against V?

V	0	0.5	1	1.5	3	6
I				18		

GCSE MATHEMATICS M3 AND M7

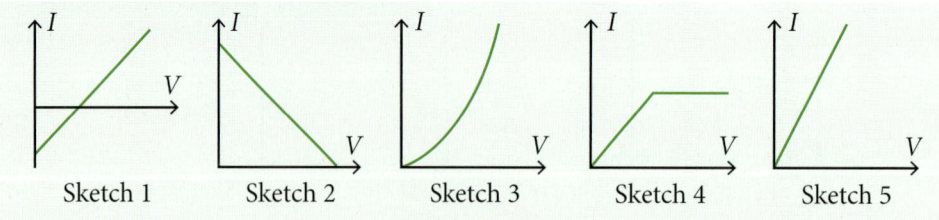

Sketch 1, Sketch 2, Sketch 3, Sketch 4, Sketch 5 (graphs of I against V)

3. P is directly proportional to Q^3. Which of the sketches below could represent the graph of P against Q?

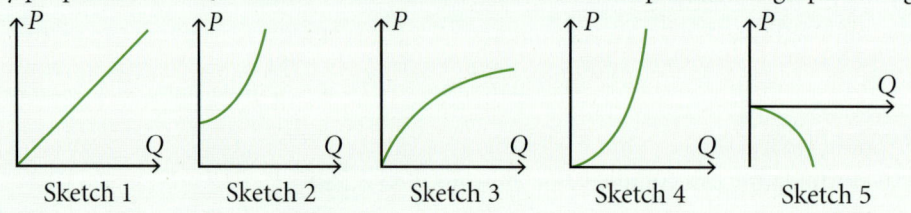

Sketch 1, Sketch 2, Sketch 3, Sketch 4, Sketch 5 (graphs of P against Q)

4. Given that $L \propto M^2$, what is the effect on L when:
   (a) M is doubled?
   (b) M is divided by 3?

5. P varies directly as $(Q + 2)$. When $P = 4$, $Q = 18$.
   (a) Find P when $Q = 8$.
   (b) Find Q when $P = 6$.

6. H is directly proportional to $(B + 10)^2$.
   (a) When $B = 2$, $H = 36$.
       Find a formula for H in terms of B.
   (b) Copy and complete the table of values shown.

B	−10	−4	−2	0	2		10
H	0		16		36	49	100

7. c is proportional to $\sqrt{d + 4}$.
   When $d = 21$, $c = 40$.
   (a) Find a formula for c in terms of d.
   (b) Find c when $d = 96$.
   (c) Find d when $c = 56$.

8. Y is proportional to X.
   (a) Write down a formula for Y in terms of X and a constant of proportionality, k.
   (b) When $Y = a$, $X = 2$
       When $Y = 2a - 5$, $X = 6$
       Write down two simultaneous equations in a and k.
   (c) Solve the simultaneous equations to find the values of a and k.
   (d) Hence write down the formula for Y in terms of X.

9. The heat, H (measured in joules) generated by a lamp varies with the square of the current, I passing through it (measured in amps). Gavin carries out an experiment and obtains these results for I and H. One of the values for H has been incorrectly recorded. Which one? Hint: find the constant of proportionality for each pair of values.

I	1	2	4	5	10
H	80	320	1280	4000	8000

10. y varies as x^2
    When $x = 5$, $y = 25$
    When $x = p$, $y = 2p + 3$, where p is a positive integer.
    (a) Show that $p^2 - 2p - 3 = 0$
    (b) Hence find the value of p.

33.5 Summary

In this chapter you have learnt about:
- Direct proportion. If one variable, y is directly proportional to another, x, they increase and decrease at the same rate. This can be expressed algebraically using the formula $y = kx$
- More complex relationships, for example y could be proportional to the square of x, the cube of x, or the square root of x. These relationships can also be expressed algebraically, for example $y = kx^2$
- Calculating the constant of proportionality, k and then to use the formula to calculate one variable from the value of the other.
- Recognising the graph representing the relationship between two variables that are in proportion.

Progress Review
Chapters 28 – 33

This Progress Review covers:

Chapter 28: Formulae
Chapter 29: Sequences
Chapter 30: Non-Linear Graphs
Chapter 31: Real-Life Graphs
Chapter 32: Graphical Solutions
Chapter 33: Proportion and Variation

1. Rearrange these formulae to make a the subject:
   (a) $u = \frac{1}{2}av^2$
   (b) $y = \frac{z}{\sqrt{a}}$

2. Rearrange this formula: $a = 3(b + c^2)$
   (a) to make b the subject,
   (b) to make c the subject.

3. Rearrange these formulae to make x the subject:
   (a) $g^2 - 7x = h^2$
   (b) $Jx + K - L = 0$

4. Rearrange these formulae to make y the subject:
   (a) $C = 4 - Dy$
   (b) $E = fyg$

5. Rearrange this formula to make a the subject: $6(a + D) = E$

6. Rearrange these formulae to make a the subject:
   (a) $M = \frac{1}{2(6a - b)}$
   (b) $V = \frac{b}{1 - a}$
   (c) $R = \frac{S}{T + a} - 3R$

7. Rearrange these formulae involving powers and roots to make b the subject:
   (a) $\frac{4}{\sqrt{b}} = 2U$
   (b) $W = b^3 v^2$

8. Using differences, find the next two terms in the following sequences:
   (a) 5, 13, 21, 29, …
   (b) 67, 63, 59, 55, …
   (c) 21, 32, 43, 54, …
   (d) 7, 1, –5, –11, …

9. For each of the following sequences describe, in words, the rule for finding the next term, and find the next term:
   (a) –2, –5, –8, –11, …
   (b) 59, 64, 69, 74, …
   (c) 154, 143, 132, 121, …
   (d) $3\frac{1}{8}, 4\frac{1}{2}, 5\frac{7}{8}, 7\frac{1}{4}, \ldots$

10. Using differences, state whether each of the following sequence is quadratic or linear, and find the next term:
    (a) 3, 6, 9, 12, 15, …
    (b) 3, 6, 10, 15, 21, …
    (c) 29, 27, 23, 17, 9, …
    (d) 31, 27, 23, 19, 15, …

11. Is the triangular number sequence (1, 3, 6, 10) linear or quadratic? Give a reason.

12. The number of squares in each of the frames shown on the right forms a sequence. State whether it is it linear or quadratic, and work out how many squares are in the next frame in the sequence.

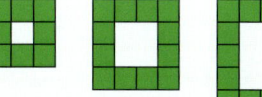

13. Find the nth term of each of these linear sequences:
    (a) 210, 190, 170, 150, 130, …
    (b) 31, 38, 45, 52, 59, …
    (c) 395, 408, 421, 434, …
    (d) 9, 3, –3, –9, –15, …

GCSE MATHEMATICS M3 AND M7

14. Find the first four terms, and the 50th term, in the sequences whose n^{th} terms are given by:
    (a) $(n-7)(n+1)$
    (b) $\dfrac{n^2+n}{2}$
    (c) $1+n+n^2$
    (d) $\dfrac{n^3-n}{6}$

15. Which of (a) – (d) is the nth term for the sequence: 1, 0, 1, 4, …
    (a) $n-1$
    (b) n^2-1
    (c) n^2-n
    (d) $(n-2)^2$

16. Identify the nth terms of these sequences as expressions involving n^2
    (a) 2, 5, 10, 17, …
    (b) −1, −4, −9, −16, …
    (c) 11, 14, 19, 26, …
    (d) 49, 46, 41, 34, …

17. Identify the nth terms of the following sequences:
    (a) $\dfrac{1}{2}, \dfrac{2}{3}, \dfrac{3}{4}, \dfrac{4}{5}, \ldots$
    (b) $\dfrac{7}{1}, \dfrac{7}{4}, \dfrac{7}{9}, \dfrac{7}{16}, \ldots$
    (c) $\dfrac{1}{4}, \dfrac{4}{7}, \dfrac{7}{12}, \dfrac{10}{19}, \dfrac{13}{28}, \ldots$
    (d) $\dfrac{2}{6}, \dfrac{8}{5}, \dfrac{18}{4}, \dfrac{32}{3}, \ldots$

18. A quadratic curve has the equation $y = x^2 + x - 6$
    (a) Copy and complete the table of values on the right, using the equation of the curve.

x	−4	−3	−2	−1	0	1	2	3
y	6		−4	−6		−4	0	

    (b) Using your table of values, plot the graph of the curve.
    (c) Using your graph:
        (i) Estimate the two values of x on this curve for which $y = 1$
        (ii) Estimate the minimum value of y

19. A quadratic curve has the equation $y = -x^2 - 4x + 1$
    (a) Copy and complete the table of values on the right, using the equation of the curve.

x	−5	−4	−3	−2	−1	0	1
y	−4		4	5		1	−4

    (b) Using your table of values, plot the graph of the curve.

20. A cubic curve has the equation $y = x^3 - 3x + 1$
    (a) Copy and complete the table of values on the right, using the equation of the curve.

x	−2	−1	−0.5	0	0.5	1	2
y	−1		2.375	1		−1	3

    (b) Using the values in your table, plot the graph of the curve.
    (c) Using your graph, estimate the three solutions to the equation $x^3 - 3x + 1 = 2$

21. A cubic curve has the equation $y = -x^3 + x^2 + 4x - 4$
    (a) Copy and complete the table of values below, using the equation of the curve.

x	−3	−2	−1	−0.5	0	0.5	1	1.5	2	3
y	20	0		−5.63	−4		0	0.875		−10

    (b) Using the values in your table, plot the graph of the curve.
    (c) Using your curve, write down the three solutions to the equation $-x^3 + x^2 + 4x - 4 = 0$

22. A reciprocal curve has the equation $y = \dfrac{5}{x}$
    (a) Copy and complete the table of values below using the equation of the curve.

x	−10	−5	−4	−2	−1	−0.5	0.5	1	2	4	5	10
y	−0.5	−1		−2.5	−5		10	5		1.25	1	

    (b) Using the values in your table, plot the graph of the curve.
    (c) Using your graph, estimate the solution to the equation $\dfrac{5}{x} = 4$

23. A reciprocal curve has the equation $y = -\dfrac{60}{x}$
    (a) Copy and complete the table of values below using the equation of the curve.

x	−20	−15	−12	−10	−6	−5	−4	−3	3	4	5	6	10	12	15	20
y		4	5	6		12	15	20	−20		−12	−10	−6		−4	−3

    (b) Using the values in your table, plot the graph of the curve. It may be helpful to use x and y values from −20 to 20, going up in steps of 5

24. Match the curve sketches below with an appropriate equation.

Curve 1   Curve 2   Curve 3   Curve 4

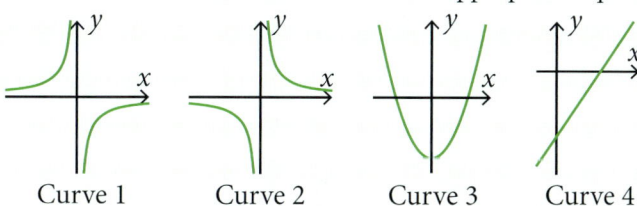

Equations

A: $y = \dfrac{3}{x}$

B: $y = x^2 - 2$

C: $y = x - 2$

D: $y = -\dfrac{2}{x}$

25. Match the curve sketches below with an appropriate equation.

Curve 1   Curve 2   Curve 3   Curve 4

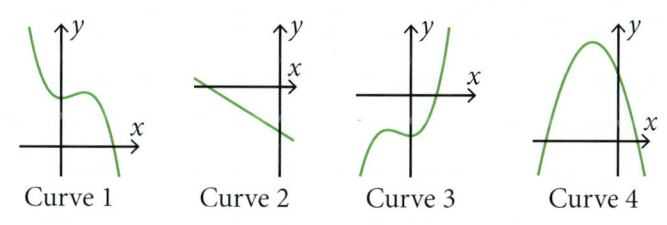

Equations

A: $y = -\dfrac{1}{2}x - 1$

B: $y = -x^3 + x^2 + 1$

C: $y = -x^2 + 2x + 2$

D: $y = x^3 + x^2 - 1$

26. The graph on the right shows the conversion between miles and kilometres. Using the graph, answer the following questions.
  (a) How many kilometres is 25 miles?
  (b) Roughly how many miles is 50 km?
  (c) How many kilometres is 100 miles?
  (d) What is the gradient of the line?
  (e) Interpret the gradient of the line in the context of the question.

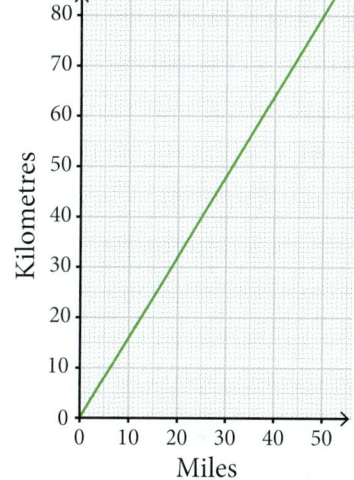

27. A lemon costs 24 p.
  (a) Copy and complete the table on the right.

Number of lemons	1	2	3	4	5
Cost (pence)	24				

  (b) Draw a straight line graph to show this information. The horizontal axis should represent the number of lemons and the vertical axis the total cost.
  (c) Using the graph, how much would it cost to buy 50 lemons?

28. The Johnston family are going camping. They leave home in the morning and the graph on the right shows their distance from home during the journey.
  (a) After they had been travelling some time, Mrs Johnston realises they have forgotten to bring the tent. They return home to pick it up. What time did Mrs Johnston realise this?
  (b) Later, the Johnston family stopped to buy some drinks. What time did they do this?
  (c) The Johnston family arrived at the camp site at 12:30 pm. How far from their home is the camp site?

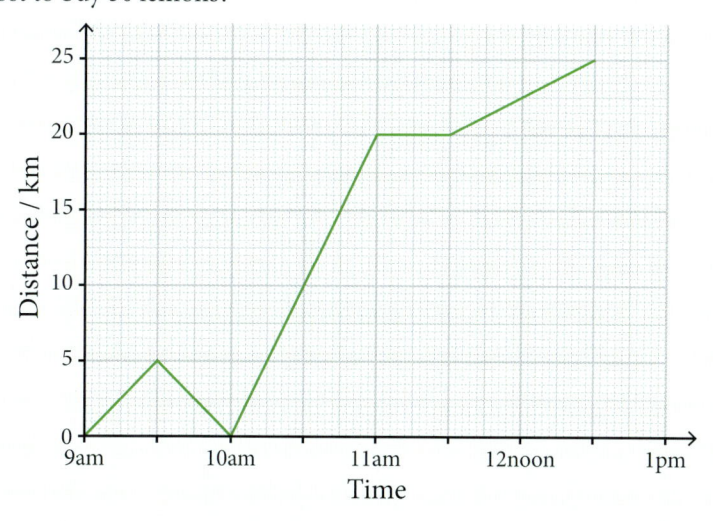

29. Clodagh climbs to the top of a mountain in the Sperrins, then back down to ground level. The graph on the right shows her height above ground level throughout the day.
   (a) At what time did Clodagh start climbing the mountain?
   (b) When Clodagh reached the top, how long did she stop to look at the view?
   (c) How high above ground level was Clodagh at 4 pm?

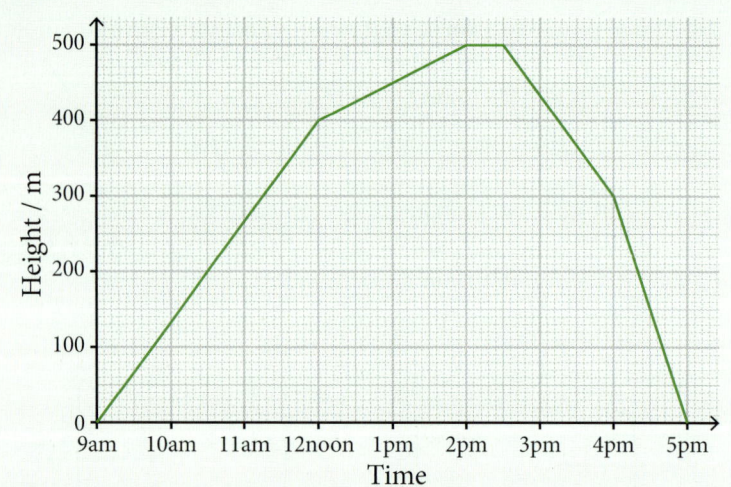

30. Leonard lives in London. He walks to a house called Howards End in the countryside outside London. The graph on the right shows his journey.
   (a) When does Leonard leave London?
   (b) How long does he spend at Howards End?
   (c) On the way back Leonard stops for a rest. How far is Leonard from London when he stops?
   (d) How long does he stop for?
   Charles drives from London to Howards End at a steady speed. He leaves at 1:15 pm and arrives at 1:45 pm.
   (e) Copy the graph. Add a line representing Charles' journey.
   (f) At what time does Charles pass Leonard? What is Leonard doing at that time?
   (g) How far is Charles from Howards End when he passes Leonard?

31. A, B and C are three stops on a bus route. The travel graph on the right shows the journey of a bus from A to C via B.
   (a) (i) When does the bus leave stop A?
       (ii) When does the bus arrive at stop B?
       (iii) For how long does the bus stop at B?
       (iv) What is the average speed of the bus between A and B?
       (v) What is the distance between stops B and C?
   (b) A second bus leaves C at 10:10. It travels at a steady speed to B, arriving at 10:45
       (i) Copy the travel graph and show the second bus's journey on it.
       (ii) When do the two buses pass each other?
       (iii) Roughly how far apart are the buses at 10:40?

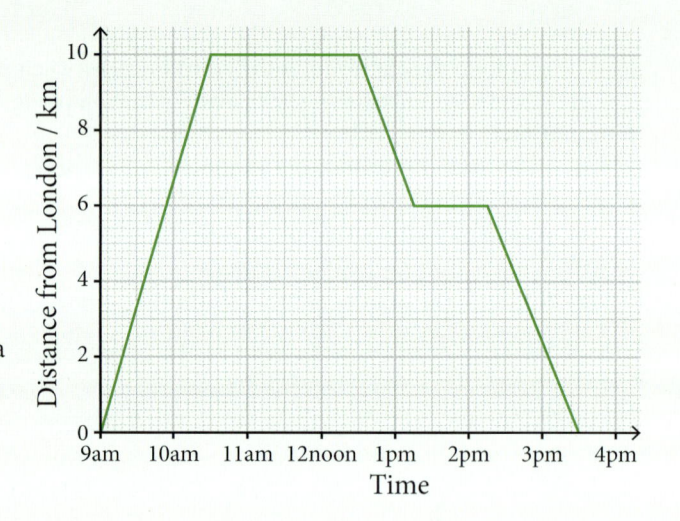

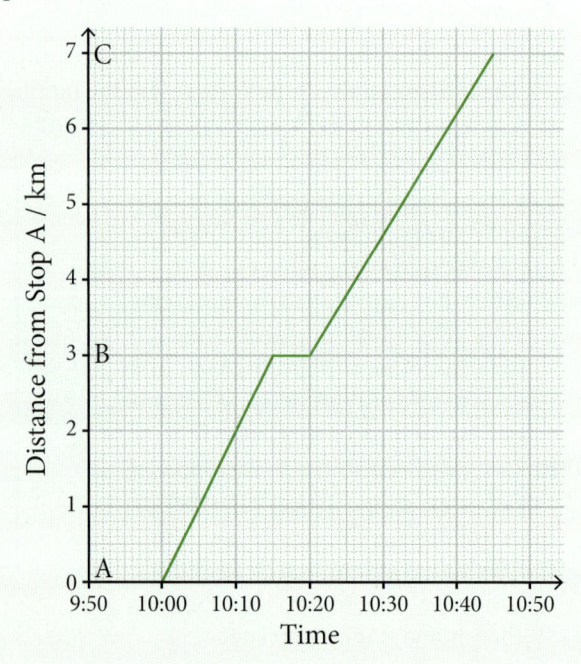

32. To film a scene for an action movie, a remote-controlled car is driven off a 50 metre cliff, as shown on the right. The height of the car can be calculated using this quadratic function:

$y = 50 - \frac{1}{5}x^2$

where y is the car's height and x is its horizontal distance from the cliff. x and y are both measured in metres.

(a) Copy and complete the table of values below.

x (metres)	0	2.5	5	7.5	10	12.5	15
y (metres)	50				30		

(b) Copy and complete the graph on the right using the data in your table. Join the points on your graph with a smooth curve. Extend the curve so that it touches the x-axis.

Answer parts (c) – (e) using your graph:

(c) Roughly how high above the ground is the car when it is 8 metres from the cliff?

(d) Roughly how far from the cliff is the car when it is at a height of 10 metres?

(e) Roughly how far from the cliff does the car land?

33. The graph on the right shows two straight lines:
$y = -x + 4$; and
$y = 1.5x - 1$

Using the graph, write down x and y values that are the solutions to the simultaneous equations:
$y = -x + 4$
$y = 1.5x - 1$

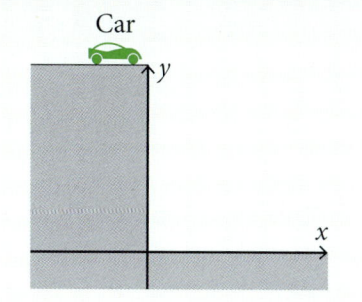

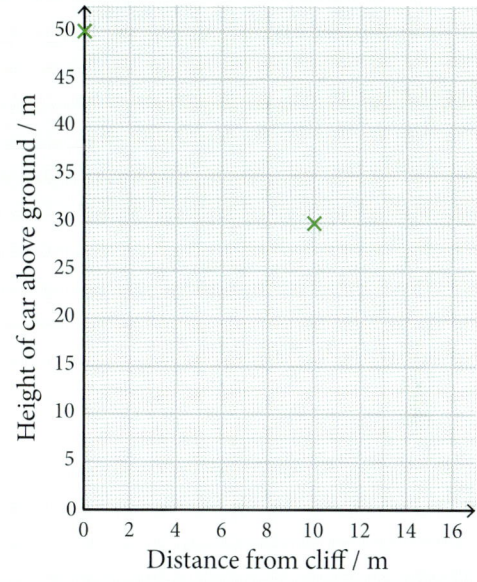

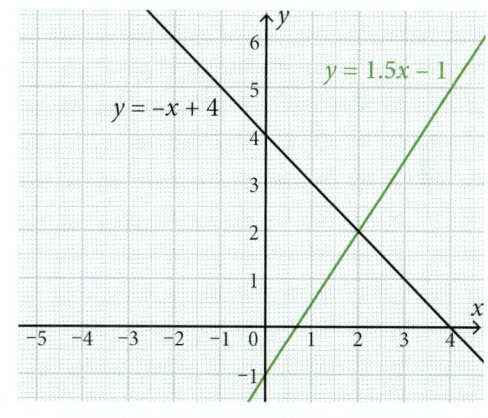

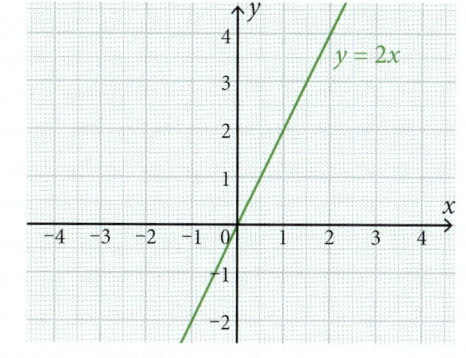

34. The graph on the right shows the straight line $y = 2x$

(a) A second straight line has the equation $y = -x + 3$
Copy and complete the table of values below using the equation of this line.

x	−1	0	1	2	3	4
y	4		2		0	

(b) Using your table of values, plot the line $y = -x + 3$

(c) Using your graph, solve the following pair of simultaneous equations:
$y = 2x$
$y = -x + 3$

35. Look at the graph on the right showing a straight line.
    (a) What second straight line would you need to plot on the graph to solve this pair of simultaneous equations?
    $y = -2x + 3$
    $y = -\frac{1}{2}x - 1$
    (b) Plot the second line using a table of values. Label both lines on your graph.
    (c) Using your graph, estimate solutions to the simultaneous equations above.

36. The graph on the right shows the straight lines with equations: $y = -2x + 3$ and $y = 8x + 1$
    Use the graph to estimate the solutions to the simultaneous equations:
    $y = -2x + 3$
    $y = 8x + 1$
    Estimate solutions for x and y to one decimal place.

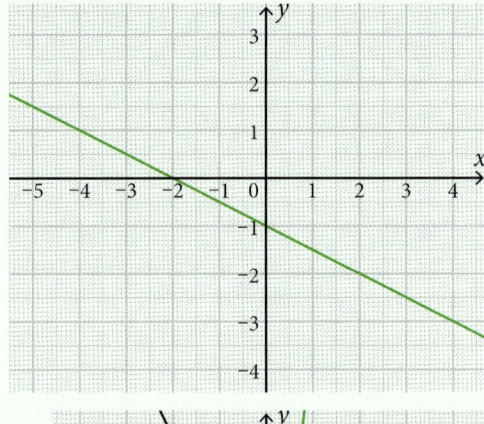

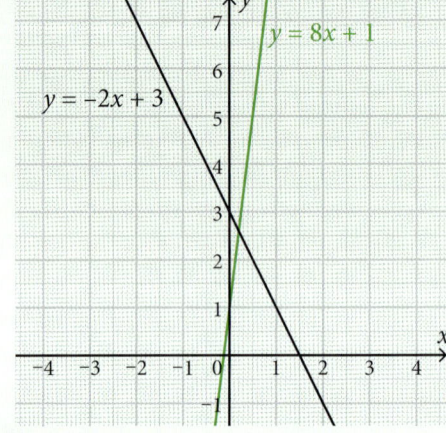

37. A curve has the equation $y = x^2 + 3x + 1$
    (a) Copy and complete the table of values on the right using the equation of the curve.
    (b) Hence plot the curve.
    (c) On the same graph, add the line with equation $y = 2x + 3$
    (d) Write down the coordinates of the two points of intersection of the curve and line.

x	−4	−3	−2	−1	0	1
y	5		−1	−1		5

38. The curve shown on the right has the equation $y = -x^2 + 5x$
    Copy the graph carefully.
    (a) On the same graph, add these two straight lines:
    $y = 3x + 1$
    $y = -x + 9$
    You may use a table of values for each line.
    (b) Write down the coordinates of the single point of intersection of the curve $y = -x^2 + 5x$ and the straight line $y = 3x + 1$
    (c) Write down the coordinates of the single point of intersection of the curve $y = -x^2 + 5x$ and the straight line $y = -x + 9$
    (d) Using your graph, solve this pair of simultaneous equations:
    $y = 3x + 1$
    $y = -x + 9$

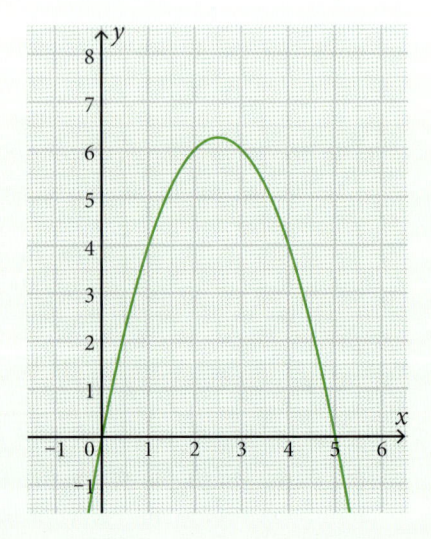

39. The graph on the right shows the quadratic curve with the equation:

    $y = \frac{1}{2}x^2 + x - 8$

    Copy the graph carefully.

    By adding the appropriate straight lines to the graph, find or estimate the coordinates of any intersection points between the curve and each of these straight lines:
    (a) $y = -4$ (b) $y = -8$ (c) $y = 0$ (d) $x = 1$

40. L varies directly as R
    (a) When $L = 4$, $R = 7$
        Find a formula for L in terms of R
    (b) Find L when $R = 5$
    (c) Find R when $L = 10$

41. J is directly proportional to M
    When $M = 18$, $J = 3$
    (a) Find a formula for J in terms of M
    (b) Find J when $M = 12$
    (c) Find M when $J = 5$

42. The variable Y is proportional to the square of the value of a second variable Z
    Given that $Y = 4000$ when $Z = 80$:
    (a) Find a formula for Y in terms of Z
    (b) Find: (i) the value of Y when $Z = 16$ (ii) the value of Z when $Y = 24$ (Give your answer to 2 decimal places.)

43. The kinetic energy E (measured in joules) of an object varies with the square of its speed v (measured in m/s). When the object is moving with a speed of 9 m/s, it has a kinetic energy of 405 joules.
    (a) Find a formula for E in terms of v
    (b) Find the kinetic energy of this object when its speed is 4 m/s.
    (c) Find its speed when its kinetic energy is 20 joules.

44. Mobile phone masts transmit and receive millions of signals from phones that are within a certain coverage area. The radius, r of this coverage area varies with the square root of the height, h of the mast. At one location, a mast with a height of 25 m covers an area with a radius of 2 km.
    (a) Find a formula for r in terms of h
    (b) Find the radius of the area covered by a mast of height 9 m.
    (c) A mast is to cover an area with a radius of 1.6 km. Find the minimum height of this mast.

45. Match each graph below with one of the formulae and find a value for the constant of proportionality in each case.

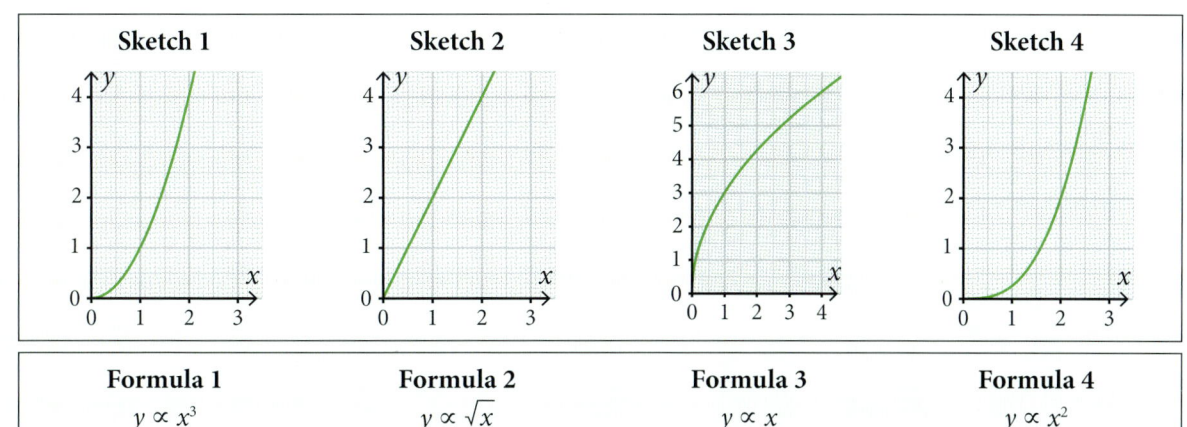

Sketch 1	Sketch 2	Sketch 3	Sketch 4
Formula 1	Formula 2	Formula 3	Formula 4
$y \propto x^3$	$y \propto \sqrt{x}$	$y \propto x$	$y \propto x^2$

46. y varies as x, and when $x = 10$, $y = 2$
    Find x when $y = 2x + 3$

47. p is proportional to the square of q, and when $p = 27$, $q = 9$
    Find p and q when $p = 2q - 3$

Chapter 34
Transformations

34.1 Introduction

A **transformation** moves an object according to a rule. Three types of transformation are discussed in this chapter: translations, rotations and reflections. In the next chapter we discuss enlargements.

Key words
- **Translation**: A translation moves an object without changing its shape, size or orientation.
- **Reflection**: A reflection reflects an object in a mirror line.
- **Rotation**: A rotation turns an object about a point.
- **Image**: The shape that results from a transformation.

Before you start you should:
- Recognise **line symmetry** and **rotational symmetry** in shapes.
- Know how to complete a shape so that it has line symmetry.
- Be able to state whether a shape has rotational symmetry and, if so, state the **order of rotation**.

In this chapter you will learn how to:
- Perform the following transformations on 2D shapes: translations, reflections and rotations.
- Describe a transformation of a 2D shape.
- Describe and transform 2D shapes using combined transformations.

Example 1 (Revision)

Copy the shape on the right.
(a) Draw the vertical line of symmetry.
(b) Copy the shape again. Shade two more squares to give the shape a horizontal line of symmetry. Add the line of symmetry to your drawing.

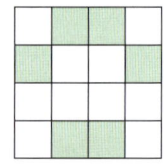

(a) The vertical line of symmetry is shown on the right.

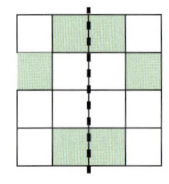

(b) The shape is redrawn with two more squares shaded so that there is an additional horizontal line of symmetry:

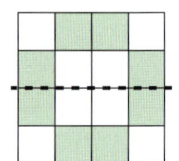

Some letters of the alphabet have rotational symmetry, while others do not.

Example 2 (Revision)

Consider some rotations of the capital letter N.
(a) By looking at the four pictures below, state whether the capital letter N has rotational symmetry. Explain your answer.

Original | 90° clockwise rotation | 180° clockwise rotation | 270° clockwise rotation

Note: The point the shape turns about is called the **centre of rotation**. The centre of rotation has been marked with a dot on these diagrams.

(b) If the letter has rotational symmetry, state its **order of rotational symmetry**.

CHAPTER 34: TRANSFORMATIONS

(a) After a 180° clockwise rotation, the capital N is identical to the original, so it **does** have rotational symmetry.
(b) The capital N has rotational symmetry of order 2, since there are two positions in which it appears the same as the original.

Note: When stating the order of rotational symmetry, remember to include the original position.

Exercise 34A (Revision)

1. Copy the shape shown on the right.
   (a) Shade one extra square so that the shape has one line of symmetry.
   (b) Add the line of symmetry to your diagram.

2. Consider the shape shown on the right. Shade one additional square so that this shape has rotational symmetry of order 4

3. The shape shown on the right has rotational symmetry order 2. Write down the coordinates of the centre of rotation.

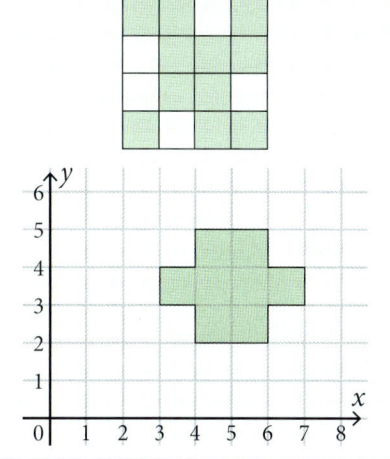

34.2 Translations

One type of transformation is a **translation**. In a translation, an object moves without turning or changing its size or shape. In the diagram below, object ABC is translated to the image A'B'C'.

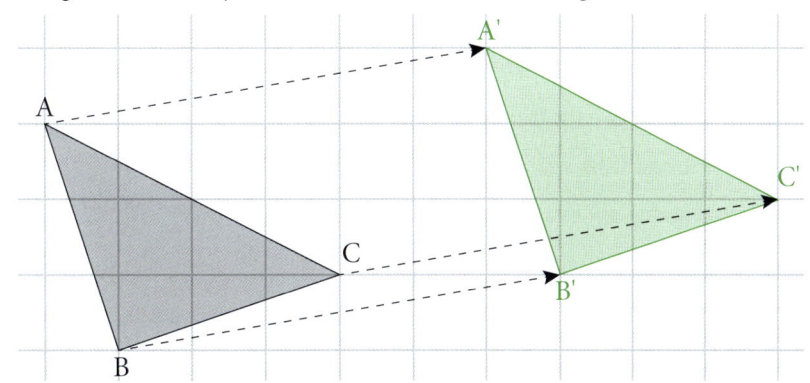

To describe a translation, count the units along and up through which a point moves.

The translation shown in the diagram is described as 6 units right and 1 unit up.

It can also be written as $\begin{pmatrix} 6 \\ 1 \end{pmatrix}$

This way of writing a translation is called a **vector**.

In the vector, we use positive numbers for movement to the right and upwards; negative numbers are used for movement to the left and downwards.

The original object and the image are congruent. (They are the same shape and the same size.)

307

Example 3

Consider the shapes shown on the right. Describe the translation that maps:
(a) P to Q   (b) Q to R   (c) R to P

(a) The translation mapping P to Q is 6 units right and 1 unit up. Using vector notation: $\begin{pmatrix} 6 \\ 1 \end{pmatrix}$

(b) The translation from Q to R is 3 units left and 5 down, or $\begin{pmatrix} -3 \\ -5 \end{pmatrix}$

(c) The translation from R to P is 3 units left and 4 up, so $\begin{pmatrix} -3 \\ 4 \end{pmatrix}$

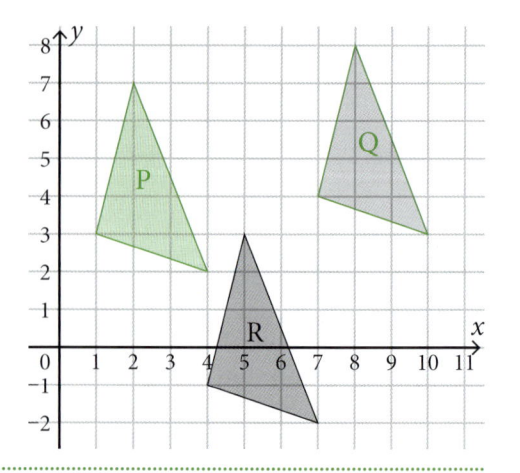

You may be asked to translate an object using a vector.

To translate a shape, translate each of the vertices (corners) and then join them up.

Example 4

(a) Draw an *x*-axis from –1 to 8 and a *y*-axis from 0 to 8
Plot and label these points: (2, 1), (2,4), (5, 4), (6, 2)
Join the points to form a quadrilateral and label it A.

(b) Translate shape A by $\begin{pmatrix} 2 \\ 4 \end{pmatrix}$ Label the image B.

(c) Translate shape B by $\begin{pmatrix} -5 \\ 0 \end{pmatrix}$ Label the image C.

(d) Describe the single transformation that maps A to C.

(a) – (c) To translate the shape A using the vector $\begin{pmatrix} 2 \\ 4 \end{pmatrix}$, translate each of the four vertices.

The point (2, 1) is moved 2 units to the right and 4 up. So the image point is (4, 5).

Repeat this process with each of the four vertices. Join the four image points and label the shape B.

Repeat the whole process above for the translation of shape B by $\begin{pmatrix} -5 \\ 0 \end{pmatrix}$, labelling the resulting shape C.

The three shapes A, B and C are shown on the right.

(d) The transformation from A to C is a translation $\begin{pmatrix} -3 \\ 4 \end{pmatrix}$

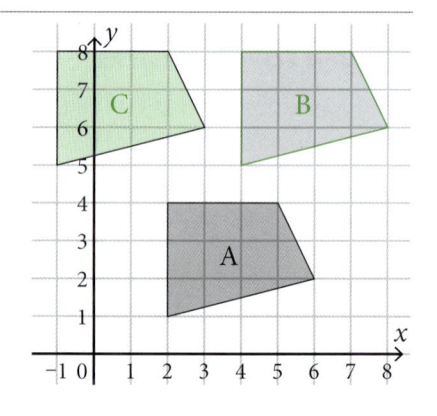

Exercise 34B

1. Copy the diagram on the right.
   Draw the triangle after:
   (a) A translation of 2 to the right and 1 up.
   (b) A translation of 1 to the left and 1 up.
   (c) A translation of 4 down.
   (d) A translation of 6 to the right and 2 down.

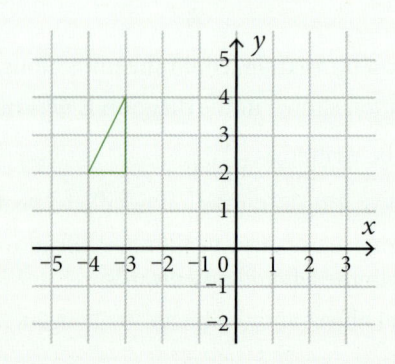

2. Copy the diagram on the right. Then carry out the transformations given in the following table. Label each of the images.

Translation	Left/right	Up/down
W→X	3 right	1 up
X→Y	1 left	7 down
Y→Z	4 left	1 up

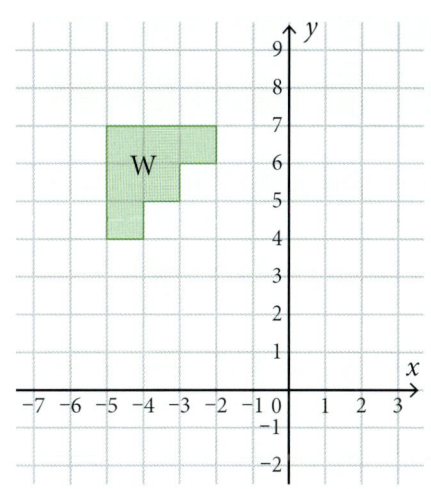

3. Look at the shapes on the right. Copy the following table and complete it to describe each translation.

Translation	Translation Vector
A→B	$\binom{8}{1}$
A→C	
B→C	
B→A	
C→A	
C→B	

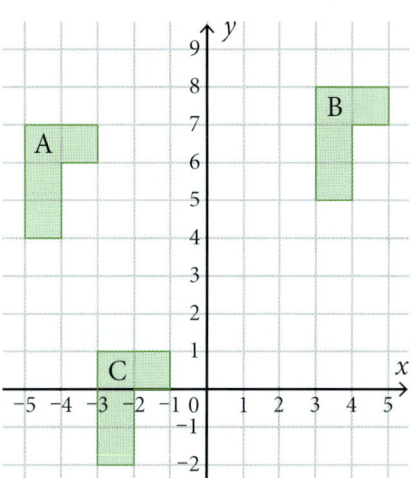

4. Draw an x-axis using values of x from -5 to 4 and a y-axis using values of y from -1 to 8
   (a) Plot the following points and join them to make a triangle: $(-5, 4), (-4, 8), (0, 5)$
       Label this triangle A.
   (b) Translate triangle A using the vector $\binom{3}{-4}$ Label the new triangle B.
   (c) Describe fully the transformation that maps B to A.
   (d) Are shapes A and B congruent? Explain your answer briefly.

5. (a) Write down the translation vector where the shape does not move.
   (b) Write down a sequence of three different translations where the shape returns to its original position.

34.3 Reflections

When an object is reflected in a line, its image is formed on the other side of the line. The line is called the **mirror line** or the **line of reflection**.

The image and the original object are congruent.

A common type of question involves reflecting a shape on a coordinate grid. The mirror line can be:

- The x-axis or the y-axis.
- Any vertical or horizontal line, e.g. $x = 3$ or $y = -2$
- The line $y = x$ or $y = -x$
  These are diagonal lines passing through the origin.

When reflecting a shape, reflect each vertex (corner) in turn and then join them up.

A point and its image are always the same distance from the mirror line.

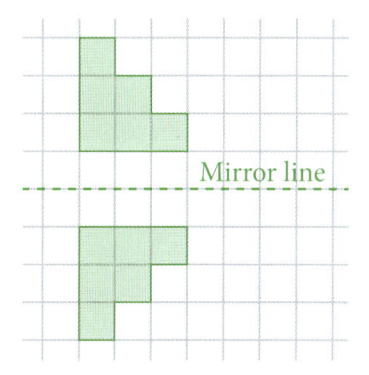

GCSE MATHEMATICS M3 AND M7

Example 5

Copy the diagram on the right. Draw the image of the triangle when it is reflected in the *y*-axis.

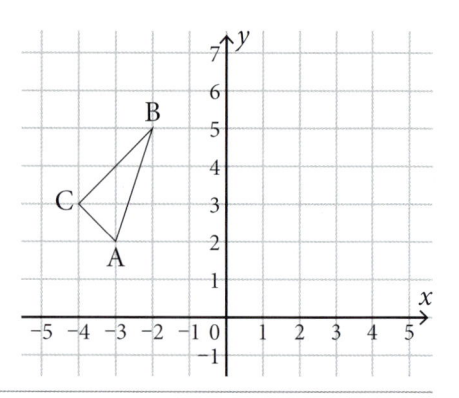

For each vertex, count the number of squares from the vertex to the mirror line. Then count the same number of squares on the opposite side.

For example, vertex A is 3 squares from the mirror line. We count 3 squares from the mirror line on the right-hand side and draw the image of point A at (3, 2).

> **Note:** It is common for the image of point A to be labelled A', etc.

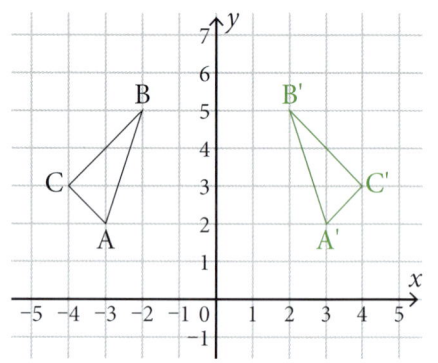

Describing and transforming 2D shapes using reflections in the lines $y = \pm x$

In the M7 module you may be asked to reflect shapes in the line $y = x$ or the line $y = -x$
These are diagonal lines passing through the origin.

Example 6

Look at triangle ABC on the upper right.
(a) Reflect the triangle in the line $y = x$
(b) Reflect the triangle in the line $y = -x$

(a) The mirror line $y = x$ is shown in the diagram on the lower right. For each vertex, count the number of squares to the mirror line, diagonally, meeting the mirror line at right angles.

The construction line for point A is shown. From point A, you move across 2½ squares diagonally to the mirror line. Move the same distance on the other side of the mirror line to find the position of the image point A'

Repeat for each vertex of the original triangle. Finally join points A', B' and C' to form the image triangle.

> **Note:** After a reflection in the line $y = x$ the coordinates of each point are reversed. In the example above, the point (−1, 4) is mapped to (4, −1), etc.

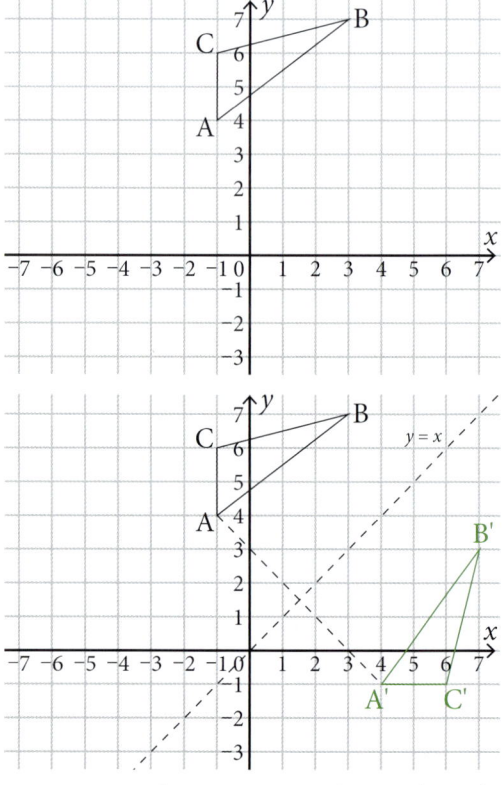

(b) The mirror line $y = -x$ is shown in the diagram on the next page. For each vertex, count the number of squares to the mirror line, diagonally, meeting the mirror line at right angles.

310

CHAPTER 34: TRANSFORMATIONS

The construction line for point A is shown on the right. From point A, you move across 1½ squares diagonally to the mirror line. Move the same distance on the other side of the mirror line to find the position of the image point A".

Repeat for each vertex of the original triangle. Finally join points A", B" and C" to form the image triangle.

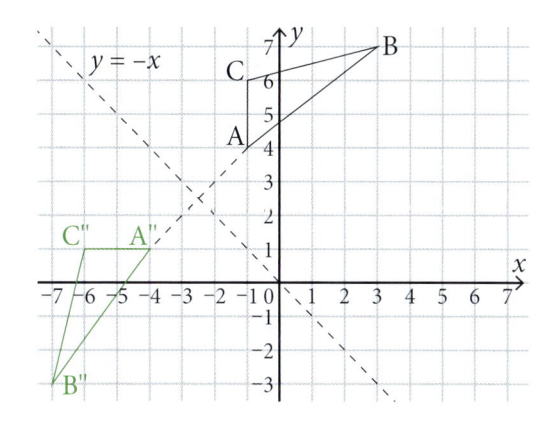

Note: After a reflection in the line $y = -x$ the coordinates of each point are reversed and the signs are changed. In the example above, the point $(-1, 4)$ is mapped to $(-4, 1)$, etc.

Exercise 34C

1. Copy the diagram on the right. Reflect the triangle shown in the *y*-axis.

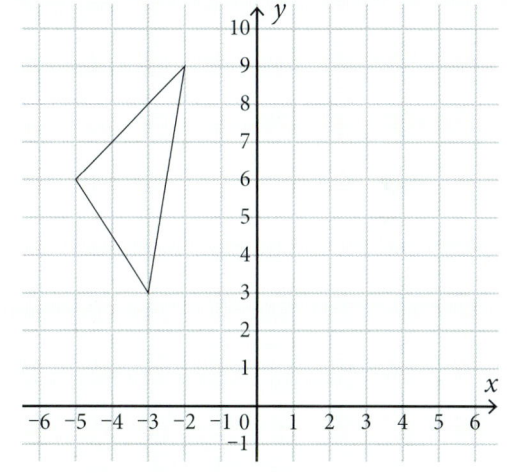

2. Draw *x*- and *y*-axes from −7 to 7
   (a) Plot these points and join them to form a triangle:
   $(-6, -1), (-1, -4), (-4, -5)$
   Label the triangle A.
   (b) Reflect A in the *x*-axis and label the image B.
   (c) Reflect B in the *y*-axis and label the image C.
   (d) Reflect C in the *x*-axis and label the image D.
   (e) What transformation maps triangle D onto triangle A?
   (f) Are shapes A and D congruent?
   Explain your answer briefly.

3. (a) Copy the diagram on the right and reflect the shape in the *y*-axis. 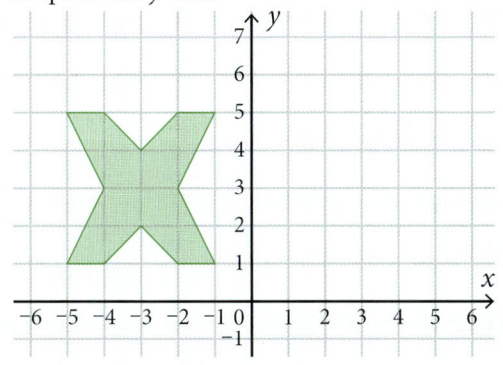
   (b) Copy the diagram on the right and reflect the shape in the *y*-axis.

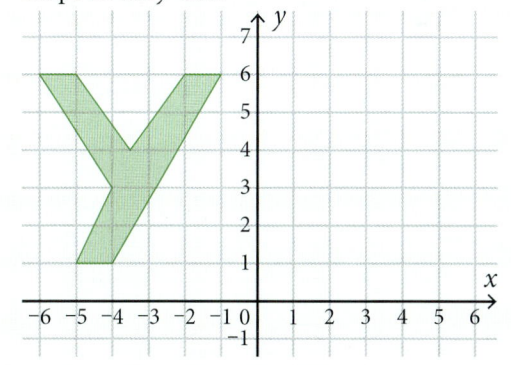

   (c) Why is the letter X the 'right way round' after the reflection, but the letter Y is not?

4. Copy the diagram shown on the right.
   It contains a shape made of two straight lines.
   (a) Reflect the shape in the *y*-axis.
   (b) Now reflect both shapes in the *x*-axis.
   (c) Describe the shape you have drawn.
   (d) Rotate trapezium A by a quarter turn anti-clockwise, centre the origin. Label the image D. What are the coordinates of the vertices of D?
   (e) What single transformation would map C onto D?
   (f) Are trapeziums A and D congruent?
   Explain your answer briefly.

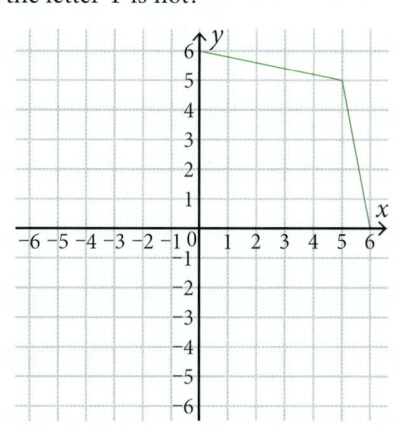

5. The diagram on the right shows a parallelogram ABCD.
   (a) Reflect ABCD in the line $y = x$
   (b) Reflect ABCD in the line $y = -x$

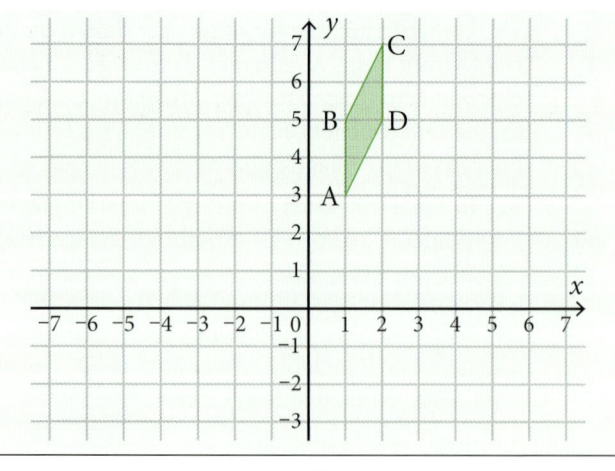

34.4 Rotations

Another type of transformation is a rotation. A rotation is performed by turning an object.

A full turn is 360°, so:
- a quarter turn is a 90° rotation,
- a half turn is a 180° rotation,
- a three-quarters turn is a rotation through 270°

In the diagram on the right the green flag is the image of the white flag when it is rotated through 90° clockwise about the point (1, 0).

The grey flag is the image of the white one when it is rotated through 180° about the point (1, 0).

A rotation of 180° can be clockwise or anti-clockwise.

After any rotation, the image shape is always congruent to the original shape.

Using tracing paper is the easiest way to rotate a shape. Follow these steps:

1. Trace the **original** shape **and** some of the grid lines onto tracing paper.
2. Without moving the tracing paper, put the point of your pencil on the centre of rotation to fix that point.
3. Rotate the tracing paper. After a 90° or 180° rotation the grid lines on the tracing paper should line up with those on the page.
4. Copy the shape back onto the page as the **image**.

The next example demonstrates rotating a shape about the origin.

Example 7

Look at the shape shown on the right.
(a) Rotate shape A by 90° in a clockwise direction, centre (0, 0). Label the resulting image B.
(b) Rotate shape A by 180°, centre the origin. Label the resulting image C.
(c) Why does the direction of rotation in part (b) not matter for this rotation?

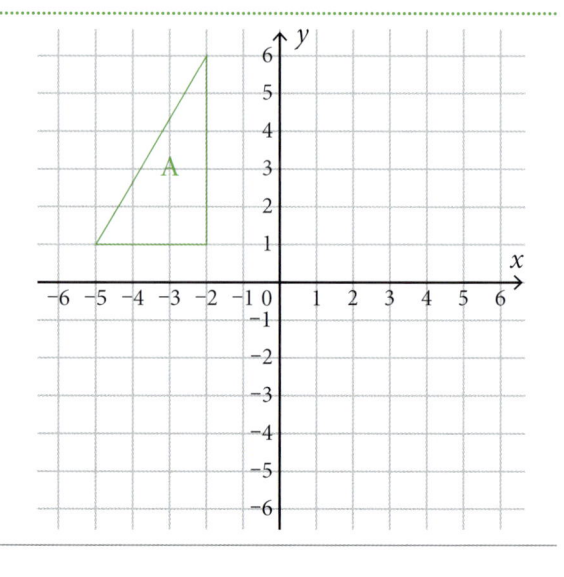

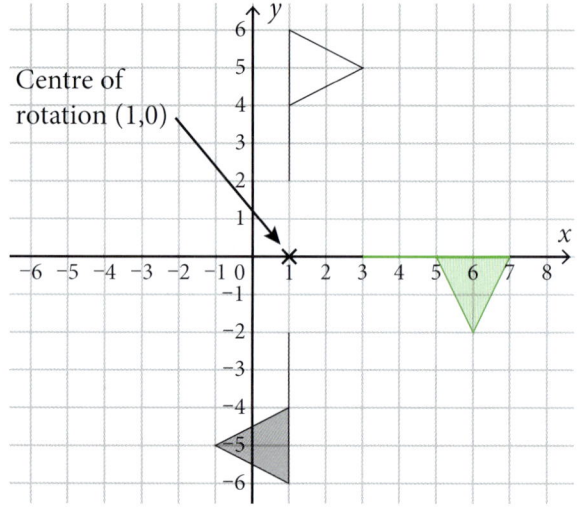

312

(a) and (b) Shapes B and C are shown on the diagram on the right.
(c) The direction does not matter because a turn of 180° is the same whichever direction you turn.

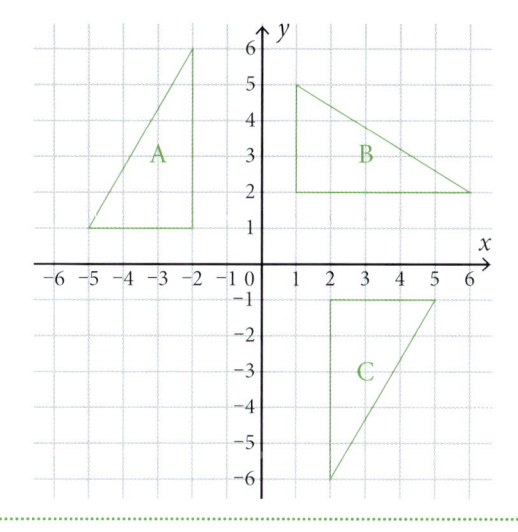

Describing a rotation

The next example shows how to **describe a rotation**.

Example 8

Look at the diagram on the right showing three arrowheads.
(a) Describe the transformation from shape A to shape B.
(b) What are the coordinates of the four vertices of shape B?
(c) Describe the transformation from shape A to shape C.

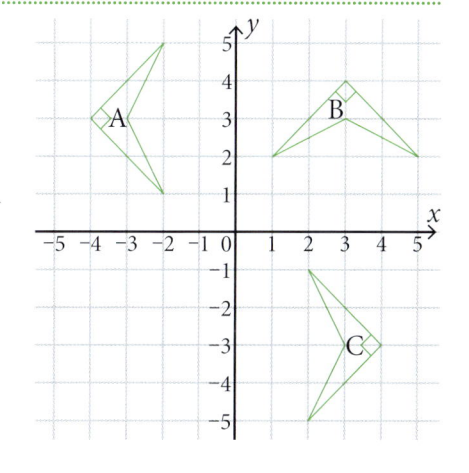

(a) Remember, to describe a rotation you must give the angle, the direction and the centre of rotation. This is a rotation through 90° clockwise about (0, 0).
(b) The coordinates of the vertices of B are: (1, 2), (3, 4), (5, 2) and (3, 3).
(c) This is a rotation through 180° about (0, 0). To describe a rotation of 180° you don't need to give the direction. Both clockwise and anti-clockwise are correct.

Exercise 34D

Reminder: You can use tracing paper in all questions on rotations.

1. Copy the diagram on the right.
   (a) The triangle A is rotated through a quarter turn in a clockwise direction about the origin. Draw its image and label it B.
   (b) The triangle A is rotated through a quarter turn in an anti-clockwise direction about the origin. Draw its image and label it C.
   (c) Describe the rotation that maps triangle B onto triangle C.

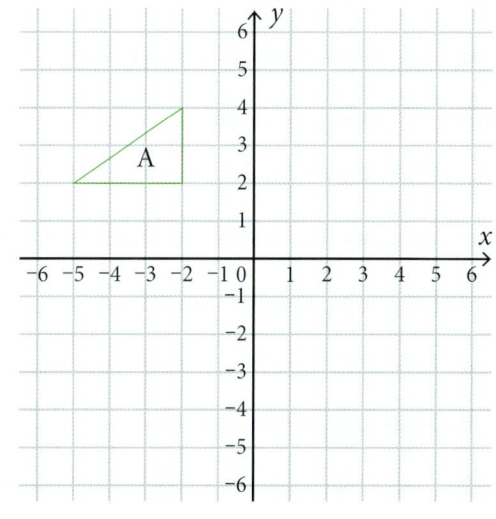

2. (a) Draw x- and y-axes from −7 to 7
   Plot the points: A(6, 1), B(2, 6) and C(4, 1).
   Join them to form a triangle.
   (b) Rotate the triangle through 90° clockwise about the origin. Label the image A'B'C' and write down the coordinates of each vertex.
   (c) Rotate A'B'C' by a further 180° clockwise about the origin. Label the image A"B"C" and write down the coordinates of each vertex.

3. Draw a grid using an *x*-axis from −4 to 4 and a *y*-axis from −4 to 4
   (a) The vertices of triangle A are (0, 0), (1, 0) and (0, 3). Draw the triangle on the grid.
   (b) Draw the image of A when it is rotated through 90° in a clockwise direction about the origin. Label it B.
   (c) Draw the image of A when it is rotated through 180°, centre the origin. Label it C.
   (d) Draw the image of A when it is rotated through 270° in a clockwise direction about the origin. Label it D.
   (e) Describe the effect of rotating A through 360° about the origin.

4. (a) Draw *x*- and *y*-axes from −5 to 5
       Plot these coordinates and join them to make a trapezium: (1, 1), (1, 4), (2, 4), (4, 1)   Label it A.
   (b) Rotate trapezium A by a quarter turn clockwise, centre the origin. Label the image B. What are the coordinates of the vertices of B?
   (c) Rotate trapezium A by a half turn, centre the origin. Label the image C. What are the coordinates of the vertices of C?
   (d) What single transformation would map C onto D?
   (e) Are trapeziums A and D congruent? Explain your answer briefly.

5. Look at the diagram on the right.
   (a) Describe each of these rotations:
       (i)   A to B
       (ii) A to C
       (iii) B to A.
   (b) Draw and label *x*- and *y*-axes from −4 to 4 Copy shape A onto your diagram. Rotate A through 180° about the origin. Label the image D.

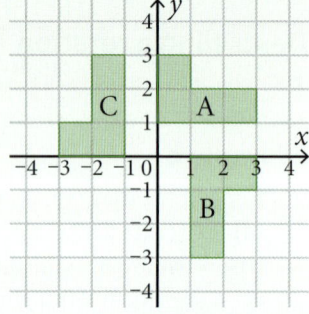

6. Look at the arrow on the diagram on the right. Copy the diagram.

   Draw the arrow after each of the transformations given below. In each part the image should have a length of 3 units.

   (a) A rotation 90° clockwise, centre (1, 1).
   (b) A rotation 270° anti-clockwise, centre (1, 1).
   (c) A rotation 180° clockwise, centre (1, 1).
   (d) A rotation 180° anti-clockwise, centre (1, 1).

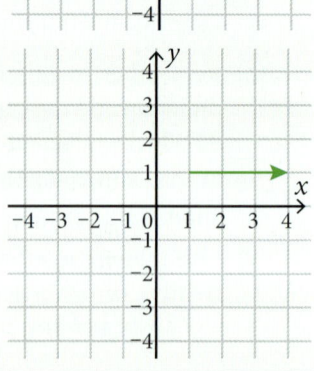

34.5 Combined Transformations

You may be asked to transform 2D shapes using a combination of transformations.

You may also be asked to describe one or more of these transformations, or describe a single transformation that would have the same effect.

Example 9

The diagram on the right shows a triangle T1.
(a) Rotate T1 by 90° clockwise about the point (0, 0). Label the image T2.
(b) Reflect triangle T2 in the *y*-axis. Label the image T3.
(c) What single transformation maps T1 onto T3?

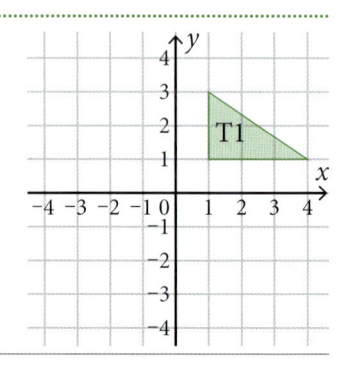

(a) – (b) The two transformations are shown in the diagram on the right.
(c) The single transformation that maps T1 onto T3 is a reflection in the line $y = -x$

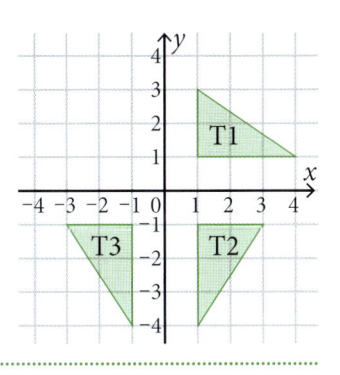

Exercise 34E

1. The diagram on the right shows a triangle labelled T1.
   (a) Rotate T1 anti-clockwise 90° about the origin. Label the image triangle T2.
   (b) Reflect T2 in the y-axis. Label the image T3.
   (c) What single transformation maps T1 to T3?

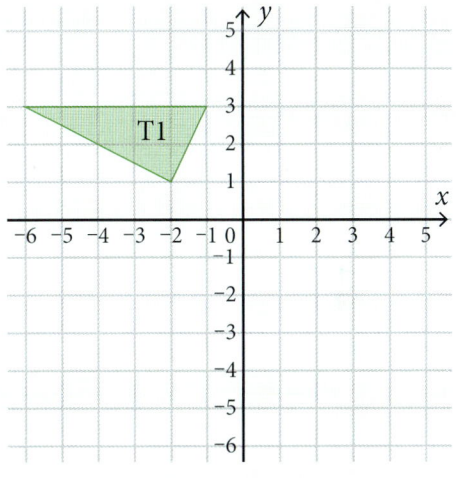

2. Shape A shown on the grid on the right is a trapezium.
   (a) Translate A using the vector $\begin{pmatrix} 2 \\ 4 \end{pmatrix}$. Label the image B.
   (b) Reflect shape B in the line $x = 0$. Label the image C.
   (c) Write down a single transformation that maps A to C.
   (d) What single transformation maps C to A.

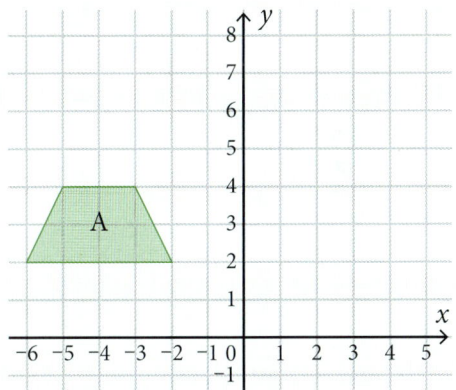

3. Look at shape A on the right.
   (a) Translate A using the vector $\begin{pmatrix} -4 \\ 2 \end{pmatrix}$. Label the image B.
   (b) Rotate shape B 180° about the point (3, 6). Label the image C.
   (c) State **two** single transformations that map A onto C.

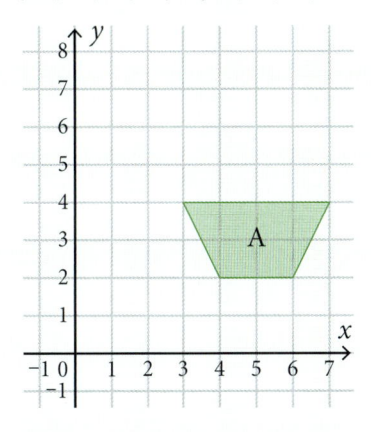

4. The diagram on the right shows an arrowhead, W and point C(3, 3).
   (a) Rotate W 180° about point C. Label the image X.
   (b) Reflect shape X in the mirror line $x = 3$
       Label the image shape Y.
   (c) Translate X by the vector $\begin{pmatrix} 0 \\ -2 \end{pmatrix}$
       Label the image shape Z.
   (d) State a single transformation that would take shape W to shape Y.
   (e) State a single transformation that would take shape W to shape Z.

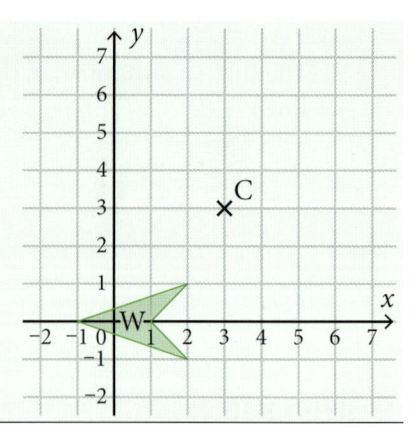

34.6 Summary

In this chapter you have learned:

- About three types of transformation: translations, rotations and reflections.
- That you may be asked to draw the **image** of a shape after any of these transformations.
- That you may also be asked to describe a transformation. In questions like this, state the type of transformation **and** give all the required information:
  - For a translation: give the vector.
  - For a reflection: give the equation of the mirror line.
  - For a rotation: give the coordinates of the centre of rotation, the angle and the direction.
- That you may be asked to carry out more than one transformation (combined transformations) to obtain the final image shape.

Chapter 35
Enlargement and Similarity

35.1 Introduction

A **transformation** moves an object according to a rule. In Chapter 34, three types of transformation were discussed: translations, rotations and reflections. In this chapter, a fourth type of transformation is introduced: **enlargements**. You will learn about enlargements in Section 35.2.

After an enlargement, the image shape is mathematically **similar** to the original. Similar figures have the same shape but different sizes.

In Section 35.3 you will learn how to prove that two shapes are similar.

In Section 35.4 you will use similarity to find properties of shapes, for example side lengths.

In Section 35.5 you will see some examples of harder problems involving both proof and similarity.

In Section 35.6 you will learn about the relationship between the lengths and areas of similar shapes.

In Section 35.7 we discuss the effect of enlargement on the volume of a 3D shape.

Key words
- **Enlargement**: An enlargement increases the size of a shape without changing its shape.
- **Image**: The shape that results from a transformation.
- **Similar**: Shapes that are similar have the same shape but different sizes. If one shape is an enlargement of another, the two shapes are similar.
- **Congruent**: Two shapes are congruent if they are exactly the same size and shape.

Before you start you should:
- Know how to simplify a ratio.
- Recall the sum of angles in a triangle and a quadrilateral.
- Recall the angle properties relating to parallel lines.
- Understand that two shapes are **congruent** if they are exactly the same size and shape.

In this chapter you will learn:
- How to carry out and describe an enlargement of a 2D shape.
- About mathematical similarity.
- How to prove that two shapes are similar.
- How to use similarity to find unknown side lengths or angles.
- How to use the relationship between the ratios of lengths and areas of similar 2D shapes.
- To understand and use the effect of enlargement on the volume of a solid.

Congruence

Two shapes are congruent if they are exactly the same size and shape. Their orientation does not matter.

Example 1 (Revision)

The shapes shown in the diagram are drawn on a 1 cm grid.
Which two of the shapes are congruent?

Shapes D and E are congruent as they are exactly the same size and shape. The difference in the orientation (which way they are facing) does not matter.

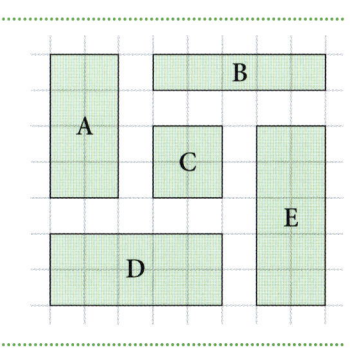

GCSE MATHEMATICS M3 AND M7

Exercise 35A (Revision)

1. Simplify these ratios.
   (a) 14:2   (b) 20:15   (c) 2:6:10

2. (a) In the isosceles triangle ABC, angle A is 110°
   Find the size of angles B and C.
   (b) In a quadrilateral, PQRS, angle P is 64°, angle Q is 101° and angle S is 84°
   Find angle R.

3. Find the unknown angles in the following diagrams.
   (a) 72°, a
   (b) b, 108°
   (c) c, 64°
   (d) d, 135°

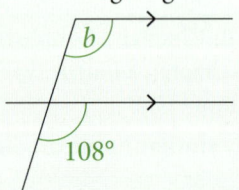

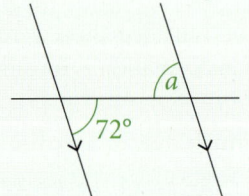

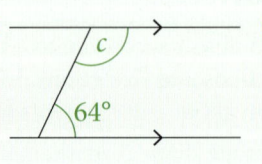

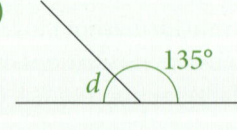

4. Look at these 8 triangles. There are four congruent pairs.
   Copy and complete the following:
   (a) Triangle A is congruent to triangle _____
   (b) Triangle _____ is congruent to triangle _____
   (c) Triangle _____ is congruent to triangle _____
   (d) Triangle _____ is congruent to triangle _____

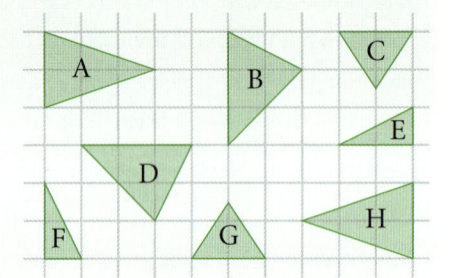

35.2 Enlargements

Enlarging 2D shapes

You may be asked to enlarge a 2D shape using a positive whole number scale factor.

Example 2

Look at the shape shown on the right.
Enlarge this shape using a scale factor of 3

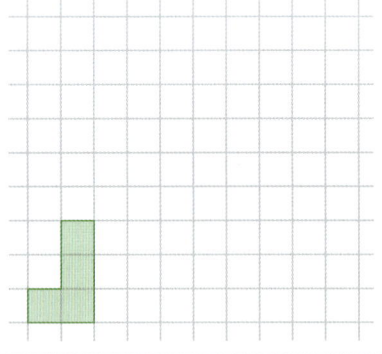

All the side lengths are multiplied by the scale factor 3
The orientation of the shape remains the same.
One possible solution is shown on the right.

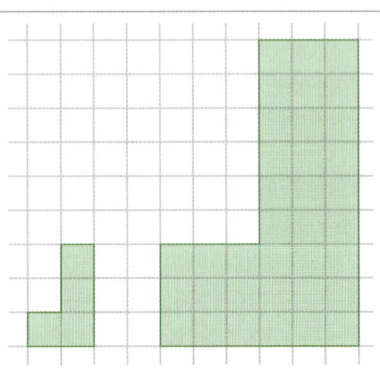

Sometimes you will be asked to use a **centre of enlargement**.

CHAPTER 35: ENLARGEMENT AND SIMILARITY

Example 3

Enlarge the shape PQRS, shown on the upper right, using a scale factor of 2 and a centre of enlargement C(1, 1). Label the image P'Q'R'S'.

For each vertex of the original shape:
- Calculate the vector from the centre of enlargement to the vertex. For example, from the centre C(1, 1) to P(3, 2), we move 2 to the right and 1 up, giving a vector $\begin{pmatrix} 2 \\ 1 \end{pmatrix}$.
- Multiply both numbers in the vector by the scale factor. In this case, the scale factor is 2, giving $\begin{pmatrix} 4 \\ 2 \end{pmatrix}$.
- Use this vector to move from the centre of enlargement to the image point. Moving from (1, 1) using the vector $\begin{pmatrix} 4 \\ 2 \end{pmatrix}$ takes us to P'(5, 3).
- Repeat this for each vertex to get the positions of all four vertices in the image.
- Join the vertices in the image to form the enlarged shape, as shown in the diagram on the lower right.

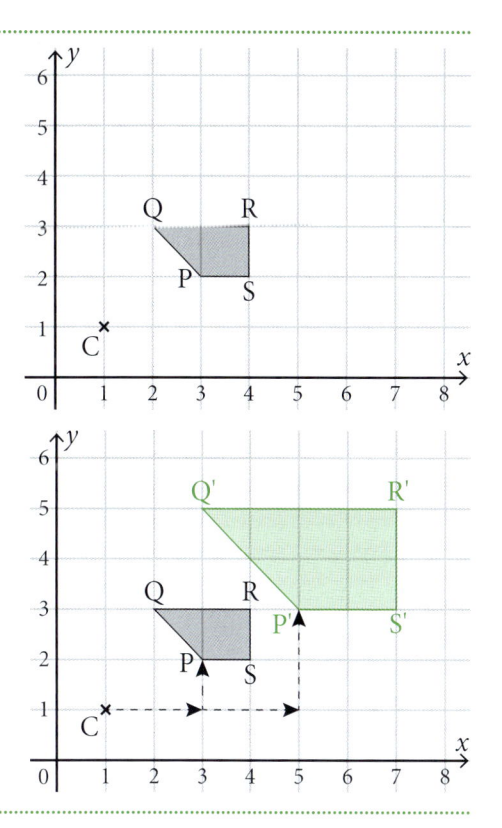

Enlargements using fractional scale factors

You may be asked to enlarge a 2D shape using a fractional scale factor. If the scale factor is between 0 and 1, the image shape is smaller than the original. The same method is used as in the previous section.

Example 4

The diagram on the right shows a quadrilateral PQRS. Enlarge this shape using a scale factor of $\frac{1}{2}$ and centre of enlargement C(1, 2).

For each vertex of the original shape:
- Calculate the vector from the centre of enlargement to the vertex. For example, from the centre C(1, 2) to P(7, 6), we move 6 to the right and 4 up, giving a vector $\begin{pmatrix} 6 \\ 4 \end{pmatrix}$.
- Multiply both numbers in the vector by the scale factor. In this case, the scale factor is $\frac{1}{2}$, giving $\begin{pmatrix} 3 \\ 2 \end{pmatrix}$.
- Use this vector to move from the centre of enlargement to the image point. Moving from (1, 2) using the vector $\begin{pmatrix} 3 \\ 2 \end{pmatrix}$ takes us to P'(4, 4).
- Repeat this for each vertex to get the positions of all four vertices in the image.
- Join the vertices in the image to form the image shape, as shown in the second diagram on the right.

Note: The image is smaller than the original because the scale factor of $\frac{1}{2}$ is between 0 and 1

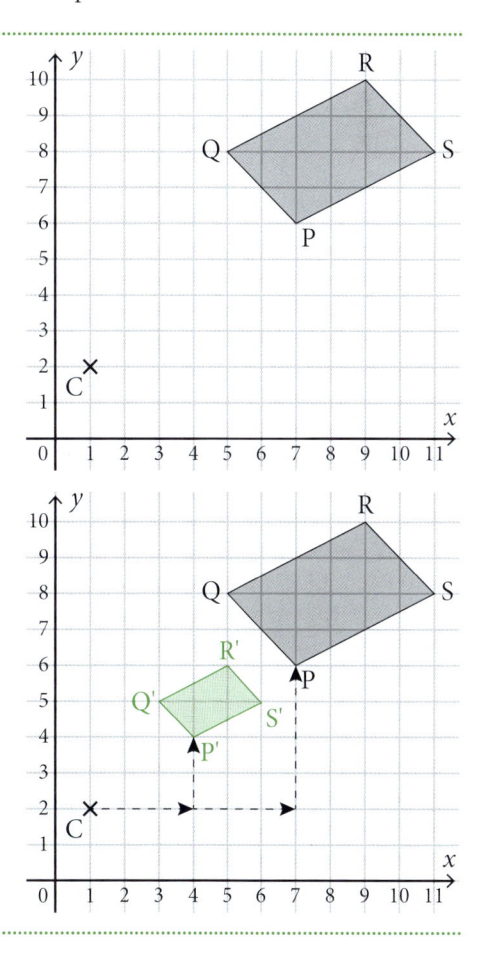

GCSE MATHEMATICS M3 AND M7

Describing enlargements of 2D shapes

You may be given a diagram of the original shape and the image and asked to describe the enlargement. If there is no centre of enlargement, you should just state the scale factor.

Example 5

The two shapes shown on the right are drawn on a 1 cm grid.

Shape B is an enlargement of shape A. What is the scale factor?

The length of each side in shape B is two times the length of the corresponding side in shape A. For example, the base of shape B is 4 cm and the base of shape A is 2 cm.

Therefore, the scale factor is 2

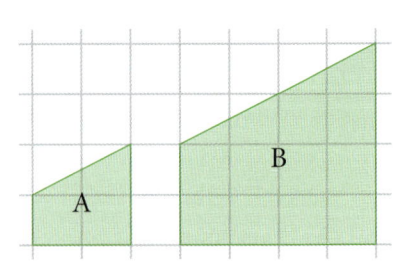

If the two shapes are drawn on a grid with coordinate axes shown, you should also give the centre of enlargement.

Example 6

Look at the diagram on the right.

Copy and complete the following:

Shape B is an _____ of shape A using a scale factor of _____ and a centre of enlargement (___, ___).

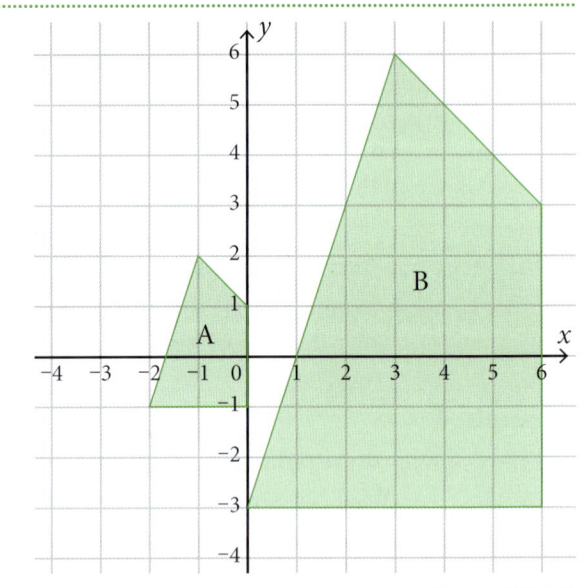

The original, A, and image B have the same shape, but different sizes, so this is an enlargement.

To find the scale factor, note that the base of shape B is 6 units, while the base of shape A is 2 units, so the scale factor is 3

To find the centre of enlargement, connect all 4 pairs of corresponding vertices, as shown in the diagram on the right. The lines you draw should intersect at the centre of enlargement.

These four lines intersect at (−3, 0), so this is the centre of enlargement.

The completed sentence is:

Shape B is an ___enlargement___ of shape A using a scale factor of __3__ and a centre of enlargement (_−3_, _0_).

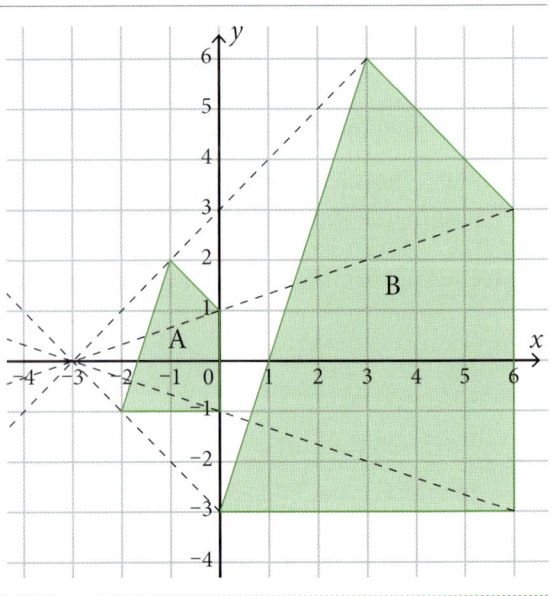

The following example shows how to describe an enlargement by a fractional scale factor.

320

CHAPTER 35: ENLARGEMENT AND SIMILARITY

Example 7

The diagram on the right shows a kite PQRS and its image P'Q'R'S' after a transformation.

Describe the transformation taking PQRS to P'Q'R'S'.

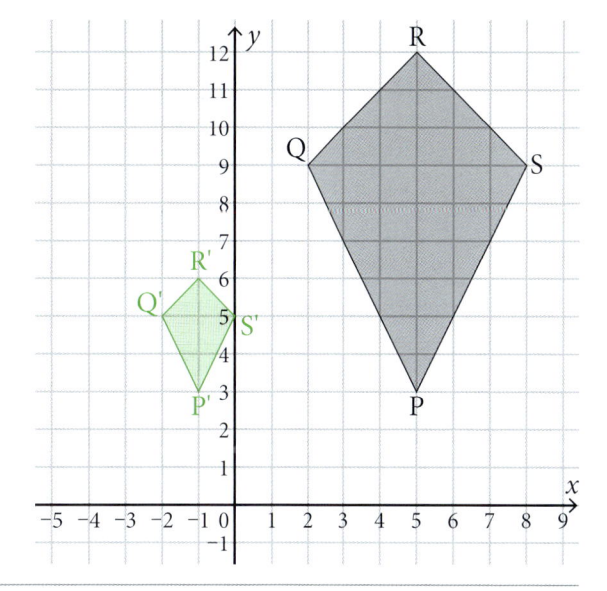

The original PQRS and image P'Q'R'S' have the same shape, but different sizes, so this is an enlargement.

To find the scale factor, note that the height of the original is 9 units, while the height of the image is 3 units, so the scale factor is $\frac{3}{9}$ or $\frac{1}{3}$.

To find the centre of enlargement, connect all 4 pairs of corresponding vertices, as shown in the diagram on the right. The lines you draw should intersect at the centre of enlargement, C.

These lines intersect at (−4, 3), so this is the centre of enlargement.

The complete description is: **Enlargement using a scale factor of $\frac{1}{3}$ and a centre of enlargement (−4, 3).**

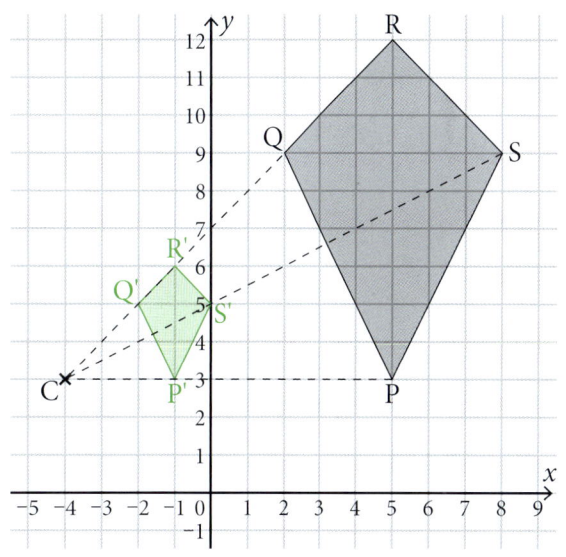

Exercise 35B

1. Copy the shape below onto squared paper. On your grid, enlarge this shape using a scale factor of 2

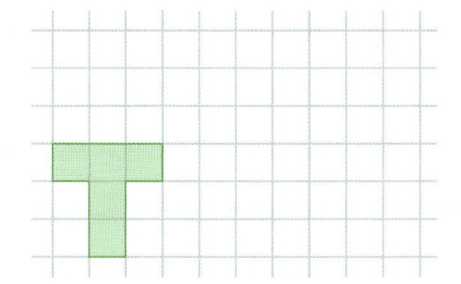

2. Copy the shape below onto squared paper. On your grid, enlarge the shape using a scale factor of 2

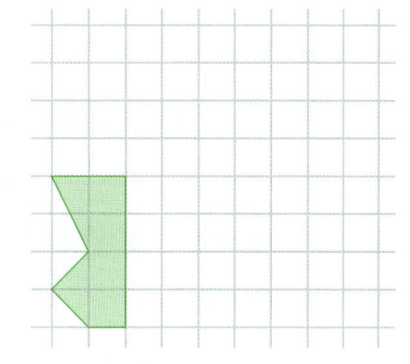

3. Copy the grid and the kite shown in the diagram below. Enlarge the kite using a centre of enlargement (0, 0) and a scale factor of 3.

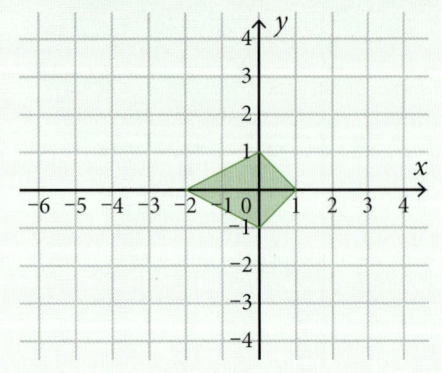

4. Copy the grid and the triangle shown in the diagram below. Enlarge the triangle using a scale factor of 3 and a centre of enlargement (2, 2).

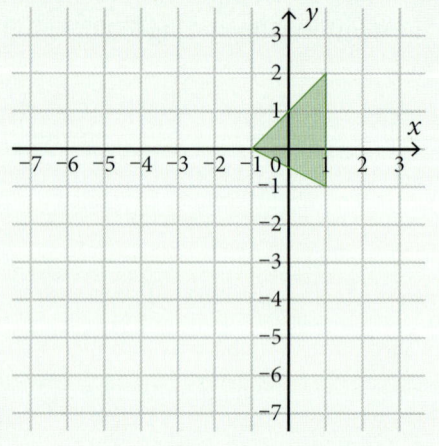

5. After an enlargement using a scale factor of 2, are the original shape and the image congruent? Explain your answer.

6. The two shapes shown below are drawn on a 1 cm grid. Shape B is an enlargement of shape A. What is the scale factor?

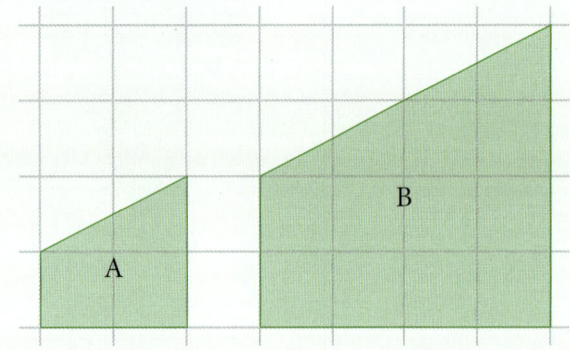

7. In the diagram be;pw, shape B is an enlargement of shape A. Find the scale factor and the centre of enlargement.

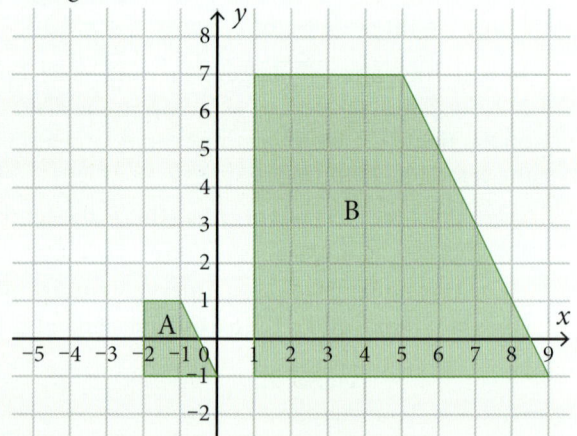

8. Look at the diagram on the right. Describe fully the transformation that maps shape A to shape B.

   Hint: to find the scale factor you could think about the distances PT and P'T'

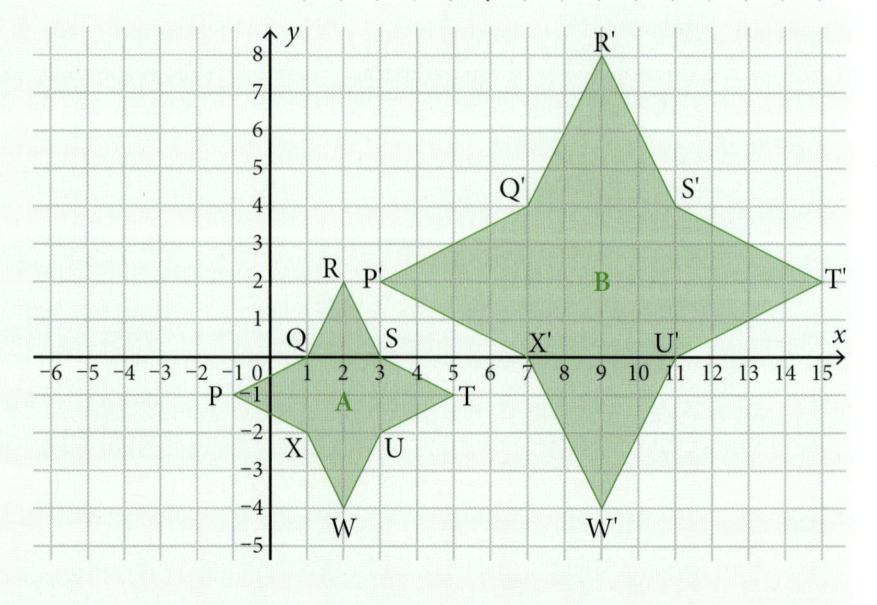

322

CHAPTER 35: ENLARGEMENT AND SIMILARITY

9. The diagram on the right shows a quadrilateral WXYZ. Copy the diagram and draw the image of this shape after an enlargement with scale factor $\frac{1}{2}$ and centre of enlargement (−4, 1).

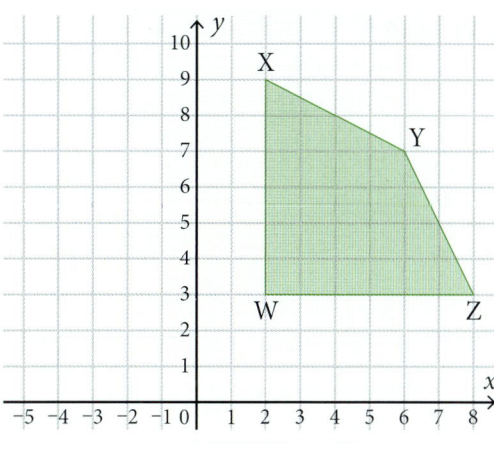

10. Under a single transformation, the triangle T, shown in the diagram on the right, is mapped to triangle T'. Describe in full the transformation.

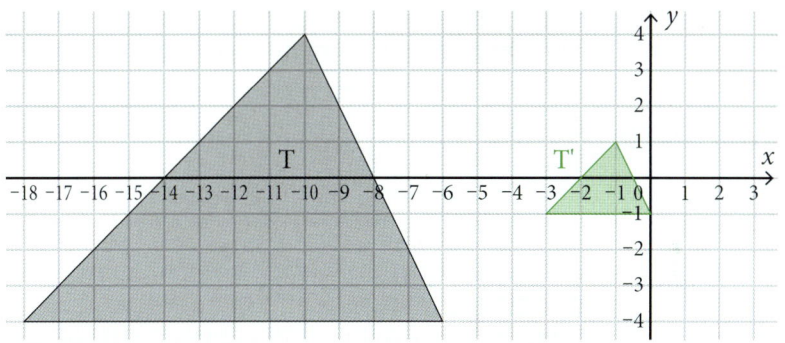

35.3 Proving Similarity in 2D Shapes

Shapes that are **similar** have the same shape but different sizes.

In two similar shapes:
- Each angle in one shape is equal to the corresponding angle in the second shape; **and**
- The corresponding sides in each shape are in the same ratio.

Proving two triangles are similar

For two triangles, if one of these conditions is true, then the second one is automatically true. Therefore, to prove that two triangles are similar, we only need to show that **one** of the conditions above is true. So, we must show that **either**:
- Each angle in one shape is equal to the corresponding angle in the second shape; **or**
- The corresponding sides in each shape are in the same ratio.

You can use either method. Example 8 below uses angles to prove that two triangles are similar, while Example 9 uses side lengths.

Example 8

Look at the diagram on the right. Prove that triangles ABC and DEF are similar.

We must calculate angle C and angle E.

In triangle ABC, since the three angles add up to 180°:
C = 180 − 33 − 59 = 88°

In triangle DEF:
E = 180 − 33 − 88 = 59°

So: A = D; B = E and C = F

Therefore, the triangles are similar.

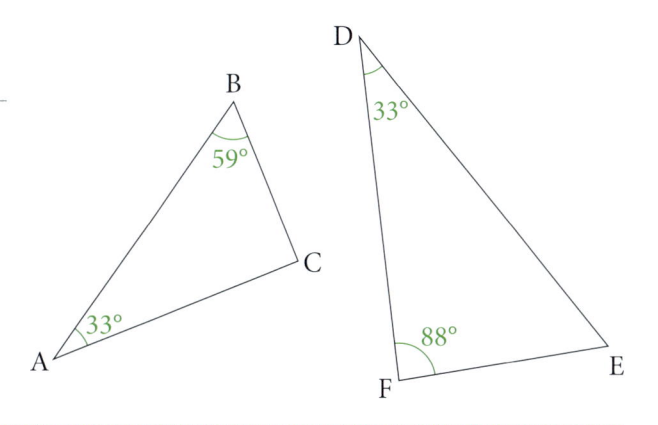

323

GCSE MATHEMATICS M3 AND M7

Example 9

Look at the diagram on the right. Prove that triangles ABC and DEF are similar.

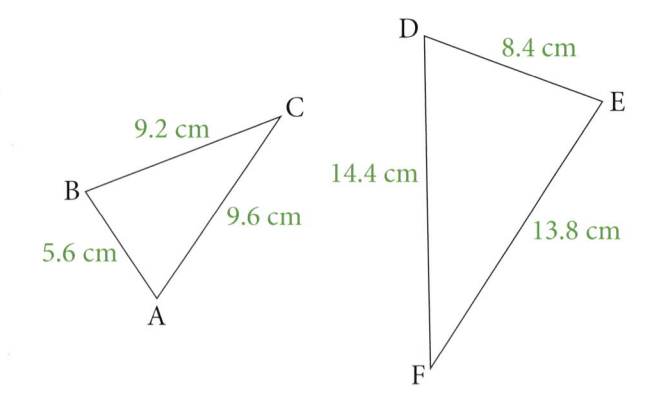

Consider the longest side in each triangle, AC and DF:
$\frac{DF}{AC} = \frac{14.4}{9.6} = 1.5$

The next longest pair is BC and EF:
$\frac{EF}{BC} = \frac{13.8}{9.2} = 1.5$

The shortest sides are AB and DE:
$\frac{DE}{AB} = \frac{8.4}{5.6} = 1.5$

Since each pair of sides is in the same ratio, the triangles ABC and DEF are similar.

Exercise 35C

1. In each part (a) – (f) prove that each pair of triangles is similar.

   (a) 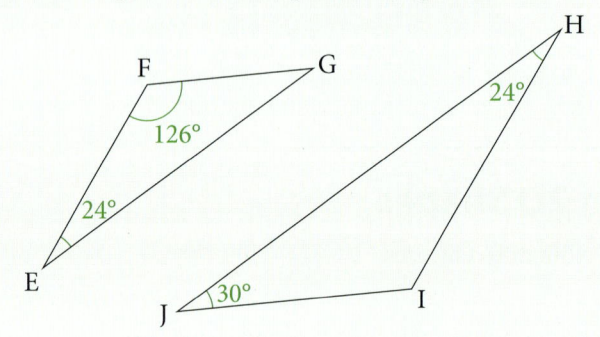

   (b) 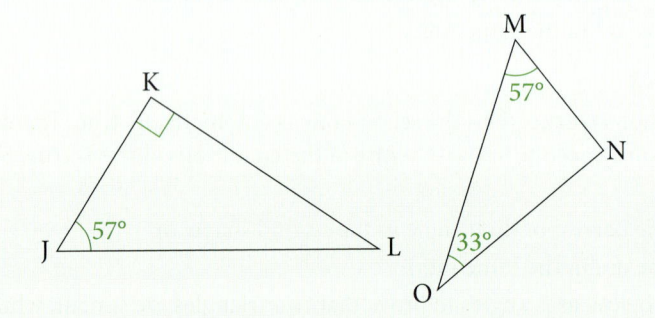

   (c) 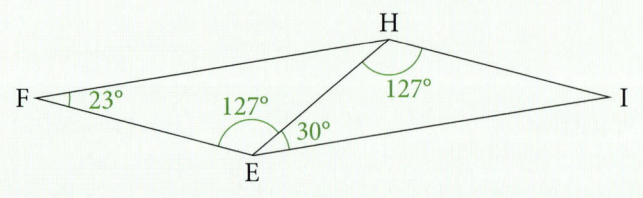

   (d)

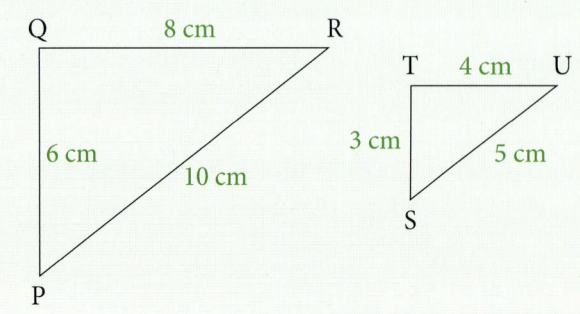

CHAPTER 35: ENLARGEMENT AND SIMILARITY

(e)

B 10 cm C
7 cm
14 cm
A

F
12 cm
16.8 cm
E
8.4 cm
D

(f)

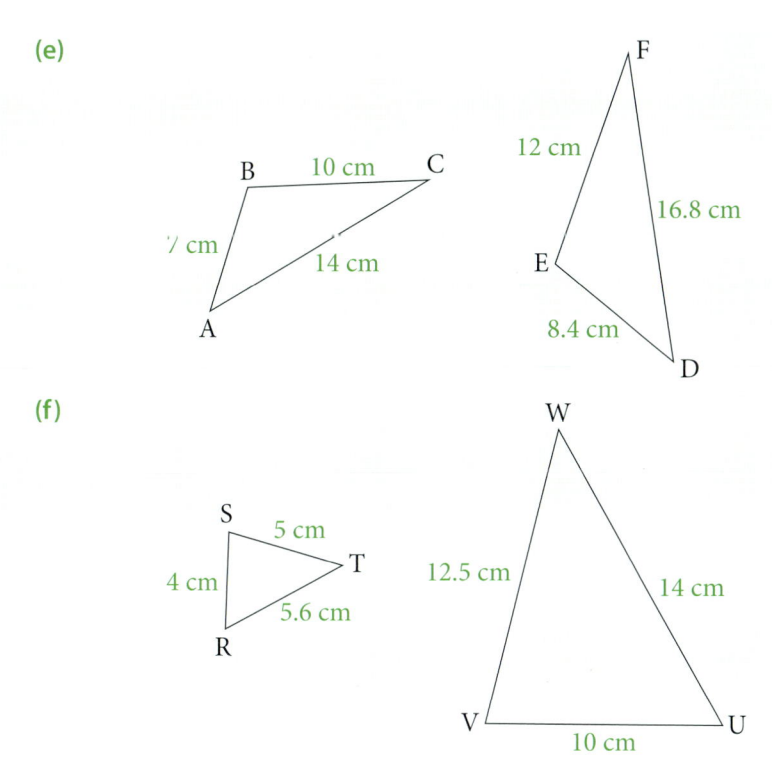

S
5 cm
4 cm   T
5.6 cm
R

W
12.5 cm   14 cm
V 10 cm U

2. Triangle T has sides of length 10, 15 and 20 cm. Triangle U has sides of length 15, 22.5 and 25 cm. Are triangles T and U similar? Show all your working.

Proving similarity in other shapes

To prove similarity in non-triangular shapes, we must prove that **both** conditions are true:
- Each angle in one shape is equal to the corresponding angle in the second shape; **and**
- The corresponding sides in each shape are in the same ratio.

Example 10

Prove that the two quadrilaterals shown on the right are mathematically similar.

Since angles in a quadrilateral add up to 360°:

Angle B = 360 − (78 + 127 + 84)
= 71°

Angle G = 360 − (78 + 71 + 84)
= 127°

So the angles are the same in both quadrilaterals.

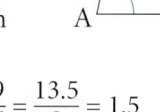

For the side lengths: $\frac{15}{10} = \frac{11.1}{7.4} = \frac{9}{6} = \frac{13.5}{9} = 1.5$

So the sides are in the same ratio, 1:1.5 or 2:3

Therefore, the two shapes are similar.

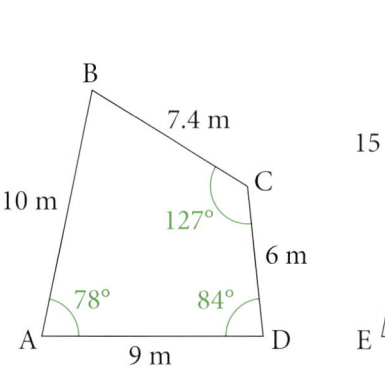

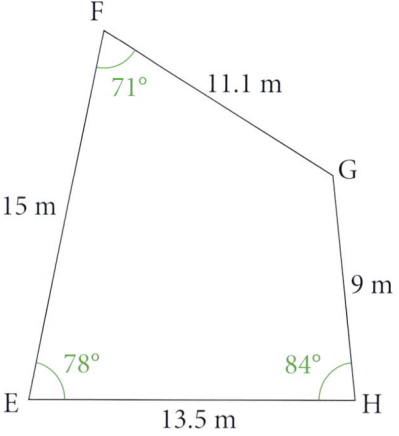

325

GCSE MATHEMATICS M3 AND M7

Exercise 35D

1. Prove that the following pairs of shapes are mathematically similar.

   (a)

   (b)

   (c)

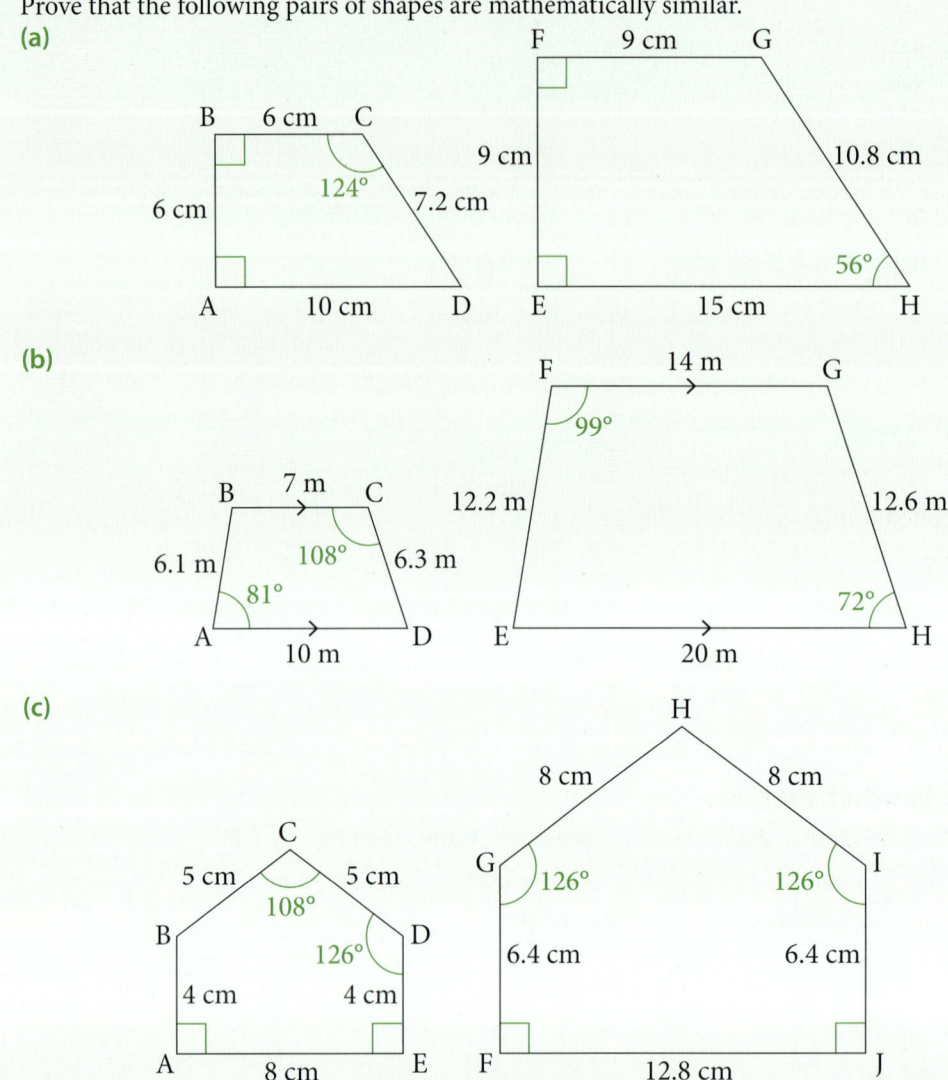

2. Two regular hexagons have sides of length 4 cm and 6 cm. Prove that the two hexagons are similar.

3. In kite K, the two shorter sides are 3 cm long and the two longer sides are 7 cm long. In kite L, the two shorter sides are 6 cm long and the two longer sides are 14 cm long. Which of the following statements is true?
   - Statement 1: Kites K and L are similar.
   - Statement 2: Kites K and L are not similar.
   - Statement 3: We do not have enough information to decide whether kites K and L are similar.

4. A rectangle R has a length of $(3x + 2)$ cm and a width of $(x - 2)$ cm. A second rectangle S has a length of $(9x + 6)$ cm and a width of $(3x - 6)$ cm. Which of the following statements is true:
   - Statement 1: Rectangles R and S are similar.
   - Statement 2: Rectangles R and S are not similar.
   - Statement 3: We do not have enough information to decide whether rectangles R and S are similar.

35.4 Using Similarity in 2D Shapes

Sometimes you may be told that two shapes are similar. You can use this information to find unknown angles or sides.

CHAPTER 35: ENLARGEMENT AND SIMILARITY

Example 11

The two trapezia shown on the right are mathematically similar.
(a) Find the size of angle B.
(b) Find the length of the side marked x.
(c) Find the length of the side marked y.
(d) Write down the sizes of all four angles in trapezium PQRS.

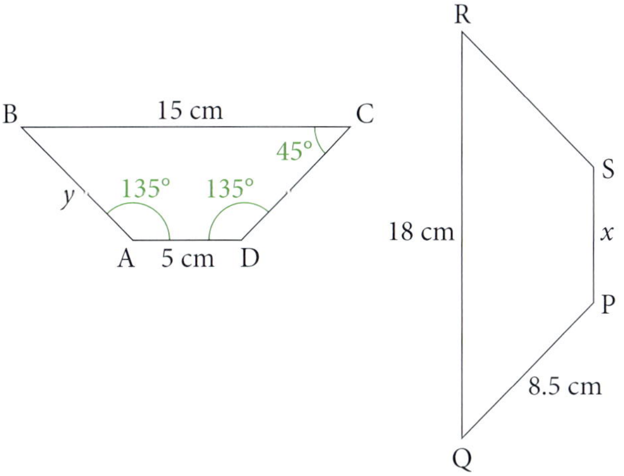

(a) In trapezium ABCD, the angles add up to 360°, so:

B = 360 − (135 + 135 + 45)
B = 45°

(b) Side QR in PQRS corresponds to side BC in ABCD.
Side PS in PQRS corresponds to side AD in ABCD.

Corresponding sides are in the same ratio. So: $\dfrac{x}{5} = \dfrac{18}{15}$

Cross-multiply: $15x = 90$

Giving: $x = \dfrac{90}{15} = 6$ cm

(c) Side PQ in PQRS corresponds to side AB in ABCD.

So: $\dfrac{8.5}{y} = \dfrac{18}{15}$

Cross–multiply: $18y = 127.5$

Giving: $y = \dfrac{127.5}{18} = 7.1$ cm (1 d.p.)

(d) Since the two shapes are similar, the angles in PQRS are the same as those in ABCD. So:
P = 135°, Q = 45°, R = 45°, S = 135°

Exercise 35E

1. Calculate the sides marked x and y in the following pairs of similar shapes. Round your answers to 1 decimal place where necessary.

(a)

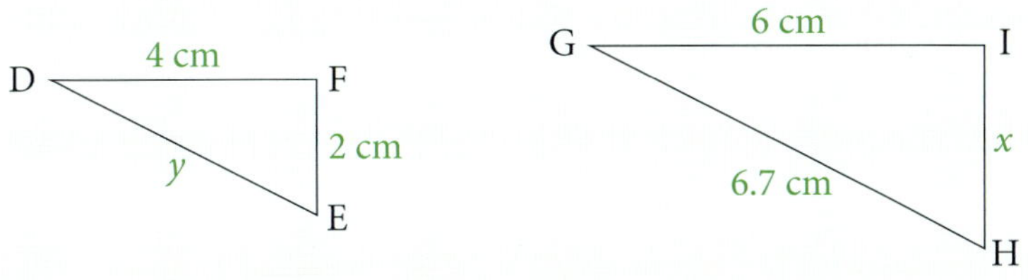

(b)

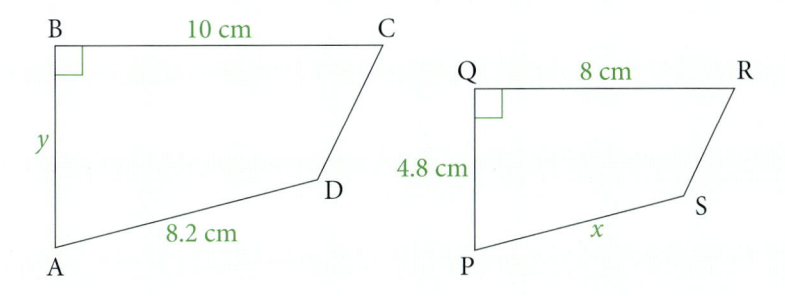

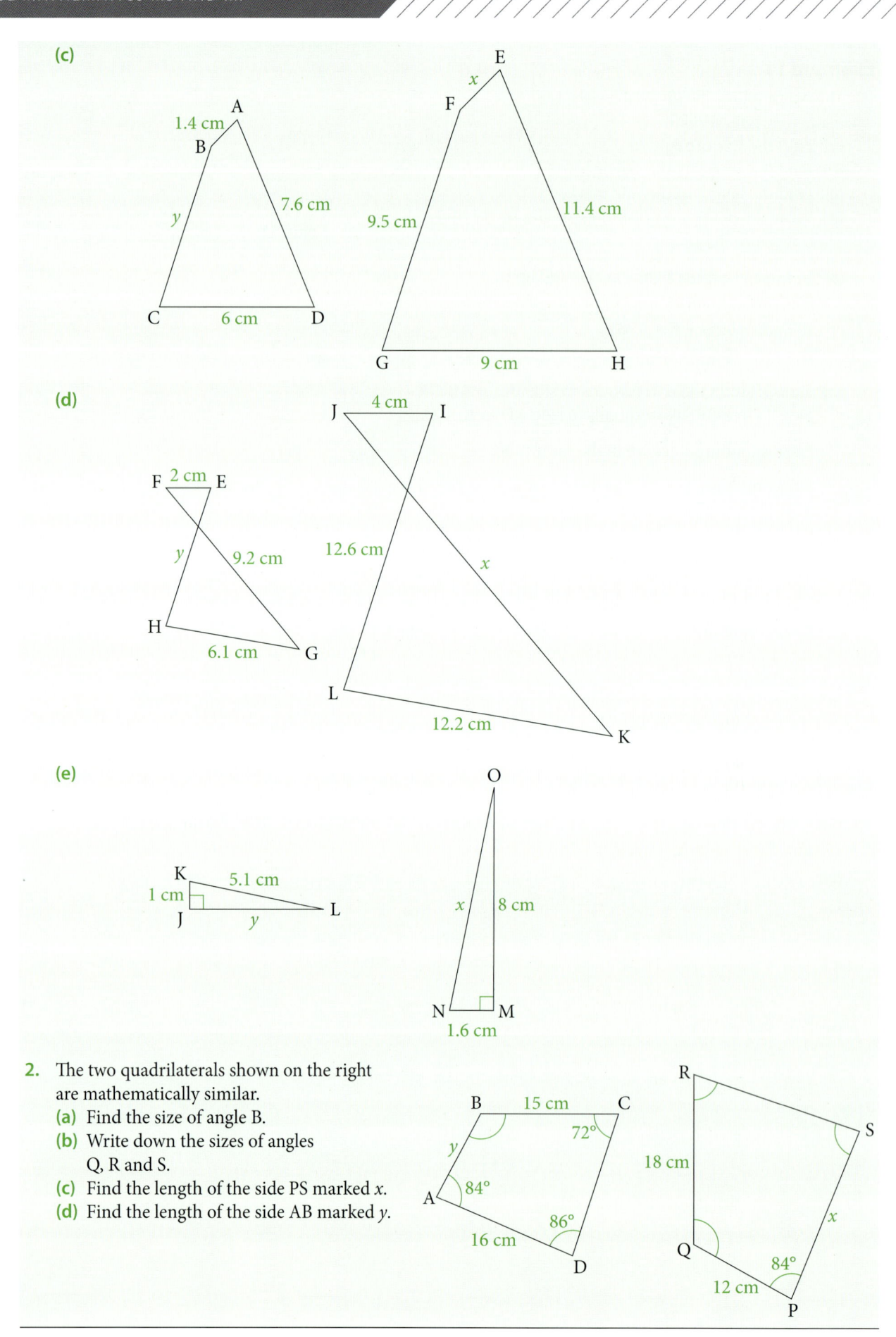

(c)

Triangle ABCD: AB = 1.4 cm, BD = 7.6 cm, CD = 6 cm, AC = y

Triangle EFGH: EF with angle x at E, FG = 9.5 cm, EH = 11.4 cm, GH = 9 cm

(d)

Triangle EFGH: FE = 2 cm, EG = 9.2 cm, HG = 6.1 cm, FH = y

Triangle IJKL: JI = 4 cm, JK = 12.6 cm, IK = x, LK = 12.2 cm

(e)

Triangle JKL: JK = 1 cm, KL = 5.1 cm, JL = y, right angle at J

Triangle MNO: NM = 1.6 cm, MO = 8 cm, NO = x, right angle at M

2. The two quadrilaterals shown on the right are mathematically similar.
  (a) Find the size of angle B.
  (b) Write down the sizes of angles Q, R and S.
  (c) Find the length of the side PS marked *x*.
  (d) Find the length of the side AB marked *y*.

Quadrilateral ABCD: BC = 15 cm, angle C = 72°, angle A = 84°, angle D = 86°, CD = 16 cm, AB = y

Quadrilateral PQRS: RQ = 18 cm, QP = 12 cm, angle P = 84°, SP = x

CHAPTER 35: ENLARGEMENT AND SIMILARITY

35.5 Harder Problems

Certain problems may involve a combination of the skills you learnt in sections 35.3 and 35.4: you will prove that two shapes are similar, and then use this fact to find an unknown side or angle.

You may also have to use Pythagoras' Theorem, corresponding, alternate or vertically opposite angles.

The next example demonstrates a proof involving corresponding angles. Note that, in this example, it is possible to show that the angles are equal without knowing the sizes of any of them.

Example 12

In the diagram on the right, AB is parallel to DE. Prove that the triangles ABC and DEC are mathematically similar.

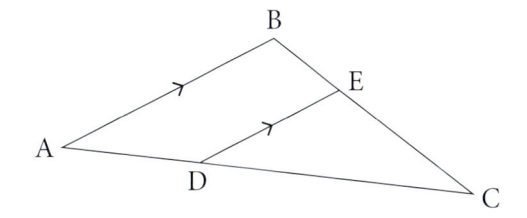

Identify equal angles in the diagram.

Triangle ABC		Triangle DEC	
∠BAC	=	∠EDC	since they are corresponding angles.
∠ABC	=	∠DEC	since they are corresponding angles.
C	=	C	since this is an angle that is common to both triangles.

Since the three angles in ABC are equal to the three angles in DEC, the two triangles are similar.

The following example uses the ratio of the side lengths to show that the two triangles are similar. However, some of the side lengths must be calculated using Pythagoras' Theorem.

Example 13

Prove that triangles ABC and BCD are similar in the diagram on the right.

First use Pythagoras' Theorem in triangle ABC to find the length BC:
$BC^2 = 8.1^2 + 10.8^2$
$BC^2 = 182.25$
$BC = \sqrt{182.25} = 13.5$ cm

Now use Pythagoras' Theorem in triangle BCD to find the length BD:
$BD^2 = 13.5^2 + 10.125^2$
$BD^2 = 284.765625$
$BD = \sqrt{284.765625} = 16.875$ cm

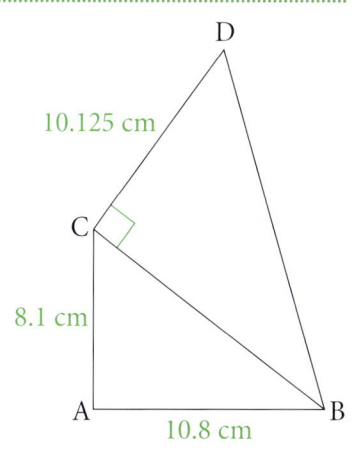

Now show that corresponding sides are in the same ratio:

$\frac{BD}{BC} = \frac{16.875}{13.5} = 1.25$ $\frac{BC}{AB} = \frac{13.5}{10.8} = 1.25$ $\frac{CD}{AC} = \frac{10.125}{8.1} = 1.25$

Since the pairs of sides are all in the same ratio, the triangles are similar.

GCSE MATHEMATICS M3 AND M7

Exercise 35F

1. Prove that triangles ABC and BCD, in the diagram below, are similar.

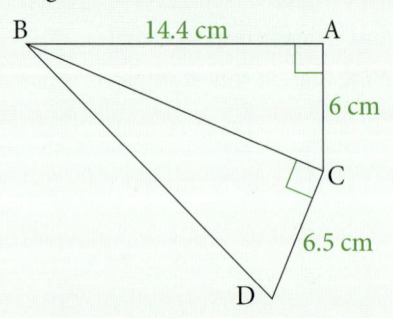

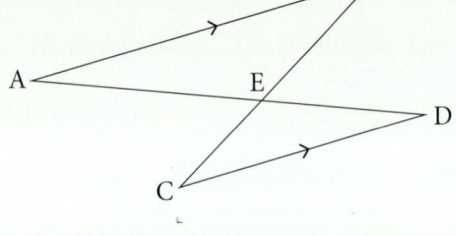

2. Prove that triangles MNP and QRP, in the diagram below, are similar.

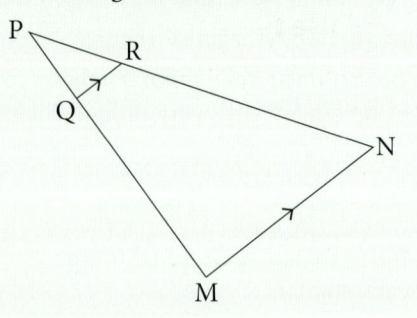

3. Prove that triangles ABE and CDE, in the diagram below, are similar.

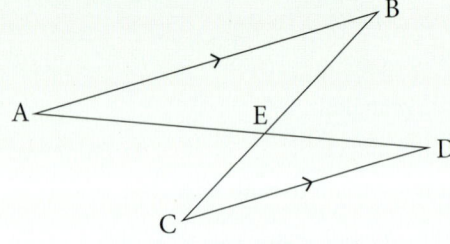

4. ABCD, shown below, is a rectangle. Prove that triangles ADE and BCE are similar.

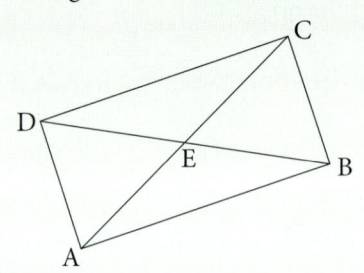

5. Triangle ABC, shown on the right, is right-angled at A. AD is the perpendicular line from BC to point A. Show that triangles ABC and ADC are similar.

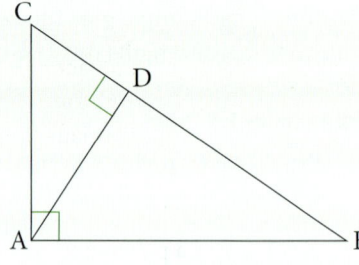

6. In the diagram on the right, lines AB and CD are parallel.
   (a) Prove that triangles ABE and CDE are mathematically similar.
   (b) Find the length of AE, marked x.
   (c) Find the length of CD, marked y.

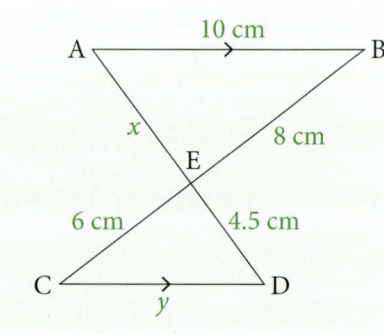

7. In the diagram on the right, lines PQ and RS are parallel. Lines PS and QR intersect at point T. Copy and complete the following sentences.
   (a) **Angle PTQ = 113° because** _____
   (b) **Angle PQT = 28° because** _____
   (c) **Angle SRT = 28° because** _____
   (d) **Angle RST = 39° because** _____
   (e) **Triangles PQT and RST are similar because** _____

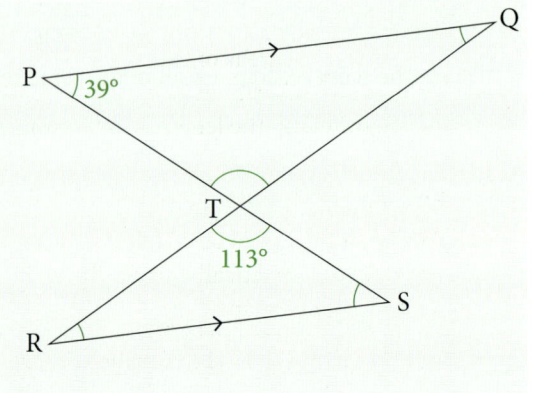

CHAPTER 35: ENLARGEMENT AND SIMILARITY

35.6 The Relationship Between the Ratios of Lengths and Areas in 2D Shapes

Understand and use the effect of enlargement on perimeter and area of shapes

We have already seen that, for an enlargement with scale factor 2, each length in the image is 2 times the corresponding length in the original shape. However, in this case the area of the image is 4 times the area of the original.

If the scale factor is 3, then the area of the image is 9 times the area of the original.

> The **Area Scale Factor** is the square of the **Length Scale Factor**.

Example 14

The shape shown on the right is enlarged with a scale factor of 2
(a) If the base of the original has a length of 4 cm, find the length of the base of the image.
(b) If the area of the original shape is 18 cm², what is the area of the image?

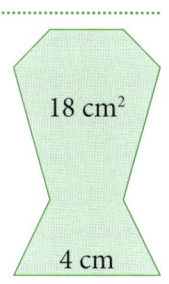

18 cm²

4 cm

> **Note:** Since the shape is enlarged, the image shape and the original are similar shapes.

(a) The length scale factor is 2
So, the length of the base of the image is 2 × 4 = 8 cm

(b) The length scale factor is 2
The area scale factor is $2^2 = 4$
The area of the image is 4 × 18 = 72 cm²

Example 15

A shape has a perimeter of 40 cm and an area of 15 cm²
It is enlarged with a scale factor of 2
(a) What is the perimeter of the enlarged shape?
(b) How many times bigger is the area of the enlarged shape than the area of the original?
(c) Find the area of the enlarged shape.

(a) The perimeter is a length. In the enlarged shape (or image), the perimeter is 2 times the perimeter of the original = 2 × 40 = 80 cm

(b) For this enlargement, the area scale factor is $2^2 = 4$
The area of the enlarged shape is 4 times the area of the original.

(c) The area of enlarged shape = 4 × 15 = 60 cm²

You may be given the areas of two similar shapes and one length, then asked to find the corresponding length in the other shape. In this situation, find the area scale factor first. The length scale factor is the square root of the area scale factor. Using the length scale factor, the unknown length can be calculated.

Example 16

Shape P has a width of 3.5 cm and an area of 46 cm²
Shape Q is mathematically similar to shape P and has an area of 414 cm²
Find the width of shape Q.

The area of shape Q is 414 cm² and the area of shape P is 46 cm²

We can calculate the area scale factor as $\frac{414}{46} = 9$

331

The length scale factor is the square root of the area scale factor.

So the length scale factor is $\sqrt{9} = 3$

This means the width of Q is 3 times the width of P.

So the width of Q = $3 \times 3.5 = 10.5$ cm

Exercise 35G

1. A company makes Christmas decorations in two different sizes: small and large, which are mathematically similar. The small Christmas decorations have a height of 20 cm and a surface area of 300 cm². The large Christmas decorations have a height of 120 cm. Find the surface area of the large Christmas decorations.

2. The octagon shown on the right is enlarged with a scale factor of 2
   The original octagon has an area of 28 cm²
   Its height is 6 cm and its base is 2 cm, as shown.
   (a) Find the height the enlarged shape.
   (b) Find the length of the base of the enlarged shape.
   (c) Find the area of the enlarged shape.

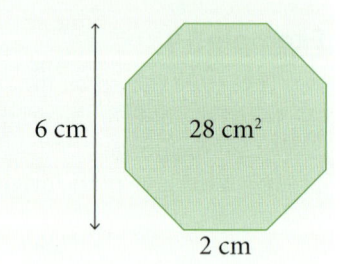

3. A shape has a width of 11 cm and an area of 50 cm²
   It is enlarged with a scale factor of 3
   (a) What is the width of the enlarged shape?
   (b) How many times bigger is the area of the enlarged shape than the area of the original?
   (c) Find the area of the enlarged shape.

4. Shape B is an enlargement of shape A. Shape A has a length of 3 cm and Shape B has a length of 6 cm.
   (a) What is the scale factor for the lengths of the two shapes?
   (b) What is the scale factor for the areas of the two shapes?
   (c) Copy and complete the table on the right.

	Shape A	Shape B
Length	3 cm	6 cm
Area	5 cm²	

5. Shape Q is an enlargement of shape P.
   Copy and complete the table on the right.

	Shape P	Shape Q
Length	5 m	
Area	8 m²	72 m²

6. ABCD and AEFG, shown on the right, are mathematically similar trapezia. Trapezium ABCD has an area of 64 cm² Find the area of the **unshaded** region on the diagram.

7. At a zoo, there is a scale model of the zoo in the visitor's centre. The length of the zoo is 200 times the length of the model. How many times larger is the area of the zoo than the area of the model?

8. A 2D shape has a perimeter of 18 cm and an area of 8 cm²
   The shape is enlarged using a scale factor of 3
   Find:
   (a) The perimeter of the enlarged shape.
   (b) The area of the enlarged shape.

9. The diagram on the right shows two pentagons P and Q.
   Pentagon Q is an enlargement of pentagon P. Pentagon P has an area of 12.5 cm² and pentagon Q has an area of 50 cm²
   (a) What is the area scale factor? (In other words, how many times larger is the area of Q than the area of P?)
   (b) What is the length scale factor?
   (c) The width of pentagon P is 5 cm. What is the width of pentagon Q?
   (d) The perimeter of pentagon Q is 28.8 cm. What is the perimeter of pentagon P?

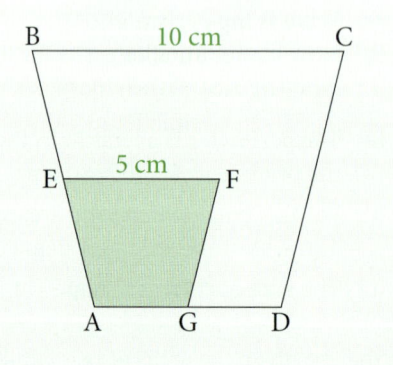

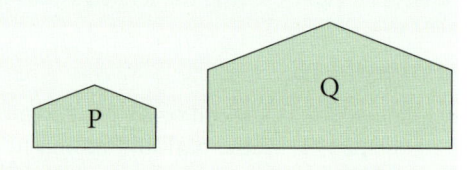

35.7 The Effect of Enlargement on Volume

You should understand the effect of enlargement on the **volume** of solids.

> If a 3D shape is enlarged by scale factor k, the volume will be enlarged by a factor of k^3

Example 17

Simon makes a model of the Eiffel Tower in Paris. His model is 5 cm high and has a volume of 20 cm³ Gretta makes a similar model that is 15 cm high. What is the volume of Gretta's model?

Gretta's model is 3 times taller than Simon's, so the length scale factor is 3

Therefore the volume scale factor is 3^3, or 27

To calculate the volume of Gretta's model, multiply the volume of Simon's model by 27:

27 × 20 = 540 cm³

Exercise 35H

1. Copy and complete the following paragraph.

   Object A has a length of 10 cm and a volume of 100 cm³

   Object B has a length of 20 cm and is mathematically similar to object A.

   Object B is an enlargement of Object A with a length scale factor of _____

   The volume scale factor is the cube of the length scale factor.

   So the volume scale factor is ____³ which is ____

   The volume of Object B is ____ × 100 = ____ cm³

2. Cuboid Q is an enlargement of cuboid P, using a scale factor of 4
   If Cuboid P has a volume of 10 cm³, find the volume of cuboid Q.

3. The diagram on the right shows two hexagonal prisms. The large prism is an enlargement of the small prism. The smaller prism has a height of 2 cm and the larger prism has a height of 4 cm. If the volume of the smaller prism is 7 cm³, find the volume of the larger prism.

4. A flat box has a height of 1.5 cm, as shown on the right. Its volume is 37.5 cm³
   A second box is an enlargement of the first box. It has a height of 3 cm. Find the volume of the second box.

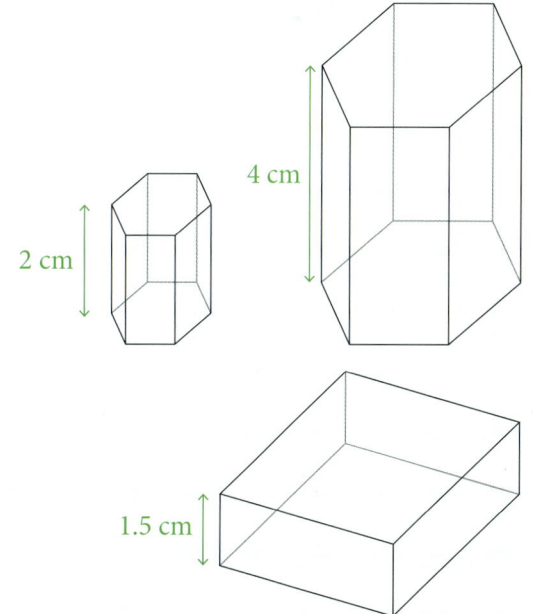

5. A shop sells two sizes of beach ball. The larger beach ball is an enlargement of the smaller one. One has a diameter of 20 cm and the other has a diameter of 40 cm. The smaller beach ball has a volume of 4200 cm³ What is the volume of the larger beach ball? **Hint**: you do not need to use the formula for the volume of a sphere to answer this question.

6. A twelve-sided coin is shown on the right. It has a height of 2 mm and is made of 0.25 cm³ of metal. A second, larger coin is an enlargement of the first and has a height of 3 mm. How much metal is needed to make each of the larger coins? Give your answer in cm³ to 2 decimal places.

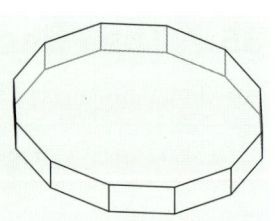

35.8   Summary

In this chapter you have learnt:
- About a fourth type of transformation: enlargements.
- That you may be asked to enlarge a 2D shape using a positive scale factor greater than 1, either with or without a centre of enlargement.
- That you may be also be asked to enlarge a 2D shape using a fractional scale factor less than 1, in which case the image is smaller than the original.
- To describe an enlargement, give the coordinates of the centre of enlargement and the scale factor.
- That under an enlargement, all the lengths of the original are multiplied by the scale factor. So, for example, if the scale factor is 2, a side length of 4 cm becomes a side length of 8 cm in the enlarged shape.
- That shapes that are **similar** have the same shape but different sizes. In two similar shapes:
  - Each angle in one shape is equal to the corresponding angle in the second shape; **and**
  - The corresponding sides in each shape are in the same ratio.
- That to prove that two triangles are similar, you only need to show that **one** of the conditions above is true.
- That to prove similarity in other shapes, you must prove that both conditions are true.
- That if you are told that two shapes are similar, you can use this information to find unknown angles or sides. In harder problems, you may need Pythagoras' Theorem, corresponding, alternate or vertically opposite angles.
- That for similar 2D shapes, areas are multiplied by the square of the scale factor.
  So, using a length scale factor of 2, the area scale factor is 2^2 or 4
  If the original 2D shape has an area of 10 cm², the enlarged shape has an area of 40 cm²
- That when enlarging a 3D shape, volumes are multiplied by the cube of the scale factor.
  So, using a length scale factor of 2, the volume scale factor is 2^3 or 8
  If the original 3D shape has a volume of 10 cm³, the enlarged shape has a volume of 80 cm³

Chapter 36
Constructions and Loci

36.1 Introduction

In this chapter, you will learn how to draw accurate constructions. Constructions of triangles and polygons in Section 36.2 require a ruler, a pair of compasses and a protractor. The standard constructions in Section 36.3 should be done using a ruler and compasses only. You will also learn how to draw loci. A locus is a set of points, defined by some rule. A locus can be a line or a curve or an area.

All this work requires you to make accurate drawings.

> **Note:** In an examination, you will usually be given the marks if lines are within 2 mm of the required length and angles are within 2 degrees of the required size.

Key words
- **Construction**: An accurate drawing.
- **Locus**: A set of points. The plural of locus is **loci**.

Before you start you should know:
- How to draw and measure lines and angles.
- How to work with metric units of length.
- How to draw and measure bearings.
- How to use a scale to calculate distance.
- The names of different types of triangle.
- How to draw triangles and other 2D shapes accurately using a ruler and protractor.

In this chapter you will learn how to:
- Use a ruler, a pair of compasses and a protractor to construct triangles and other polygons.
- Use a ruler and pair of compasses to do constructions.
- Construct loci.

Exercise 36A (Revision)

1. The map on the right shows the locations of Armagh, Banbridge and Craigavon, marked with the letters A, B and C respectively. The scale of the map is 1 cm = 5 km.
   (a) If a bird flies directly from Armagh to Banbridge, what is the actual distance it travels?
   (b) What is the bearing of Craigavon from Armagh?
   (c) What is the bearing of Craigavon from Banbridge?
   (d) Dromore is 16.7 km from Craigavon on a bearing of 101°.
   Copy the map and mark the position of Dromore with the letter D.

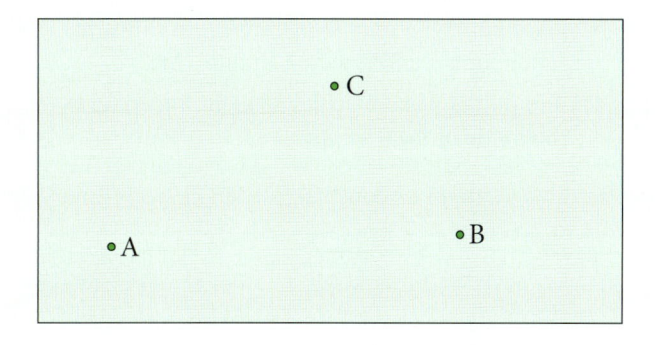

36.2 Constructing Triangles

A construction is different from a sketch. All lines and angles in a construction should be drawn exactly.

You may be asked to do a construction of a triangle using a ruler and protractor.
You should learn three such constructions:

- Construction of a triangle given one side length and two angles.
- Construction of a triangle given two side lengths and one angle.
- Construction of a triangle given all three side lengths.

You should also be able to construct regular polygons, for example a regular pentagon.

In the first example we construct a triangle, given one side length and two angles.

Example 1

Construct the triangle ABC where AB = 7.2 cm, angle BAC = 55° and angle ABC = 37°.

Follow these steps:
1. Draw the line AB as the base of the triangle, 7.2 cm long.
2. At point A, measure the angle 55° between the base and the side AC.
3. At point B, measure the angle 37° between the base and the side BC.
4. Extend the sides AC and BC until they meet at point C.

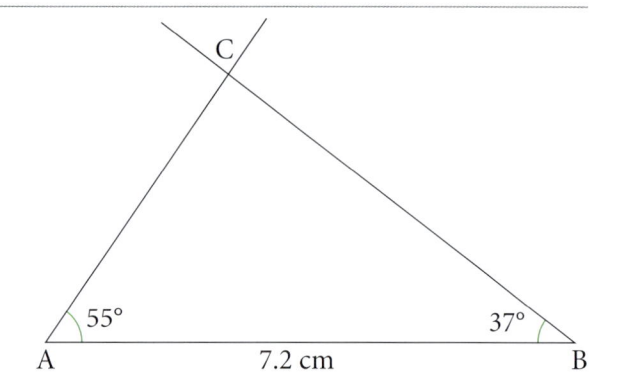

In the next example we are given two sides and one angle.

Example 2

Construct the triangle ABC, where AB = 6.4 cm, AC = 5.2 cm and angle BAC = 30°.

Follow these steps:
1. Draw the line AB as the base of the triangle, 6.4 cm long.
2. At point A, measure the angle 30° between the base and the side AC.
3. Place point C 5.2 cm from point A in this direction. Join points A and C with a straight line.
4. Join points B and C with a straight line to form the third side of the triangle.

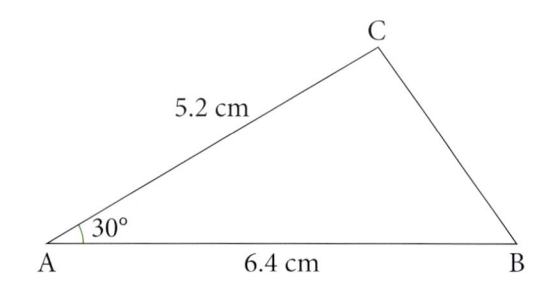

In the third example we construct a triangle from three side lengths.

Example 3

Construct the triangle ABC where AB = 8 cm, AC = 5 cm and BC = 6 cm.

Follow these steps:
1. Draw the line AB as the base of the triangle, 8 cm long.
2. Using your compasses, draw a circle with a radius of 5 cm centred on point A. Point C is 5 cm from point A, so it must lie on this circle.
3. Using your compasses, draw a circle with a radius of 6 cm centred on point B. Point C is 6 cm from point B, so it must lie on this circle.
4. Since point C lies on both circles, it must be the point where the circles intersect. Label this point C and join it with straight lines to points A and B.

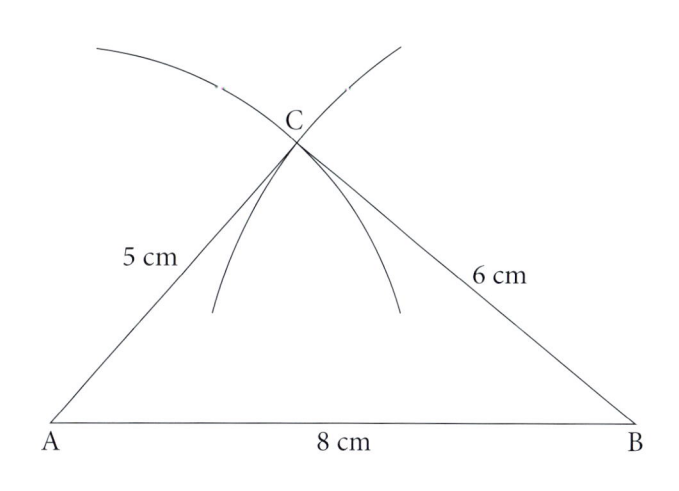

Note: You don't need to draw full circles, as long as the arcs you draw intersect. However, you should leave the arcs as a part of your working – don't rub them out!

Exercise 36B

1. Construct the triangle ABC where AB = 6 cm, angle BAC = 60° and angle ABC = 24°.
2. Construct the triangle ABC, where AB = 9.2 cm, AC = 7.5 cm and angle BAC = 15°.
3. Construct the triangle PQR where PQ = 5.5 cm, PR = 7.9 cm and QR = 6.9 cm.
4. Construct the isosceles triangle WXY, where WX = XY = 7.4 cm and WY = 5.2 cm.
5. A plane flies 600 km due north from airport A to airport B and then flies 500 km to airport C on a bearing of 140°.
   (a) Using a scale of 1 cm = 100 km, construct a scale drawing showing the plane's journeys.
   (b) Using your diagram, find:
      (i) The direct distance from A to C in kilometres.
      (ii) The bearing of C from A.
6. Construct an equilateral triangle XYZ with sides of length 6.5 cm.
7. Construct a regular pentagon ABCDE with sides of length 5 cm. (Hint: Calculate the size of each interior angle. Draw the base, then draw the next side at the correct angle to the base. Continue working your way around the pentagon in this way.)
8. The diagram on the right shows a sketch of shop floor. Make an accurate scale drawing, using a scale of 1 cm to 2 m.

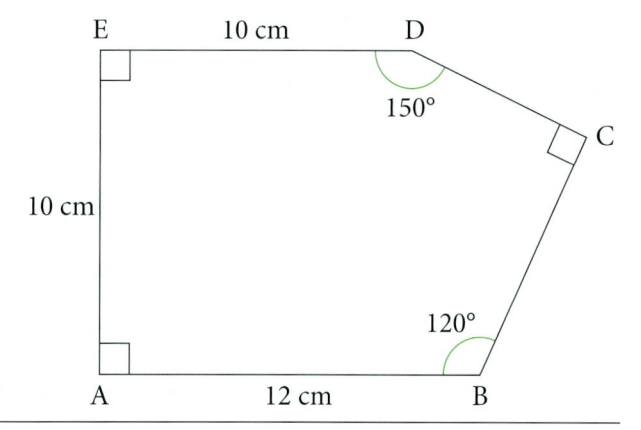

36.3 Other Standard Constructions

In this section we discuss the standard constructions that you can do with only a ruler and a pair of compasses.

The perpendicular bisector of a straight line

A perpendicular bisector is a straight line that cuts another straight line in half at right angles.

> **Note:** After completing a construction, it is very important to leave all your construction lines on your diagram. These are an important part of your solution.

Example 4

Draw a line PQ, which is 5 cm long. Construct the perpendicular bisector of PQ.

The completed construction is shown on the right.
To draw it, follow these steps:

1. Draw the line PQ, 5 cm long.
2. With the compass point at P and a radius bigger than half of 5 cm, draw an arc above and below the line PQ.
3. With the compass point at Q and keeping the radius the same, draw an arc that intersects the first arc in two places. Label these two points A and B.
4. Draw a straight line joining A and B.

Leave the two arcs on the page, as well as the bisector. Do not rub them out as these are an important part of your solution.

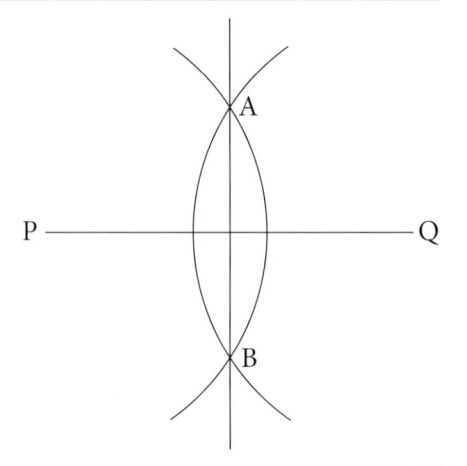

> **Note:** When you have completed your construction, check that the bisector splits PQ into two equal line segments, in this case 2.5 cm each. You can also measure the angle at which the two straight lines meet – this should be 90°

The angle bisector

Bisecting an angle means cutting it in half. You can use a ruler and a pair of compasses to bisect an angle.

Example 5

Copy the diagram below, and construct the bisector of the angle ABC.

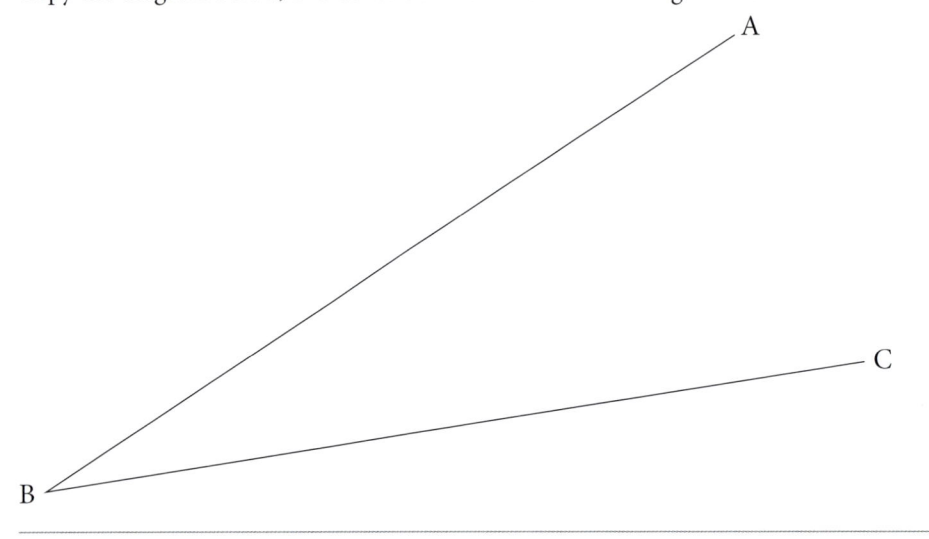

CHAPTER 36: CONSTRUCTIONS AND LOCI

Follow these steps:
1. With the compass point at B, draw an arc cutting both lines AB and CB at points D and E, as shown below.
2. Move the compass point to D. Draw a second arc.
3. Move the compass point to E and using the same radius draw a third arc, making sure the second and third arcs intersect. Label the point of intersection F, as shown.
4. Join points F and B with a straight line. This line should bisect the angle.

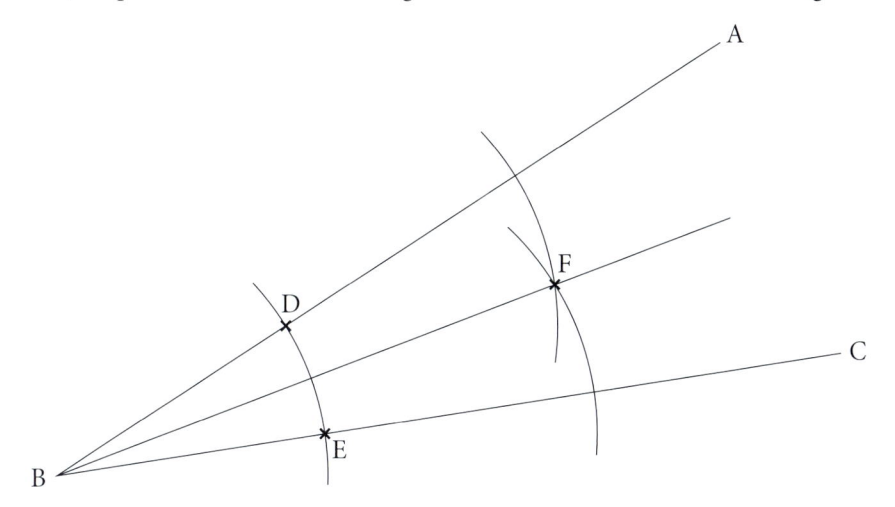

Note: The line FB lies exactly halfway between lines AB and CB. So, a question could ask you to construct the line that is the same distance (or **equidistant**) from two lines that meet at a point. This is the same as constructing the angle bisector.

Perpendicular from a point to a line

A perpendicular to a line is another line at right angles to it.

Example 6

Copy the diagram on the right. Construct the perpendicular from point A to the line PQ.

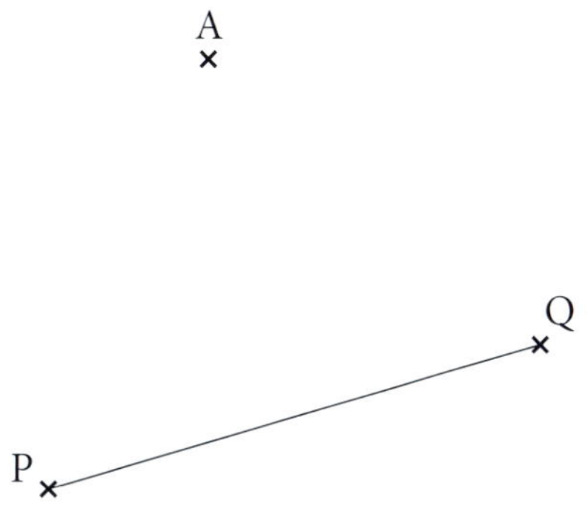

Follow these steps:
1. Extend the line PQ in both directions.
2. With compasses, draw an arc centred on A. Make sure it passes through the extended line in 2 places, labelled X and Y in the diagram on the right.
3. Putting the compass point at X, draw a second arc.
4. Putting the compass point at Y, draw a third arc **with the same radius** as the second arc. Make sure the second and third arcs intersect, as shown on the right, at point E.
5. Join points A and E with a straight line. This line is the perpendicular from A to the line PQ.

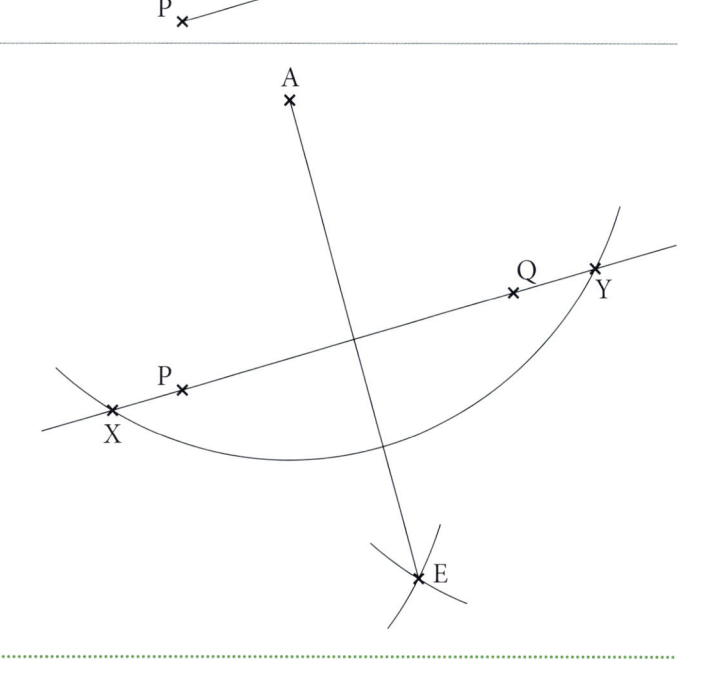

GCSE MATHEMATICS M3 AND M7

Exercise 36C

1. Copy the line segment AB shown on the right.
   Construct the perpendicular bisector of AB.

2. Draw a line 7 cm long.
   Construct the perpendicular bisector of the line.

3. Copy the diagram shown on the right.
   Construct the angle bisector for angle PQR.

4. Draw an angle of 50°
   Construct the angle bisector.

5. Draw a line CD 8.4 cm long.
   Construct the perpendicular bisector of the line.

6. Draw the angle PQR, which is 150°
   Construct the angle bisector.

7. Draw two straight lines LM and MN, which meet at point M at 80° Construct a straight line that is exactly equidistant from LM and MN.

8. Copy the diagram shown on the right. Construct the perpendicular from point A to the line PQ.

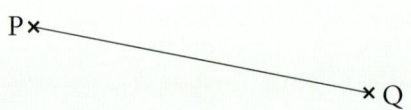

9. (a) Construct a triangle XYZ by following these steps:
     (i) Draw the base XY, which is 8 cm long.
     (ii) Draw a line that makes an angle of 40° with the base at the point X.
     (iii) Draw a line that makes an angle of 45° with the base at the point Y.
     (iv) The point of intersection of these two lines is the third vertex of the triangle. Label it Z.
   (b) Construct the perpendicular from Z to the line XY.

10. (a) Using a ruler and protractor, construct a rhombus ABCD such that:
      - Each side is 5.6 cm long; and
      - Angle DAB is 42°
    (b) Construct the perpendicular bisector of AB.
    (c) Construct the bisector of the angle BCD.

36.4 Loci

Identify the loci of points, including real life problems

A locus is a set of points. The locus is described by some mathematical rule or set of conditions. A locus may be a line or a curve, or it may be an area.

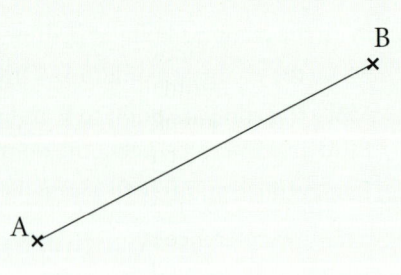

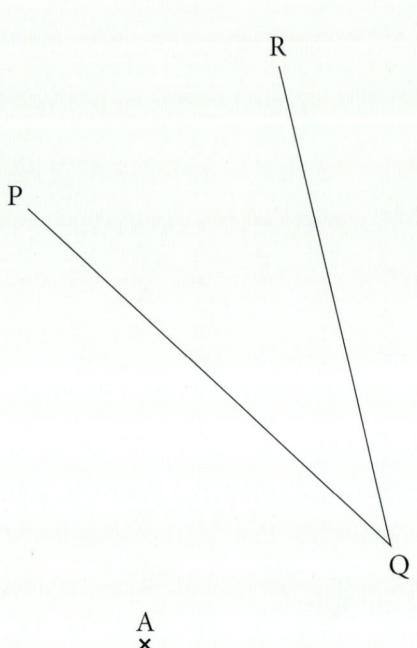

CHAPTER 36: CONSTRUCTIONS AND LOCI

Example 7

O is a fixed point. Draw the locus of points that lie:
(a) Exactly 3 cm from O
(b) **Less than or equal to** 3 cm from O
(c) Between 2 and 3 cm from O

(a) The points that lie 3 cm from a fixed point form the circumference of a circle with a radius of 3 cm, as shown below.

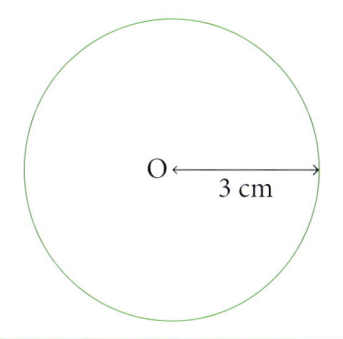

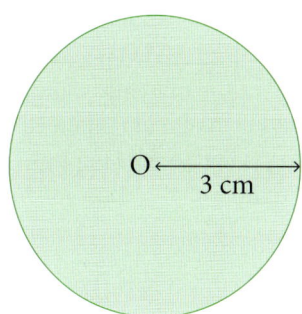

(b) All of the points inside the circle are less than or equal to 3 cm from O, so this locus is an area, as shown below.

Note: You can shade the area in any way, as long as you make it clear that the locus is the circle and everything inside it.

(c) The locus of points lying between 2 and 3 cm from O is shown in the diagram on the right. It is a doughnut shape, between two circles of radius 2 cm and 3 cm.

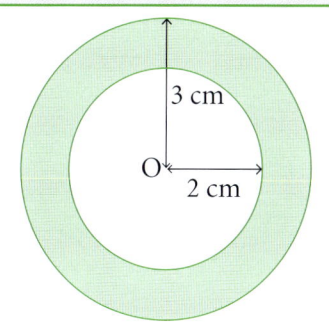

Example 8

Show the locus of points that lie exactly 1 cm from the fixed line shown on the right. The line extends indefinitely in both directions.

The locus of points 1 cm from a straight line is a pair of parallel lines 1 cm either side of the original line. Every point on each of the parallel lines is exactly 1 cm from the original line.

The locus is shown as the green parallel lines in the diagram on the right.

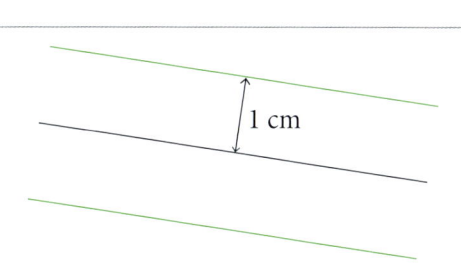

In Example 5 above, the original (shown in black) is a **line** and not a **line segment**. That means it goes on for ever in both directions, as do the two parallel lines that form the locus (shown in green).

The next example shows that the locus is a different shape if the original is a **line segment**.

Example 9

Draw the locus of points that lie 1 cm from the line segment AB shown.

341

GCSE MATHEMATICS M3 AND M7

Two lines run parallel to the line segment, 1 cm above and below it.

At each end there is a semicircle. Each point on the left-hand semicircle lies 1 cm from point A. Each point on the right-hand semicircle lies 1 cm from point B.

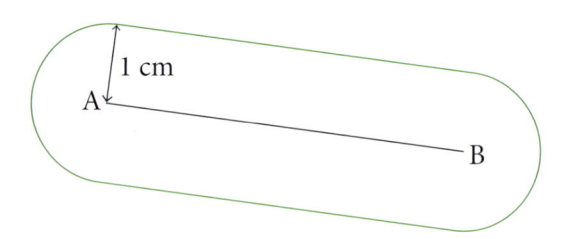

You may be asked to draw a scale drawing and find a locus on it.

Example 10

A lamppost is 30 cm from a long building.
(a) Draw a scale drawing, showing the positions of the lamppost and the building. Let 1 cm on the drawing represent 10 cm in real life.
(b) A dog is tied to the lamppost. The dog's lead allows him to move 50 cm from the lamppost. On your diagram, show the locus of points representing the area that the dog can move in.

(a) The scale drawing is shown on the right.
(b) The locus representing the area the dog can move in is shown shaded in green on the diagram.

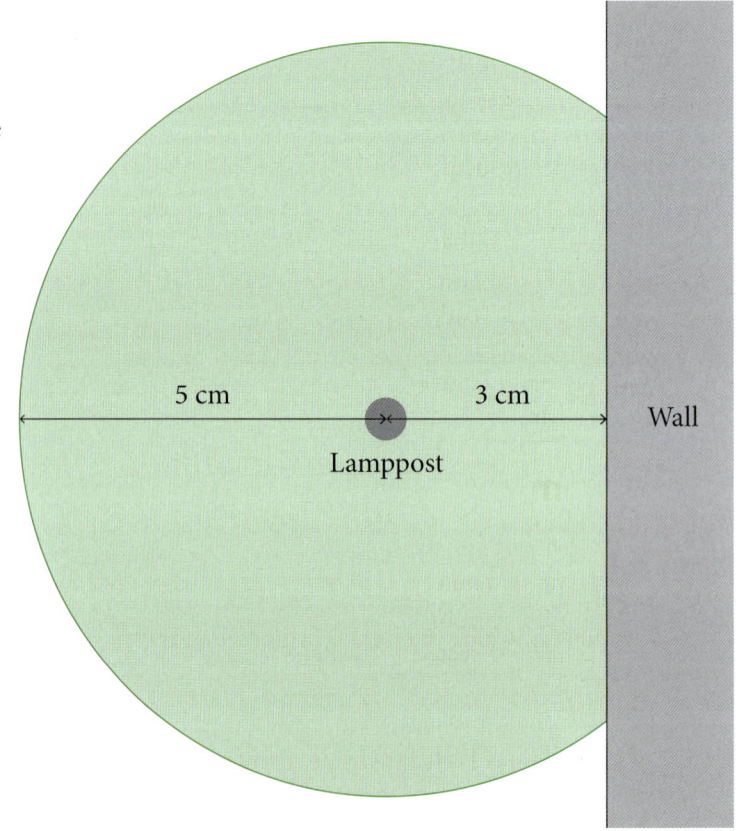

Exercise 36D

1. Show with shading on a diagram the locus of points that are **more than** 2 cm from a fixed point O.
2. Shade on a diagram the locus of points that are less than 5 cm from a fixed point O.
3. Draw the locus of points 1.5 cm from a line AB that is 5 cm long.
4. (a) Draw a straight line across your page, going all the way from one side to the other.
   (b) Draw the locus of points that are 2 cm from the original line.
5. Construct the locus of points that are:
   (a) Exactly 1.8 cm from a line segment AB.
   (b) Less than 1.8 cm from a line segment AB.
6. (a) Draw a horizontal line XY, which is 6.4 cm long.
   (b) Construct the locus of points that are **equidistant** (the same distance) from points X and Y.

7. Copy the diagram shown on the right. Construct the locus of the points that are equidistant from PQ and QR.

8. Construct a rectangle ABCD, where AB is 8 cm long and BC is 5 cm long. Construct the locus of points that lie equidistant from points A and C.

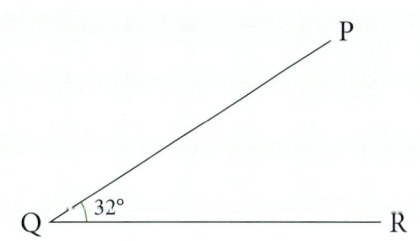

9. (a) Construct a triangle ABC with AB = 9 cm, BC = 7 cm and AC = 6 cm.
   (b) Construct the locus of points that are inside the triangle and lie closer to AB than to AC.

10. A mobile phone transmitter is located at point O. Construct the locus of points that lie within 2.5 km of the transmitter. Use a scale of 1 cm = 1 km for your drawing.

11. A gas pipeline FG runs in a straight line, 7 km long. After a leak in the pipe, the authorities wish to evacuate all properties within 4 km of the pipeline.
    (a) Draw a scale drawing of the pipeline, using a scale of 1 cm = 1 km.
    (b) Construct the locus of points that lie within 4 km of the pipeline.

12. Point M is 64 miles east of point L.
    (a) Show points L and M on a scale diagram, using a scale of 1 cm = 10 miles.
    (b) Construct the locus of points that are equidistant from L and M.

13. (a) Construct the triangle PQR, where PQ = 6.5 cm, QR = 8.5 cm and RP = 10 cm.
    (b) Construct the locus of points that are equidistant from the points P and Q.
    (c) Construct the locus of points that are equidistant from the lines PQ and QR.

14. Construct the rectangle DEFG where DE = 7 cm and EF = 4.5 cm. Construct the locus of points that are **both** (i) nearer to D than F; **and** (ii) more than 3.5 cm from G.

15. Adam likes to go out on his bike. His parents say he must stay within 2 km of their house. They also ask him not to cross a busy road, which runs from north to south 1.5 km to the east of the house.
    (a) Draw a scale drawing, showing the house and the road.
        Let 1 cm on the drawing represent 1 km in real life.
    (b) On your drawing, shade the area within which Adam can travel on his bike.

36.5 Summary

In this chapter you have learnt:
- How to construct accurate drawings of triangles and other polygons using a ruler, compasses and a protractor.
- About other standard constructions that you can do with only a ruler and compasses. These are:
  - the perpendicular bisector of a straight line,
  - the angle bisector,
  - the perpendicular from a point to a line.
- About loci. A locus is a set of points, defined by some rule. Typical loci include the set of points that lie:
  - a certain distance from a point,
  - a certain distance from a straight line,
  - a certain distance from a line segment.

Chapter 37
Probability

37.1 Introduction

In this chapter we will study what is meant by probability and the probability scale. We will look at how to list outcomes for events in a systematic way, how to use the product rule for counting and how to use relative frequency as an estimate of probability. We will also consider what is meant by mutually exclusive outcomes, independent events, how to calculate the probability of an event happening and how to compare experimental and theoretical probabilities. We will also look at probability tree diagrams. You should also understand that using a large sample size generally gives better estimates of probability.

Key words

- **Event**: An event is something that may happen
- **Probability**: The probability is the chance of an event taking place. The probability of an event happening is a number between 0 and 1, with 0 being impossible and 1 being certain
- **Fair**: A fair dice is one in which every outcome (1 to 6) has the same probability. For a fair coin, there is the same probability for heads and tails.
- **Biased**: Biased is the opposite of fair. For a biased coin, there may be a greater probability of getting heads than tails, for example.
- **Mutually exclusive**: Mutually exclusive events cannot both happen. For example, if a single dice is rolled once, the events 'getting a 1' and 'getting a 6' are mutually exclusive.
- **Exhaustive**: Events are exhaustive if one of them must happen. For example, if a coin is tossed, the events 'getting heads' and 'getting tails' are exhaustive. No other events are possible.
- **Independent**: With independent events, if one event happens it does not affect the probability of the other event happening. For example, getting heads when tossing a coin and getting a 6 when rolling a dice.
- **Relative frequency**: An estimate of probability based on experimental or observational data.

Before you start you should:

- Understand and use the vocabulary of probability, including the terms 'fair', 'random', 'evens', 'certain', 'likely', 'unlikely ' and 'impossible'.
- Understand and use the probability scale from 0 to 1
- Be able to work out probabilities expressed as fractions or decimals from simple experiments.
- Understand that the probability of an event not occurring is one minus the probability that it occurs.
- Be able to use probabilities to calculate expectation.

Notation

In this chapter we will use the following notation:

- For the probability of an event A happening, we use the mathematical notation **P(A)**. For example, we can write: P(C), P(X = 4) or P(rain tomorrow).
- For the probability of two events A and B happening one after the other, we write **P(A, B)**. For example: P(rain tomorrow, cycling).

In this chapter you will learn:

- How to list all outcomes for single events, and for two successive events, in a systematic way.
- The product rule for counting combinations.
- About **mutually exclusive**, **exhaustive** and **independent** events. You will learn when to add or multiply two probabilities.
- How to work with probability tree diagrams for independent events.
- How to work with relative frequency.

Exercise 37A (Revision)

1. (a) Explain what is meant if a coin is described as **fair**.
   (b) What is the opposite of fair?

2. Pair up these events with a word describing the probability.

Event	Probability
A caterpillar turning into an elephant	Unlikely
Pressing the light switch in my bedroom and the light coming on	Impossible
Spring following winter	Certain
Meeting the Pope on the way to school	Likely

3. The fair spinner shown on the right is spun. It can land on grey, light green, dark green or white.
   (a) Find the probability that the spinner:
      (i) Lands on grey, giving your answer as a fraction.
      (ii) Lands on dark green, giving your answer as a decimal.
   (b) If the spinner is spun 200 times, find the expected number of times it would land on grey.

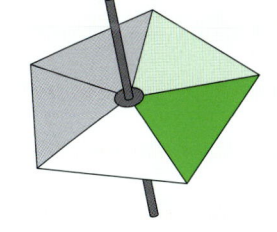

4. Copy the probability scale below.

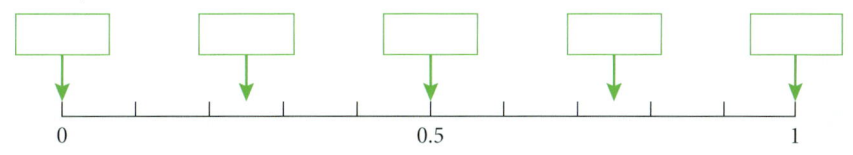

   0   0.5   1

   Place each of the following 5 words in the correct box above the line:
   Likely   Certain   Unlikely   Impossible   Evens

5. There are 10 socks in a drawer: 6 are black and 4 are white. A sock is taken from the drawer at random.
   (a) Find the probability that the sock is:
      (i) Black   (ii) White   (iii) Black or white   (iv) Blue
   (b) Which of the four events above is **unlikely**?

37.2 Mutually Exclusive Events

If events A and B cannot both happen, they are called **mutually exclusive** events. For example, when rolling a single dice, the two events 'getting a 3' and 'getting an even number' are mutually exclusive events. To find the probability of getting a 3 **or** an even number, you add the probabilities.

Example 1

A single dice is rolled once. Event A is getting a 3. Event B is getting an even number.
(a) Find P(A).
(b) Find P(B).
(c) Are events A and B mutually exclusive?
(d) Find the probability of A or B happening.

(a) When rolling a single dice, there is just one way of getting a 3 out of 6 possible outcomes.
So P(A) = $\frac{1}{6}$

(b) There are three even numbers: 2, 4 and 6.
So the probability of getting an even number is $\frac{3}{6}$ or $\frac{1}{2}$
So P(B) = $\frac{3}{6}$ or $\frac{1}{2}$

(b) Events A and B are mutually exclusive because they cannot both happen.

(c) To find the probability of A or B happening, we add P(A) and P(B).

$$P(A \text{ or } B) = \frac{1}{6} + \frac{1}{2} = \frac{2}{3}$$

Example 2

A single biased dice is rolled once. Copy and complete this table using the following information:

Number	1	2	3	4	5	6
Probability				0.15	0.22	

The probability of getting a 2 is equal to the probability of getting a 5
The probability of getting an even number is 0.56
The probability of getting a prime number is 0.6

> **Note:** Biased means the probability of each outcome is **not** $\frac{1}{6}$
> Some of the probabilities are bigger and some smaller than $\frac{1}{6}$

Since $P(2) = P(5)$, we know $\quad P(2) = 0.22$

Since $P(\text{even}) = 0.56$, we know
$P(2) + P(4) + P(6) = 0.56$
So: $0.22 + 0.15 + P(6) = 0.56$
$P(6) = 0.56 - 0.22 - 0.15 = 0.19$

Since $P(\text{prime}) = 0.6$, we know
$P(2) + P(3) + P(5) = 0.6$
$0.22 + P(3) + 0.22 = 0.6$
$P(3) = 0.6 - 0.22 - 0.22 = 0.16$

All six probabilities must add up to one because the six outcomes in the table are mutually exclusive (i.e. only one of them can happen) and they are exhaustive (i.e. no other outcomes are possible). To find $P(1)$, we add the other five probabilities and subtract from one:

$$P(1) = 1 - (0.22 + 0.16 + 0.15 + 0.22 + 0.19)$$
$$P(1) = 0.06$$

The completed table is shown below.

Number	1	2	3	4	5	6
Probability	0.06	0.22	0.16	0.15	0.22	0.19

Example 3

In each part (a) and (b) below:
(i) State whether events A and B are mutually exclusive.
(ii) Write down $P(A)$ and $P(B)$.
(iii) Find the probability of A or B happening.

(a) A bag contains 15 black counters, 10 red counters and 5 white counters. A single counter is taken at random. Event A is getting a black counter; event B is getting a white counter.
(b) A fair six-sided dice is rolled. Event A is getting a prime number; event B is getting a 3

(a) (i) Events A and B are mutually exclusive, because they cannot both happen.
   (ii) There are 30 counters altogether, so $P(A) = \frac{15}{30} = \frac{1}{2}$; $P(B) = \frac{5}{30} = \frac{1}{6}$
   (iii) Since A and B are mutually exclusive, $P(A \text{ or } B) = P(A) + P(B) = \frac{1}{2} + \frac{1}{6} = \frac{2}{3}$

(b) (i) There are three prime numbers on the dice: 2, 3 and 5
       The events A and B are not mutually exclusive, because they could both happen if a 3 is rolled.
   (ii) $P(A) = \frac{3}{6} = \frac{1}{2}$; $P(B) = \frac{1}{6}$
   (iii) Since A and B are not mutually exclusive, it is not possible to find $P(A \text{ or } B)$ by adding $P(A)$ and $P(B)$. Instead, we list the outcomes in A or B: they are 2, 3, 5
       $P(A \text{ or } B) = \frac{3}{6} = \frac{1}{2}$

Exercise 37B

1. The probability that it will rain tomorrow is $\frac{2}{3}$. What is the probability that it won't rain?

2. There are some blue, red, green and black balls in a bag, and no others. One ball is chosen at random. The probabilities of choosing a blue, red or green ball are shown in the table. Find the probability of a black ball being chosen.

Blue	Red	Green	Black
0.2	0.4	0.3	

3. A fair six-sided dice is rolled. Which of these pairs of events are mutually exclusive and which are not?
   (a) The number is even and the number is a multiple of 3
   (b) The number is odd and the number is a multiple of 2
   (c) The number is odd and square.

4. There are some blue, red, green and yellow balls in a box. A ball is chosen at random. The probabilities of choosing a blue, red and green ball are shown in the table.

Blue	Red	Green	Yellow
0.1	0.3	0.4	

   (a) Find the probability of choosing a yellow ball.
   (b) Find the probability of choosing a ball that is either red or blue.
   (c) Find the probability of choosing a ball that is not blue.

5. A fair, six-sided dice is rolled. Event A is 'getting a 6'. Event B is 'not getting a 6'. Event C is 'getting a 5'.
   (a) Find P(A)
   (b) Find P(B)
   (c) Find P(C)
   (d) Are events A and B mutually exclusive? Give a reason for your answer.
   (e) Are events A and B exhaustive? Give a reason for your answer.
   (f) Are events A and C mutually exclusive? Give a reason for your answer.
   (g) Are events A and C exhaustive? Give a reason for your answer.
   (h) Find P(A or B)
   (i) Find P(A or C)

6. Nine counters numbered 1 to 9 are put into a tin. One counter is selected at random.
   (a) What is the probability of getting a counter with:
      (i) The number 4?
      (ii) An odd number?
      (iii) Not an odd number?
      (iv) A prime number?
      (v) A square number?
      (vi) A multiple of 3?
   (b) Event A is 'getting a square number'. Event B is 'getting a prime number'. Are events A and B mutually exclusive? Give a reason.

7. Izzy is going on holiday, and she can choose one activity to take part in.
   The probability that she chooses archery is 0.4
   (a) Work out the probability that Izzy doesn't choose archery.
   (b) The probability that Izzy chooses kayaking is 0.35. Find the probability that Izzy chooses archery or kayaking.
   (c) Find the probability that Izzy doesn't choose archery or kayaking.

8. State whether these pairs of events are mutually exclusive or not:
   (a) A single dice is rolled. Event A is 'throwing a number less than 4'; Event B is 'throwing a number greater than 4'.
   (b) Event A is 'eating toast for breakfast'; Event B is 'eating chips for dinner'.

9. Which of the pairs of mutually exclusive events in question 8 are also exhaustive?

10. Here are the probabilities of some events happening.
    In each part, find the probability of the event not happening.
    (a) P(A) = 0.4
    (b) P(B) = 0.26
    (c) P(C) = 97%
    (d) P(D) = $\frac{3}{5}$

11. A hotel owner conducts a survey of guests staying at her hotel. The table shows some of the results of her survey. A guest is chosen at random.

Type of guest	Probability
Man	0.45
Woman	0.55
From Northern Ireland	0.62
Married	0.65
Not married	0.35
Vegetarian	0.27

    (a) From the table find these probabilities:
        (i) The guest is not from Northern Ireland.
        (ii) The guest is not married.
        (iii) The guest is not a vegetarian.
    (b) Which two types of guest form a mutually exclusive pair? There may be more than one possible answer.
    (c) Do the events 'The guest is married' and 'The guest is not married' form an exhaustive group? Explain your answer.

37.3 Independent Events

If two events are **independent**, then one event happening does not affect the probability of the other event happening.

Questions in this section usually feature two successive trials or experiments, meaning that one follows the other, for example throwing a dice and then tossing a coin.

If A and B are independent events, the probability of both A and B occurring is P(A) × P(B)

Sample space diagrams

In a sample space diagram, all possible outcomes from two experiments are shown in a table. Sample space diagrams are mostly used when all the outcomes of each experiment have equal probabilities, such as for a fair dice or a fair coin.

Example 4

A fair dice is thrown and a fair coin is tossed.
(a) Draw a sample space diagram to show all the possible outcomes.
(b) Find the probability of getting tails on the coin and a 2 on the dice.
(c) Find the probability of getting a head on the coin and an odd number on the dice.

(a) The sample space diagram is as follows:

		\multicolumn{6}{c}{Dice}
		1	2	3	4	5	6
Coin	H	(H, 1)	(H, 2)	(H, 3)	(H, 4)	(H, 5)	(H, 6)
	T	(T, 1)	(T, 2)	(T, 3)	(T, 4)	(T, 5)	(T, 6)

(b) There are 12 possible outcomes.
Since the coin and dice are both fair, each outcome has the same probability of $\frac{1}{12}$
So P(T, 2) = $\frac{1}{12}$

> **Note:** Because getting tails on the coin and a 2 on the dice are independent events, you could calculate P(T, 2) in this way: P(T, 2) = P(T) × P(2) = $\frac{1}{2} \times \frac{1}{6} = \frac{1}{12}$

(c) There are 12 possible outcomes. Three of them involve a head on the coin and an odd number on the dice: (H, 1), (H, 3) and (H, 5).
So P(heads and odd number) = $\frac{3}{12} = \frac{1}{4}$

CHAPTER 37: PROBABILITY

Exercise 37C

1. Two fair dice are rolled and the **sum** of the two numbers is calculated.
   (a) Copy and complete this sample space diagram.

+	Dice 1					
	1	2	3	4	5	6
1	2	3	4			
2	3	4	5			
3	4	5				
4	5					
5						
6						

   (Dice 2 labels the rows)

   (b) Find the probability of getting a total of 9

2. Two coins are tossed.
   (a) Draw a sample space diagram to show the possible outcomes.
   (b) Using your sample space diagram, calculate the probability of getting:
      (i) one head and one tail;   (ii) at least one tail.

3. Colm spins this fair spinner and tosses a fair coin.
   (a) He records the **number** that the spinner lands on (1 to 6), as well as the outcome of the coin toss. Construct a sample space diagram showing the possible outcomes for these two events.
   (b) Find the probability of Colm getting a tail on the coin and 4 on the spinner.
   (c) Find the probability Colm gets a head on the coin and an even number on the spinner.
   (d) Colm's sister Jo spins the spinner and tosses the coin. She records the **colour** that the spinner lands on (green or white), as well as the outcome of the coin toss. Construct a sample space diagram showing the possible outcomes.

4. Two fair dice are rolled and the **product** of the two numbers is calculated.
   (a) Copy and complete this sample space diagram.

×	Dice 1					
	1	2	3	4	5	6
1	1	2	3			
2	2	4				
3	3	6				
4	4					
5						
6						

   (Dice 2 labels the rows)

   (b) Find the probability of getting a product of 6
   (c) Find the probability of getting a product of 9
   (d) If the two dice are rolled together 120 times, how many times would you expect to get a product of 4?

37.4 Tree Diagrams

Probability tree diagrams can be used to represent successive (usually two or three) experiments or trials. For the M7 module, these experiments are independent.

The leftmost branches in the diagram represent the first experiment and you can follow the branches from left to right according to the outcome of each one.

In this section you will learn when to add or multiply two probabilities.

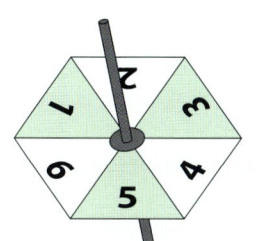

Example 5

A fair 6-sided dice is rolled and a fair coin is tossed. Draw a tree diagram to find the probability of getting a 6 on the dice and tails on the coin.

The tree diagram is shown on the right.

The combination of getting a 6 on the dice and tails on the coin is highlighted in the tree diagram.

We multiply the probabilities when moving from left to right through the tree diagram.

So P(6, Tails) = $\frac{1}{6} \times \frac{1}{2} = \frac{1}{12}$

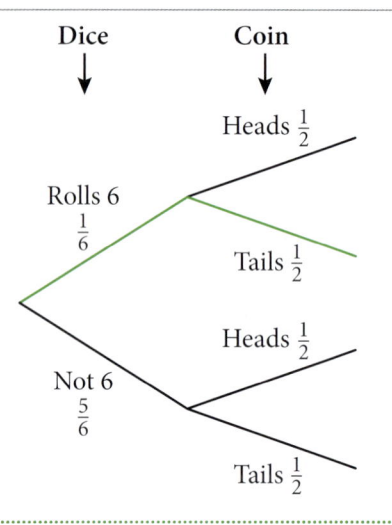

Example 6

(a) Draw a tree diagram to show all the possible outcomes when a fair coin is tossed 3 times.
(b) Use your tree diagram to find the probability of getting three heads.

(a) The tree diagram is shown on the right.

(b) The combination heads, heads, heads has been highlighted in the tree diagram.

When following a set of branches through the diagram, you multiply the individual probabilities together.

So P(heads, heads, heads) = $\frac{1}{2} \times \frac{1}{2} \times \frac{1}{2} = \frac{1}{8}$

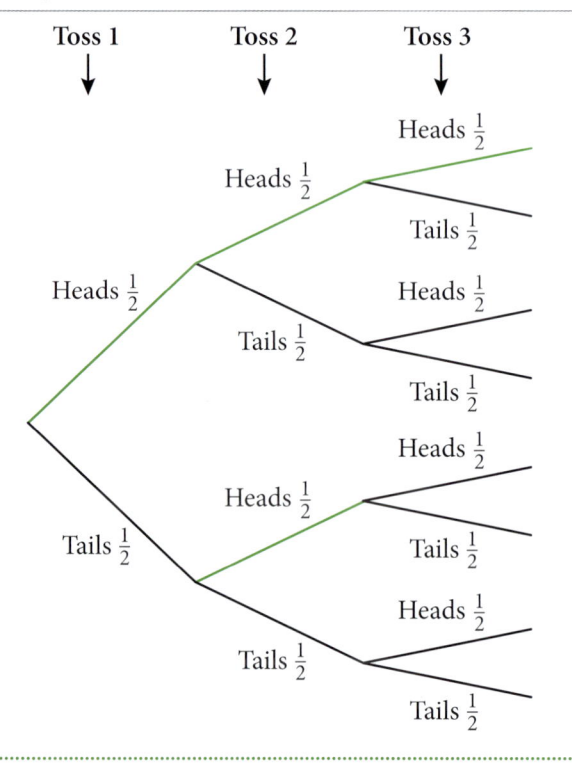

CHAPTER 37: PROBABILITY

Example 7

In Class 11F there are 13 boys and 17 girls. In Class 11G there are 14 boys and 11 girls. To represent the school at a public event, a pupil is chosen at random from Class 11F and another pupil at random from Class 11G. Find the probability that:
(a) Two girls are chosen.
(b) One boy and one girl are chosen.

(a) There are 30 pupils altogether in Class 11F.

Since there are 13 boys, the probability of a boy being chosen from Class 11F is $\frac{13}{30}$

The other probabilities in the diagram on the right are calculated in a similar way.

The combination Girl, Girl is highlighted on the tree diagram.

To calculate the probability that two girls are chosen:

$P(\text{Girl, Girl}) = \frac{17}{30} \times \frac{11}{25} = 0.249$ (3 s.f.)

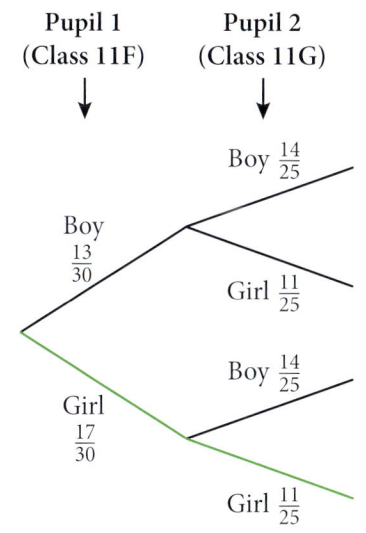

(b) There are two ways to select one boy and one girl. The tree diagram is shown again, with the relevant combinations highlighted and ticked.

We work out the probability of each combination and then add the results:

$P(\text{Boy, Girl}) = \frac{13}{30} \times \frac{11}{25} = 0.1906\ldots$

$P(\text{Girl, Boy}) = \frac{17}{30} \times \frac{14}{25} = 0.3173\ldots$

$P(\text{One boy and one girl})$
$= 0.1906\ldots + 0.3173\ldots$
$= 0.508$

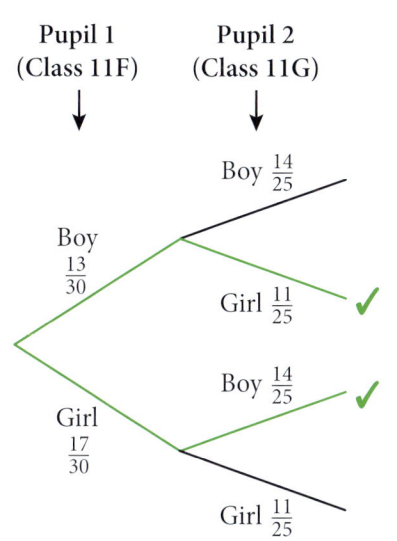

Selection with replacement

In some cases, an item is taken at random from a collection and then put back, before another item is chosen. This is known as **selection with replacement**.

Example 8

In a game, a player must choose a letter tile at random from a bag and then put it back. A second letter tile is then chosen at random. The bag contains 2 letter A tiles, 3 letter B tiles and 5 letter C tiles.

(a) Draw a tree diagram to show the possible outcomes.

Use the tree diagram to find the probability that:

(b) Two letter A tiles are chosen.
(c) One letter A tile and one letter B tile are chosen.
(d) At least one letter A tile is chosen.
(e) No more than one letter A tile is chosen.

(a) There are 10 tiles in total. The completed tree diagram is shown on the right.

(b) Following the A branches in the diagram:
$$P(A, A) = \frac{2}{10} \times \frac{2}{10} = \frac{1}{25}$$

(c) There are two ways to get one A tile and one B tile:
$$P(A, B) = \frac{2}{10} \times \frac{3}{10} = \frac{2}{25}$$
$$P(B, A) = \frac{3}{10} \times \frac{2}{10} = \frac{2}{25}$$
$$P(\text{one A and one B}) = \frac{2}{25} + \frac{2}{25} = \frac{4}{25}$$

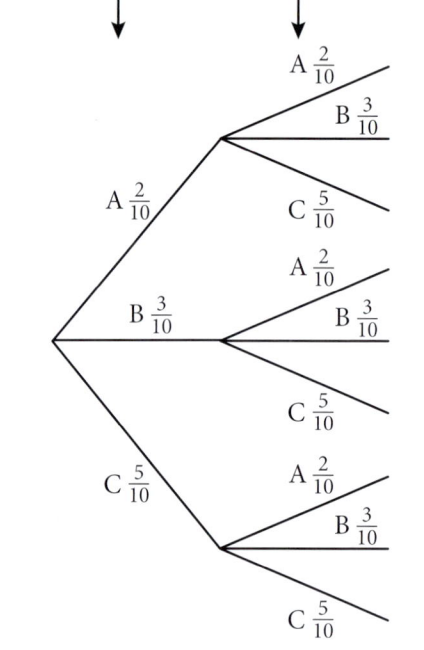

(d) To find the probability of getting at least one A tile, it is helpful to simplify the tree diagram. We can consider the two events 'getting A' and 'not getting A'. The simplified diagram is shown on the right, with all combinations involving at least one A ticked.

The probabilities of all four possible combinations add up to 1

So, calculate P(not A, not A) and subtract from 1:
$$P(\text{not A, not A}) = \frac{8}{10} \times \frac{8}{10} = \frac{16}{25}$$
$$P(\text{at least one A}) = 1 - \frac{16}{25} = \frac{9}{25}$$

(e) Getting no more than one A means getting either zero or one letter A tile. From the simplified tree diagram A, A is the only combination we do not need. Calculate this and subtract it from 1:
$$P(A, A) = \frac{2}{10} \times \frac{2}{10} = \frac{4}{25}$$
$$P(\text{no more than one A}) = 1 - \frac{4}{25} = \frac{21}{25}$$

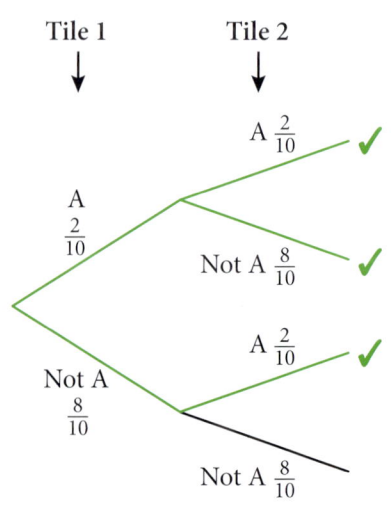

CHAPTER 37: PROBABILITY

Example 9

Caoimhe is offered a sweet from a bag. The bag contains 4 red sweets and 8 green sweets, but Caoimhe doesn't like the green ones. Caoimhe takes a sweet at random. If it is red, she eats it. If it is green, she puts it back in the bag and takes another one at random.

(a) Draw a tree diagram to show all the possible outcomes.
(b) Use the tree diagram to find the probability that Caoimhe still hasn't taken a red sweet after taking two sweets.

(a) The tree diagram is shown on the right.

(b) We are interested in the combination green, green which is highlighted on the diagram. The probability that Caoimhe takes two green sweets is:

P(green, green) = $\frac{8}{12} \times \frac{8}{12} = \frac{4}{9}$

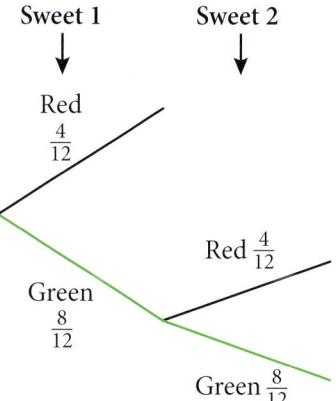

> **Note:** When multiplying two numbers between 0 and 1, the answer is always smaller than each of the individual numbers.
>
> For example, $\frac{1}{4} \times \frac{1}{2} = \frac{1}{8}$
>
> The answer of $\frac{1}{8}$ is smaller than both $\frac{1}{2}$ and $\frac{1}{4}$
>
> When dealing with two independent events, we multiply two probabilities. Probabilities are numbers between 0 and 1, so the probability of them both happening is less than the probability of either event happening individually.
>
> This is always the case, unless both probabilities are 0 or both 1

Exercise 37D

1. There are 10 wooden bricks in a box. Seven of them are blue and the rest are yellow. A child chooses a brick at random, puts it back in the box and then chooses a second brick at random.
   (a) Copy and complete the tree diagram shown on the right.
   (b) Use your tree diagram to find:
      (i) The probability that two yellow bricks are chosen.
      (ii) The probability that two bricks of the same colour are chosen.

2. The probability that Alan goes to the gym on Saturday is $\frac{3}{10}$
   The probability that Tess goes to the gym on Saturday is 0.45
   (a) Copy and complete the tree diagram shown on the right.

   Use your tree diagram to answer the following questions.
   (b) What is the probability Alan and Tess both go to the gym this Saturday?
   (c) What is the probability neither Alan nor Tess goes to the gym this Saturday?
   (d) What is the probability only one of them goes to the gym this Saturday?

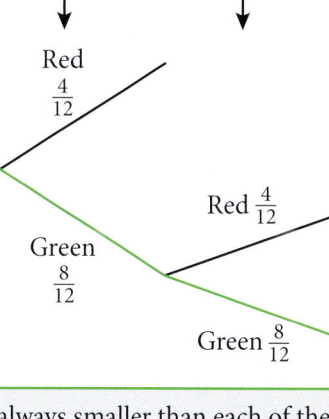

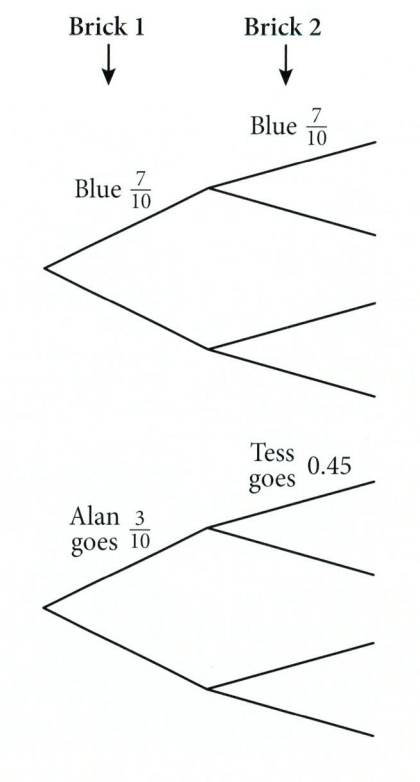

3. There are 20 coins in a jar. Twelve of them are 10p coins, seven of them are 50p coins and one is a £1 coin. A coin is chosen at random and put back in the jar. Then, a second coin is chosen at random.
   (a) Copy and complete the following tree diagram on the right to show all possible outcomes.

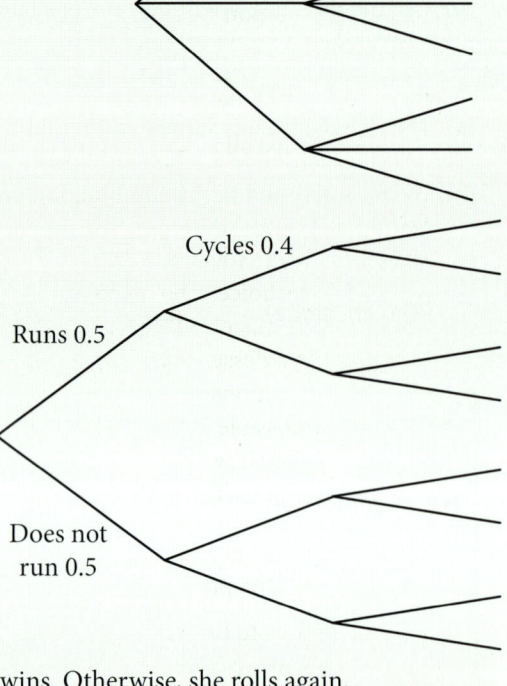

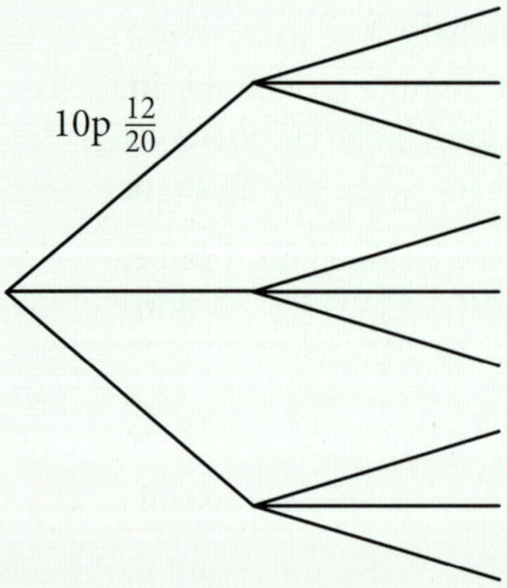

   Use your tree diagram to find the probability that:
   (b) A 10p coin is chosen twice.
   (c) The £1 coin is chosen twice.
   (d) The two coins chosen are the same.
   (e) The total value of the two coins chosen is not £1

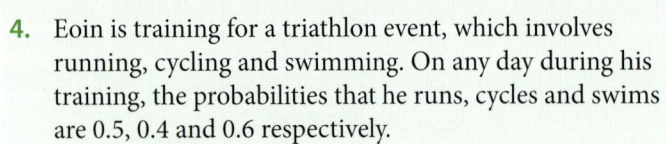

   10p $\frac{12}{20}$

4. Eoin is training for a triathlon event, which involves running, cycling and swimming. On any day during his training, the probabilities that he runs, cycles and swims are 0.5, 0.4 and 0.6 respectively.
   (a) Copy and complete the tree diagram on the right to show the possible outcomes for one of Eoin's training days.

   Use your tree diagram to find the probability that on any day:
   (b) Eoin trains for all three sports.
   (c) Eoin trains for two of the three sports.
   (d) Eoin trains for at least one sport.
   (e) Eoin swims, but does not cycle.

   Cycles 0.4
   Runs 0.5
   Does not run 0.5

5. Sue plays a game involving rolling a dice. If she rolls a 6 she wins. Otherwise, she rolls again. If she does not roll a 6 after 3 rolls of the dice, she loses.
   (a) Draw a tree diagram to show the possible outcomes of the game.
   (b) Use your tree diagram to find the probability that Sue wins the game on the second roll.

6. A bucket contains 100 marbles, 35 of which are red with the rest white. A marble is chosen at random and put back in the bucket. A second marble is then chosen at random.
   (a) Using a tree diagram, find the probability that:
      (i) Both marbles chosen are red.
      (ii) One red and one white marble are chosen.
   (b) Philip says 'The probability of choosing two white marbles is smaller than the probability of choosing at least one red marble'. Is Philip correct? Show all your working.

7. When Darcy cycles to school, there are 2 sets of traffic lights on the route.
   The probabilities of the lights being green are $\frac{3}{5}$ and $\frac{7}{10}$ respectively.
   Using a tree diagram or otherwise, find the probability that Darcy:
   (a) Does not have to stop.
   (b) Has to stop at one set of lights.

8. In a small town, there is a probability of $\frac{1}{100}$ that a man chosen at random is called Paul. In the same town, there is a probability of $\frac{1}{200}$ that a man chosen at random has the last name Thompson. Using a tree diagram, or otherwise, find the probability a man chosen at random has the name Paul Thompson.

9. Raquel goes to a speed dating event. For each person she talks to, there is a probability of $\frac{1}{4}$ that they agree to meet up again. Raquel decides that if somebody agrees to meet up again, she leaves the event. She also decides to talk to a maximum of 3 people.
   (a) Draw a tree diagram showing the possible outcomes from the event. Label the branches 'Yes' and 'No' ('Yes' representing an agreement to meet up again).
   (b) Use your tree diagram to find:
      (i) The probability that Raquel arranges to meet again with the second person she talks to.

(ii) The probability that Raquel arranges to meet somebody again.

10. Archie puts 20 beads in a box, some of them black and some of them white. His sister Molly chooses one at random, puts it back and chooses another one. Both Molly's beads were black. Archie says 'The probability of that happening are only $\frac{4}{25}$'.

   (a) Let x be the number of black beads in the box. Use a tree diagram to show that: $\frac{x^2}{400} = \frac{4}{25}$
   (b) Solve this equation to find the number of black beads.

37.5 Listing or Counting Outcomes For Successive Events

You may be asked to list all the possible outcomes for two events.

Example 10

On a menu there are four choices for a main course and three choices of dessert. Harry finds it very difficult to make decisions. He chooses a main course and a dessert by pointing his fork at random at both sections of the menu.
(a) How many possible combinations are there?
(b) Find the probability Harry chooses chicken followed by apple crumble.

Menu

Main Course	Dessert
Beef	Chocolate Brownie
Chicken	Ice Cream
Fish and Chips	Apple Crumble
Vegetarian Pasta	

(a) We can list all the possible outcomes in a systematic way:

- Beef, Chocolate Brownie
- Beef, Ice Cream
- Beef, Apple Crumble
- Chicken, Chocolate Brownie
- Chicken, Ice Cream
- Chicken, Apple Crumble
- Fish and Chips, Chocolate Brownie
- Fish and Chips, Ice Cream
- Fish and chips, Apple Crumble
- Vegetarian Pasta, Chocolate Brownie
- Vegetarian Pasta, Ice Cream
- Vegetarian Pasta, Apple Crumble

There are 12 possible outcomes.

(b) There are 12 possible outcomes and each one is equally likely. So:

P(chicken, apple crumble) = $\frac{1}{12}$

The product rule for counting combinations

You should know how to use the product rule for counting combinations. If there are m ways of doing one task and for each of these, there are n ways of doing another task, then the total number of ways the two tasks can be done is $m \times n$

Example 11

If a building has two doors to enter and three to exit, how many ways are there to enter and exit the building?

With two entrance doors and three exits, there are 2×3 ways you could enter and exit.
$2 \times 3 = 6$

Example 12

On a restaurant menu, there are 3 options for a starter, 4 for a main course and 5 choices for dessert. Lauren chooses to have three courses. How many combinations are there to choose from?

There are $3 \times 4 \times 5$ different combinations altogether.
$3 \times 4 \times 5 = 60$

GCSE MATHEMATICS M3 AND M7

Exercise 37E

1. On a menu there are 8 choices for a main course and 7 choices for a dessert. How many different combinations of main course and dessert could be chosen?

2. In a Taekwondo club competition, Radhika can choose one of 5 club members as her first round opponent, and then one of 7 different opponents in the second round. How many different possible combinations are there altogether?

3. Emma wants to visit one country in Asia, one country in Africa and one country in South America in the next three years. She has narrowed her choices down, so that she is choosing one of 10 countries in Asia, one of 12 countries in Africa and one of 5 in South America. How many combinations of countries does she have to choose from?

4. Pete has 5 shirts and 4 pairs of trousers.
   (a) In how many different ways could he choose a shirt and a pair of trousers to wear?
   (b) Pete is meeting his girlfriend. He knows she doesn't like his pink shirt. His black trousers are being cleaned. Without the pink shirt and the black trousers, in how many different ways could he choose a shirt and a pair of trousers?

5. The Laverty family, who live in Northern Ireland, are going for a short break in County Donegal. There are 6 main roads into County Donegal. They would like to travel back to Northern Ireland by a different route.
   (a) How many combinations of routes are available to them on a normal day?
   (b) On the morning of their departure, the Laverty family learn that the A38 road into County Donegal is closed due to repairs. It will still be closed when the family are making their return journey. How many combinations of routes are available to them now?

6. In SubsRUs, you can make your own sub sandwiches. There are 5 choices of protein, 4 different salads and 3 different types of bread, as shown below.

   ## SubsRUs

Protein	*Choose one of these:* Chicken, roast beef, tuna, cheddar cheese, egg
Salad	*Choose one of these:* Green salad, Asian salad, coleslaw, house special salad
Bread	White, Brown or Wholemeal

   (a) How many different choices of Sub are there altogether?
   (b) When Torin visits SubsRUs, the Asian salad and the roast beef are not available. By what percentage is the number of combinations decreased?

7. Jenni must choose one after-school activity for Mondays and one for Thursday. Her options are listed on the right.
   (a) By listing all the possible combinations, find how many different combinations Jenni must choose from.
   (b) If Jenni chooses at random, find the probability she chooses Hockey and Netball.
   (c) Find the probability Jenni **doesn't** choose Lifeguard Training.
   (d) The following term, football is moved from Monday nights to Thursday nights. Jenni must make her choices again. How many possible combinations must Jenni choose from now?

Monday
Cross Country
Cookery
Hockey
Sewing
Football

Thursday
Netball
Lifeguard Training

8. In a car showroom there are electric cars, petrol cars and diesel cars. There are also four colours to choose from: black, silver, white and red.
   (a) Annie is buying a new car from this showroom. State how many combinations are available to her and list them. The first one has been done for you:
   - Electric, black
   - ...
   (b) Annie decides she does not want a black car. After browsing, she approaches a salesperson. He tells Annie that there is a waiting time of one year for electric cars and that there are no white cars

available. Annie must buy a car today. List the combinations that are still available to her given the new information.
(c) If Annie chooses at random from the options still available to her, what is the probability she chooses a red car with a petrol engine?

37.6 Relative Frequency

Relative frequency as an estimate of probability

Relative frequency (RF) is an estimate of probability, based on observations.

Consider a football team playing a game. The possible outcomes are win, lose and draw. However, these are not all equally likely to happen. The outcome of the match depends on various factors such as the current form of the team and the quality of the opposition.

We can use relative frequency as an estimate for the probability that the team wins, draws or loses.

The formula for relative frequency is:

$$RF = \frac{\text{number of times event happens}}{\text{number of times experiment takes place}}$$

Example 13

A football team plays 40 games in a season. The outcomes of these games are shown in the table on the right.
(a) Estimate the probability that the football team wins their next game.
(b) Explain why the answer to part (a) is only an estimate of the true probability.

Win	Draw	Lose
16	10	14

(a) $RF = \dfrac{\text{number of wins}}{\text{total number of games}}$

$RF = \dfrac{16}{40} = \dfrac{2}{5}$

So, the best estimate of the probability of the team winning their next game is $\dfrac{2}{5}$ or 0.4
Or using mathematical notation, we can estimate that P(win) = 0.4

(b) The answer to part (a) is an estimate because:
- It is only based on 40 games. If we took a larger sample, we may get a better estimate.
- The next game may be affected by special circumstances. For example, the opposition may be a very strong team; there may be players injured, etc.

In some cases, there may be a list of results over various periods of time. The best estimate for the true probability is made by using the biggest sample size. Look for the largest number of times the experiment took place.

> **Note:** Increasing the sample size generally leads to better estimates of probability.

Example 14

An experiment is carried out with a drawing pin to estimate the probability of it landing point up when thrown. It is thrown onto a desk 200 times. The table on the right shows the number of times the pin lands point up after 50, 100, 150 and 200 throws. Estimate the probability of the drawing pin landing point up.

Number of times pin landed point up	Total number of throws
37	50
67	100
115	150
144	200

The best estimate for the probability is found using the final row of the table, since this row contains the largest number of times the experiment was carried out, i.e. 200 times.

GCSE MATHEMATICS M3 AND M7

$$RF = \frac{\text{number of times event happens}}{\text{number of times experiment takes place}}$$

$$RF = \frac{144}{200} = 0.72$$

We can estimate that P(pin lands point up) = 0.72

Comparing relative frequency data with theoretical probabilities

In the following example, we work out the relative frequencies for each number on a six-sided dice. By comparing these with theoretical probabilities, we can decide whether this is a fair dice.

Example 15

A six-sided dice is thrown 120 times. Look at the table on the right. It shows the number of times each number from 1 to 6 was rolled.

Number	1	2	3	4	5	6
Frequency	19	18	34	14	19	16

(a) Find the relative frequency for each number.
(b) Rolling a **fair** six-sided dice 120 times, how many times would you expect to roll a six?
(c) Do you think this is a biased dice?
(d) Using the relative frequencies you have calculated, estimate the number of sixes you would expect if you rolled this dice 1000 times.

(a) Copy the table and add another row for the relative frequency.

Number	1	2	3	4	5	6
Frequency	19	18	34	14	19	16
Relative Frequency	$\frac{19}{120} = 0.158$	$\frac{18}{120} = 0.15$	$\frac{34}{120} = 0.283$	$\frac{14}{120} = 0.117$	$\frac{19}{120} = 0.158$	$\frac{16}{120} = 0.133$

(b) For a fair dice, the probability for each number would be $\frac{1}{6}$
To calculate the expected number of sixes from 120 rolls:
$120 \times \frac{1}{6} = 20$

(c) After 120 rolls the number 3 was rolled far more than any other number.
This dice is probably biased towards the number 3

(d) To calculate the expected number of sixes after 1000 rolls of this dice:
$1000 \times 0.133 = 133$

Exercise 37F

1. A women's hockey team plays 30 games in a season. The outcomes of these games are shown in the table on the right.

Win	Draw	Lose
15	10	5

   (a) Estimate the probability that the team wins their next game.
   (b) Explain why the answer to part (a) is only an estimate of the true probability.

2. Tilly is in an archery club. She fires 20 arrows at the target. The table on the right shows the number of times Tilly hits the centre of the target after 5, 10, 15 and 20 arrows.

Number of times Tilly's arrow hits the centre of the target	Total number of arrows
1	5
3	10
4	15
5	20

   (a) Estimate the probability Tilly hits the centre.
   (b) How many times would you expect Tilly to hit the centre of the target if she fires 48 arrows?

3. A six-sided dice is thrown 60 times. Look at the table on the right. It shows the number of times each number from 1 to 6 was rolled.

Number	1	2	3	4	5	6
Frequency	9	8	8	10	9	16

   (a) Find the relative frequency for each number.

(b) Rolling a **fair** six-sided dice 60 times, how many times would you expect to roll a six?
(c) Do you think this is a biased dice? Explain your answer.
(d) Using the relative frequencies you have calculated, estimate the number of sixes you would expect if you rolled this dice 1000 times.

4. There are 90 beads in a box. Each bead is either red or black. A bead is taken from the box at random. Its colour is noted and then the bead is put back. The table shows the number of red beads withdrawn after 20, 40, 60, 80 and 100 times.
   (a) Copy the table and complete the relative frequency column. Round your answers to 3 decimal places where appropriate.

Total number of draws	Number of times a red bead is withdrawn	Relative frequency
20	17	0.85
40	35	
60	40	
80	55	
100	72	

   (b) Which of the values in the relative frequency column is the best estimate for the true probability of taking a red bead? Briefly explain your answer.
   (c) Using your estimate for the probability of choosing a red bead, estimate the number of red beads and the number of black beads in the box.

5. For a film, a stuntman must jump across a river on a motorbike. During filming on Monday he attempts the jump 30 times. The director records the number of times he succeeds in the table on the right.
   (a) Copy the table and complete the relative frequency column.
   (b) Which of the values in the relative frequency column is the best estimate for the probability of success? Briefly explain your answer.
   (c) Final rehearsals for the film take place on Tuesday. The stuntman has to attempt the jump another 20 times. Estimate how many of these are successful.

Number of jumps	Number of successes	Relative frequency
5	3	
10	7	
15	9	
20	13	
25	17	
30	21	

6. Shane, Cian, Jo and Clare are working in a group to test whether a coin is biased. They each toss the coin a number of times. The table on the right shows their results.
   (a) Find the total number of times the coin was tossed and the total number of heads.
   (b) Copy and complete the table as follows:
      (i) Enter the two totals you calculated in part (a).

Name	Number of times coin tossed	Number of heads	Relative frequency
Shane	30	15	
Cian	54	24	
Jo	45	21	
Clare	60	17	
Total			

      (ii) Calculate and enter the relative frequency for each person
      (iii) Calculate and enter the relative frequency for the total.
      Round your relative frequencies to 3 significant figures where appropriate.
   (c) Which of the relative frequency figures you have calculated is the best estimate for the probability of getting heads? Explain your answer.
   (d) What should the group conclude? Is the coin biased? Explain your answer.
   (e) Estimate the number of heads you would expect if the coin is tossed 500 times.

7. Rowan counts the number of questions he gets right for each 10 questions in a spelling test. There are 100 questions. The results are shown in the table on the right.
   (a) Copy and complete the table. Round the relative frequency values to 3 significant figures where appropriate.
   (b) What is the best estimate for the probability Rowan gets a question right?
   (c) The following week Rowan does another spellings test. This time the test has 40 questions. Using your answer to part b, estimate the number of questions Rowan gets right.

Question numbers	Number correct	Total correct so far	Relative frequency
1 – 10	5	5	0.5
11 – 20	4	9	0.45
21 – 30	5		
31 – 40	6		
41 – 50	5		
51 – 60	6		
61 – 70	8		
71 – 80	7		
81 – 90	6		
91 – 100	4		

8. A gym owner records the number of times each facility is used during one week. The results are shown in the table on the right.
   (a) What is the total number of times the facilities were used?
   (b) Find the relative frequency for each facility. Round your answers to 3 significant figure where appropriate.
   (c) Using your answer to part b, estimate the number of times the exercise bikes are used on a day that the facilities are used 250 times.

Facility	Number of times used
Treadmills	55
Weights	35
Pool	42
Exercise bikes	28

37.7 Summary

In this chapter you have learnt:
- How to identify different mutually exclusive outcomes and know that the sum of the probabilities of all these outcomes is 1
- That the probability of an event not occurring is one minus the probability that it occurs.
- That a sample space diagram can be used to put the outcomes of two events into a table and calculate probabilities.
- How to systematically list all outcomes for single events and for two successive events.
- How to use the product rule for counting combinations.
- When to add or multiply two probabilities: if A and B are mutually exclusive, then the probability of A or B occurring is P(A) + P(B). If A and B are independent events, the probability of A and B occurring is P(A) × P(B).
- How to use tree diagrams to represent successive events that are independent.
- That relative frequency is an estimate of probability, based on observations.
  The formula for relative frequency is:
  $$RF = \frac{\text{number of times event happens}}{\text{number of times experiment takes place}}$$
- About comparing experimental data and theoretical probabilities, for example to determine whether a coin or dice is biased.
- That increasing the number of times an experiment takes place generally leads to better estimates of probability.

Progress Review
Chapters 34 – 37

This Progress Review covers:

Chapter 34: Transformations

Chapter 35: Enlargement and Similarity

Chapter 36: Constructions and Loci

Chapter 37: Probability

1. Look at the four congruent shapes A, B, C and D in the diagram on the right.
   (a) Describe fully the transformation that maps C to D.
   (b) Describe fully a **reflection** that map A to B.
   (c) Describe fully a **translation** that map A to B.
   (d) Describe fully a **reflection** that map A to C.
   (e) Describe fully a **rotation** that map A to C.
   (f) Explain why there is no **translation** mapping A to C.
   (g) Describe fully a transformation mapping A to D.

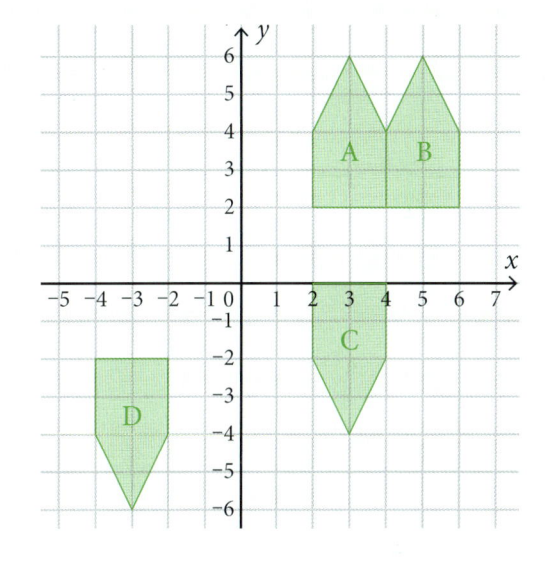

2. Draw and label axes from −6 to 6
   (a) Plot the following points and join them to form a triangle: (3, 2), (5, 4), (4, −2)
       Label the triangle A.
   (b) Reflect triangle A in the x-axis and label the image B.
   (c) Write down the coordinates of the vertices of triangle B.
   (d) Translate triangle B by $\begin{pmatrix} -6 \\ -2 \end{pmatrix}$ and label the image C.
   (e) Reflect triangle C in the x-axis. Label the image D.
   (f) Describe the transformation that maps D on to A.

3. Look at the images A, B, C and D on the grid on the right.
   (a) Describe fully the transformation from shape A to shape B.
   (b) Describe fully the transformation from shape A to shape D.
   (c) Describe fully the transformation from shape B to shape C.

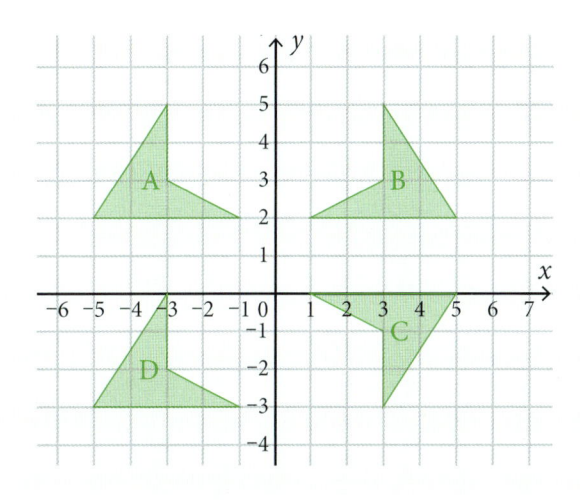

361

4. Look at the shapes in the diagram on the right.
   (a) Describe each of these translations:
       (i) A to B
       (ii) A to C
       (iii) A to D
       (iv) C to D
   (b) Explain briefly why there is no translation that maps A to E.

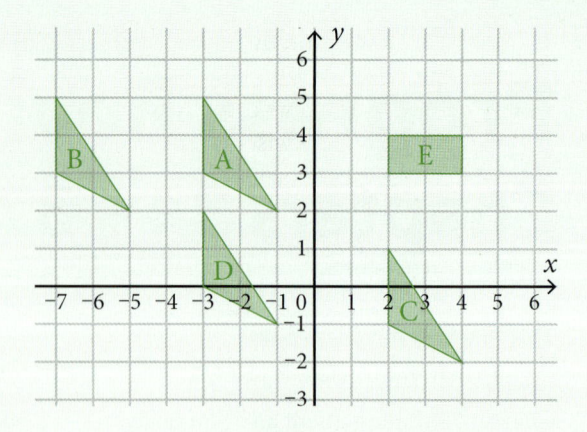

5. Look at the diagram on the right. Match each of the following mappings with the transformation described.

Mapping	Transformation
A → P	Rotation 180° about (3, 3)
A → Q	Reflection in line $y = 3$
A → R	Translation by vector $\begin{pmatrix} 3 \\ 3 \end{pmatrix}$
A → S	Reflection in line $x = 3$

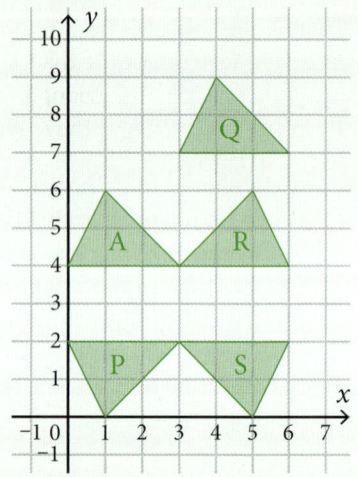

6. (a) You only need one piece of information to describe a translation. What is it?
   (b) You need three pieces of information to describe a rotation fully. What are they?

7. Look at the diagram on the right.
   (a) What **translation** could be used for each of these transformations?
       (i) A to E
       (ii) A to C
       (iii) E to C
       (iv) B to D
   (b) What **rotation** could be used for each of these transformations?
       (i) A to B
       (ii) A to C
       (iii) B to D
       (iv) B to E

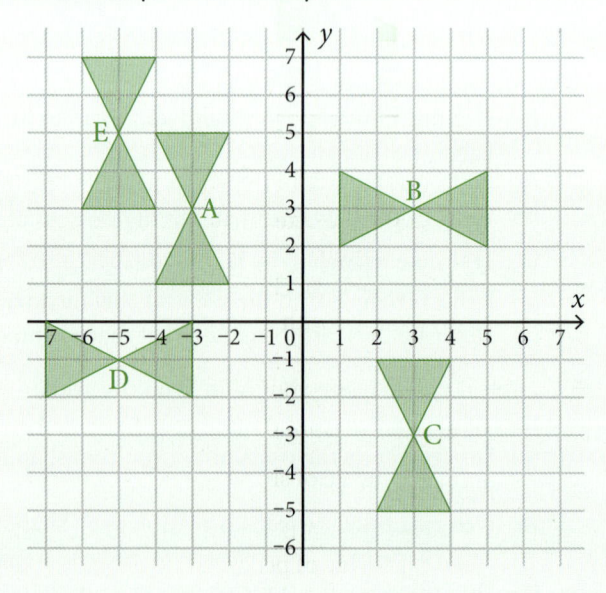

PROGRESS REVIEW: CHAPTERS 34–37

8. Copy the diagram on the right. Rotate the flag about point the origin (0, 0):
   (a) a quarter turn anti-clockwise;
   (b) a half turn clockwise;
   (c) a three-quarter turn clockwise.

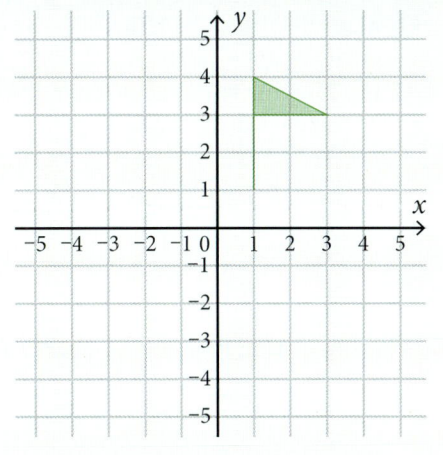

9. The triangle A in the diagram below is rotated through 90° clockwise with centre of rotation P(−3, 2). Its image is B. The triangle B is then rotated through 90° clockwise with centre of rotation Q(−1, 2). Its image is C. Find the single transformation that maps A onto C.

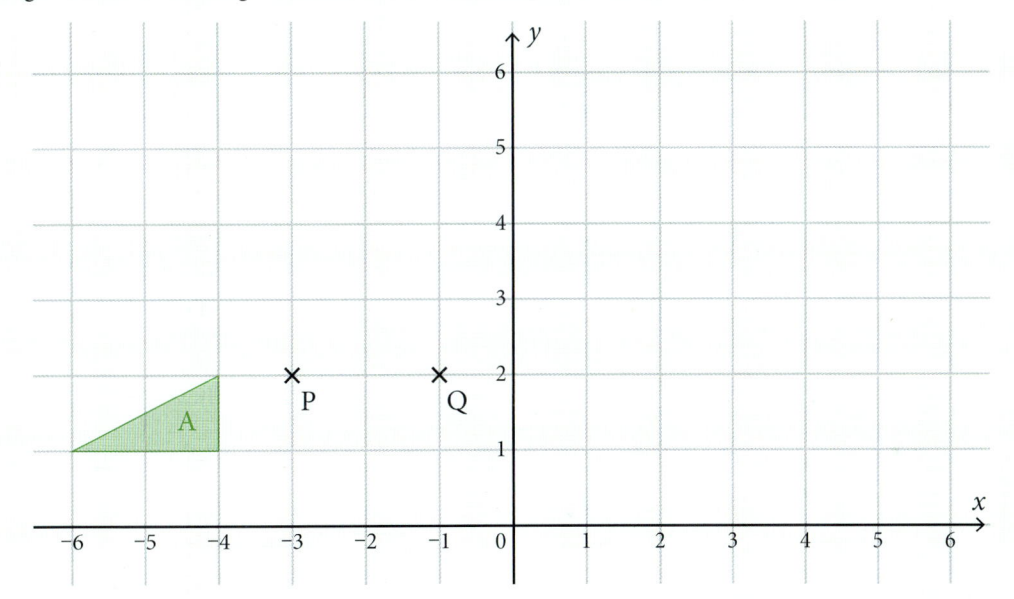

10. Copy the coordinate axes shown on the right.
    (a) Plot these points:
       A(−3, 1), B(−3, 3), C(−1, 1) and D(−1, 3).
    (b) Join the points with three straight lines: from A to B, from B to C and from C to D.
    (c) What letter have you drawn?
    (d) Translate the entire shape using the vector $\begin{pmatrix} 4 \\ 0 \end{pmatrix}$.
       Label the image of shape ABCD as A'B'C'D'.
    (e) Rotate the lines A'B' and C'D' by 90° clockwise, using a centre of rotation P(4, 0). 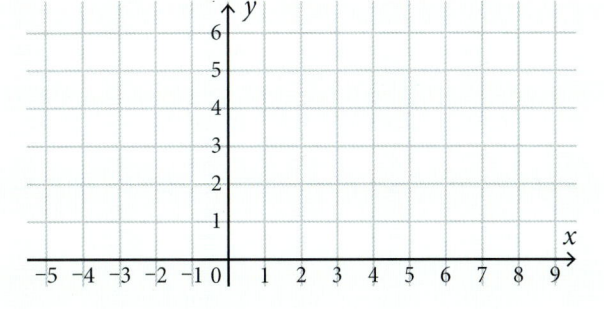
       Label the image of point A' as A". Label the images of the other three points B", C" and D".
    (f) Reflect the diagonal line B'C' in the mirror line $x = 4$
       What letter have you drawn?

363

GCSE MATHEMATICS M3 AND M7

11. Copy the diagram on the right. Shape B is an enlargement of shape A.
    (a) Find the scale factor of the enlargement from shape A to shape B.
    (b) The point P is the centre of enlargement. Find the coordinates of P.
    (c) Describe the transformation fully.

12. A circle has a **diameter** of 16 cm. Its area is 201 cm² to the nearest whole number. The circle is enlarged with a scale factor of 2
    (a) Find the **radius** of the enlarged circle.
    (b) Calculate the area of the enlarged circle to the nearest whole number.

13. Calculate the sides marked x and y in the pair of similar shapes shown on the right. Round your answers to 1 decimal place where necessary.

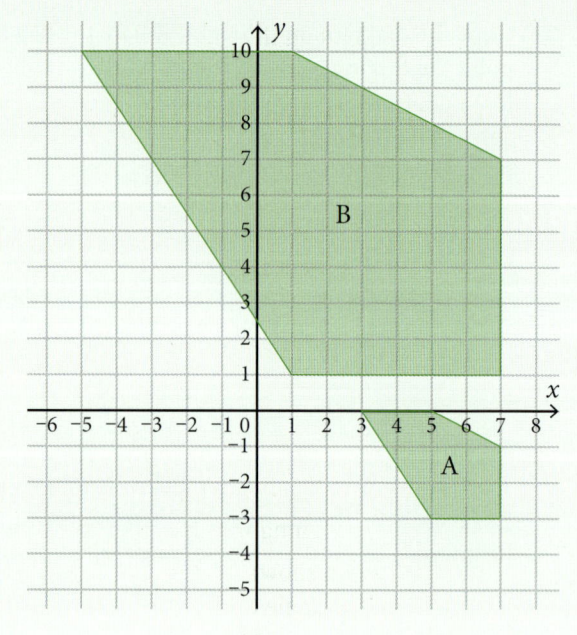

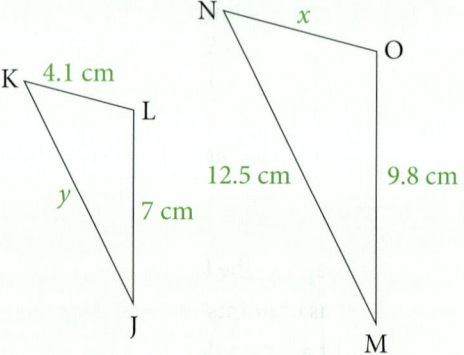

14. The two quadrilaterals shown on the right are mathematically similar.
    (a) Find the size of angle B.
    (b) Write down the sizes of angles Q, R and S.
    (c) Find the length of the side PS marked x.
    (d) Find the length of the side AB marked y.

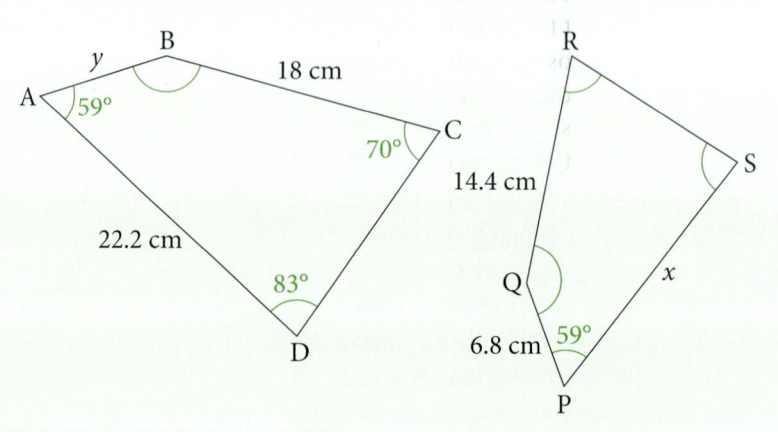

15. The diagram on the right shows a rectangle R1. Copy the diagram. On your copy:
    (a) Enlarge R1 using a scale factor of $\frac{1}{2}$ and centre of enlargement C(3, 0). Label the image rectangle R2.
    (b) Translate R2 using the vector (4, 0). Label the image rectangle R3.
    (c) What single transformation takes rectangle R3 to rectangle R1?

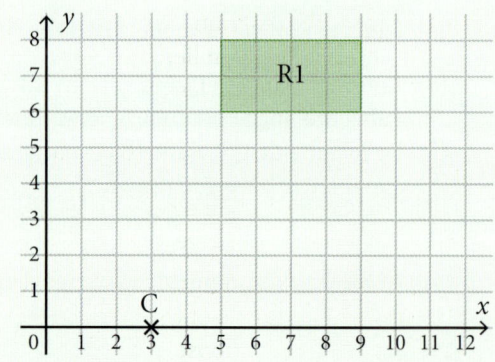

16. Sam thinks that all regular hexagons are congruent. Paul says that all regular hexagons are similar. Who is correct?

PROGRESS REVIEW: CHAPTERS 34–37

17. A pack of cheese sold in the supermarket has a length of 10 cm and a volume of 400 cm³.

    A mathematically similar pack of cheese is introduced with a length of 15 cm. What volume of cheese does the new pack contain?

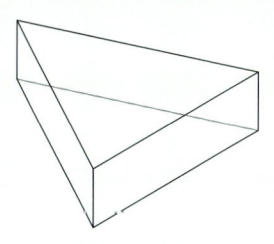

18. Copy the line segment DE shown below. It is 10.6 cm long. Construct the perpendicular bisector of DE.

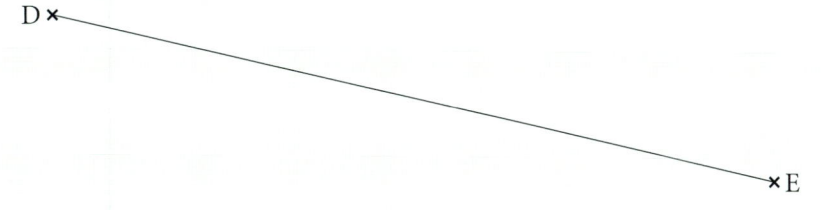

    D ×
    × E

19. Draw a line 8.8 cm long. Construct the perpendicular bisector of the line.

20. Copy the diagram shown on the right. Construct the angle bisector for angle CAB.

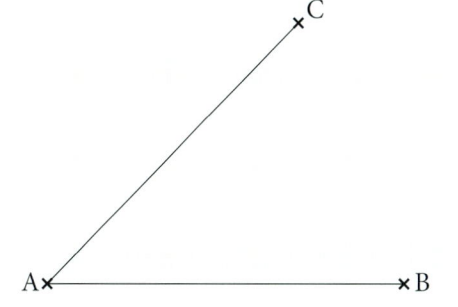

21. A straight road is 7.4 km long. A second road passes through the first one at 90° exactly halfway along. Using a scale of 1 cm = 1 km, construct a scale drawing showing both roads.

22. Draw an angle of 132° and construct the angle bisector.

    × C
    A × ———————— × B

23. Shade on a diagram the locus of points that are less than 4.5 cm from a fixed point O.

24. Draw the locus of points that lie **more than** 2.5 cm from a line FG that is 6 cm long.

25. (a) Construct a rectangle TUVW, where TU = 12 cm and UV = 8 cm.
    (b) Construct the locus of points that lie inside the rectangle, **and**:
        (i) are closer to point U than to point W.
        (ii) are closer to the line VW than to the line UV.
        (iii) are less than 5 cm from point V.
    (c) Construct the locus of points that satisfy all three of the conditions in part (b).

26. Grace and Henry live at opposite ends of a straight road, which is 0.5 km long. Grace is allowed to go 400 m from her house. Henry is allowed to go 300 m from his house. Draw the road as a straight line, using a scale of 1 cm = 100 m. Mark on your diagram Grace's house and Henry's house. By constructing a locus, shade the area in which they are allowed to meet up.

27. (a) Construct a triangle XYZ with XY = 11 cm, XZ = 6 cm and YZ = 7 cm.
    (b) Construct the locus of points that are inside the triangle and lie closer to XZ than to YZ.

28. Two points X and Y are 6 cm apart, as shown on the right. Copy the diagram. On your diagram, construct the locus of points that lie:
    • More than 3 cm and less than 4 cm from point X;
      and
    • More than 3 cm and less than 4 cm from point Y.

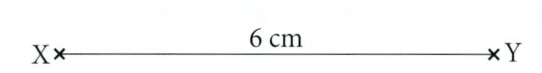

    X × ———— 6 cm ———— × Y

29. For a randomly chosen day in February, the probability of rain in Northern Ireland is 0.68
    Find the probability there is no rain.

30. Every Saturday morning, James has the choice of going to chess club, swimming or badminton. He always goes to one of these clubs.
    The probability he chooses chess club is 0.4 and the probability he chooses swimming is 0.25
    Find the probability that James chooses to go to badminton.

GCSE MATHEMATICS M3 AND M7

31. Which of these pairs of events (a) to (d) are mutually exclusive?
  (a) Winning a football match and drawing the same match.
  (b) Wearing one red sock and one blue sock.
  (c) Eating toast for breakfast and eating chips for dinner.
  (d) Being on time for school and being late for school on a particular day.

32. There are some black, red, white and blue beads in a bag. A bead is chosen at random. The probabilities of choosing a black, red and blue bead are shown in the table on the right.

Black	Red	White	Blue
0.15	0.25		0.35

  (a) Find the probability of choosing a white ball.
  (b) Find the probability of choosing a ball that is either black or red.

33. In each part (a) to (d) are the probabilities of some events happening. In each case, find the probability of the event not happening.
  (a) P(A) = 0.54   (b) P(B) = 30%   (c) P(C) = 54.5%   (d) P(D) = $\frac{11}{15}$

34. In each part (a) to (c) state whether events A and B are mutually exclusive or not.
  (a) A coin is tossed. Event A is 'getting heads'; event B is 'getting tails'.
  (b) Two teenagers are chosen at random from a group of two girls and three boys. Event A is 'choosing two girls'; event B is 'choosing two boys'.
  (c) A football team plays a match. Event A is 'the team winning'; event B is 'the team drawing the match'.

35. Which one of the pairs of mutually exclusive events in question 34 are also exhaustive?

36. Alex and Bella play a game of noughts and crosses. From looking at previous games, the probability of Alex winning is 0.29, the probability Bella will win is 0.33. What is the probability of a draw?

37. A spinner consists of an outer ring of coloured sections and an inner circle of numbered sections, as shown.
  (a) The probability of getting 2 is $\frac{1}{4}$. The probabilities of getting 1 or 3 are equal. What is the probability of getting 3?
  (b) The probability of getting grey is $\frac{1}{4}$. The probability of getting white is also $\frac{1}{4}$. The probability of getting black is $\frac{3}{8}$. What is the probability of getting green?
  (c) Which of these pairs of events are mutually exclusive?
     (i) Getting 3 and getting 2
     (ii) Getting 3 and getting black.
     (iii) Getting 1 and getting grey.
     (iv) Getting grey and getting white.

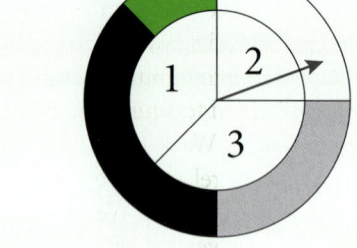

38. An operation has a 70% success rate the first time it is attempted. If it is unsuccessful, it can be repeated, but with a success rate of only 40%.
  (a) Draw a tree diagram showing this information. Use fractions or decimals on the branches rather than percentages.
  (b) Using your tree diagram, find the probability that the operation will fail twice. Give your final answer as a percentage.

39. Two fair coins are tossed.
  (a) Copy and complete the sample space diagram on the right showing the possible outcomes.
  (b) Using your completed sample space diagram, find the probability of getting at least one tail.

		Coin 1	
		H	T
Coin 2	H		(T, H)
	T		

40. Two fair, six-sided dice are thrown and the sum of the two numbers is calculated.
    (a) Copy and complete the sample space diagram on the right to show the possible outcomes.
    (b) Use your sample space diagram to find the probability that the sum of the two numbers is:
        (i) A multiple of 4
        (ii) A prime number.

		Dice 1				
+	1	2	3	4	5	6
1	2	3				
2	3					
3						
4						
5						
6						

(Dice 2 labels the rows)

41. Granny Valerie is in a muddle with her passwords. She knows the password for her computer is the name of one of her children. She has four children. She guesses the computer password at random from these four names. She knows the password for her email is the name of one of her grandchildren. She has seven grandchildren. She guesses the email password at random from these seven names. Find the probability that Granny Valerie guesses both passwords correctly first time.

42. Luke is a darts player. His results for his last 40 games are shown in the table on the right.
    (a) Estimate the probability that Luke wins his next game.
    (b) Explain why the answer to part (a) is only an estimate of the true probability.

Win	Lose
25	15

43. On a factory production line, a robot squirts cream into plastic tubs. Usually, the cream goes into the tub successfully, but sometimes it misses. The table on the right shows the number of times the robot was successful and unsuccessful during Working Week 1.
    (a) Find the total number of times the robot attempted to squirt cream into a tub during Working Week 1.
    (b) Find the relative frequency of success.
    (c) During Working Week 2, the robot makes a further 40 000 attempts to squirt cream into tubs. Using your answer to part (b), estimate how many attempts are successful during Working Week 2.

Day	Successful	Unsuccessful
Monday	7900	100
Tuesday	8700	300
Wednesday	6000	250
Thursday	7500	500
Friday	4700	300

44. A six-sided dice is thrown 300 times. Look at the table below. It shows the number of times each number from 1 to 6 was rolled.

Number	1	2	3	4	5	6
Frequency	42	90	48	39	45	36

(a) Find the relative frequency for each number.
(b) Rolling a **fair** six-sided dice 300 times, how many times would you expect to roll a two?
(c) Do you think this is a biased dice? Explain your answer.
(d) Using the relative frequencies you have calculated, estimate the number of times you would expect to roll a 2 if you rolled this dice 1000 times.

45. On a beach, water samples are taken to decide whether the water is safe to swim in. Over the course of a week, 100 samples are taken and are shown in the table below.

Total number of samples	Total number of times water quality is acceptable	Relative frequency
20	16	
40	35	0.875
60	54	
80	66	
100	85	

(a) Copy and complete the table.
(b) From the numbers in your table, write down an estimate of the probability that the water is safe to swim in.
(c) Explain briefly how you arrived at your answer in part (b).
(d) The following week, a further 200 samples are taken. Assuming the water quality remains the same, estimate the number of samples that would be acceptable.

46. At a small youth club one week there are 3 boys and 2 girls. Jake and Lucy are two of the members and they are brother and sister. One boy and one girl are chosen at random for an activity.
(a) How many different combinations of one boy and one girl are there?
(b) What is the probability that both Jake and Lucy are chosen?

The following week the same people are at the youth club, as well as an extra boy and an extra girl. Again, one boy and one girl are chosen at random. This week:
(c) How many combinations of one boy and one girl are there?
(d) What is the probability that both Jake and Lucy are chosen?
(e) Find the probability that **either** Jake or Lucy are chosen.